AGROFORESTRY

SYSTEMS AND PRACTICES

ABOUT THE EDITORS

Dr. Sunil Puri, NAAS Fellow, is a trained forest scientist from Oxford Forestry Institute, U.K. and has a wide research experience of 27 years in both tropical and temperate regions of India. He has played a key role in bringing forestry education within the University system in India. He was the first to start M. Sc. Agroforestry programmes at Haryana Agricultural University, Hissar in 1989 when he was Head of Forestry Department and later in 1997 at Indira Gandhi Agricultural University, Raipur. Dr. Puri is actively working on tree crop interaction under agroforestry practices as well as development of wasteland/ degraded lands. He has standardized the technology of propagating vegetatively many MPTs of temperate and tropical regions and thereby raise quality seedlings in mass. Dr. Puri is a member of many international and national committees on forestry. He has over 200 publications in scientific journals and books. He has organized many workshops, seminars, summer institutes and has traveled widely in Europe, Asia, Africa and Pacific. He has been director of many courses and trainings conducted for the upliftment of rural masses both at farmers and officers' level. Presently he is on the Editorial Board of the Agroforestry Systems since 1997 and Coordinator of Tropical Agroforestry of IUFRO Division 1 (Silviculture), since 2003 and of Silviculture & Management in Arid & Semi-Arid Regions, IUFRO, since 2005.

Born in Nangal, India on 17 March 1953. Educated at Govt. Boys High School, Nangal, 1965-69; DAV College, Jalandhar, 1969-71; Punjab Agricultural University, Ludhiana, 1971-76; Guru Nanak Dev University, Amritsar, 1976-79; Oxford Forestry Institute, U.K, 1985-86; Ph.D. 1979.

Professor & Head of Forestry, Indira Gandhi Agricultural University, Raipur, 1994-2004; Associate Professor of Forestry, Haryana Agricultural University, Hisar, 1985-94; Head Forestry, HAU, Hisar, 1988-91; Director Instructions, Indira Gandhi Agricultural University, Raipur, 2004.

Fellow of National Academy of Agricultural Sciences since 2006 and was awarded AAAS, USA International Diploma of Honour in 1998. He served as Vice President (2003-05) of Indian Society of Agroforestry.

Dr. Pankaj Panwar is a forest scientist with a research experience in both temperate and sub-tropical climates. Dr. Panwar was born in Kasuali, Himachal Pradesh and did his senior secondary from Jawahar Navodaya Vidyalaya, Theog (Shimla) followed by his B.Sc (forestry), M.Sc (forestry) and Ph.D (forestry) from Dr. Y.S. Parmar University of Horticulture and Forestry, Solan (Himachal Pradesh). He has to his credit 45 research publications along with two books. He is life member of Indian Journal of Agroforestry, Range Management and Agroforestry. At present Dr. Panwar is working on tree crop interaction under agroforestry practices, biodiesel production from jatropha and on bamboos. He is handling projects on agroforestry, bamboo and jatropha from funding agencies like Ministry of Rural Development, DST and DBT. He is Incharge - Biofuel Development Cell of the University. He has guided one M.Sc (forestry) student. Dr. Panwar was awarded "2000 millennium award" in 8th World Environment Congress. Dr. Panwar is referee of "Seed Research" and "Journal of Indian Society of Remote Sensing".

AGROFORESTRY

SYSTEMS AND PRACTICES

Editors :

PROF. SUNIL PURI
Ph.D (Bioscience), PDF (Oxford)
Director Instruction
Indira Gandhi Agricultural University
Raipur - 492 006, (Chattisgarh)

&

DR. PANKAJ PANWAR
B.Sc (For.), M.Sc (For.), Ph.D (For.)
Department of Forestry
Uttar Banga Krishi Viswavidyalaya
Pundibari, Distt. Cooch Behar - 736 165
(West Bengal)

New India Publishing Agency
Pitam Pura, New Delhi- 110 088

Published by
Sumit Pal Jain *for*
New India Publishing Agency
101, Vikas Surya Plaza, CU Block, L.S.C. Mkt.,
Pitam Pura, New Delhi- 110 088, (India)
Phone: 011-27341717, Fax: 011-27341616
E-mail: newindiapublishingagency@gmail.com
Web: www.bookfactoryindia.com

ISBN : 978-81-89422-62-2

Composed and Designed by NIPA

INDIAN COUNCIL OF AGRICULTURAL RESEARCH
Krishi Anusandhan Bhawan-II, Pusa
New Delhi - 110 012
Tel. : 91-11-25848364, 25842262 (Ext. 1101)
Fax : 91-11-25848366, 23387293
E-mail : jssamra2001@yahoo.com

भाकृ अनुप
ICAR

Dr. J.S. Samra
Dy. Director General
(Natural Resource Management)

Foreword

Agro-forestry systems have always been looked upon to diversify livelihood, income, employment and environmental services with all inclusive equity and sustainability. These systems are based on the principles of cycling and recycling of residues, nutrients and energy fluxes to minimize loading of the environment and optimize overall utilization efficiencies. They also insure regular flow of income with diversified risks and preoccupation of the human resources throughout the year. Trees including fruits, bushes, grasses and their permutations and combinations in space and time nave gained importance in mitigating global warming. Agro-forestry is becoming a commodity for carbon trading. Bio-fuels, bamboo cultivation, private wood industries, partnerships for pulp, synthetic wood and fuel for captive power plants have the potential of attracting investments by the industry especially in the wastelands. Fish and banana based agro -forestry systems have tremendous opportunities in high rainfall regions of eastern India, coastal belts, waterlogged canal commands and other shallow water table agro-ecologies. Primary productivity of the arid and semi-arid region is about two times less as compared to humid and sub-humid region and is limited by the moisture availability. Therefore, enhancing income in the low rainfall region can be realized by cultivating endemic medicinal and aromatic plants of low volume and high value. In the current scenario of growth in population, urbanization and industrialization with diminishing per capita resources, agro-forestry needs reprioritization, harmonization with the market demands, reorientation and revamping research and development in the country. I am sure this book will stimulate rethinking process which in any case has to be a dynamic exercise for addressing ever emerging new issues. I congratulate the authors for their hard work in compiling, editing and publishing the information.

(J.S.Samra)
Deputy Director General (NRM)
ICAR, Krishi Anusandhan Bhawan-II
Pusa, New Delhi - 110 012

Dr. S.K. Dhyani
Director
National Research Centre for Agroforestry (ICAR)
Jhansi-284 003 (Uttar Pradesh)

Message

Retaining trees on the farm fields had been an age old practice for fulfilling the fuel, fodder, timber and other minor product requirements. Trees had been a part in festivals, fairs and religious rites. However, with the greed of producing more from the same patch of land the monocropping of agriculture started which led to vanishing of the trees from the farmlands. After a long period of monocropping of agriculture crops it was realized that this had resulted in increased soil erosion, less fertility of soil, scarcity of tree related produces, increased insect pest incidence, decreased biodiversity thus leading to threatening of the existence of the life on the mother earth. To produce the crops sustainably it was emphasized to increase the biodiversity of the farmlands and incorporating trees in scientific manner with crops was considered to be the best viable option. Since trees may show variable results with different crops and in different agroclimatic conditions coupled with the fact that different farm owners have liking towards different trees, a need was felt to develop and\or standardize region based agroforestry systems through a chain of agroforestry centres working under one umbrella. This lead to initiation of All India Coordinated Research Project (AICRP) on Agroforestry by the ICAR in 1983. At present there are 36 centres located at different agro-climatic zones of the country. Valuable research had been conducted and many agroforestry systems had been standardized by these Centres. The need is to share the results and information generated with the scientists\academicians\students\farmers and other end users for their ultimate adoption.

This book entitled **"Agroforestry: Systems and Practices"** is a novel effort of Dr.Sunil Puri and Dr.Pankaj Panwar in bringing literature on agroforestry from different parts of the country in a single book. The book will certainly benefit the researchers, extension workers, policy makers, progressive farmers and student.

(S.K.Dhyani)

Preface

Agroforestry is unique in many respects, both as a science and as a practice. It encompasses a set of land use practices which aims to realize the benefits of growing woody and herbaceous species together. World over agroforestry is practiced since time immemorial but it has emerged as a focal theme in late 1970's. Today the term agroforestry has become a widely used 'term of art' in agriculture and forest policy, despite the fact that it has no universally agreed definition. It also has become a widely used political and public relation tool. The Indian Planners in mid 1970's while preparing Seventh Plan Document recommended to include agroforestry as a core subject in curricula of State Agricultural Universities (SAU's). National Agriculture Policy 2000 of India has emphasized the need for diversification in agriculture through agroforestry. Similarly, the Task Force on Greening India for Livelihood Security and Sustainable Development, cover Joint Forest Management and Agroforestry are the viable options.

Today agroforestry has become an important land use strategy and has emerged as one of the sustainable option of land management. Presently there is a great variety of perspectives and a lively debate on both its practical and scientific meaning. India being a large and diverse country, with reference to its climatic, edaphic and socio-economic conditions, the need is to develop land use systems, which fit well in these conditions. Lot of research with regards to complimentary and supplementary effect of perennials on annual/associated crops has been done with a support from agencies like ICAR or by individual efforts of SAU's/ Institutes/ Scientists. These research efforts were able to recommend some well to do location specific agroforestry systems/ practices. The need is to share these location specific sustainable systems, which had been developed with great efforts, with rest of the scientists/ farming communities working elsewhere. Usually this is done in the form of annual reports or seminars. However, due to lack of funds these reports are circulated to only few centers and SAU's, leave aside their reach to students, which are budding scientist of future.

Taking this into account, it was thought to publish a book on Agroforestry in order to club all the agroforestry systems and practices prevailing and/or researched in every nook and corner of the country. This will act as a ready reference of literature for students, teachers, researchers, planners and other people who are or would like to be involved in this Herculean task of greening the nation and alleviating poverty through approach of agroforestry in different physiographic conditions of country. For the sake of convenience, the book has been divided into four broad sections i) Agroforestry in India ii) Agroforestry system and practices in North, West and Central India iii) Agroforestry systems and practices in East, North-East and Southern India iv) Allied topics related to Agroforestry. The states have been taken as a unit. As far as possible from each state a senior author was identified and requested to compile the agroforestry achievements obtained in the state. In totality there are 37 papers covering almost all the states of the country, besides covering some of the important topics like women, tree improvement, modeling, canopy management, bamboo, mined areas and medicinal plants where agroforestry can play a dominant role.

We feel that the authors have done justice in compiling the research and technology developed in their respective state and regions. We appreciate the commitment of the contributing authors and take the responsibility for those important issues still missing in the book for an inquisitive reader.

We wish to thank all the contributors for complying with the time schedule and bearing with the persistent requests and queries made by us. We would like to extend our deep appreciation to Prof. M.K. Mazumdar, Vice Chancellor and Prof. S.K. Ghosh, Dean, Faculty of Horticulture of Uttar Banga Krishi Viswavidyalaya, Cooch Behar (West Bengal) for their continuous guidance and support. Their deep insight on aspects of integration of trees with crops and/or animals into the present land use systems has been a guiding principle in undertaking research activities on agroforestry in Uttar Banga Krishi Viswavidyalaya.

Prof. Sunil Puri
Dr. Pankaj Panwar

Content

Section 1

Agroforestry in India

1

How Can We Promote Sustainability and Development of Agricultural Lands through Agroforestry Practices — Some Experiences from India

SUNIL PURI* AND PANKAJ PANWAR#

*Indira Ghandhi Agricultural University, Raipur - 492 006

Department of Forestry, # Uttar Banga Krishi Viswavidyalaya

Pundibari - 736 165 (West Bengal)

ABSTRACT

In India agroforestry practices are ancient but the story of organized agroforestry research began in 1979 at Imphal. Considerable efforts have been expended towards promoting and implementing agroforestry practices in the country by establishing centers for agroforestry research in almost all agro-climatic zones. Though almost 30 years has passed since systematic research started however, adoption of agroforestry is still in patches. This is because of myths attached with integrating trees with crops, market condition, rules regarding timber movements, allelopathy, long duration and so on. This paper examines the various arguments for and against agroforestry practices in India and presents thc opportunity and challenges in adopting agroforestry as a sustainable land use practice.

INTRODUCTION

In India the green revolution led to increase in production through enhanced productivity. Unfortunately the population growth is increasing at a rapid rate and there is an urgent need to accelerate agricultural growth to address issues on food security, nutrition adequacy, rural income generation, employment and poverty. Despite higher production, the per capita availability of food over time has not increased significantly. About 70 to 80 per cent of our people depend on agriculture for their livelihood. Contribution of agriculture to GDP is declining; this affects the rural poor most. The urban –

rural divide and regional disparities are on the increase. To achieve a desirable rate of 7 per cent growth in the economy, agriculture must register a growth not less then 4 per cent. Now it is around 2 per cent and is declining (Singh, 2002). These continuous improvements in productivity are essentials but these must be capable of being maintained in perpetuity. In other words, an ever-green revolution rooted in the principles of ecology, economics, social and gender equity, energy conservation, employment generation and social auditing, is essential in major farming systems of India. Land, water and vegetation care constitute the foundation for building up such an evergreen revolution movement.

A major opportunity to meet the challenge exist, if only we are able to break the traditional sectoral divide between 'agriculture' and 'forestry', and recognize 'agroforestry' as farmer-led efforts to meet livelihood needs on a limited land base without categorical distinctions between 'perennials' and 'annuals' components of their enterprise. In India agroforestry practices are ancient but the story of organized agroforestry research began in 1979 at Imphal, the capital of Manipur when a seminar on Agroforestry was organized by Indian Council of Agricultural Research, New Delhi to accumulate and compile the data on the research and development of agroforestry research in India.

In India, agroforestry has been a way of life. Its evolutionary process could be understood through the historical development. The process of human evolution has been from forests when man learnt the art of domesticating plant and animals after leaving hunting and gathering habit. Thus agroforestry is an age old practice whose origin is believed to be since the Vedic period. The period of Indian history during which two great epics (Ramayana and Mahabharata) were composed by Maharishi Valmiki, there is an illustration of Ashok Vatika, a form of home garden. It comprised of plants encircling the sarovar (water tank) in three rings. Forest trees (like *Ashok, Neem, Pipal, Bargad*) in the outer ring, horticultural trees (jack fruit, mango, banana and guava) in the middle ring and flowers (*champa, chameli, juhi, bela* etc) in the inner ring. The tribal dances performed around trees are symbolic of the intimate ethnic bias to protect trees. Trees had considerable socio, ethnic and religious backing, but their use as agroforestry component was limited.

Despite the considerable efforts that have been expended towards promoting and implementing agroforestry practices (e.g. the recent National Agriculture Policy of 2000), the farmers are reluctant to take up. Moreover, during recent years agroforestry practices have sparked off some controversies.

Critics argue that such activities are taking over land which previously provided them with the means to feed themselves and earn a living. The planting of large areas of eucalyptus, acacias, pines and poplars under the agroforestry/ social forestry has sparked off controversies. Some claim that the practice will destroy the environment and ruin small farmers. Other say they will help protect natural forest and provide economic growth. Most of the public does not know what to believe.

This paper examines the various arguments for and against agroforestry practices in India. This is a complex topic. Sometimes following agroforestry practices is an excellent way to use the land; sometimes it is not. In one location a plantation of fast growing eucalyptus might have a profoundly negative impact on the crop or on the amount of water available to other users. Yet a similar agroforestry practice elsewhere might do little or no harm to water resources. A plantation of acacias or poplars under agroforestry might produce significant social and economic benefits. Yet a similar plantation elsewhere might lead to changes that hurt local communities.

Our principal objectives is to separate fact from fiction, science from speculation. In some cases established scientific findings are at odds with popular perceptions and we highlight these. But science cannot provide answers to complex questions that have an ethical, political and social dimension. Rather, science provides information, and we must do our best to use the information as an aid to better decision- making, rather then as the sole basis for it. We hope that this paper will make a constructive contribution to the debate on agroforestry myths and provide a useful guide to policy makers. Strength and weakness of agroforestry in India are highlighted. The paper also discusses technology adoption issues in the context of agroforestry systems in India and what policy makers might do (if anything) to expand and promote agroforestry.

MYTHS AND REALITIES

Competition

There is a general belief amongst the farming community that when trees are grown in combination with agricultural crops, the latter are adversely affected in the vicinity of trees due to competition. Competition between trees and crops is often viewed as an inherently negative aspect of agroforestry systems, which one has to avoid or reduce by appropriate management protocols. Indeed, this view has led to the conclusion that one should shy away altogether from integrating trees with crops, especially in drier climatic zone and rather focus on fallow rotations (Sanchex, 1995). It remains debatable whether competition is so inherently negative that the above conclusion is warranted. No doubt it depends on ones perspective and on the way one evaluate the different components of the system.

An example of the above, farmers in the sub-humid India are following traditionally a parkland agroforestry system which includes a highly competitive, fast growing native species, *Acacia nilotica,* because it provides a wide range of products for household needs as well as generating income. It is valued principally because its high quality timber is termite resistant. At the same time the tree provides a variety of non-timber products (gums, seeds) and its leaves and fruits provide fodder for goats, sheep and cattle. Typically, a farmer establishes 10 to 20 trees/ ha, mainly from wilding, on a farm. By practicing agroforestry (rice + tree) system, farmer gets higher cash returns on a short term (10 year) harvest cycle of trees (Vishwanath *et al.*, 2000)Thus this species in sub-humid zone is considered as compatible with the agroecosystem even though it is highly competitive with crops. In contrast to this, the farmers of Indo Gangetic Plains and of North India have no preference for this tree as it reduces crop productivity by 40 to 60 per cent and is not a remunerative tree as agroforestry (Sharma, 1992; Puri *et al.*, 1995; Chaturvedi and Das, 2002). Under these circumstances it is regarded as a competitive system.

Early studies of agroforestry begin with the general assumption that complementarity was the norm as tree are believed, as a rule, to be more deeply rooted than crops. The myth is that where tree can exploit resources that are unavailable to crops, agroforestry can increase productivity per unit area of land through the efficient use of resources; in other words through spatial complementarity. The distribution of the root system through space and time is usually influenced by both the genetic character of the species and localized soil conditions (Schroth, 1999). Many studies have inferred competition between tree and crop roots for soil resources due to shallow tree roots (e.g. Puri *et al.*, 1994b in *Populus deltoids;* Swamy *et al.*, 2002 in *Gmelina arborea*). Whereas, there is a tacit assumption that the shallow –rooted trees can compete with associated crops and this may lead to yield depressions and contribute to economic failure of land-use-system. This can be over come either by management of root system (a tedious job) or by selection of deep rooted trees, like *Acacia nilotica, Prosopis cineraria* etc. It suggests that to reduce the negative effects of competition, tree spacing should be increased to minimize competition (Singh *et al.*, 1989). An extension of this idea is that to further minimize negative effects of trees on crops they should be matched to appropriate niches within farm. Thus the agroforestry systems can increase the efficient use of resources by combining species with different patterns of resource uptake.

Crop Yield

The general perception is that crop yield decreases when grown in association with trees. However, when plants grow in proximity to each other they interact in positive ways (complementarity) or in negative ways (competition). Crop yields under trees may be unaffected during the early years of tree growth, but could increase or decrease when tree are large. Any positive influence of trees on crop yields through soil fertility improvement would not be expected for many years, as trees modify the soil environment slowly. It has been postulated that the soil beneficial effects of *Prosopis* on crop yields will start only after the trees are more then 10 years old (Puri and Kumar, 1995). On the other hand, the negative effect of trees because of competition for growth resources can be notified within few years. The slow-growing trees such as *Prosopis* may not influence crop yields for many years after their establishment (Puri and Bangarwa, 1992), but the fast-growing trees such as eucalyptus reduce crop yields fairly quickly, within two to three years after their establishment (Ralhan *et al.*, 1992). At Hissar 3.5 year old *Eucalyptus tereticornis* on field boundaries reduced crop yield by 41 per cent up to 10 m from the tree row (Malik and Sharma, 1990).

There are simultaneous agroforestry systems that work remarkably well: the *Prosopis cineraria* parklands in the arid and semiarid regions (Shankarnarayan *et al.*, 1987; Puri *et al*, 1994a). Millet and Sorghum crops growing within a 5 to 10 m radius around mature *P. cineraria* tree yield two to three times more then adjacent crops growing in open. Its foliage and pods are used as fodder for ruminant animals during the dry season. Animals enjoy the shade, urinate and defecate more often under the *P. cineraria* trees. Birds congregate on tree branches, adding nutrient inputs to the soil under the tree canopy. *P. Cineraria* also fixes nitrogen during the dry season with nodules located on roots several meters below the soil surface near to the water table (Tejwani, 1994).

There is little question about the positive tree-crop interaction in *P. cineraria* parklands, but it does not always happen (Yadav and Blyth, 1996). The processes involved need to be fully quantified at well - characterized sites. How densely can the *P. cineraria* trees be planted and still provides the positive effects? How long does it takes for the tree to start exerting such positive effects?

It is frequently claimed that when trees and crops are brought together they complement rather than compete with each other in their capture growth resources. The biophysical bottom line of agroforestry is how to manage the interaction for natural resources (light, water and nutrients) between the tree component and the crop and/or livestock components for the benefit of the farmers (Rao *et al.*, 1998). Agroforestry literature on tree-crop interactions is largely based on the studies with young trees, with the exception of studies of parkland system. There is considerable evidence that mature scattered trees in such environment are associated with improved soil nutrient supply in parkland system, but it is uncertain that whether certain trees actually increase the total pool of nutrients in the system or simply re-distribute existing soil nutrients more efficiently across the landscape (Kumar *et al.*, 1999), Nair *et al.*, 1999). If higher nutrients under the trees are due to redistribution, no net gain in production at the landscape scale should result. However, this will still enable the farmer for less input of nutrients. Unfortunately, few studies have sampled soil nutrients randomly over sufficiently large areas to test this hypothesis. For example, Puri and Kumar (1995) and Pandey *et al.*, 2000 reported increased soil fertility under mature scattered trees but their study only compared under-storey soil to open soil. Tree roots are known to spread laterally well beyond their canopies (e.g. up to 4 times the canopy dimensions of *Prosopis cineraria* and *Gmelina arborea* as reported by Puri *et al.*, 1994b; Swamy *et al.*, 2002) so it would remain uncertain as to whether the contrast in soil fertility resulted from redistribution or truly from addition of nutrients.

Quick growing trees with spreading crowns generally impede light infiltration thereby adversely affecting under-storey crop productivity. Crop yield declines, relative to yields in open- treeless fields, are noted mostly under the canopies of large, evergreen unmanaged trees. Wheat yields were reduced by up to 60 per cent (Puri and Bangarwa, 1992) and mustard (*Brassica spp.*) yields by up to 65 per cent (Yadav *et al.*, 1993) under *Acacia nilotica* trees. In environment where seasonal rainfall (above approximately 800 mm during the cropping season) or irrigation provides adequate water for crop growth, reduction of crop yield under trees is mainly due to reduced light.

Light resource availability in agroforestry systems can be altered through managerial interventions like tree canopy management, broad spatial geometries of tree planting and growing crops when tree become deciduous (e.g. poplars) or trees with light/sparse crown (like *Ceiba pentandra*). Alternately, if light infiltration is impeded due to tree growth/thick canopy cover, shade loving crops like *Zingiber officinale* and *Curcuma longa*, be grown (Kumar, 1999). Shade tolerant medicinal crops like *Pogostemon cablin*, *Kaempferia galangal*, etc are also suitable intercrops (Kumar *et al.*, 1999).

It must be recognized that in spite of large yield increases or decreases under scattered trees and near boundary plantings, the overall effect of trees on crop yields could be small because only a small proportion of the area is subjected to tree/crop interaction (Rao *et al.*, 1998). Consider a scattered - tree system that contains 20 tree ha^{-1} with each tree affecting crops over $100m^2$ area. If the trees increase or decrease crop yield by 50 per cent over that area, the overall yield of the system will be only 10 per cent higher or lower than that of sole crop. Similarly, if a boundary planting of a 100 m tree row reduces crop yield by 50 per cent over a 10 m distance on either side, then the yield of the system as a whole over one hectare increases or decreases by only 10 per cent (or 50 per cent to each neighboring farmer).

Water

The relation between trees and water continues to be subject of confusion in the public debate, but the fact that young tree plantations, especially of evergreen species tend to use more water then established, deciduous forests or agricultural (non-irrigated) lands has gained attention in the form of the *Eucalyptus* debate. While there is no reason to single out *Eucalyptus* species in this regards, the high water use of fast growing, evergreen trees can be a concern in areas with a shortage of groundwater of subsurface flows of water. In other areas such interception of subsurface flows can be seen as the basis of an environmental service function, where it prevents salt movement in groundwater flows. 'Environmental service' perceptions will thus depend on the local agro-ecosystem.

The hydrological effects of agroforestry, and particularly the effects of trees on water yields, have been the subject of much myth-making. Almost invariably, whenever there is water scarcity the cry goes up that it has been caused, or aggravated, by deforestation. And this is generally accompanied by the claim that the obvious way to prevent a recurrence is to plant trees in the water catchments, the idea being that they will soak excess water. There are often good reasons for planting trees in agroforestry, but they can seldom guarantee an even flow of water, whatever the climatic conditions.

Understanding the links between agroforestry and water resources requires a basic knowledge of hydrology and plant physiology, not to mention an understanding of various environmental processes. It is a complex matter since the real contribution to water regime will vary from one site to another. Topography, soil type, local climate, the type of agroforestry practice involved and a variety of other factors will exert their own particular influence.

Water competition between trees and crops can occur both within the rainy season and through the build-up of a soil water deficit caused by tree transpiration during the dry season. Inter-seasonal water deficit can be particularly severe when there is a succession of droughts and when recharge is small and infrequent (McIntyre *et al.*, 1996, Govindarajan *et al.*, 1996). Thus, temporal complementarity is only possible if residual water is available after the crop harvest or if rainfall occurs when there is no scope for cropping. For example, in Hyderabad, 20 per cent (152 mm) of the annual rainfall occurs outside the normal cropping season and even the traditional intercrop of sorghum/pigeon pea utilize only 41 per cent of the annual rainfall. The reminder of the rain is lost as runoff (26 per cent) or deep percolation (33 per cent). In such environments, agroforestry practices increase rainfall utilization by extending the growing season. Thus, the total rainfall utilization of perennial pigeon pea/groundnut systems was 84 per cent, about twice the value reported from the most productive intercrop systems in Hyderabad (Ong *et al.*, 1992). Thus competition for below-ground resources is inevitable and it is best managed by manipulating the interface between species in a mixture.

The most remarkable example of temporal complementarity in water use is the unusual phenology of *Faidherbia albida* which is in leaf in the dry season and sheds its leaves in rainy season. It retains this phenology even when planted in the Deccan plateau of India, where the water -table is too deep for access by tree roots. One of the few experiments in which *F. albida* was compared with *Albizia lebbek*, a tree with conventional leaf phenology, is that reported by Ong *et al.*, 1992 at Hyderabad. Comparison of sap flow rates of *F. albida* begins when the understorey crops have developed full canopy later in the rainy season. In contrast, *A. lebbek* produces a full canopy well before the onset of the rains and sheds its leaves when the *F. albida* starts to develop its canopy. Thus tree phenology can be selectively used for temporal complementarity in which trees like *F. albida* make growth demands at a different time from that of the crops.

The hydrological impact of fast growing exotic trees has caused concern to local people in India, especially in semiarid areas, due to the eucalyptus controversy. In southern India, eucalypt plantations have been reported to not only extract all the rainfall that enters the soil, but also utilize an additionally 100 mm of water from each meter depth of soil that the roots penetrate (Calder *et al.*, 1997). This is a matter of concern as studies in Karnataka have shown that eucalyptus roots can reach to a depth of 8 m in three years (Calder *et al.*, 1997). Consequently, when planning trees in association with crops, it is essential to consider the implications of increased water use on the medium and longer-term water budgets. Special consideration should be paid to the source of water used by trees, the rate of water depletion below the crop rooting zone and the prospects for deep recharge as using water at faster than replacement rates demonstrates a serious ecological consequence of not mimicking the resource use patterns of locally adapted species (Ong and Leakey, 1999).

It is advocated that trees harvest resources from their surroundings and accumulate them under trees. As the positive influences of woody plants on productivity are due to the redistribution of resources within landscape it will always be difficult to develop agroforestry without risking the depletion of resources at a distance from the tree. An intriguing possibility is the prospect of capitalizing on this phenomenon of resource capture at a distance through strategic use of 'hydraulic lift'. This is the process in which water taken up by roots moist zones of soil is transported through the root system and released into drier soil. Rainfall captured through stem flow, especially by woody canopy, can be stored deep in the soil close to roots and be returned to the topsoil beneath canopy by hydraulic lift for later use with associated benefit for under story species (Ong and Leakey, 1999).

There is strong perception that soil water beneath trees is high. Rhoades, 1995 reported increased soil water (4 - 53 per cent greater than in open) in the crop root zone beneath *Faidherbia albida* canopies. In theory, trees can increase soil water content underneath their canopies, if the water is saved by shading effects on soil evaporation and rainfall redistribution (e.g. funneling of intercepted rainfall as stem flow), exceeds that removed by the root systems beneath tree canopies. At high tree densities, the proportion of rainfall lost as interception by tree canopies and used for tree transpiration would exceed that saved by shading and stem flow, resulting in drier soil below the tree canopy.

It is frequently said that trees can regulate water flow by retaining water during the wet season and releasing it slowly during the dry season. Many scientists see some conceptual validity in this (Cossalter and Pye -Smith, 2003). After all, root activity leads to better soil structure and increases the capacity of the soil to store water. In theory, these soils should release water slowly, rather than rapidly. However, this is rarely borne out by empirical analyses, and what little evidence exists to support this view is anecdotal.

SOIL EROSION AND NUTRIENTS

Soil degradation has become an increasingly serious problem where many soils are inherently poor in nutrients and at high risk of erosion. The main causes of soil degradation are poor agricultural practices, deforestation and overgrazing. As a rule, erosion tends to increase during site preparation for crop cultivation, when the soil is exposed to wind and water. The amount of erosion depends on slope, soil type, rainfall, the duration of the barren phase when the soil is directly exposed to wind and water, and the nature of the tree canopy. Once the trees are established, erosion may still be an issue, especially where there is little ground cover, on sloping ground and in trees where leaf size encourages the creation of large water drops. Trees planted for short-rotation will experience more erosion than long-rotation trees when established under identical conditions.

Keeping the soil resources in place is one of the main sustainability issues. Many agroforestry systems help keep the soil resource in place by biological means. Closely spaced trees on slopes reduce soil erosion by water through two main processes: first, as a physical barrier of stems, low branches, superficial roots and leaf litter against running surface water; second, as sites where water infiltrates faster because of generally better soil structure under trees than on adjacent land. The scientifically acceptable evidence is overwhelming (Rao *et al.*, 1998). Trees, however, do not provide these functions until they are well established and have developed litter layer. Sequential agroforestry systems that include crops or ground cover while trees are small overcomes this limitation.

As a general principle, multiple rotations of trees (e.g. eucalyptus) will have a more significant impact on soil fertility then trees (e.g. teak in taungya) that are harvested after a long period of time (Nair *et al.*, 1995). The quantity of nutrients removed will be greater, and so, consequently, will be the requirements for fertilizers. Nutrient recycling processes vary considerably from one soil type to another, and with different harvesting and replanting procedures. Nutrient losses are generally most significant during site preparation and harvesting. Some nutrient losses also occur as result of erosion.

It is important to stress that trees are generally much less degrading to the soil then many commercial agricultural crops. FAO cites a study that found that the amount of nitrogen removed by cereal crop was two and half times more than the amount removed by eucalyptus plantation. In the case of phosphorous it was fifteen times more (FAO, 2000). A study comparing fast-growing pine with grains, potatoes and alfalfa found that the latter removed 15 to 35 times more nitrogen (with the exception of nitrogen-fixing alfalfa), 80 to 250 times more phosphorous, and up to 10 times more calcium than the pines (Cossalter and Pye-Smith, 2003). Acacia trees, like alfalfa, fix atmospheric nitrogen, so unlike eucalyptus they increase the amount of nitrogen in the soil. It goes without saying that good management will help to minimize soil loss and decline in fertility, regardless of whether the land is devoid to trees and /or annual agricultural crops.

The relationship between trees, crop and the soil is extremely complex. Short rotation trees are more likely to experience higher rates of erosion and a greater decline in fertility than longer-rotation trees, but they are generally much less degrading to the soil than many agricultural crops. In some situations they can have beneficial influence on the soil. Agroforestry soils are especially liable to erosion during harvesting and crop raising. Ground vegetation, with or without trees, is much more important than the canopy when it comes to reducing soil erosion (Nair *et al.*, 1999).

The nutrient resource base for agricultural production can be accomplished through trees in the agroforestry system. There are some indications of positive interactions or at least additive effects between soil organic matter (provided by trees) and inorganic fertilizers (Breman and Kessler, 1997; Palm *et al.*, 1997), but it is not yet clear whether agroforestry has a major role to play in the soil replenishment initiatives which are currently underway (Sanchez *et al.*, 1997). Trees may indeed increase nutrient and water availability for annual food crops and pasture production, but only if trees:

- have the right properties (can generate stem flow of water, are deep rooted, are unpalatable and of low 'quality', and thus able to contribute more to long term accumulation of soil organic matter rather than rapid mineralization, exhibit slow growth rate and have a long life span);
- are used at an appropriate tree density
- are properly managed by pruning, lopping and controlled harvesting;
- are growing on appropriate sites and soil types in which the above goals can be realistically achieved.

Trees with the best set of properties, such as *Prosopis cineraria*, may increase grain yield in the Thar desert, provided they have an average tree canopy cover of about 20 per cent and can gain access to the water-table (Puri *et al.*, 1994a). A general 'grow more trees' approach is not likely to meet such conditions.

If the beneficial effects of parklands trees on soil properties are linked to trees with a high proportion of woody aboveground structures, then it would take a long time before the beneficial effects can be realized, since investment in woody structure slows tree growth. Therefore, it is not surprising that the positive interactions between mature, deep-rooted and widely spaced *Prosopis cineraria* trees would require 10 – 15 years. Such long time scales are well beyond the planning horizon of many farmers for the relatively small benefit in crop productivity and may well explain why farmers rarely plant these trees, even though they are well aware of tree species which improve soil and climate.

ALLELOPATHY

This refers to the production of certain chemicals popularly known as allelochemicals by one or more plant species when grown as mixed communities which affect, alter and/or modify the growth and production of associated species. Allelopathy is one of the widely considered limitations for promotion and adoption of agroforestry at the field scale.

Allelopathy is discussed here as applicable to all agroforestry systems. If allelopathy exist, its effects could be more important in simultaneous systems than in sequential systems because of the permanent presence of trees and direct exposure of crops to the continuous release of chemicals. Some studies have reported the deleterious allelopathic effects of certain agroforestry trees on crops (Puri and Khare, 1991; Nandal *et al.*, 1994). Most allelopathy literature is based on laboratory studies or bioassays conducted under controlled conditions, using unrealistic concentrations of aqueous leachates or extracts of tree parts and it is not clear to what extent these results have practical value. Even though certain trees may release harmful chemicals into soil through leaf litter or root secretion, they may have only temporary or no effect on crops under field conditions because of rapid elimination of chemicals by soil microorganisms and /or leaching, particularly in the humid environments. Moreover huge soil volume acts as a sink for decomposition of allelochemicals. The practical significance of allelopathy cannot be ascertained unless the effect of allelochemicals on crop yields is demonstrated under field conditions, independent of the competition of trees for water and nutrients, and independent of the temporary immobilization of plant nutrients due to addition of low quality tree liter (Sanginga and Swift, 1992). To prove weather allelochemicals accumulate in soil under the tree canopies, Singh *et al.*, 1998 took 30 cm surface soil under the 25 year old plantations of *Eucalyptus tereticornis*, *Acacias nilotica*, *Prosopis cineraria*, *Terminalia arjuna* and *Albizia lebbek* and filled in pots. Soil was also taken from adjacent field where trees were never associated with agricultural crops. Wheat and oats were grown as test crops in these pots. The average yields of these crops were significantly higher in plantation soils than in the control soil. This indicated that nothing like allelochemicals accumulated in the soil even after 25 years of tree occupancy which could affect the yield of two most popular agricultural crops of the region.

BIODIVERSITY

Agroforests are typically less diverse than native forest (but more diverse than agriculture), they do contain much greater number of plants and animal species than forest plantation (Murdiyarso *et al.*, 2002). This diversity can, at time, provide ecological resilience and contribute to the maintenance of beneficial ecological functions. Similar to plantation forests, agroforests are 'working forests' and they can help relieve some of the pressure to harvest native forests (although their presence as such is not a sufficient condition for protection of old growth forests) Angelsen and Kaimowitz, 2001. If natural forest is cleared to make way for agroforestry practices, there will be a loss of biodiversity. Yet a similar practice, established on degraded land, might bring about an increase in biodiversity. In other words, the impact on biodiversity will be a function of what they replace. Other factors of importance include the location, its size, length of tree rotation and species composition. The issue of contiguity is also important. If agroforestry practices (such as shifting cultivation) are sited close to existing natural forests, they may benefit from their biodiversity: animals, birds and insects will be readily available to invade. However, if no such reservoir of biodiversity exists, then the chance of being invaded by wildlife from outside become remote.

However, there are two sides to this story. In some situations agroforestry practices may have a positive impact on biodiversity. When the vegetation has already been destroyed or seriously damaged-for example, by unsustainable agriculture of overgrazing-tree plantations under agroforestry practices may help to restore some biodiversity, either by sheltering relics of the original flora and fauna and allowing them to proliferate, or by creating a new ecosystem, albeit with a different mix of species from the one which originally occupied the site. In many parts of India, trees have been established on barren lands or abandoned agricultural land. On the whole, this has been good for biodiversity, although growing trees on land no longer capable of sustaining farm crops might result in mediocre wood yields.

Some trees will support more biodiversity than others, and the same species will invariably shelter and attract more wildlife in its natural habitat than when it is planted as an exotic. Native Australian eucalypts, for example, support a rich diversity of animals and plants. These have evolved with the trees. However, when planted as an agroforestry component in India, the same eucalypt species tend to have little or no undergrowth and support relatively few species (Lal *et al.*, 1997).

GLOBAL WARMING

Rising concentrations of greenhouse gases, in particularly carbon dioxide released by the burning of fossil fuels, are leading to global warming. According to the Intergovernmental Panel on Climate Change (IPCC), if current trends continue, temperatures will rise by 1.4 to 5.8 degrees C over the next century. In the worst-case scenario, sea levels will rise by just under a meter, and costal cities and deltas will be inundated. There are two main ways of tackling global warming. One involves conserving carbon in its solid state; the other - and this is where agroforestry practices come in - involves its sequestration.

The science of carbon sequestration is simple enough: trees and crops convert carbon dioxide into soil carbon and stores it in the form of wood or crop biomass. Carbon cannot be stored in crops for long duration. Much of the opportunity to store carbon through afforestation will occur on agricultural lands due to the vast land area devoted to agriculture in India. Obviously, the nature of trees (fast or slow growing) and the uses to which its timber are put will determine whether or not it is beneficial from an environmental and social point of view. If the timber is used to make high-quality furniture with a long shelf life, then the carbon produced by the trees will be locked up for an appreciable period of time. If, on the other hand, they produce wood that is pulped to make paper, much of which will be burnt or quickly discarded, they will fail to serve their true purpose of carbon sequestration. Critics of carbon sequestration fear that sequestering carbon in one area-by, for example, planting trees- could lead to pressure on forests elsewhere, thus neutralizing the benefits of the scheme. They also suggest that carbon sequestering projects involving trees will have a very short-term impact on carbon budgets. Recently research related to carbon sequestration in agroforestry practices has been initiated in India. At 6 years, *Gmelina arborea* based agrisilviculture system sequestered 31.37 Mg ha^{-1} carbon (Swamy *et al.*, 2003). In a silvopastoral system carbon flux in net primary productivity increased due to integration of *Prosopis juliflora* and *Dalbergia sissoo* with grasses (Kaur *et al.*, 2002).

IMPACT ASSESSMENT

Although agroforestry has been a way of life in India, it was in mid 1960's when planting of eucalyptus in agricultural fields brought revolution in the economic activities of farmers (Pathak and Pateria, 1999). In 1970 - 75 only 7 per cent of the farmers in Punjab were planting eucalyptus and fetched good prices at the harvest resulting in increase in adoption of eucalyptus (46 % farmers in 1980 - 85). But when these plantations became harvestable (in late 1980's to early 1990's), the prices crashed due to glut in the market of eucalyptus timber and low consumption by paper/pulp mills. This resulted into sharp decline in the percentage of eucalyptus growing farmers (5 %) in 1990-91. This situation diverted the farmer's attention to *Populus* plantation which increased from 3 per cent (in 1970 - 75) to 48 per cent (in 1990-95). Large scale poplar plantations are confined to above 28^0N latitude as it is unable to withstand harsh climatic conditions. It also require good soil moisture (higher water table) and is very adaptable to agrisilvicultural practices

(Joshi, 1996). However, in late 1990's a few clones have also been introduced in Central parts of India, i.e up to 21°N latitude (Puri *et al.*, 2002). Presently it is grown on large scale and is replacing eucalyptus in the irrigated belt. Easy marketability of this species has encouraged its wide adoption. In 6-8 years, it produce 10 – 30 m^3/ha/yr (under intensive management 46.92 m^3/ha/yr). It also helps enriching the soil, works as wind break in fruit orchards and thus used on the farm boundary. Limited plantations of other trees species, such as, *Dalbergia sissoo, Acacia nilotica, Gmelina arborea*, bamboos, etc have been undertaken by farmers. Farmers have also developed agrisilvicultural and agrisilvihorticultural systems involving many fruit trees.

The quantifiable impact of agroforestry research/technologies generated in the past is visible as evidenced under:

i) The poplar and eucalyptus based agroforestry has been adopted in a big way by the farmers of Punjab, Haryana and western Uttar Pradesh particularly the Terai belt. There are reports that by raising poplar with crops, the farmers in these regions are getting higher income per ha from the sale of poplar after 6-8 years of planting, compared to that of agriculture produce (Joshi, 1996). Punjab, having only 2- 3 per cent area under forests, now has become a major wood surplus state and is presently supplying wood to Himachal Pradesh for making boxes for fruit packaging.

ii) Agroforestry played a major role in the recent past in the development of wastelands such as deserts, salty lands, ravines and gully affected areas and in the control of water and wind erosion. For example, by using the agroforestry technologies developed at Central Soil Salinity Research Institute, Karnal, the state forest departments, NGO's, National Wastelands Development Board (NWDB), New Delhi and other developmental agencies have rehabilitated salt-affected soils particularly the village community lands, government lands, absentee lands, area along road sides, canals and railways tracks (Singh *et al.*, 1994). Seeing the large scale success of these technologies at the grass root level, the international funding agencies such as World Bank and European Economic Community funded pilot projects on afforestation/agroforestry in several states in the past.

iii) Indian Farm Forestry Development Co-operative Limited has implemented agroforestry in the states of Uttar Pradesh, Madhya Pradesh, Chhattisgarh and Rajasthan. This has been done by people's participation through the formation of Co-operative Societies. Similarly, Tree Growers Co-operative in Gujrat has come in a big way to implement agroforestry on farm lands and wastelands.

iv) Agroforestry got a major boost in north of India through farmers friendly policies of M/s WIMCO Ltd. This company has not only motivated the farmers to plant trees, such as poplar, on their farm lands but also created a big market on the economy of these states by creating assured market for poplar wood (Joshi, 1996).

v) Owing to increased supply of wood in the market, there has been a significant increase in the number of factories/industries dealing with wood and wood based ventures in the near past. Such industries have promoted agroforestry (through eucalypts) and contributed significantly in increasing area under farm forestry. Notable examples are ITC Bhadrachalam Paperboards and J & K Paper Mills; Newsprint Factory has a buy back deal with the farmers for their agroforestry produce.

vi) Seeing agroforestry as a viable venture many corporate bodies, NRI's and Limited Companies such as Kitply Industries, Golden Forests, Green Gold, etc. have entered this business in a big way and initiated agroforestry activities in collaboration with farmers at a large scale.

OPPORTUNITIES AND CHALLENGES

In India, the sectoral divide between forestry and agriculture is particularly pronounced. The Indian constitution places the control of natural resources in the hands of the states and states that these must be managed for the benefits of the people. Foresters have experimented for more than a century with ways to get local farmers to participate in their efforts to plant and manage trees, in various forms of 'taungya', 'agroforestry' or 'social forestry'. Most of these efforts have been a hard way to learn a simple lesson: unless farmers share substantially in the long-term benefits of forest plantation efforts, the interaction between the 'agro' and the 'forestry' component remains a competitive one. Because of land scarcity, large-scale plantations and smallholder development programmes tend to be mutually exclusive. What is needed is that foresters start to participate in farmers' tree planting efforts, rather than expecting farmers to participate in foresters' efforts.

Presently the success of industrial or government schemes related to plantation is widely questioned. The reasons for the failure are numerous, but the key reason is non-involvement of local people. In general these plantations are established by technicians and contract laborers who have no post-planting responsibility, concern or expectations of future benefits. Central planning of plantation schemes often assumed that local people would protect the newly established trees. However, having been excluded from the planning process, local people feels no sense of ownership of the plantation and no incentive to protect the trees. Plantations are often heavily damaged or completely destroyed by fires, grazing, or appropriation of the site for other uses.

In contrast, the loss of local forest resources often leads to increased incentives for spontaneous expansion of smallholder tree husbandry. Farmers protect and plant more trees on their own farms to provide tree products for household needs and market demands. Smallholders have limited time and financial resources. The trees they plant represent a conscious investment for which other options have been forfeited. Farmers generally restrict planting to the number of trees that can be maintained. They integrate tree growing with their crop and animal production activities. The management practices undertaken to assure good food crop yields- cultivation, weed control and fertilization - also benefit their trees. The available land, labor, and other resources are allocated according to the farmers objectives. Because landholdings are small, farmers select the farm niches most appropriate for tree production. The combination of limited resources, small individual planting, and intimate familiarity with the planting site result in high tree survival and good growth rates.

In India, smallholders form the majority farming community. Smallholders with diverse, risk-averse farms that include a significant tree components ('agroforestry') are seen to be the agroforest producers of the future. Thus, a paradigm shift is required in the agroforestry sector, to include the more flexible farmer-led approach in wood and fiber production as part of the solution to achieve sustainable land use objectives. There are, however, a number of challenges that need to be addressed before the full potential of this new evergreen revolution can be realized.

Usually, smallholders start timber production systems by planting short-rotation species to meet household and local market needs. As more farmers begin producing timber, supply meets or exceeds demand and prices decline. At this point lead farmers either stop or diversify into long-rotation, premium-quality timbers (Saxena, 1995). However, the dynamics of tree product supply, market demand, and marketing channels at the small holder level are poorly understood by farmers.

Market agents perform the important roles of linking farmers to processors and manufacturers who transform the raw materials (commodities) into finished goods. Local and regional dealers serve very important roles - collecting, sorting, grading and transporting raw materials. One of the largest risks reported by middlemen is unreliable quality and quantity of small holders products. This uncertainty, plus time and expense required to interact with numerous smallholders, are usually cited as the reason dealers pay low rates to individual farmers. The absence of price incentives at farmer level for higher quality products, however, maintains the status quo on quality. This constraint on the contribution of agroforestry can be overcome, if public domain information access on market conditions improves.

Worse still, the policy measures aimed at protecting forests also are applied to agroforestry systems that are managed sustainably by small-scale farmers. The Unintended result of treating all timber alike- regardless of its origin in forests or on farms - is that smallholders who plant and tend trees are unfairly penalized. They are effectively denied the opportunity to produce timber, a product that could provide them with a much-needed source of income.

STRATEGIES AND POSSIBLE SOLUTIONS

Political will : It is imperative to create a strong political will and commitment at the level of Central and State Government to enable Forest Departments to ensure sustainable development and scientific management of forest resources. Prevention of further degradation of well stocked forests and restoration of their growing stock and productivity to optimum levels through adequate financial investments and appropriate silvicultural and sustainable management practices should receive highest priority. This will be not only cheaper and cost effective compared to reforestation, but also ecologically and environmentally more desirable.

Improvement in productivity : Substantial improvement in productivity of forest resources on sustainable basis is critically important for meeting the national needs and conserving the forest resources. Intensively managed agroforestry practices based on genetically improved planting stock and sound silvicultural practices can achieve productivity levels of 20 - 30 m^3/ha. Therefore national policy must accept agroforestry as the choicest strategy for reforestation of degraded forest lands and marginal agricultural lands. Priority for tree plantation should be given for areas with good soils but devoid of tree cover as it is counter productive to tackle most difficult sites first meanwhile allowing the better site to degrade to the point of no return.

Corporate sector participation : As massive financial investments will be required for reforestation of degraded land supplemented with genetically improved planting stock and competent professional management with accountability for results, corporate sector and all sections of society should be encouraged to participate in the national effort of reforestation through innovative policy changes, fiscal incentives and tax benefits.

Local community co-operation : Co-operation of the local communities and all political parties for sustainable development of forest resources is absolutely essential. However, rights and responsibilities must go together. Unregulated grazing and illicit cutting of trees even for firewood will have to be stopped. State Governments can direct Forest Departments/ Forest Development Corporations to supply requisite quantity of firewood to the local communities ex-depots at concessional rates or free provided the concerned State Governments make suitable budgetary allocations for compensating the Forest Department/Forest Development Corporation for the subsidies involved in supply of firewood and controlled/ regulated grazing.

Liberalization of ceiling limits/restrictions : Statutory ceiling limits on agricultural land holdings for wastelands should be liberalized. Tree planting and farm forestry plantations should be treated on par with plantations of commercial crops like rubber, tea and coffee etc. Long term finance, fiscal incentives and tax benefits should be provided to promote technology based reforestation and farm forestry projects. Restrictions on felling of farm growing trees and movement of wood from farm plantations to markets/factories should completely be abolished in all states.

CONCLUSIONS

Due to complexity and multidisciplinary mix in agroforestry, its adoption by the Indian farmers is slow. The situation is aggravated either by the lack of information or due to myths in the minds of farmers. In the present paper attempt has been made to sort fact from fiction, truth from misinformation, as far as agroforestry is concerned. Our understanding of agroforestry practices in India has advanced considerably during the last few years. Scientists have sometimes exaggerated the impact of agroforestry. The truth is that in some situations agroforestry is undesirable; in other situations, it can yield benefits not just for the economy, but for the environment and local communities. It is emphasized that the sooner subsidies to agriculture are phased out, or at least reduced, the better. Subsidies create economic distortions and make agriculture viable in situations where other land uses might make better economic and environmental sense.

The relationship between trees, crops and the soil is extremely complex. Short rotation trees are more likely to experience higher rates of erosion and a greater decline in fertility than longer-rotation trees, but they are generally much less degrading to the soil than many agricultural crop. In some situations they can have a beneficial influence on the soil. Ground vegetation, with or without trees, is much more important than the canopy when it comes to reducing soil erosion.

Competition of trees can be managed to some extent by pruning, maintaining the appropriate density and sequential thinning; however, the scope for management of belowground competition is limited to the choice of trees which have deep and compact root systems. Whether tree density can be increased in parklands where crop yields are higher under trees than in the open (e.g. under *P. cineraria* trees) depends on the nature of processes that contribute to increased yields.

On the whole, in addition to promoting indigenous agroforestry models, it appears that a great deal of research needs to be done to identify other short-rotation, high-value species which suit the farmers' requirements of planting on marginal lands and bunds. Species which complement agricultural production, rather than reduce it, would be welcomed by the farmers as their primary land use continues to be crop production. It would probably be more realistic to select trees which could provide direct cash benefits to farmers through their products, and to accept that in longer term they will provide indirect environmental benefits arising from a complex agro ecosystem. Economic gains from marketable products could compensate for any loss in crop yield.

It is believed that it is essential that governments adopt a landscape approach to agroforestry development. Investment in trees, especially on community lands, should not be considered, and permission for private companies to establish plantations should not be given, if it can be demonstrated that the trees will prevent the delivery of a full range of tree goods and services at the landscape level. For example, if agroforestry practice are likely to adversely disrupt the hydrological cycle or reduce water quality, then it should not be followed. Likewise, trees should not be planted if they have an adverse effect on local communities; if , for example, they are likely to lead to a net loss of employment or to local communities being deprived of firewood, grazing land and other goods and services on which they depend. All these factors should be considered together, not independently, as there may be trade-offs that are acceptable. In any case, local communities, like other stakeholders, need to be involved at the earliest stage of planning and development.

The other question is that of ensuring that farmers get a fair price for their trees. Instead of taking over the regulatory and management functions of private plantations, the forest bureaucracy should be working in area where their services are needed more. Forest farmers' cooperatives can go a long way in giving farmers remunerative prices. The forest bureaucracy could work towards creating such cooperatives so that wood prices are determined by the farmers themselves. It could also perform market surveys and advise farmers on the wood market and on what trees are in demand.

All these issues are examined in this paper, but it is worth bearing in mind that generalizations about the impact of agroforestry on local communities, are often misleading. The problems related to agroforestry are often site-specific, and the way in which they are planned and managed is of paramount importance.

REFERENCES

Angel sen, A. and D. Kaimowitz, D. (2001) Agricultural Technologies and Tropical Deforestation. Wallingford, CABI Publishing, U.K.

Breman, H. and Kessler, J.J (1997). The potential benefits of agroforestry in the Sahel and other semiarid regions. *European Journal of Agronomy* 7: 25 – 33.

Calder, I.R; Rosier, P.T.W and Prasanna, K.T (1997) Eucalyptus water use greater than rainfall input-a possible explanation from southern India. Hydrology and Earth System Science.

Chaturvedi, O.P and Das, D.K. (2002) Studies on rooting patterns of 5 - years - old important agroforestry tree species in north Bihar, India. *Forests, Trees and Livelihood* 12: 329 - 339.

Cossalter, C. and Pye-Smith, C. (2003). Fast Wood Forestry – Myths and Realities. CIFRO, Bogor, Indonesia. Pp 50.

FAO (2000) The global outlook for future wood supply from forest plantations. Working paper GFPOS/WP/03. Rome, Italy.

Follis, M.B and Nair, P.K.R (1994) Policy and institutional support for agroforestry : An analysis of two Ecuadorian case studies. *Agroforestry Systems* 27: 223 – 240.

Govindarajan, M; Rao, M.R; Mathuva, M.N and Nair, P.K.R. (1996) Soil-water and root dynamics under hedgerow intercropping in semi-arid Kenya. *Agronomy Journal* 88: 513 - 520.

Joshi, B.C (1996) Poplar cultivation under agroforestry. In : Bach I(ed) Environmental and social issues in Poplar and Willow cultivation and utilization. FAO Proceedings 20th Session of the International Poplar Commission. Budapest, Hungary. Pp 409 – 423.

Kaur, B; Gupta, S.R and Singh G (2002) Carbon storage and nitrogen cycling in silvipastoral system on a sodic soil northwestern India. *Agroforestry Systems* 54: 21 – 29.

Kumar, B.M (1999) Agroforestry in the Indian tropics . Indian Journal of Agroforestry 1: 47 – 62.

Kumar, S.S; Kumar, B.M; Wahid, P.A; Kamalam, N.V and Fisher, R.F (1999). Root competition for phosphorous between coconut, multipurpose trees and kacholam (*Kaempferia galangal* L) in Kerala, India. *Agroforestry Systems* 46: 131- 146.

Lal, P; Kulkarni, H.D; Srinivas, K; Venkatesh, K.R and Santakumar, P. (1997) Genetically improved clonal planting stock of Eucalyptus - a success story from India. *India Forester* 123 : 117- 1138.

Malik, R.S and Sharma, S.K (1990) Moisture extraction and crop yield as a function of distance from row of *Eucalyptus tereticornis*. *Agroforestry systems* 12: 187 – 195.

McIntyre, B.D; Riha, S.J and Ong, C.K (1996). Light interception and evapotranspiration in hedgerow agroforestry systems. *Agriculture and Forest Meterology* 81:31 – 40.

Murdiyarso, D; Van Noordwijk, M; Wasrin, U.R; Tomich, T.P and Gillisom, A.N (2002) Environmental benefits and sustainable land-use options in the Jambi transect, Sumatra, Indonesia. *Journal of Vegetation Science* 13: 429 – 438.

Nair, P.K.R (1993) An introduction to Agroforestry. Kluwer Academic Publishers, Dordrecht, The Netherlands. Vol. 1 pp 499.

Nair, P.K.R; Buresh, R.J; Mugendi, D.N and Latt, C.R. (1999) Nutrient cucling in tropical agroforestry systems: myths and science. In : Buck L.E, Lassoie, J.P and Fernandes, E.C.M (eds) Agroforestry in Sustainable Agricultural Systems. Pp 1-31. CRC Press, Boca Raton FL.

Nair, P.K.R; Kang, B.T and Kass, D.C.L (1995). Nutrient cycling and soil-erosion control in agroforestry systems. In : Juo, A.S.R and Freed, J.D (eds) Agriculture and the Environment: Bridging Food Production and the Environment Protection in Developing Countries. American Society of Agronomy, Madison, W.I pp 117-138.

Nandal, D.P.S; Bisla, S.S;Narwal, S.S and Kaushik, J.C (1994) Allelopathic interactions in agroforestry systems. In : Narwal, S.S and Taura P. (eds) Allelopathy in Agriculture and Forestry. Pp 92-139. Scientic publishers, Jodhpur, India.

Ong, C.K and Leakey, R.R.B. (1999) Why tree crop interactions in agroforestry appear at odds with tree-grass interactions in tropical savannahs. *Agroforestry Systems* 45: 109 – 129.

Ong, C.K; Odongo, J.C.W; Marshall, F and Black, C.R (1992) Water use of agroforestry systems in semiarid India. *In* : calder, I.R; Hall, L.R and Adlard, P.G (eds) Growth and wtare use of plantations. Proceedings of the international Symposium on the Growth and water use of Forest Plantations, Banglore pp 347 – 358. Chichester, Wiley.

Palm, C.A; Myers, R.J.K and Nandwa, S.W (1997) Combined use of organic and inorganic nutrient sources for soil fertility maintenance and replenishment. In: Buresh, R.J; Sanchez, P.A and Calhoun, F (eds) Replenishing soil fertility in Africa. SSSA Spec. Pub. Nr. 51 pp 193 – 217. Soil Science Society of America, Madison, Wisconsin, USA.

Pandey, C.B; Singh, A.K and Sharma, D.K (2000) Soil properties under Acacia nilotica trees in a traditional agroforestry system in central India. *Agroforestry Systems* 49: 53 - 61.

Pathak, P.S and Pateria, H.M (1999) Agroforestry in the Indo-Gangetic Plains: a analysis. *Indian Journal of Agroforestry* 1: 15 - 36.

Pathak, P.S and Solanki, K.S (2002) Agroforestry Technologies for different Agroclimatic Regionas of India. National Research center for Agroforestry, ICAR, New Delhi. 42 pp.

Puri, S and Bangarwa, K.S (1992) Effects of trees on the yield of irrigated wheat crop in semiarid regions. *Agroforestry Systems* 20: 229 - 241.

Puri, S and Khara, A. (1991) Allelopathic effects of *Eucalyptus tereticornis* on *Phaseolus vulgaris* seedlings. *International Tree Crops Journal* 6: 287 - 293.

Puri, S and Kumar, A. (1995) Establishment of *Prosopis cineraria* (L.) Druce in the hot desert of India. *New Forests* 9: 21-33.

Puri, S; Bangarwa, K.S and Singh, S (1995) Influence of multipurpose trees on agricultural crops in arid region of Haryana, *Indian Journal Arid Environment* 30: 441 - 451.

Puri, S.; Kumar, A and Singh, S (1994a) Productivity of *Cicer arietinum* (chickpea) under *Prosopis cineraria* based agroforestry system in arid regions of India. *Journal Arid Environment* 27: 85 - 98.

Puri, S.; Singh, V.,; Bhushan, B. and Singh S (1994b) Biomass production and distribution of roots in three stands of *Populus deltoids. Forest Ecology and Management* 65: 135 - 147.

Puri, S; Swamy, S.L and Jaiswal, A.K (2002) Evaluation of *Populus deltoids* clones under nursery, field and agrisilviculture system in subhumid tropics of Central India. *New Forests* 23: 45 - 61.

Ralhan, P.K; Singh, A and Dhanda, R.S (1992) Performance of wheat as intercrop under poplar (*Populus deltoids* Bartr.) plantation in Punjab (India). *Agroforestry Systems* 19: 217 - 222.

Rao, M.R; Nair, P.K.R and Ong, C.K (1998) Biophysical interactions in tropical agroforestry systems. *Agroforestry Systems* 38: 3 - 50.

Rhoades, C (1995) Seasonal pattern of nitrogen, mineralization and soil moisture beneath *Faidherbia albida* (syn. *Acacia albida*) in central Malawi. *Agroforestry Systems* 29: 133 - 145.

Sanchez, P.A (1995) Science in agroforestry. *Agroforestry Systems* 30 : 5 - 55.

Sanginga, N.K and Swift, M.J (1992) Nutritional effect of Eucalyptus litter on the growth of maize (*Zea Mays*). *Agriculture, Ecosystems and Environment* 41: 55 - 65.

Schroth, G (1999) A review of belowground interaction in agroforestry focusing on mechanisms and management options. *Agroforestry Systems* 43: 5-34.

Shankaranarayan, K.A; Harash, L.N and Kathju, S (1987) Agroforestry in the aird zones of India *Agroforestry Systems* 9 : 259 – 274.

Sharma, K.K (1992) Wheat cultivation in association with *Acacia nilotica* (L) Willd ex. Del. Field bund plantation – a case study. *Agroforestry Systems* 17 : 43 – 51.

Singh, G; Singh, H and Bhojvaid, P.P (1998) Sodic soil ameliotration by tree plantations for wheat and oats production. *Land Degradation and Development* 9 : 453-462.

Singh, g; Singh, N.T and Abrol, I.P (1994) Agroforestry techniques for the rehabilitation of degraded salt-affected lands in India. *Land Degradation and Rehabilitation* 5: 223 – 242.

Singh, P. (2002) Realising an agricultural dream. In : ravi N (ed) Hindu Survey of Indian Agriculture. The Hindu, Chennai. Pp 15 – 21.

Singh, R.P, Ong, C.K and Saharan, N (1989) Above and below ground competition in alley cropping in semi-arid India. *Agroforestry Systems* 9: 259 – 274.

Swamy, S.L; Mishra, A and Puri, S (2002) Biomass production and root distribution of *Gmelina arborea* under an agrisilviculture system in subhumid tropics of central India. *New Forests.*

Swamy, S.L; Puri, S and Singh, A.K (2003) Growth, biomass, carbon storage and nutrient distribution in Gmelina arborea Roxb. Stands on red lateritic soils in Central India. *Bioresource Technology* 90(2) : 109 – 126.

Tejwani, K.G (1994) Agroforestry in India. Oxford & IBH Publishing Co., New Delhi. 233 pp.

Viswanath, S; Nair, P.K.R; Kaushik, P.K and Praksasam, U (2000) *Acacia nilotica* trees in rice fields: a traditional agroforestry system in central India. *Agroforestry Systems* 50: 157 – 177.

Yadav, J.P and Blyth, J.B. (1996) Combined productivity of *Prosopis cineraria*-mustard agrosilviculture practice. *International Tree Crop Journal* 9: 47-58.

Yadav, J.P; Sharma, K.ak and Khanna, P (1993). Effect of *Acacia nilotica* on mustard crop. *Agroforestry Systems* 21: 91 – 98.

• • •

2

Agroforestry Research and Extension Efforts at National Level

PANKAJ PANWAR
Department of Forestry, Uttar Banga Krishi Viswavidyalaya
Pundibari- 736 165 (West Bengal)

ABSTRACT

Agroforestry research and development in scientific mode in India started about thirty years ago with few individual scientists working on the aspect. However, concentrated efforts started with the establishment of a network under co-ordinated mode throughout the country in the year 1983. Since then lot of research has been done and recommendations made. The present paper compiles the chronological sequences of the starting of AICRP on AF along with its co-ordinating partners. Suitable perennial species and agroforestry systems recommended for respective agroclimatic zones of the country are also compiled. National seminars and group meeting conducted for popularizing and extending agroforestry to every corner of the country had also been provided.

INTRODUCTION

Agroforestry is an age old practice and was a way of life since the settled agriculture evolved from forest. However, with passage of time the cropping system was converted to monocropping and in the way the benefits of integrating the perennials with annuals were forgotten. In the late 70's of 20th century, benefits of integrating perennials in cropping system was again realized and efforts were diverted towards scientifically associating the perennials and annuals. Dr. M.S. Swaminathan an eminent scientist while addressing first agroforestry seminar at Imphal in 1979 stated that agroforestry lies the key solution of some of the problems of forestry "what is now important is to give a scientific context and better planning like a good architect who can plan the use of space more efficiently". This seminar was a spark in the history of scientific agroforestry in India. Taking into account the vide variation in edaphic and climatic factors in India along with the perennial nature of the systems, it was felt that working in isolation would not yield fruitful results and therefore, an association of scientist and institutes of different agroclimatic regions in form of a network was felt imperative and a viable wayout and hence the birth of AICRP on Agroforestry. All India Co-ordinated Research Project on Agroforestry (AICRPAF) was initiated vide ICAR letter number 21(1)/80-SW & DF dated 7th February, 1983 at 20 centers of which 12 centers were located in State Agricultural Universities and eight at the ICAR institutes. Later on more co-ordinating centers were added and at present there are 37 AICRPAF co-ordinating centers with 10 centers in ICAR institutes and 27 centers located in State Agriculture Universities covering all the major agro-ecological zones of the country. The coordinating unit of the project was initially located at ICAR, Krishi Bhawan, New Delhi, which was shifted to National Research Center for Agroforestry, Jhansi w.e.f. 1st April, 1997 under the administrative control of Director, NRCAF.

Mandate of AICRPAF

- Screening and genetic upgrading of different plant species for their compatibility under different agroforestry systems.
- To study the effect of the trees on legumes/trees/suitable orchard species, while growing mixed with suitable species of grasses and fodder commercial plants.
- Performance of the local predominant grasses/ crops and
- Refining the technology for proper propagation and management of the species.

Objectives of AICRPAF

- Diagnostic survey and appraisal of existing farming system and agroforestry practices and farmers preference.
- Collection and evaluation of promising tree species/cultivars of fuel, fodder and small timber (tree and shrubs).
- Studies on management practices of agroforestry systems such as agrisilviculture, boundary plantation, silvipasture, silvihorticulture, agrisilvihorticulture, multistory, homestead etc.
- Analysis of social, anthropological and economical relation of agroforestry systems with particular emphasis to women, children, tribal and landless farmers.

- Environment, wasteland and community land development, pollution, pest, diseases, fire, soil conservation, saline, alkaline and watershed development by agroforestry interventions.
- Studies on post harvest technology or forest product utilization, wild life, recreation and animal science, fishery, apiculture, lac etc. in relation to agroforestry systems.

Locations of AICRP centers

A) Centers in State Agricultural Universities

S.No.	Agro-climatic center	Name of the center and place	Year of establishment
1.	Himalayan Region	Dr. Y.S. Parmar University of Horticulture and Forestry, Solan (H.P)	1983-84
		Sher-E-Kashmir Univ. of Agri. & Tech., Srinagar (J&K)	1983-84
		Assam Agriculture University, Jorhat (Assam)	1983-84
		Chaudhaury Sharwandev Himachal Pradesh Krishi Viswavidyalaya, Palampur (H.P)	2001
2.	Indo-Gangetic Region	Panjab Agriculture University, Ludhiana (Punjab)	1983-84
		G.B. Pant Univ. of Agri. & Tech., Pantnagar (Uttaranchal)	1983-84
		Rajendera Agriculture University, Pusa (Bihar)	1985
		Narendra Dev Univer. Of Agri. & Tech., Faizabad (U.P)	1985
		Chander Shekhar Azad Univ. of Agri. & Tech., Kanpur (U.P)	1985
3.	Humid and Sub-Humid	Regional Research Station, Jhargram Bidhan Chandra Krishi Viswavidyalaya, Kalyani (West Bengal)	1983-84
		Orissa Univ. of Agri. & Tech., Bhubaneshwar (Orissa)	1983-84
		Birsa Agriculture University, Ranchi (Jharkhand)	1985

Contd...

Table Contd...

		Indira Gandhi Krishi Viswavidyalaya, Raipur (Chhattisgarh)	1996
4.	Arid and Semi-Arid	Sardar Krishinagar Dantiwada Agriculture University, S.K. Nagar (Gujrat) *formerly* Gujarat Agriculture University	1983-84
		Rajasthan Agriculture University, Fatehpur-Shekhawati (Rajasthan)	1983-84
		Chaudhary Charan Singh Harayana Agriculture University, Hisar (Haryana)	1985
		Mahatama Phule Krishi Vidyapeeth, Rahuri (Maharashtra)	1985
		Dr. PanjabRao Deshmukh Krishi Vidyapeeth, Nagpur (Maharastra)	1985
		Acharya N.G. Ranga Agri. Univ., Hyderabad (A.P)	1985
		Jawaharlal Nehru Krishi Viswa Vidyalaya, Jabalpur (M.P)	1985
5.	Tropical	University of Agriculture Sciences, Dharwar (Karnataka)	1983-84
		Tamil Nadu Agriculture University, Coimbatore (T.N)	1983-84
		Kerala Agriculture University, Thrissur (Kerala)	1983-84
		Marathwada Agri. Univ., Parbhani (Maharashtra)	1985
		Tamil Nadu Veterinary and Animal Science University, Kattupakkam (T.N)	1996
		Dr. Bala Saheb Konkan Krishi Vidyapeeth, Dapoli (Maharashtra)	2001
		University of Agriculture Sciences, Banglore (Karnataka)	2001

* In the year 2001 two new AICRPAF centers viz. Palampur and Banglore were opened in place of Kanpur and Parbani which were winded off in the year 2000.

B) Centers at ICAR institutes

S.No.	Agro-climatic center	Name of the center and place	Year of establishment
1.	Himalayan Region	Central Soil and Water Conservation Research and Training Institute, Dehra dun (Uttaranchal)	1983-84
		ICAR NEH Region Complex, Barapani (Meghalaya)	1983-84
		ICAR NEH Region Complex, Imphal (Manipur)	1983-84
		ICAR NEH Region Complex, Gangtok (Sikkam)	1983-84
2.	Indo-Gangetic Region	Central Soil Salinity Research Institute, Karnal (Haryana)	1985
3.	Humid and Sub-Humid	ICAR NEH Region Complex, Agartala (Tripura)	1983-84
4.	Arid and Semi-arid	Central Arid zone Research Institute, Jodhpur (Rajasthan)	1983-84
		Indian Grassland and Fodder Research Institute, Jahansi (U.P)	1983-84
		Central Research Institute for Dryland Agriculture, Hyderabad (A.P)	1985
5.	Tropical	Central Agricultural Research Institute, Port Blair (A & N Islands)	1983-84

Source: Anonymous, 2002

In addition to the above centers there are voluntary center of AICRPAF namely Central Plantation Crop Research Institute, Kasargod (Kerala); Central and Sheep and Wool Research Institute, Avikanagar (Rajasthan); Central Institute for Research on Goats, Makhdoom (U.P)

To exchange the ideas among the scientist working in agroforestry and to evaluate the progress made in agroforestry research and to train the scientists in establishing and managing the agroforestry systems regular group meeting and workshops were organized by AICRPAF either alone or in association with ICAR or other agencies. Table 2.1 gives list of such events.

Annual meeting of AICRPAF is a regular feature. Universities are also conducting conferences/seminar, short term training, winter/ summer schools through the fund support from ICAR/ICFRE/NABARD and other funding agencies to disseminate agroforestry to the farmers field. Fund support for large scale agroforestry plantations is being provided under Technology Development, Extension and Training (TDET), Ministry of Rural Development, Govt. of India.

Looking into account the multi-dimensional nature of agroforestry it was imperative to associate experts from all agriculture sciences viz agronomist, forester, horticulturist etc different post were sanctioned to different AICRPAF centers taking into account their major mandate. To make the working easier sufficient supporting staff was also provided so that the project can be run smoothly without any burden to institutes/ universities. In some of the centers vehicle was also provided. Table 2.2 gives the sanctioned staff strength in different AICRPAF centers.

Research breakthrough obtained under AICRPAF

Since perennial component is an important part in any agroforestry system hence choosing prominent perennial component among the available species of the region was the major thrust area of the research in AICRPAF. In all the AICRPAF centers number of tree species both local, indigenous and exotics were tried to evaluate their performance. The top five promising species of different centers are provided in table 2.3 in increasing order according to their growth performance.

For getting profit from agroforestry on sustainable basis it is important that the perennial and annuals are chosen in such a way that they have complimentary effect or at least supplementary effect on each other and in no case restrict the growth and reproductive capacity of each other. Various trials were laid in AICRPAF in different centers to find out the most compatible crops which will give overall best economic and environmental benefits. Table 2.4 provides some of the recommended agroforestry systems along with the perennial and annual crops which can be associated.

Acknowledgement

The author is thankful to Director, NRCAF, Jhansi for allowing us to publish the information. Thanks are also due to Dr. Anil Handa for providing information regarding recent changes in the AICRP centers.

Table 2.1. Workshops/seminar/group meetings organized by AICRPAF and other organization for developing and disseminating agroforestry research in India

Title of the seminar/workshop/meeting	Place	Year
1st Agroforestry seminar	Imphal	1979
1st Group meeting	Sikkam	1983
2nd Group meeting	Jhansi	1984
1st Workshop-cum-seminar	Dharwad	1985
2nd Workshop-cum-seminar	Hyderabad	1986
3rd Workshop-cum-seminar	Karnal	1988
4th Biennial Workshop-cum-symposium	Navsari	1990
1st Annual meeting-cum-symposium	Madras	1991
2nd Annual Group Meeting	Nagpur	1992
6th Biennial workshop	Coimbatore	1994
Group meeting for Himalayan and Gangetic plains	Solan	1995
Group meeting for Arid and semiarid regions	Raipur	1996
Group meeting for Humid and subhumid regions	Coimbatore	1996
Annual group meeting	Hyderabad	1997
Annual group meeting- cum- symposium in Agroforestry	Thrissur	1998
Biennial workshop of AICRPAF	Kumarganj (Faizabad)	1999
Annual group meeting	Solan	2000
Annual group meeting	Ludhiana	2001
Symposium on mortality in Shisham and Kikar in North India	Bhatinda	2002
National Symposium on Agroforestry in 21st Century	Ludhiana	2003
Agroforestry and sustainable production	Jhansi	2003
Workshop on AICRPAF	Banglore	2004
Workshop on AICRPAF	Karnal	2005
Agroforestry for livelihood security, environment protection and biofuel production	Jhansi	2006

Table 2.2. Sanctioned staff strength for AICRPAF centers

Agroclimatic zone	AICRPAF center	Scientist	Assistant scientist	Sr. Tech. Assistant
Himalayan	SKUAT, Srinagar	1	2	2
	YSPUHF, Solan	1	2	2
	HRS, Kahikuchi (AAU, Jorhat)	1	2	2
	CSHPKV, Palampur	1	2	2
	Subtotal	*4*	*8*	*8*
Indo-Gangetic	PAU, Ludhiana	1	2	2
	G.B.PUAT, Pantnagar	1	2	2
	RAU, Pusa	1	2	2
	NDUAT, Faizabad	1	2	2
	CSAUAT, Kanpur	1	2	2
	Subtotal	*5*	*10*	*10*
Humid and Subhumid	BAU, Ranchi	1	2	2
	OUAT, Bhubhaneshwar	1	2	2
	RRS, Jhargram (BCKV, West Bengal)	1	2	2
	IGAU, Raipur	1	1	2
	Subtotal	*4*	*7*	*8*
Arid and Semiarid	CCSHAU, Haryana	1	2	2
	ANGRAU, Hyderabad	1	2	2
	RAU, Rajasthan	1	2	2
	GAU, Gujarat	1	2	2
	PDKV, Nagpur	1	2	2
	MPKV, Rahuri	1	2	2
	JNKVV, Jabalpur	1	2	2
	Subtotal	*7*	*14*	*14*
Tropical	TNAU, Mettupalayam (Tamil Nadu)	1	2	2

Contd...

Table 2.2. Contd...

	KAU, vellanikkara	1	2	2
	BSKVV, Dapoli	1	2	2
	UAS, Dharwad	1	2	2
	MAU, Parbani	1	2	2
	TNVAU, Kattupakkam	1	2	2
	UAS, Banglore	1	2	2
Subtotal		7	***14***	***14***
Grand Total		27	53	54

Source: Anonymous, 2002

In addition to Research and technical staff, supporting staff viz. Typist/clerk, Field man, beldar, driver and messengers were also sanctioned for administrative support.

Table 2.3. Promising tree species at different AICRPAF centers

Agroclimatic zone	AICRPAF center	Tree species in order of their performance				
		1st	2nd	3rd	4th	5th
A) State Agricultural Universities						
Himalayan	SKUAT, Srinagar	*Ulmus wallichiana*	*Ailanthus excelsa*	*Catalpa bignonoides*	-	-
	YSPUHF, Solan	*Melia composita*	*Morus alba*	*Grewia optiva*	*Robinia pseudoacacia*	*Grevillea robusta*
	HRS, Kahikuchi (AAU, Jorhat)	*Terminalia chebula*	*Amoora wallichii*	*Terminalia belerica*	*Embilica offcinalis*	*Syzygium cumini*
	CSHPKV, Palampur	-	-	-	-	-
Indo-Gangetic	PAU, Ludhiana	*Eucalyptus tereticornis*	*Dalbergia sissoo*	*Albizia lebbeck*	*Melia azedarach*	*Leucaena leucocephala*
	G.B.PUAT, Pantnagar	*Casuarina equisetifolia*	*Eucalyptus citriodora*	*Anthocep-halus cadamba*	*Terminalia belerica*	*Albizia lebbeck*
	RAU, Pusa	*Pithecellobium dulce*	*Acacia auriculifor-mis*	*Erythrina indica*	*Acacia catechu*	*Peltophorum ferrugineum*
	NDUAT, Faizabad	*Casuarina equisetifolia*	*Eucalyptus hybrid*	*Dalbergia sissoo*	*Leucaena leucocephala*	*Bambusa spp.*

Contd...

Table 2.3. Contd...

	CSAUAT, Kanpur	*Leucaena leucocephala*	*Eucalyptus hybrid*	*Acacia nilotica*	*Azadirachta indica*	*Albizia procera*
Humid and Subhumid	BAU, Ranchi	*Eucalyptus tereticornis*	*Acacia auriculifor-mis*	*Acacia mangium*	*Leucaena leucocephala*	*Terminalia arjuna*
	OUAT, Bhubhanes-hwar	*Acacia mangium*	*Acacia auriculifor-mis*	*Cassia siamea*	*Gliricidia sepium*	*Casuarina equisetifolia*
	RRS, Jhargram (BCKV, West Bengal)	*Leucaena leucocephala*	*Acacia auriculifor-mis*	*Eucalyptus hybrid*	*Gmelina arborea*	*Dalbergia sissoo*
	IGAU, Raipur	*Dalbergia sissoo*	*Moringa olerifera*	*Albizia procera*	*Madhuca latifolia*	*Artocarpus heterophyllus*
Arid and Semiarid	CCSHAU, Harayana	*Leucaena leucocephala*	*Prosopis juliflora*	*Acacia nilotica*	*Dalbergia sissoo*	*Azadirachta indica*
	ANGRAU, Hyderabad	*Acacia leucocephala*	*Eucalyptus camaldulens*	*Faidherbia albida*	*Dalbergia sissoo*	*Acacia auriculifor-mis*
	RAU, Rajasthan	*Dalbergia sissoo*	*Acacia tortilis*	*Cordia myxa*	*Albizia lebbeck*	*Prosopis cineraria*
	GAU, Gujarat	*Albizia lebbeck*	*Acacia tortilis*	*Eucalyptus tereticornis*	*Cassia siamea*	*Faidherbia albida*
	PDKV, Nagpur	*Dalbergia sissoo*	*Acacia nilotica*	*Albizia lebbeck*	*Leucaena leucocephala*	*Eucalyptus tereticornis*
	MPKV, Rahuri	*Acacia cupressiformis*	*Anogeissus latifolia*	*Khaya grandiflora*	*Eucalyptus spp*	*Leucaena leucocephala*
	JNKVV, Jabalpur	*Eucalyptus tereticornis*	*Dalbergia sissoo*	*Leucaena leucocephala*	*Albizia lebbeck*	*Acacia nilotica*
Tropical	TNAU, Mettupala-yam (Tamil Nadu)	*Eucalyptus camaldulens*	*Eucalyptus tereticornis*	*Casuarina equisetifolia*	*Azadirachta indica*	*Acacia nilotica*
	KAU, vellanikkara	*Acacia auriculiformis*	*Peraseriant-hes falcataria*	*Casuarina equisetifolia*	-	-
	BSKVV, Dapoli	*Acacia auriculiformis*	*Eucalyptus citriodora*	*Gliricidia sepium*	*Gmelina arborea*	*Acacia catechu*
	UAS, Dharwad	*Casuarina equisetifolia*	*Eucalyptus tereticornis*	*Albizia lebbeck*	*Grevillea robusta*	*Dalbergia sissoo*
	MAU, Parbani	*Leucaena leucocephala*	*Albizia lebbeck*	*Eucalyptus spp.*	*Dalbergia sissoo*	*Acacia nilotica*
	TNVAU, Kattupakkam	*Leucaena leucocephala*	*Erythrina indica*	*Gliricidia sepium*	-	-

Contd...

Table 2.3. Contd...

	UAS, Banglore	*Leucaena leucocephala*	*Casuarina equisetifolia*	*Anogeissus latifolia*	-	-
B) ICAR Institutes						
Himalayan Region	CSWCRTI Dehra dun (Uttaranchal)	-	-	-	-	-
	ICAR NEH RC, Barapani (Meghalaya)	*Parkia javanica*	*Alnus nepalensis*	*Paraseriant-hes falcataria*	*Gmelina arborea*	*Michelia oblonga*
	ICAR NEH RC, Imphal (Manipur)	*Acacia auriculiformis*	*Paraseriant-hes falcataria*	*Samanea saman*	*Albizia procera*	*Quercus serrata*
	ICAR NEH RC, Gangtok (Sikkam)	*Leucaena leucocephala*	*Alnus nepalensis*	*Erythrina arborescence*	*Evodia fraxinifolia*	*Terminalia myriocarpa*
Indo-Gangetic Region	CSSRI, karnal (Haryana)	-	-	-	-	-
Humid and Sub-Humid	ICAR NEH RC, Agartala (Tripura)	*Acacia auriculifor-mis*	*Eucalyptus hybrid*	*Gmelina arborea*	*Tectona grandis*	*Sesbania sesban*
Arid and Semi-arid	CAZRI, Jodhpur (Rajasthan)	*Dichrostachys cinerea*	*Acacia tortilis*	*Tamarix articulata*	*Colophos permummo-phane*	-
	IGFRI, Jhansi (U.P)	*Eucalyptus camaldulens*	*Acacia nilotica*	*Albizia lebbeck*	*Prosopis juliflora*	*Acacia tortilis*
	CRIDA, Hyderabad (A.P)	-	-	-	-	-
Tropical	CARI, Port Blair (A & N Islands)	*Cocos nucifera*	*Areca catechu*	*Musa paradisiaca*	*Ceiba pentendra*	-

Adopted from Pathak *et al*, 2000 with modification

Table 2.4. Recommended agroforestry systems under different agroclimatic regions

Agroclimatic zone	AICRPAF center	Agroforestry system	Perennial and annual crops
A) State Agricultural Universities			
Himalayan	SKUAT, Srinagar	a) Agrisilviculture	a) *Ulmus wallichiana* + pea and maize followed by oats-mung bean

Contd...

Table 2.4. Contd...

		b) Hortipasture	b) Almond + red & white clover, *Festuca paratense, Dactylis glomerata*
	YSPUHF, Solan	a) Agrisilviculture (for low and mid hills)	a) *Grewia optiva/Morus alba/ Populus deltoids* + wheat/maize/ barley/pulses
		b) Hortisilviculture (for low, mid and high hills)	b) *Pyrus malus/Mangifera indica/* Citrus/*Prunus domestica/Prunus armeniaca* + *Grewia optiva/Morus alba/Populus ciliata/Toona ciliate*
		c) Silvipasture	c) *Morus alba/Grewia optiva/Robinia pseudoacacia/Populus ciliata/ Quercus leucotrichophora* + *Setaria anceps/Chloris guyana*/Hybrid napier/*Dactylis glomerata*/Red & white clover/ *Stylosanthes hamata*
	HRS, Kahikuchi (AAU, Jorhat)	a) Agrisilviculture	a) *Anthocephala cadamba/ Azadirachta indica + crops; Gmelina arborea* + turmeric; *Areca catechu* + turmeric/ginger/pineapple
Indo-Gangetic	PAU, Ludhiana	a) Agrisilviculture and / as Farm boundary	*Populus deltoids/Eucalyptus/Acacia nilotica* + sugarcane for 2 years followed by wheat/oat/barley
	G.B.PUAT, Pantnagar	a) Agrisilviculture and/as random trees on farm	*Populus deltoids/Melia azedarach/ Dalbergia sissoo/Albizia lebbeck* + wheat/turmeric/potato/ radish/coriander
		b) Block plantation	
		c) Hortisilviculture	
	RAU, Pusa	a) Agrisilviculture and/as farm boundary/ shelterbelts	a) *Populus deltoids/Leucaena leucocephala/Sesbania sesban* + wheat/til/tori/rajmash/ coriander/oats
	NDUAT, Faizabad	a) Agrihortisilvi-culture	a) *Casuarina equisetifolia* + *Emblica offcinalis/Psidium guajava / Zizyphus mauritiana*
	CSAUAT, Kanpur	a) Agrisilviculture and/as boundary trees	a) *Casuarina* on salt affected soils.
		b) Silvipasture	b) *Leucaena leucocephala/Albizia amara* + napier/stylo
Humid and Subhumid	BAU, Ranchi	a) Silvihorticulture	a) *Acacia auriculiformis* + *Artocarpus heterophyllus*

Contd...

Table 2.4. Contd...

		b) Agrihorticulture	b) sugarcane+ *Acacia nilotica/ Phoenix dactylifera/Borassus flabellifer*
	OUAT, Bhubhaneshwar	a) Agrisilviculture	a) Ginger/soybean/wheat + *Tectona grandis/Dalbergia sissoo/ Santalum album/Perocarpus marsupium*
		b) Agrihorticulture	b) Ginger/soybean/wheat + *Psidium guajava*
	RRS, Jhargram (BCKV, West Bengal)	a) Farm boundary	a) *Annona squamosa/Embilica offcinalis/Zizyphus mauritiana/ Punica granatum/ Madhuca latifolia* + agriculture crops
		b) Homestead	b) *Annona squamosa/Embilica offcinalis/Zizyphus mauritiana/ Punica granatum/ Madhuca latifolia* + agriculture crops+ *Leucaena leucocephala/Eucalyptus spp./ Terminalia arjuna*
	IGAU, Raipur	a) Agrisilviculture	a) *Leucaena leucocephala/Ceiba pentandra/Gmelina arborea* + soybean
		b) Silviculture on wastelands	b) *Gmelina arborea/Pongamia pinnata/Eucalyptus/Emblica officinalis/Tectona grndis/Dalbergia sissoo/Bambusa bamboos/ Dendrocalamus strictus.*
Arid and Semiarid	CCSHAU, Harayana	a) Agrisilviculture and/as farm boundary	a) *Populus deltoids* + crops (irrigated condition) *A. excelsa/Prosopis cineraria* + crops (rainfed condition)
		b) Agrihorticulture	b) *Embilica officinalis/Zizyphus mauritiana* + crops (rainfed condition)
	ANGRAU, Hyderabad	a) Agrisilviculture	a) *Acacia albida* + jowar, *Leucaena leucocephala*+ *stylo*, *Dalbergia sisoo* + sorghum, *Cocus nucifera* on field boundaries
		b) Hortipasture	b) fodder crops like maize/ sorghum/cowpea + *Psidium guajava/Syzygium cumini/Annona squamosa*
		c) Silvipasture	c) *Acacia leucophloea* + *Cenchrus ciliaris*, cowpea + *Gmelina arborea/ Tectona grandis*

Contd...

Table 2.4. Contd...

	RAU, Rajasthan	a) Agrisilviculture	a) *Acacia nilotica/Acacia cupressiformis/Ailanthus excelsa* + wheat/ mustard/ cumin (irrigated condition); *Prosopis cineraria/Tecomella undulata/Acacia Senegal/Acacia tortilis/Prosopis juliflora* + pearl millet/ cluster bean/ moth bean (rainfed condition)
		b) Silvipasture	b) *Zizyphus nummularia/ Calligonum polygonides/Capparis deciduas* + *Cenchrus ciliaris/ Panicum turgidum/L. sindicus/ Clitoria ternatea*
	GAU, Gujarat	a) Agrisilviculture	a) *Prosopis cineraria*+ crops
		b) Agrihorticulture	b) *Zizyphus mauritiana/Embilica officinalis* + crops
		c) Hortisilviculture	c) *Zizyphus mauritiana/Embilica officinalis/Citrus spp./Punica granutum* + *Ailanthus excelsa/ Hardwika binnata/Dalbergi sissoo*
	PDKV, Nagpur	a) Hortiagriculture	Oranges and vegetable, Sapota and vegetables, Mango and vegetables
		b) Silvipasture / Hortipasture	*Hardwikia binnata/ Albizia lebbeck* + *Cenchrus ciliaris/ Stylo hamata; Zyziphus zujuba, Annna sqamosa* + Dicantheu/ cowpea/ sorghum; *Mangifera indica/ Emblica officinalis* + Napier/ Cluster bean/ cowpea
	MPKV, Rahuri	a) Silvipasture	
		b) Agrihortisilviculture	a) *Chrysopogon fulvus/Cenchrus ciliaris* + *Schima*
			b) *Mangifera indica/ Annona squamosa*+ *Azadirachta indica*+ crops
	JNKVV, Jabalpur	a) Block plantation	a) *Dalbergia sissoo/Acacia nilotica/ Albizia lebbeck/Leucaena leucocephala* (distantly located farms of big farmers)
		b) Agrisilviculture	b) *Leucaena leucocephala* + mustard, *Acacia nilotica* + linseed

Contd...

Table 2.4. Contd...

		c) Agrihorticulture	c) *Psidium guvajava* + paddy/ potato/spinach
Tropical	TNAU, Mettupalayam (Tamil Nadu)	a) Agrisilviculture	a) *Ceiba pentandra* + medicinal plants
	KAU, vellanikkara	a) Homegarden	*Cocos nucifera/ Areca catechu/Hevea brasilensis* + *Mangifera indica/ Artocarpus heterophyllus/ Anacardium occidentale* + crops
		b) Agrisilviculture	b) *Hevea brasiliensis* + banana/ tapioca/yams/ vegetables ; *Areca catechu* + cowpea /sorghum / tapioca/ elephant foot yam / turmeric /ginger
	UAS, Dharwad	a) Agrisilviculture	a) bamboo + crops (in drylands); *Acacia auriculiformis* + ragi
		b)agrihortisilviculture	b) Teak+ papaya+crops
	MAU, Parbani	a) Boarder planting	a) *Mangifera indica/Zizyphus mauritiana/Tectona grandis/ Eucalyptus*
		b) Agrisilviculture	b) *Tectona grandis/ Azadirachta indica* + jowar/cotton/ground nut
		c) Agrihorticulture	c) *Mangifera indica/ Annona squamosa* + jowar/ cotton/ ground nut
		d) Silvipasture	d) *Tectona grandis/ Annona squamosa/ Dalbergia sissoo* + stylo
	TNVAU, Kattupakkam	a) Silvipasture	a) *Leucaena leucocephala/ Albizia lebbek/Gliricidia sepium* (having high coppicing ability) + grasses
B) ICAR Institutes			
Himalayan Region	CSWCRTI Dehra dun (Uttaranchal)	a) Agrihorticulture	a) *Mangifera indica/Litchi chinensis/Citrus spp*/peach + *Curcuma longa/Zingiber officinale*
		b) Agrisilviculture	b) *Grewia optiva/Bauhinia purpurea/Morus alba /Eucalyptus* + rice/wheat/maize.
	ICAR NEH RC, Barapani (Meghalaya)	a) Silvipastoral	a) *Alnus nepalensis/Schima wallichii/Michelia oblonga* + Setaria spp/guinea grass

Contd...

Table 2.4. Contd...

		b) Agrisilviculture	b) *Alnus nepalensis/Michelia champaca/Paraserianthes falcataria/ Parkia roxburghii* + soybean-linseed/Ginger/turmeric; *Morus alba*+ ground nut- mustard
		c) Agrihorticulture	c) *Psidium guajava* + ginger/ turmeric/ground nut/soybean
		d) Homegardens	d) *Psidium guajava/Musa paradisiaca* + groundnut/beans/ chillies
	ICAR NEH RC, Imphal (Manipur)	a) Agrislivipasture	a) *Bauhinia variegata/Pinus roxburghii/Schima wallichii* + maize+ napier grass
		b) Hortipasture	b) *Citrus rotundifolia* +pineapple + napier
		c) Silvipasture	c) *Eucalyptus spp/Leucaena leucocephala* + *Pannicum pedicellatum*
	ICAR NEH RC, Gangtok (Sikkam)	a) Silvipasture	a) *Ficus hookerii* + *Thysanoleana maxima*/napier
		b) Agrisilviculture	b) ginger/turmeric/large cardamom (upto 11-15 m away) and rice/soyabean/stylo (beyond 15 m away) from bamboo
Humid and Sub-Humid	ICAR NEH RC, Agartala (Tripura)	a) Agrisilvihorticul-ture	a)*Tectona grandis*+ assam lemon+ turmeric
		b) hortisilviculture	b) Pineapple+ *Acacia auriculiformis/ Morus alba/Gliricidia sepium*
		c) Agrisilviculture	c) Til + *Acacia auriculiformis*
Arid and Semi-arid	CAZRI, Jodhpur (Rajasthan)	a) agrisilviculture	a) *Prosopis cineraria* + bajra/til/ guar
		b) Silvipasture	b) Jodhpur grass + *Acacia tortilis/ Zizyphus mauritiana*
	IGFRI, Jhansi (U.P)	a) Agrisilviculture	a) *Acacia nilotica/Leucaena leucocephala/Dalbergia sissoo* + paddy/wheat/mustard/gram (irrigated condition) ; *Azadirachta indica/Acacia nilotica/Hardwickia binata/Dalbergia sissoo* + groundnut/sorghum/gram/

Contd...

Table 2.4. Contd...

			lentil/soybean (rainfed condition)
		b) Agrihorticulture	b) *Mangifera indica/Psidium guajava* + paddy/wheat/ mustard/gram/sunflower (irrigated condition) ; *Zizyphus mauritiana/Annona squamosa/ Embilica offcianalis* + groundnut/ sorghum/gram/lentil/soybean (rainfed condition)
		c) Silvipasture	c) *Leucaena leucocephala/ Hardwickia binata* + *Cenchrus ciliaris/Stylosanthes hamata*
Tropical	CARI, Port Blair (A & N Islands)	a) Homesteads	a) *Areca catechu*+ *Cinnamomum zeylanicum* + lemon grass/hybrid napier/stylo; *Cocus nucifera*+ *Areca catechu*+ *Musa paradisiaca/ Carica papaya/Achras sapota/* Psydium guajava + black pepper + rice/*Vigna spp.*

Adopted from Pathak *et al*, 2000 with modification

REFERENCES

Anonymous (2002) *Annual Report of All India Co-ordinated Research Project on Agroforestry*, NRCAF, Jhansi: 35 – 36.

Pathak, P.S. Pateria, H.M. and Solanki, K.R. (2000) *Agroforestry systems in India- A Diagnosis and Design Approach.* AICRP on Agroforestry, NRCAF, Jhansi.: 106.

Section 2

Agroforestry: Systems and Practices in North West and Central India

3

An Overview of Agroforestry in Kashmir Valley

A. H. MUGHAL AND M.A.KHAN
Division of Forestry, Sher-e-Kashmir University of Agricultural Sciences and Technology - Kashmir, Shalimar
Srinagar, J & K - 191 121

ABSTRACT

The state of Jammu and Kashmir, especially the valley of Kashmir, has a rich diversity of traditional agroforestry models, which are in existence since time immemorial. Agroforestry is based primarily on natural occurrence of trees. The paper highlights the traditional agroforestry systems prevailing in the valley. A number of studies have been undertaken since the inception of AICRP-AF in University of Agricultural Sciences and Technology, Kashmir. Besides design and diagnostic survey, studies on evaluation of twelve multipurpose tree species identified during design and diagnostic survey and designing of various agroforestry models for different problematic sites have been undertaken. These studies have been tested over a period of time and have been found suitable for different sites. The out come of these studies and the recommended agroforestry systems are given in this paper.

INTRODUCTION

Land is a vital natural resource and is the basis of our existence. Due to mounting pressure of increasing human and livestock population it has been subjected to various pressures and misuse resulting in degradation of land and loss of biodiversity. Management of land is therefore, of paramount importance so that increasing needs of human and livestock are met on sustainable basis. During the last three decades agroforestry has been recognised as a distinct land management system, which is destined to play an important role in socio-economic transformation of marginal lands into more productive agrarian economy. India has a long tradition of agroforestry, farmers and land owners in different parts of the country integrate a variety of woody perennials in their crop and livestock production fields, depending upon the agro climatic conditions and local needs. Most of these practices are, however, location specific and information on these is very scanty, therefore, benefits have remained vastly under exploited and un extrapolated to other sites.

The state of Jammu and Kashmir, especially the valley of Kashmir, has a rich diversity of traditional agroforestry models, which are in existence since time immemorial. Agroforestry over here is based primarily on natural occurrence of trees. These traditional agroforestry practices have, to some extent, helped people in meeting some of their diverse needs i.e. food, fodder, fuel wood and timber. Mughal and Bhattacharya (2000) have identified following 7 different agroforestry models in the valley.

A. Traditional Agroforestry Models

(1) Boundary plantations

This oldest traditional model is practiced on paddy lands. Farmers generally do not plant any tree within the paddy fields, but if the same land lies besides a road then people plant trees on the roadside. In addition to this if a canal passes through the farmers field, then on the both sides of the canal farmers plant *Salix* trees. Other trees raised are *Populus deltoides, Populus nigra* and *Aesculus indica* and the spacing is maintained at 1.20 x 1.20 m. This model is adopted by marginal as well as small and big farmers. *Avena sativa* and *Brassica* sp. are grown during winter. These lands suffer from impended drainage and are not suitable for any other agricultural crop except paddy.

(2) Agrisilviculture on sloping lands

This system is practiced by people residing on the mid-mountains or at the foothills of mountains. These hilly areas are completely devoid of vegetation and suffer from scorching heat during summer and lack of moisture in the months from May onwards till winter sets in. People raise forest tree species around their farms, which serve as boundaries as well as provide much needed fuel and fodder. Sometimes trees may be found scattered inside the farm area also. Amongst the agricultural crops maize is raised as monoculture as it can survive under moisture stress conditions compared to other crops. During winter, mustard and at some places vegetables like brassica, carrot, reddish, turnip etc. are grown. Trees planted by the farmers in these lands are *Robinia pseudoacacia, Ailanthus altissima, Aesculus indica, Populus nigra* and *Salix alba.* Spacing of trees is maintained at 1.20 x 1.20 m. Problems of the area include lack of water in streams during summer, exposed denuded surface and low agricultural output.

(3) Agrisilviculture in plains

The land of plain areas is fertile and very well suited for agriculture. Farmers whether marginal or small keep a separate area within the paddy fields of the size of about 500-600 m^2 which is generally raised in comparison to paddy field. This area is kept surrounded by tall trees of *Populus deltoides, Populus nigra, Salix alba* and some times *Robinia pseudoacacia* and the spacing is maintained at 1.5 x 1.5 m. Some trees are kept scattered within the farm also. Vegetables are grown both in rabi as well as kharif seasons. Rabi season vegetables include knol khol, cabbage, cauliflower, turnip, radish, carrot, onion, peas, spinach, garlic etc. In the kharif season vegetables like tomato, brinjal, chillies, capsicum, french beans, cucumber, bottle gourd, bitter gourd, pumpkin, brassica, potato etc. are grown.

(4) Hortisilviculture

This system is practiced generally on the slopes with or without irrigation facility. The main purpose of the system is the production of fruits for marketing. Fruit trees are grown at regular spacing within the farm at a spacing ranging from 3.0 x 3.0 m to 3.5 x 3.5 m depending upon the fruit tree species. Forest trees are grown on all the four sides of the orchard in a single or in double rows along the boundary. The trees are planted at a close spacing of 1.20 x 1.20 m and the purpose of growing these trees is wood for packing of fruits , timber for construction and fuel wood production. Forest tree species raised are *Populus deltoides, Salix alba, Robinia pseudoacacia, Ailanthus altissima, Aesculus indica* etc. Fruit trees include *Punica granatum* (Pomegranate), *Juglans regia* (Walnut), *Prunus cerasus, Prunus amygadalus* (Almond), *Malus pumila* (Apple), etc.

(5) Hortisilvipasture

This system is also practiced on sloping lands but in some cases it has been observed in orchards located on plain lands. The system has all the components of hortisilviculture as listed above. Pasture grasses or *Avena sativa* (oats) are grown in these orchards at the time when fruit trees are in a juvenile stage or when the fruit trees are planted at a much wider space i.e. 6.0 x 6.0 m. Grasses grown include *Festuca pratense, Trifolilum repens* and *Dactylis glomerata.* In some established orchards only *Avena sativa* is grown in winter as there is not much shading effect of trees on the crop during winter.

(6) Hortisilviagriculture

This system is practiced in places where orchards are located either on terraces or flat lands with irrigation facility throughout the year. In this system people devote a little area within the orchard for raising vegetables during Rabi and Kharif seasons for domestic consumption. The vegetables are intercropped between fruit trees whose spacing is maintained at 6.0 x 6.0 m and the vegetables grown are tomato, brinjal, chillies, capsicum, french beans, cucumber, bottle gourd, bitter gourd, pumpkins, potatoes, knol khol, cabbage, cauliflower, turnip, radish, carrot, onion, peas, spinach, brassica etc. in different seasons. Trees and fruit components are same as that of horti silviculture. The problem with the system is that people raise vegetables in the small area for domestic consumption only while as the rest of the area is not utilized for agriculture.

(7) Kitchen garden

This system is practiced throughout the district and is generally common among those farmers who have a land holding of about 500-1000^2 m located just around their residence. On the piece of land people raise fruit trees and vegetables of all kinds during Rabi and Kharif season. Poultry is also reared on the same piece of land. Trees raised on the land all along the boundary at a close spacing of 1.20 x 1.20m in single or double rows are *Populus deltoides, Salix alba, Robinia pseudoacacia, Ailanthus altissima*. Fruit trees may consist of pomegranate, apple, peach, almond, queens apple, cherry etc. Sometimes all the fruit trees are found in the kitchen garden. Vegetables are grown in the inter spaces of fruit trees and may include cabbage, turnip, radish, carrot, onion, peas, spinach, brassica etc. The problem with kitchen garden is that it is established on a small piece of land and sometimes people have to hire labour for the purpose of cultivation which they can not afford as they are generally poor farmers.

Drawbacks of Traditional Agroforestry Systems

The traditional agroforestry models discussed above are not in a position to meet the demands of the farmers, because these systems suffer from a number of lacuna which have compounded the problem resulting in low yields of food, fodder and fuel wood from the unit area. Besides all these systems do not have adequate representation of all the components and hence there is no efficient utilization of space in these systems. Broad evaluation of the system also shows that only three tree species i.e. *Populus deltoids, Salix alba* and *Robinia pseudoacacia* are extensively planted.

Why Interventions are Required to Develop New Agroforestry Systems

Demand of food, fodder, fuel wood and timber like in other parts of the country is on an increase in J&K state also. The annual demand of fuel wood, timber and fodder has been put at 357 lakh quintals, 6.59 lakh M^3 and 114 million tonnes respectively and there is huge gap between the demand and supply as a result of which only Kashmir valley imports about 6 lakh quintals of fodder annually, besides imports of timber for meeting demands of constructional purposes has also gained momentum. Forests can not meet all these demands on a sustained basis because of following reasons:

1. Forests are faced with huge problem of degradation due to illegal felling.
2. 85 per cent of forests of the state either lack regeneration completely or are deficit in natural regeneration (Anonymous, 2003)
3. Nomadic culture of 'Gujjars' and 'Bakerwals' who pass through the forests without permission also affect the natural regeneration as a result of grazing, browsing and trampling of young seedlings.
4. There is an increasing incidence of encroachments on forest lands and on an average about 4,000 hectares is encroached yearly (Anonymous, 2003).

Forests as such are not in a healthy state to meet the needs of people on sustained basis. In the cultivation area major portion is under dry farming and in J&K cultivation can be resorted to any degree of slope without regard to any conservation practices, as such these areas are subject to very high rate of erosion. The barren and uncultivated wastelands in the state is about 289 thousand hectares (Anonymous,2002) and these are subject to great misuse, cutting of tree and vegetation, heavy grazing and other biotic interference resulting in heavy soil erosion due to wind, water and snow.

Under these circumstances in order to bridge the gap between demand and supply of food, fodder, fuel wood and other NWFP and at the same time reduce pressure on forest lands agroforestry models if practiced and followed on scientific lines seem to be the best option. The models will at the same time bring more and more land under green cover thereby reducing soil erosion and improving environmental conditions of the valley at micro level.

A number of studies have been undertaken since the inception of AICRP-AF in University of Agricultural Sciences and Technology, Kashmir. Besides design and diagnostic survey, studies on evaluation of twelve multipurpose tree species identified during design and diagnostic survey and designing of various agroforestry models for different problematic sites have been undertaken. These studies have been tested over a period of time and have been found suitable for different sites. Models developed will not only provide more agricultural productivity, but would increase the land capability with regard to nutrient status and also provide much needed fuel wood, fodder and small timber to the farmer on sustainable basis. Results of some studies carried out over a period of time are discussed below.

B. Scientific Agroforestry

1. Evaluation of twelve multi purpose tree species (MPTS)

Studies were conducted on the basis of design and diagnostic survey and appraisal of existing agroforestry systems in Kashmir region on twelve MPT species having agroforestry promise. The species included were *Ulmus wallichiana, Ailanthus altissima, Populus euroamericana, Morus alba, Populus ciliata, Robinia pseudoacacia, Salix alba, Alnus nitida, Catalpa bignoniodies, Prunus armeniaca, Aesculus indica* and *Fraxinus floribunda.* The studies were conducted on 'Kerewa lands', which are elevated plateaus of alluvial or lacustrine materials without irrigation, low in water retention capacity and dependent on natural precipitation. Four year data revealed that amongst various MPTS *Ulmus wallichiana* (leaves are cherished by cattle) recorded maximum survival (100 %) and it also ranked first in terms of girth and plant height. *Ailanthus altissima* ranked second though the leaves are not liked by cattle due to its offensive smell, but its survival and performance under moisture stress conditions was better. *Catalpa bignoniodies* had big leaves though slow in growth, ranked third in the species evaluation trial. The remaining MPTS did not show desirable performance (Makaya, 1994).

2. Moisture conservation in agroforestry by trench method

The experiment was designed under rain fed conditions to conserve available moisture through better water harvesting methods on 'Kerewa lands' having low water retention capacity. The tree component was *Ailanthus altissima.* Under the tree component mustard (*Brassica* spps.) and moong (*Phaseolus* spp) were raised in rotation in rabi and kharif seasons, Data at the conclusion of the experiment revealed that trench system of planting is superior to flat system of planting under medium moisture stress conditions (Anonymous, 2000).

3. Agrisilviculture (Alley cropping)

Experimental trial was laid on hill slope (which had no vegetation and was suffering from soil erosion and run off losses) of 30-40 per cent with *Ulmus wallichiana* as a tree component planted in alley across the slope with trees pruned at 3 m height. The trees were planted at three different alley widths of 1, 1.5 and 2m. Arable crops i.e. peas and beans were planted in rabi and kharif seasons respectively. Data at the conclusion of the experiment proved that there is an increase in average cumulative yield of both rabi and kharif crops when compared with the average cumulative yield of crops obtained in control. Increase in yield of peas and beans were 314.34 kg ha^{-1} and 454.67 kg ha^{-1} respectively (Table 3.1). Besides the maximum yield of 949.92 kg ha^{-1} of peas and 1164.67 kg ha^{-1} of beans was recorded in an alley width of 2m (Table 3.2). In addition to yield of peas and beans, yield from *Ulmus wallichiana* trees in the form of fodder and fuel wood yielded additional benefits. Maximum fodder (8.43 kg $tree^{-1}$ yr^{-1} was obtained in closely spaced alleys and maximum fuel wood of 14.42 tonnes $ha.^{-1}$ was obtained where alley width was maintained at 1.5m. Agrisilviculture model so devised can help in stabilizing the degraded environment and at the same time help in food and fuel wood production.

4. Hortipasture model

About 2.5 lakh ha area is under orchards of apple, almond, cherry and other stone fruits in Kashmir valley. Under hortipasture, various studies have been conducted to evaluate some temperate legumes and grasses in Kashmir valley for hortipasture models (Table 3.3). Mughal & Bhattacharya (2002) have identified hortisilvipasture and hortisilviagricultural systems in the valley, which are followed traditionally by the farmers, but lack scientific tinge.

The scientific hortipastoral system developed for the valley will help in exploiting the ground floor of the orchards, which otherwise remain unexploited thus helping in increasing fodder production and relieving pressure on forest lands. In hortipastoral trial in almond orchard four species of grasses *viz. Festuca pratense, Dactylis glomerata, Trifolium repense and Trifolium pratense* were compared with natural undergrowth by Makaya and Gangoo (1995). The grasses were harvested from the experimental plots three times during the year on 1st June, 15th August and 1st November. The undergrowth in the orchard was identified as *Plantago major, Plantago lancelula, Poa balboose, Trifolium repense* and *Indigofera articulata.* Three year data showed that all the introduced grasses and legumes recorded increase in yield over the natural under growth as is evident from Table 3.4. Grasses are the most effective soil binders. Grasses as under storey crops are usually better than crops because the forage grows taller under shade and therefore, associates with trees without much loss of yield. The preferential shoot growth (at the expense of root growth) may be desirable to avoid root competition. The data on soil analysis. revealed an almost negligible change in soil pH and EC, whereas the organic matter content, an index of available N, increased. Continuous cropping depleted the soil of available P_2O_5 & K_2O by 6.6 and 10.1 per cent respectively.

5. Silvipastoral model

A single tree experiment under rainfed conditions was laid to evaluate the tree crop interaction at Forest Nursery, Shuehama, Srinagar, SKUAST (K) following Huxley *et al.*, (1987). The soil at the site is silty gravel in texture, shallow in depth, low in water retention capacity, low in organic matter and available N, P and K with 6.8 pH and safe electric conductivity. Tree component used was *Amorpha fruticosa* while as oats *(Avena sativa)* and M.P.Chari (*Sorgum vulgare)* were used as forage crops in rabi and kharif seasons respectively. The results obtained over a period of time reveal that in both the rabi and kharif seasons yield of fodder is more in control compared to other treatments. General trend is that as we move away from the tree component, yield of grasses increase in both the rabi and kharif seasons. In oats, yield obtained around tree was 146 q ha^{-1} and at 0.5m away from the tree 153.8 q ha^{-1}. Yield of fodder crop at a distance of 1.5 m and 2m away from the tree was 157.8 and 161.7q ha^{-1} respectively. Yield of M. P. Chari obtained around tree was 163.9 q ha^{-1} and it increased to 195.3 q ha^{-1} at 2m away from tree (Table 3.5). This decrease in yield of fodder crops near the tree is presumably due to competition for moisture and nutrients by the shallow root system of *Amorpha fruticosa* (Mughal *et al.*, 2003).

Besides, fodder from forage crops, fodder though meager was also regularly harvested from the tree component and it increased from 0.45 and 0.40 q ha^{-1} in the first year in both control and combination plot to 1.50 and 1.48 q ha^{-1} in control and combination plot respectively at the end of 5th year. Diameter and height also recorded an increase, diameter increase was 117 per cent and 159 per cent in combination plots and sole tree plot (control) respectively, while as height increase was 346 per cent in combination plot and 320 per cent in control over a period of 5 years. This shows that good yield of forage crops and fodder from trees can be obtained from otherwise unproductive waste and degraded lands if scientific silvipasture models are adopted on such lands. Mughal and Makaya (2000) have also identified silvipasture model for degraded lands as the best option to tide over fodder scarcity and at the same time stabilize degraded lands.

The intercropping of trees *(Amorpha fruticosa)* and forage crops (Oats and M.P.Chari) exhibited land equivalent ratio (LER) of more than 1.00 (Table 3.6), which shows that the silvipastoral system is more efficient in utilizing the available resources than sole cropping thus resulting in higher productivity per unit area. Highest LER was obtained in treatment where yield of grasses was taken from a distance of 2m away from the tree, which suggests that distance of 4m should be maintained in alleys or row plantation for obtaining maximum yield from the unit area. The development of silvipastoral system on degraded lands has also a potential for providing sustainable employment to rural people in various activities.

Table 3.1 Average production of agricultural crops over a period of nine years (1989-98)

Crop	Yield outside alley kg ha^{-1}	Yield within alley kg ha^{-1}	Increase in yield over control kg ha^{-1}	Percent increase over control
Rabi (Peas)	554	868.34	314.34	56.77
Kharif (Beans)	566	1014.67	454.67	81.19

Table 3.2. Average yield of crops, fodder and fuel wood under different alleys

Alley width	Crop yield (Kg ha^{-1})		Fodder yield Kg $tree^{-1}$ yr^{-1}	Fuel wood tonnes ha^{-1}
	Rabi	Kharif		
2m	949.92	1164.67	5.31	10.30
1.5m	845.84	1080.76	5.40	14.42
1m	819.26	798.60	8.43	13.09

Source, Mughal *et al.*, 2003

Table 3.3. Combination of grasses recommended for hortipasture in temperate Himalayas

Combinations	Seed rate kg ha^{-1}	Fertilizer rate kg ha^{-1}
Orchard grass + Red clover	5+4	90 N+60P_2O
Rye grass + Red Clover	5+4	120N+90P_2O_5
Broom grass + Red Clover	15+4	90N+60P_2O_5
Timothy + Sub Clover	4+50	90N+60P_2O_5

Source : Misri, 1988

Table 3.4. Green forage yield (tonnes ha^{-1}) of pasture and legumes

Name of grass	Year			Average yield	% Increase over natural vegetation
	1990	1991	1992		
Natural Vegetation	12.89	17.54	13.50	14.64	-
Dactylis glomerata	20.95	22.96	22.19	22.03	56
Festuca pratense	20.47	42.48	21.25	28.07	91

Contd...

Table 3.4. Contd...

Trifolium repense	22.80	34.56	17.51	24.96	70
Trifolium pratense	23.00	31.92	18.33	24.58	62
C.D at 5%	7.89	9.56	4.46	-	-

Source : Makaya and Gangoo, 1995

Table 3.5. Average yield of oats & M.P.Chari q/hac during 5 year period in different treatments

Crop	Control	Around tree	0.5m away from tree	1.5m away from tree	2.0m away from tree	C.D at 5%
Oats	199.6	146.0	153.8	157.8	161.7	7.85
M.P. Chari	251.4	163.9	181.9	184.2	195.3	23.33

Table 3.6. Land equivalent ratio (LER) for different treatments

Treatments	LER
Control	1.00
Around tree	1.54
0.5m away from tree	1.59
1.5m away from tree	1.61
2.0m away from tree	1.63

6. Silviagricultural model for southern slopes

There exists a good scope for silviagricultural system in the areas located on the southern slopes Mughal and Makaya (2000) evaluated different tree species along with agricultural crops on such degraded lands. The system identified helps in reducing scorching heat and at the same time reduces soil erosion, besides increasing production per unit area. Trees found suitable for such lands in order of merit were *Melia azedarach, Ailanthus altissima, Aesculus indica, Gliditsia inermis, Robinia pseudoacacia* and *Ulmus wallichiana.* Maize as agricultural crop was found suitable for the area. Trees on such slopes are planted along contour 3.7 m apart, while as tree to tree distance is maintained as close as possible.

CONCLUSION

Agroforestry for Kashmir Valley is not a new concept, but as discussed traditional systems are being followed by the farmers since time immemorial and people get diverse products from their farms a goal which agroforestry always longs for. The only problem with the traditional agroforestry systems is that only a few trees are being used in the systems and as such diversity among trees is lacking, besides the interspaces are not fully exploited especially in orchards. Multipurpose trees identified for agroforestry will help in increasing diversity of tree species and the models developed will help in exploiting the interspaces to its fullest which in turn will increase the production per unit area. People will have no problem in accepting these models of agroforestry as this will not be perceived as deviation from traditional land use system, but for this proper and sustained extension activities need to be undertaken. Besides, demonstration plots especially on degraded lands need to be laid so that people are able to judge by themselves the suitability and utility of such models.

REFERENCES

Anonymous (2003). Review of State Forest Policy. Jammu and Kashmir Forest Department, Srinagar , Kashmir 24th July 2003.

Anonymous. (2000). Achievements, All India Coordinated Research Project on Agroforestry. Division of Forestry, *Technical Document* No 13.

Anonymous (2002). *Digest of Statistics* 2001-02. Directorate of Economics and Statistics. Planning and Development Department. J & K Government, Srinagar.

Huxley, P. Mead, Roger and Ngugi D. (1987). *National agroforestry research proposals for Southern Africa.* ICRAF, Nairobi, pp-46.

Makaya, A.S. (1994). Evaluation trial on multipurpose tree species. *Agroforestry News Letter*. 7&8, p p 4-6.

Makaya,A.S and Gangoo, S.A. (1995). Forage yield of pasture grasses and legumes in Kashmir Valley. *Forage Research,* 21:3, 152-154.

Misri, B. (1988). Forage production in Alpine and Sub alpine regions of north western Himalayas. *In* (Eds. Punjab Singh. *Pasture and Forage Crop Research*) - a state of knowledge Report. RMSI. Jhansi: 43-55.

Mughal, A. H & Bhattacharya, P. (2002). Agroforestry models practiced in Kashmir Valley. *The Indian Forester.* 128 (8) : 846-852.

Mughal, A.H and Makaya, A.S. (2000). Suitable Agroforestry models for the degraded and waste lands of Srinagar District Kashmir. In: M.A.Khan (ed.) *Environment Biodiversity and Conservation*. APH Pub. Corp. New Delhi, 493-500.

Mughal, A.H ; Khan, M.A; Zaffar, S. N and Mughloo, J. A (2003) Restoration of Unproductive Land Resources Through a Silvipasture System. *Oriental Sciences.* ISSN 8: 0971-703X

Mughal, A.H; Qaisar, K.N and Khan, P.A. (2003). Agroforestry for conservation of degraded ecosystems and food production in Kashmir hills. In: P. Rethy, P.P.Darbal, Binay Singh and K.K. Sood (ed) *Forest Conservation and Management.* IBD, Dehradun. India pp 262-265.

• • •

4

Agroforestry for Sustainable Development of Agriculture in North-western Himalayas — With Particular Reference to Jammu Region

M. SALEEM AND L.M. GUPTA
Division of Agroforestry, Faculty of Agriculture, Sher-e-Kashmir University of Agricutural Sciences and Technology-Jammu, Main Campus Chatha, P.O. Bhour Camp, Jammu 180 009 (J&K)

ABSTRACT

Deforestation, land degradation, biodiversity decline, demand of fuelwood, fodder and timber had necessitated the planners and scientists to develop a system of cropping which would be able to revert these detrimental effect of exploitation and simultaneously provide the day-to-day needs. In present paper we have revisited our ancestors who practiced agriculture in a sustainable basis by keeping trees in the field. This paper intends to highlight some of the problems which are being faced in North-Western Himalayas and suggest different agroforestry practices which should be adopted to sustainable utilize our resources.

INTRODUCTION

Though late, the country and the society have gradually started to realize the detrimental deterioration of our ecosystem and environment. There are several factors and reasons for the same. However, land use in the country in general and hills in particular have proved to be one of the most important factors and needs much of attention for proper planning. The vital role of the forests in maintaining the balance of the ecosystem has long been acknowledged world over. In our country too, it was rightly resolved through a National Forest Policy Resolution in the years 1952 to bring at least two-third of the land in hills under the forest cover. This decision, though wise timely and farsighted, was never implemented with same spirit and there are evidences to accept the fact that effective forest cover in hilly areas is far less. The increasing need of expanding industrialization and urbanization coupled with population growth, the process of illegal and indiscriminate cutting of trees is resulting in consistent decrease of forest cover in the hills. The cultivation of land on steep slopes and hillsides is causing heavy loss of top soil and depleting the productivity of these soils besides siltation of river beds and cultivated lands down below. Coupled with this is the reckless destruction of vegetative cover by way of over-grazing which is resulting in large scale denudation and degradation of lands and thereby increasing acreage under wastelands. With this background the regular rains for which most of our country was well known have become irregular and drought have become a permanent menace and crores of rupees spent on relief work. The need for developing correct methodologies of land uses has become imperative to protect the existing forest cover, whether on public or private lands. Forest influences the rural prosperity and the human needs for the growing population have to be ensured while preserving the environment and ecological balance. A sustainable land-use system in conformity with ecological stability, economic viability and above risk aversion is desirable land use pattern in the Himalayan region.

MAJOR PROBLEMS

Land degradation

For optimum utilization of land, a suitable land capability classification is necessary. The technique helps to determine the most suitable land use for a given area. There are several systems of classification and the most commonly accepted is the one adopted by United States Department of Agriculture (USDA). It requires very detailed physical and chemical survey, which are very expensive and time consuming. For this some alternative method based on degree of soil erosion, land slope and soil depth along with its type, lands are classified into two main groups, (a) land suitable for cultivation (b) land suitable for vegetation like pastures, orchards and forests. Both the major groups comprise of class I to class VIII. By and large, in-land classes I to IV, arable farming can be practiced and beyond that arable farming is considered a risk proposition; sometimes, even the lower classes of lands are utilized for forestry and grassland management to be on safer side.

The hill people face difficulties in raising crops to meet their needs due to poor per capita arable land availability. About 33 per cent of cultivated lands in hills belong to class V, VI and VII. This faulty and unscientific cultivation practices on such lands are the major reasons for degradation. In the J & K State while the total forest area should be around 33 per cent of the total geographical area, only 15.2 per cent of the state is actually under forests. It is estimated that around 8.93 lakh hectares of land has got degraded due to soil erosion and another 1.64 lakh hectares is estimated to be culturable wasteland. Soil erosion is one of the major problems affecting land economy. The soil is carried away from the slopes, leaving them bare and unfit for cultivation. In the foothills below, a formerly regular water supply becomes spasmodic and unreliable. During heavy rainstorms, flood waters thunder down below with gravel and stones, destroying crops and causing landslides.

Overgrazing and Deforestation

Livestock grazing is one of the most widespread pressures on forest cover today. Sustained forest productivity is the terrible lack of regeneration in vast tracts of natural forests. Destruction of herb and shrub layers by overgrazing, soil compaction and their inhibiting effect on natural or artificial regeneration of tree species is a serious problem in India, particularly in hilly states. Overgrazing causes the disappearance of the best forage species and, above all, of the perennial plants which afford the best protection for the soil against erosion. Because of small holdings, farmers concentrate entirely on growing crops, and it is very rare that an area of any size can be set aside for producing forage. Therefore, land-use policy must aim at maximum development of stall feeding, increased forage crops, forage preservation, planting of fodder trees, scientific management of pasturelands and use of concentrated feed. In the state of Punjab and Haryana where production of green fodder is adequate, the severe grazing pressure is reduced to some extent by the practice of stall feeding. Until grazing is controlled it is doubtful that natural forests will ever be replaced by new growing stock. The problem must be addressed not only through closures, social fencing, physical boundaries or rotational grazing but by incorporating forage and fodder needs into multipurpose forest management design. Grasses and fodder trees in silvipastoral system may have to be developed as buffers or in patchwork along with fuel, timber and minor forest products (MFP) plantations and natural forest reserves. Agroforestry plantation must concentrate on creating alternate sources of on farm supply and enhancing common property sources of forage and fodder so that livestock can be increasingly stall fed. Controlling the livestock by replacing local breeds with suitable improved breeds may be the quickest way to ensure more sustainable yields from natural forests.

COMMUNITY NEEDS

Fuelwood

Food is universally recognized basic need, most governments as well as those they govern, put low priority on fuelwood production, if they considered it at all. It should be obvious, however, that it is pointless to produce food without also ensuring that the means for preparing it are available. None of the principal food crops is palatable unless it has been cooked first. Availability of fuel and nutritional level are interlinked and lack of it can be as much cause of malnutrition as the lack of grain or root crops. With increasing deforestation, the women upon whom the responsibility of fuel supply depends in villages, must walk 10 km or even more everyday. The time consumed for collecting firewood otherwise would have been used in taking proper care of their children or farming activities. Fuelwood shortage have led to a switch to other fuel sources such as dried animal manure and crop residues, which has in turn started a vicious circle of decreasing crop yields and environment degradation. Efforts are being rightly made to popularize non-conventional forms of energy such as *Gobar* Gas, Biogas, wind mills, solar cookers etc. or to conserve energy through improved *chulas* etc. However, they are too adequate to meet the existing or growing needs. Those who make decisions must therefore take steps either to make sure that fuel requirements can be met from old and new sources, without causing further environmental degradation. One of the ideal solutions will be intelligent land use planning and planting suitable fast growing tree species on farm lands to provide firewood and small timber enabling the farmers to stop burning dung.

Fodder

Fuel wood is not the only need. India has 15 per cent of the world's livestock population on only 2 per cent of the world's land. Livestock constitute an important component of the rural economy and size of the herd maintained many a times becomes indicative of social status of a person in rural India. There are no reliable estimates on fodder production in India because the majority of livestock graze freely on community and public lands. However, the production of fodder in the country falls far short of the requirements and the shortage of green fodder is more acute than that of dry fodder. Available resources in the country can meet only half of the present fodder requirements. The available fodder is not only insufficient but also generally poor in nutritive value. The extent of deficit varies from state to state. The agriculture area presently utilized for green fodder production is only about 4.4 per cent. Fodder production from agriculture lands has limited possibilities in view of constraints of availability of arable land and rising demands for food due to human population explosion. The gap between demand and supply of fodder will thus further widen if effective steps for increasing production are not taken. Therefore, the only hope to meet this challenge is to employ better management techniques to other potential resources such as grasslands, uncultivable lands, permanent pastures, grazing lands and cultivable wastes which are not utilized to their potential at present. So question of setting portions of the cultivable lands exclusively for fodder production do not arise. silvipasture system is of utmost importance for the marginal, degraded and wastelands which can augment fodder resources. Another useful approach will be judicious selection of fodder trees and their plantation on field's alongwith crops without any adverse effect in different agroforestry planting patterns.

AGROFORESTRY NEED

Trees are useful to man in distinct ways as producers of wide variety of goods, commonly called forest produce and as custodians of favourable environmental conditions. It would not make sense to try to qualify one of these functions as more important than any other; both are indisputably essential to the well being, indeed to the survival of man. Forests in the hills are generally located in the ecological vulnerable areas where they serve very important function of environmental protection. If these trees are not maintained in good health, agriculture production along hillsides and bottomlands would suffer considerably. But these upland forests can not just be protected by laws and decrees. No application of the whip will stop rural people to raid forests for obtaining their day to day needs at any risk, if there is no alternative source. Fortunately, the mutually reinforcing relationship between Agriculture Production System and Forest Production System is more understood now than ever before. What was stated in 1950 literally holds true today. The foresight of late K.M. Munshi, the than Agriculture Minister can not be overlooked. His reference to forest trees as a source of food deserves a special mention. It was he who, for the first time in the country, sought to recognize the inter-relationship between agriculture and forestry. This approach to land-use which we call agroforestry i.e. tree + crops together, as one of the farm management options has tremendous potential. This approach will not jeopardize the food production. In many cases, can provide continuous vegetative cover to help conserve soil, water resources and increase in crop yields as well as fuel and fodder biomass.

Various definitions have been suggested for agroforestry. However, as opined by Nair (1983), agroforestry represents an approach to integrated land use involving deliberate mixture or retention of trees and other woody perennials in the crop/animal production fields. Thus, it combines elements of agriculture, whether crops or animals, with elements of forestry in sustainable production pattern on the same piece of land, either simultaneously or sequentially. The goal is to optimize the beneficial effects of interaction of the woody components with the crop and/or animal components to obtain a production pattern that, in terms of total quantity, diversity of end-products, or sustainability, is preferable to what is usually obtained from the same resources under prevailing social, ecological and economic conditions (Lundren,1982).

AGROFORESTRY PRACTICES

This approach to land use can benefit natural biota which are essential to agriculture and forestry productivity. Important agroforestry practices and their benefits are discussed as under:

Shelterbelts and windbreaks

Establishment of shelterbelts and wind breaks help to control water and wind erosion as well as increase crop yields. Trees and shrub shelterbelts help reduce wind velocity from 65 per cent to 87 per cent and in turn decreases erosion by as much as 50 per cent (Stocking,1995). The use of shelter belts in developing countries has increased yields of various crops from 10 per cent to 74 per cent (Pimentel *et al.*, 1997). Similarly, the establishment of windbreaks (*Azadirachta indica*) in the Maia valley of Niger resulted in initial net increase of millet ranging from 854 to 1,102 kg/ha (Le Houerou,1987).

Alley Farming for Conserving Water and Soil resources

Alley farming is defined by Kang and Wilson (1987) as the growing of arable crops between hedgerows of planted shrubs and trees, preferably leguminous species, which are periodically pruned to prevent shading the companion crops. This farming system has a potential in hills where cultivation is on slopes and when hedges are planted on contours, it can be a successful soil erosion control measure. Studies conducted (Kidd and Pimentel, 1992) by alternating three rows of *Leuceana leucocephala* with three rows of maize perpendicular to the slope of 10 per cent to 15 per cent revealed that the system can control erosion if all the maize residues and the pruning weighing 2,000 and 2,500kg, respectively, are left on the surface of the land. Erosion rate under these conditions with 1000 mm of rainfall would be about one ton/ha/year (Kidd and Pimentel, 1992). A similar study with maize-*Leucaena* intercrop on steep slope of 44 per cent produced similar results. Plots protected by *Leucaena* hedgerows lost two-ton/ha/year topsoil compared with 80 ton/ha/year for an unprotected plot (Banda *et al.,* 1994).

Agrisiviculture

Trees will serve a no less useful function for crop farming. The inter-cropping arrangement of annual food crops with woody perennials is a land-use system that enhances productivity and ensures sustainability. At the same time it helps to sustain a favourable and equitable microclimate, offer protection against wind and water erosion, and ensure that plant nutrients, leached out of reach of the crops, are returned to the surface. In the case of species that produce cattle feed, the trees will further make it possible to fatten more livestock on the farmland and to obtain soil enriching manure in the bargain. They will also yield firewood, small timber, poles, thus enabling the farmer to stop burning dung and obtain some cash income from the sale of wood. It may in fact be profitable to cover as much as 5-10 per cent of the existing farmland with trees. The improvement in agriculture yields that could be expected from the beneficial influence of the trees would undoubtedly in many cases exceed the percentage of land occupied by the trees, and in addition the wood produced by the trees would help to overcome the firewood crisis.

To obtain the most satisfactory results, the tree species and the associated crops must be selected keeping in view the compatibility, soil, climate, purpose of planting and site. Example of a few suitable traditional combinations in this region is as follows:

(a) Sub-tropical Zone

Trees

Acacia nilotica, Ailanthus excelsa, Albizia lebbeck, Grewia optiva, Bauhinia variegata, Zyziphus mauritiana, Dalbergia sissoo, Dendrocalamus strictus, Melia azedarach, Morus alba, Toona ciliata, Populus deltoids, Seshania aegyptia, Leucaena leucocephala

Crops

Wheat, maize, pulses, millets and medicinal & aromatic crops like ashwagandha, sarpgandha, brahmi, solanum, mulhathi, damask rose (variety-Jwala), mints, babchi, safed musli etc.

(b) Temperate Zone

Trees

Robinia pseudocacia, Salix spp., *Celtis australis, Ulmus wallichiana, Populus ciliata, Quercus leucotrichophora, Alnus nepalenses.*

Crops

Wheat, oats,maize, pulses, vegetables and medicinal & aromatic crops like kuth, damask rose(variety- Himrose), atees, patees, valerian, dioscorea spp., chirayta, karroo etc.

(c) Cold Arid Zone

Trees

Salix spp., *Populus* spp., *Juniperus communis, J. wallichiana*

Crops

Vegetables, sorghum and medicinal & aromatic crops like seabuck thorn, artimisia etc.

Silvipastoral System and livestock Production

Agroforestry system that include livestock provide potential increased productivity and economic security. A nutritious diet for the livestock is provided by some tree species, or livestock can graze beneath the trees, once they have reached certain height, with minimal damage to the tree. Tree foliage is an important source of feed for livestock because tree provides a reliable source of feed during conditions of floods and drought. The potential of integrating trees and livestock in hills is pressing due to increasing paucity of grazing lands because of desertification. Silvipastoral system which is a mixed land-use system for forage, livestock and wood production is gaining importance in hills where certain areas can not be dedicated to conventional agriculture without producing severe, and sometimes irreversible soil degradation. The mixture of trees, grasses and legumes conserve soil and moisture. The legume benefit soil by nitrogen fixation and in the mixture they help growth of grasses and trees besides improving forage quality. Trees and shrubs besides providing fuelwood, timber, supply fodder during lean period. The system works well with improvement in land productivity by 2.5 times as compared to traditional system. After 5-7 years grazing could be allowed, thus reducing the cost of grass harvesting. The nutritional value of tree varies between tree species and type of livestock grazed. For this reason, the tree has to be chosen for both its suitability for the grass grown and the livestock to be grazed. Trees and grass species should be selected in such a way that they are compatible to each other. Ideal tree species are (i) Fast growing and compatible with grass species (ii) Highly palatable and nutritious with no side effects (iii) Multipurpose and vigorous coppicing ability. The grasses should be highly productive, good quality and longer duration with high palatability. The legume component in the pasture need to be very hardy, easily colonizing, palatable, nutritious and with strong regeneration through roots or self grown seeds. Trees, shrubs and legumes for different ecological zones of the country have been identified by Patil and Pathak (1977).

Agrihorticultural System

In this system fruit trees and crops are grown together in various ways. Depending on planting pattern and configuration, the companion crops are known as intercrops, under planting or alley cropping. In inter-cropping, agriculture crops are generally grown between rows of fruit trees. The agriculture crops meet home requirement and seasonal revenue, whereas fruit trees managed for 30-40 years provide regular returns from fruit. In some cases fuelwood and fodder is also obtained from pruning.

In Kashmir, cultivation of pulses and vegetable crops in established orchards of Apple, Peach, Almond, Plum etc. is a traditional agrihorticulture system.

In lower zone tall and perennial medicinal trees that need to be planted at a wider spacing such as *Olea cuspidata, Sapindus mukorossi, Terminalia chebula* etc. can be inter-cropped with annual crops in early years until the tree canopy covers the ground. Shade tolerant and rhizomatic crops such as ginger, turmeric, garlic etc. can be grown on a longer term basis in widely spaced plantations. The inter-crops give some income to farmers during the period when the main trees have not started production.

Homegardens

Gardens variably called homegardens, house hold gardens or kitchen gardens are also common in the hills. These home gardens are complex agroforestry systems involving many plant species characterized by different morphology, stature, biological function and utility. Fruit and timber tree species may dominate the homegardens and occupy the middle and upper strata and medicinal plants, vegetables occupy the lower stratum. Home gardens of individual holdings generally cover small pockets of land and are established around homesteads. Although these systems in the past were mostly seen to meet the home needs, increased urbanization, transport and market opportunities in recent times are helping to produce cash value crops.

CONCLUSION

It is important to keep the various components of nature in full harmony to maintain a balance in the environment, a fact which is being increasingly realized these days. Rapid population growth and degradation of arable land have contributed to the world food crisis. People in rural areas are left with decreasing supply of land and food. What land is available is quickly being degraded by erosion, leading to a decline in crop and livestock productivity.

Agroforestry practices can provide the means to increase biomass and, in turn, food crops and livestock while at the same time improving the productivity of degraded soils. In some agroforestry combinations, the grain yield may increase. With livestock in some agroforestry systems, the yield in animal products may be increased while protecting the soil from erosion. In addition, agroforestry provides fuel wood to help meet cooking and heating needs of the rural people.

Resource poor farmers are likely to adopt agroforestry techniques when they are profitable and can help protect the farmer from risk and uncertainty. Agroforestry techniques encompass a wide variety of systems and diverse array of crop, livestock, and tree species.

There is significant evidence that agroforestry techniques can help meet the needs of rural farmers (Rocheleau, 1987). In addition agroforestry is a flexible technique that can be tailored to local environments and the needs of local people.

Though agroforestry can provide added security to food and household economy, risk is the greatest obstacle to farmer adoption of agroforestry systems. It is better to build upon existing techniques than to introduce new ones. Many agroforestry practices have been in place before agroforestry's surge in popularity in the academic community (Budowski, 1993). Traditional practices are similar to those described by science. Farmers may be more willing to adopt agroforestry if they are familiar with the use of trees in agriculture and know that integrating tree and crop production has been successful for other people (Nielsen, 1994). Traditional agroforestry techniques can help lower perceived risks and real risks.

REFERENCES

Banda, A. Z.; Maghembe, J. A.; Ngugi, D. N. and Chome, V. A. (1994). Effect of intercropping maize and closely spaced Leucaena hedgerows on soil conservation and maize yield on a steep slope at Ntchew, Malawi. *Agroforestry System*. 27, 31.

Budowski, G. (1993). Agroforestry: a discipline based on traditional knowledge. *Revista-Forestal- Centroamericana*. 2: 14.

Kang, B.T. and Wilson, G. F. (1987). The development of alley cropping as a promising agroforestry technology. *In*: agroforestry: A decade of development. Steppler, H. A. and Nair, P.K.R.(Eds.). ICRAF, Nairobi, Kenya.

Kidd, C. and Pimentel, D. (1992). Integrated resources management: agroforestry for development, Academic Press, San Diego. CA.

Le Houerou, H. (1987). Indigenous shrubs and trees in silvopastoral system of Africa. *In:* Agroforestry: A decade of development. Steppler, H. A. and Nair, P.K.R.(Eds.). ICRAF, Nairobi, Kenya.

Lundgren, B. (1982). Introduction. *Agroforestry system*.1:3-6

Nair, P.K.R. (1983). Multiple land use and agroforestry. *In:* Better crops for food: CIBA foundation Symposium 97 Landon, Pitman Books. 101-15.

Neilsen, F. (1994). Indigenous agroforestry systems in southern Luwero District Uganda and their potential in agriculture development. *CDR working paper Centre for development research.* 16: 97

Patil, B.D. and Pathak, P.S. (1977). Energy plantation and silvipastural system for rural areas. *Invention Intelligence.* 12(1&): 79-87.

Pimentel, D.; Mc Nair, M; Buck, L.; Pimentel, M. and Kamil, J. (1997). The value of Forests to World Food Security. *Hum. Ecol.* 25(1):91

Rocheleau, D.E. (1987). The user perspective and the agroforestry research agenda in agroforestry: realities and potentials. . Gholz, H.L., ed., M. Nijhoff, Dordrecht, the Netherlands.

Stocking, M. (1995). Soil erosion and land degradation. *In:* Environmental Science for Env. Mgt., O' Riordan, T.(Ed.). Longman Scientific & Technical, Essex, England, Co-published with John Wiley & Sons, New York, 223.

5

Agroforestry Systems in Himachal Pradesh

K.S. VERMA, D.R. BHARDWAJ AND KRISHAN CHAND

Department of Silviculture and Agroforestry Dr. YSP University of Horticulture & Agroforestry, Nauni, Solan, Himachal Pradesh

ABSTRACT

The practice of agroforestry in the hills and mountains of Himachal Pradesh is as old as the hill agriculture itself. Farmers are retaining trees on the farmlands for the multiple benefits they provide. The selection of the trees, their planting pattern and density is chosen in such a way that it had minimum effect on the associated crop. The present paper gives the prevalent agroforestry systems with their components in different agroclimatic zones of Himachal Pradesh along with their management practices. Adoption of landuse practices depends to a large extent on status of the farmer concerned. Paper presents the association between type of farmer and the corresponding agroforestry systems they adopt. Sustainability of agroforestry systems, their role in climate change and social and economic gain of agroforestry systems in hills had also been detailed. A comprehensive list of agroforestry components had also been appended as annexures.

INTRODUCTION

Farmers throughout the Indian Himalayas covering about 16 states partially or fully have nurtured trees on their farms and in the agricultural landscape both for production and protection benefits for millennia. Mountainous region like Himachal Pradesh located in western Himalayan range present wide range of tree growing practices and systems owing to the diverse physiographic and demographic mosaic.

The state is located between 30° 22′ 40″ to 33° 12′ 40″ N latitude and 75°47′ 55″ to 79°4′ 20″ E longitude and extends over a geographical area of 55673 km^2 which is 1.69 per cent of the country's area and 10.5 per cent of the Himalayan landmass. This land locked state is bordered by Jammu and Kashmir in the north, Punjab and Haryana in the south–west and Uttaranchal in the south–east. In the north–east the state forms the international boundary with the Tibet. Agricultural landscapes are characterized by undulating topography wherein terraced fields are made on medium to steep slopes. Crop fields on sloppy lands are in general 5-10 m wide and 20-30 m long almost conical on both ends and wide in the middle. In valley areas and on lands with gentle to mild slopes (up to 40 %) large sloping fields have been developed manually by the farmers where Government has undertaken land consolidation.

Tree planting patterns also vary with land form. In fields on hill slopes trees are favoured either on boundaries or field bunds. On flat lands fruits/ forest trees those provide early income and have short rotation periods are inter planted systematically in the entire field. The planting geometry is either rectangular or square types with spacing that easily facilitate ploughing or other inter-cultural operations for cultivating annual crops.

Himachal Pradesh has a varied climate owing to its topography and altitude. Hence, following four agro-climatic zones are very conspicuous (Fig. 5.1)

Zone-i) Sub-tropical sub-montane and low hills (up to 914 m asl)

Zone-ii) Sub-temperate sub-humid mid hills (915-1523 m asl)

Zone-iii) Wet temperate high hills (1524-2472 m asl)

Zone-iv) Dry temperate high hills (>2472 m asl)

The practice of agroforestry in the hills and mountains of Himachal Pradesh is as old as the hill agriculture itself. But use of term "agroforestry" to denote this practice is new. In different agro ecological zones of the state peasants have a tradition of cultivating food crops, raising trees for fodder, fuel, fiber, medicines and fruits as well as rearing of animals together. In each agroclimatic zone examples of low external input use and socially acceptable agroforestry practices and systems in which woody perennials are retained purposefully or planted deliberately with crops and/or animal production units in order to optimize the economic and ecological benefits from the resultant interactions are numerous. It is difficult to quantify the traditional agroforestry systems because of their inherent complexities. Nevertheless, in the recent past series of attempts have been made to appraise these systems focusing primarily on their structure, functions, economic advantages and disadvantages (Kachru, 1997; Upadhyaya, 1998 and Lehria, 2004).

The presence of an agroforestry system type and its extent among the hill farmers is determined by the factors like social stratification of farm households according to their operational holdings size, access to off - farm income sources, knowledge of the use of structural components, management skills of individuals and entrepreneurship. Together all these factors have influenced the biological composition and functional units of agroforestry systems which ultimately have given rise to different subsistence, intermediate and commercial type of agroforestry systems. The overriding factors in the development of agroforestry in Himachal Pradesh however are: strong indigenous knowledge base of the farmers about the natural resource use; enhanced access to the scientific innovations; farmer friendly Government policies; encouraging entrepreneurship and; sustained use of land to enhance family income and to diversify the production. Given below are the specific agroforestry systems in vogue among the farmers in different agro-climatic regions of the state.

(i) Agrisilviculture (AS)

Agrisilviculture system refers to a combined production system integrating agricultural crops and forest trees including shrubs leading to concurrent production of food, fibre, fuel, fodder and small timber. The primary production emphasis is on food and secondary on tree products. Hence, agricultural components predominate the system in terms of land area but systems management is influenced by the growth habits as well as nature of the harvestable produce (fodder/fuel/timber) of woody perennials. This ensures efficient utilization of the available land resource to offer diverse products. The agrisilviculture system is more productive and sustainable as compared to practicing sole agricultural system. The trees grown if managed using appropriate silvicultural techniques provide security against adverse climatic factors, supplement farm income, increase the total productivity per unit area and enrich soil fertility through the addition of leaf litter and improved nutrient cycling.

In agro-climatic zone I and II the tree species viz *Toona ciliata, Ulmus villosa, Melia azedarch, Albizia lebbeck* and *Albizia chinensis* are generally retained as timber trees whereas *Grewia optiva, Morus laevigata, Anogeissus latifolia* and *Bauhinia variegata* for fodder purpose. *Acacia catechu* is retained for its multiutility viz., katha and cutch contents, small agricultural timber, fodder, fuelwood and ecological restoration (Photo 5.1).

In wet temperate high hill (zone III), *Quercus* spp., *Salix* spp., *Robinia pseudoacacia* and *Morus serrata* are the most favoured tree species. Maize, wheat, barley, buckwheat and millets are the major cereals. Beans, tomato, cabbage and potato constitutes the main vegetable crops. In dry temperate high hill (zone IV), Poplar and *Salix spp.* are the main woody perennials whereas wheat, potato, barley represents the agricultural components. *Pinus roxburghii* is retained for its edible seeds and fuel purpose.

(ii) Agrihorticulture (AH)

It is a system in which fruit trees are grown in combination with agricultural crops on the same land management unit. In the sub-tropical submontane low hills zone mango, guava, litchi and citrus fruits are grown with maize, wheat/mustard and vegetable crops. In the sub-temperate sub-humid mid hills pear, plum, peach and apricot are the major fruit crops on agricultural fields cultivated with tomato, cabbage, cauliflower and beans besides maize and wheat. Major fruit crops in the wet temperate high hills are apple and pear. The agricultural components associated with apple and pear are maize and wheat as food crops and tomato, potato, peas and beans as the vegetable crops.

(iii) Agrisilvihorticulture (ASH)

Agrisilvihorticulture system refers to the combined production system involving agricultural crops as well as forest and fruit tree species. These systems deal with the concurrent production of food, fodder, timber and fruits and are continuous interpolated systems in which the crops forms several layers or are multilayered on the basis of their vertical stratification. It has several manifestations in Himachal Pradesh. In this system the forest trees are mainly retained for household needs and services whereas the fruit trees for the purpose of cash returns. The main distinguishing feature of this system is the premise of economic enterprise. In sub-tropical submontane low hills the forest tree species viz. *Grewia optiva, Morus alba, Leucaena leucocephala, Dalbergia sissoo, Ulmus villosa* and *Toona ciliata* are interspersed with fruits viz. mango, guava, and citrus spp. in agricultural fields. In sub-temperate sub-humid mid hills *Celtis australis, Grewia optiva, Leucaena leucocephala* and *Morus alba* are planted or retained on the field bunds for fodder and fuel wood. Poplar, *Toona ciliata* and *Bombax ceiba* for wood whereas peach, plum, pear and apricot for fruits. The system is utilized at intermediate scale. In wet temperate high hills zone the farmers practice commercial type of agrisilvihorticulture system integrating *Robinia pseudoacacia, Morus serrata* and *Quercus* spp. with fruits viz. apple, kiwi, pear, etc. The choice of tree species is very limited in dry temperate high hills. Here, *Salix* and Poplar spp. serves both as fodder as well as timber. The fruit trees are almond and apricot. Amongst the agricultural components wheat as food crop whereas, peas and potato are the major vegetables.

(iv) Silvipastoral (SP)

Silvipastoral system refers to the combined production of fodder trees and grasses/ forage crops and/or animals. Under this system grazing of cattle is permitted but stall feeding is a common practice. In general, silvipastoral system maintains higher productivity in a sustained manner at relatively low input levels even on poor soil strata. Higher productivity is primarily on account of the vertical stratification of the shoot and root systems of the different components. Understory herbage yield is often regarded as a function of the photosynthetically active radiations reaching the ground, which in turn, is dependent on the nature of the tree canopy and its orientation. In short, silvipastoralism denotes landuse systems in which pasture and livestock production is integrated with woody perennials on the same land management unit (Photo 5.2).

Animal population in the state is about 5.5 million (excluding poultry, dogs and pigs). Fodder availability to feed such a huge population from the limited and gradually shrinking agricultural and pasture lands is just 50 per cent of the total requirement. Tree leaves rich in protein and other mineral nutrients constitutes one of the potential fodder resources in the hills. Mean leaf green yield and lopping time of important fodder tree species found in sub-tropical low hills and sub-temperate mid hill zones are given in Table 5.1 (Yadava, 1997).

In sub-tropical sub-montane low hills zone *Grewia optiva, Morus alba,* and *Bauhinia variegata* are the most favoured species either planted or retained in the grasslands whereas *Dalbergia sisoo, Populus deltoids* and *Albizia chinensis* are retained for timber purpose. *Acacia catechu, Emblica officinalis, Terminalia chebula, Terminalia belerica, Terminalia arjuna, Sapindus mukorossi, Santalum album and Aegle marmelos* are the non-timber trees of economic value retained on the grasslands. The common grasses of the zone are *Eragrostis spp, Apluda spp* and *Chrysopogon spp*. In sub temperate, sub humid mid hills zone, *Grewia, Celtis, Robinia, Morus* and *Bauhinia* are the fodder species, the timber species are *Dalbergia, Bombax, Albizia, Chirpine,* Poplar and Eucalyptus. The Poplars and Eucalyptus species are the only planted species. The non-timber economic trees of this zone are *Acacia catechu, Olea spp., Myrica nagi, Punica granatum* (wild pomegranate) and *Prunus armeniaca*. The fodder tree components in the wet temperate high hills are *Celtis australis, Quercus leucotrichophora, Q. dilatata, Q. semicarpifolia, Robinia pseudoacacia* and *Morus serrata*; timber - *Pinus wallichiana, Aesculus indica and Pinus roxburghii*. *Eragrostis curvula, Apluda mutica and Chrysopogon* are the important grasses. In dry temperate high hills zone, the Poplar, *Robinia* and *Sulix* spp. are main functional units of silvipasture system. These species are used both for fodder as well as timber. The grass component for this zone comprises of *Poa, Eragrostis,* etc.

Improved tree canopy management (pollarding, coppicing), soil working techniques (gradonii, Ridge-ditch trench), mixed planting of leguminous and non-leguminous fodder trees alongwith high yielding grass species in different silvipasture systems can substantially enhance the fodder production while rejuvenating the ecologically degraded sites. Data in Table 5.2 is indicative of the significant contribution of some of the above practices to enhance biomass production.

(v) Pastoralsilvihorticulture (PSH)

Pastoralsilvihorticulture means cultivation and management of grasslands supporting forest and fruit trees simultaneously on the same unit of land. The perennials in this system provide the fodder or functions as live fences around grazing land or are retained as commercial, shade, browse and fruit trees. In most cases these trees exist as scattered but also found planted systematically. The production emphasis is on fodder, so the grasses predominate the system in terms of land area and management.

In sub-tropical submontane low hills especially shivaliks hill region *Morus, Grewia, Leucaena, Anogeissus latifolia, Albizia* spp., Bamboo, *Bauhinia variegata,* etc. are the major fodder species either planted or retained by the farmers. Whereas *Dalbergia, Toona, Albizzia,* Pines and Eucalyptus are most common timber tree species. Among fruit trees mango, citrus, guava and aonla are found interspersed in grassland. The grasses viz. *Eragrostis, Apluda* and *Chrysopogon* are common. Besides *Acacia catechu, T.chebula, T.bellerica, Sapindus mukorosii* and *Santalum album* are *the* non -timber trees of economic importance in this region. Sub-temperate sub-humid mid hills zone supports *Celtis, Grewia, Morus, Robinia* and *Bauhinia* as main fodder trees. The favoured grasses are similar as in zone I. The commercial timber trees are *Bombax,* Chirpine, *Albizzia, Toona, Poplar* whereas fruit tree species are peach, plum, apricot, pomegranate, almond, etc.

Most common fodder tree species in wet temperate high hill zone are *Robinia, Celtis, Quercus* and *Morus;* while *Pinus wallichiana* is the most favourable timber species. Pear and apple are the best suited as fruit trees of this zone along with the *Eragrostis, Apluda* and *Chrysopogon* grass species. In Dry temperate high hills salix and poplar are the commonly used tree species for fodder and timber as well. Almond and apricot are the common fruit trees, whereas, Poa and *Eragrostis* species are the favourable grasses for this zone.

(vi) Energy Plantations

Biomass will continue to play a significant role in meeting energy needs of people. Its share in cumulative world energy consumption will be about 17-20 per cent and will be third important source after natural oil and gas (WEC, 2002). In India, 55 per cent of the fuel wood comes from the lands outside the forest. Hence, energy plantations are very important. The plantation of fast growing short rotation forest tree species both on fertile as well as marginal and wastelands for fuel or energy purpose is called energy plantation. Short rotation forest tree species are grown with single stem or as coppice systems with a harvest cycle of 3-30 years and with an annual wood production of at least 10 tonnes DM/ha. (Christersson and Verma, 2003). These trees are planted in high density and are managed by regular thinning, felling, coppicing to ensure high productivity and amelioration of degraded lands. Normally marginal lands viz. mined out lands, river banks, lands with impeded drainage, roadside land fills, farm boundaries and unculturable zones of agricultural fields, etc are preferred to establish energy plantation of fuelwood lots in a watershed to optimize landuse and biomass yield as well.

In many studies tree species like *Eucalyptus teretecornis, Melia azedarach, Leucaena leucocephala* (Sharma 1989, Nayak 1996), *Acacia mollissima, Eucalyptus hybrid* (Verma *et.al.* 2003) have been planted using densities 4000-27000 trees ha^{-1} to produce fuelwood and amelioration of degraded lands (Table 5.3). The most suitable fuelwood species to be planted in the state are: *Salix* spp, *Acacia mollissima, Melia azedrach, Eucalyptus hybrid, Populus deltoides, Leucaena leucocephala,* Bamboo spp, *Morus* spp, *Dalbergia sissoo, Alnus* spp, *Albizia lebbeck* and *Albizia procera,* etc.

STATUS OF AGROFORESTRY SYSTEM TYPES AMONG DIFFERENT CATEGORIES OF FARMERS

Low and mid hill of Himachal Pradesh varying in altitude from 250-1500 m asl presents a wide range of agroforestry practices and systems established on slopes ranging from 8-30 per cent. This is because of the high biological diversity in sub-tropical and sub temperate climate of low and mid hills compared to temperate climatic conditions in high hill zones. In order to implement different agricultural and rural development programmes, farmers in the state are categorized as marginal, small, medium and large, and the basis being size of the operational land holdings (Table 5.4). Component analysis of the tree-crop systems undertaken in subtropical submontane low hills (Table 5.5) and in sub-temperate sub-humid mid hills (Table 5.6) revealed variety of system types prevalent amongst various categories of farmers. These are agrisilviculture (AS), agrihortisilviculture (AHS), agrisilvihorticulture

(AHS), agrihorticulture (AH), pastoralsilviculture (PS), pastoralsilvihorticulture (PSH) and pastoralhortisilviculture (PHS).

Marginal, small and semi-medium farmers represent about 95 per cent of the total land holders and posses about 70 per cent of the land area under operational holdings (Table 5.4). These farmers meet most of their household needs from on-farm resources. They have limited access to off farm resources and non-human capital. Hence, maintains almost all forms of agroforestry system types (Table 5.5 and 5.6). Medium and large farmers however opt for cash generating agroforestry systems like agrisilvihorticulture. Pasture or animal based system types are common to all categories of farmers. This is for two reasons:

a) Animal rearing is an essential component of hill farming systems.

b) Most lands are sloppy and unirrigated where agriculture alone is unprofitable.

SUSTAINABILITY/RESILIENCE OF AGROFORESTRY SYSTEMS

Sustainability is the ability of a land use system to maintain a certain well-defined level of performance over a period of time and, if required, to enhance the same, without damaging the essential ecological integrity of the system. Because of the time factor involved and the system's responsiveness to changing requirements, it is a dynamic phenomenon. This dynamism enables the capacities of production factors, mainly biophysical resources to meet the demands for increased output without loosing their long-term potentialities. The long-term productivity and health of the landuse system is affected by the pattern and intensity of its use. So what leads to sustainability or unsustainability is the result of match/mismatch between (1) basic characteristics of the natural resource components and, (2) patterns and methods of their utilization. The latter change with the changing needs, knowledge to the farmers of the technology adopted which in turn may transform the whole resource base.

Diverse output, interlinkages among different production activities, input use, practices as well as preference to and level of consumption of the produce impart a degree of resilience or sustainability to the agroforestry systems in vogue among different categories of the farmers in an agro-climatic region. However, improved knowledge & capacity to understand interactions among different biophysical variables offer significant opportunities for the development of more resilient tree based crop production systems. Explaining the resilience of an agroforestry system type is a complex phenomenon since there may be more than one contributing factors. Nevertheless, factors like biological yield as well as net returns may be used together or singly to explain this phenomenon depending upon the needs of the intended user and the nature of the output for which an agroforestry system is exploited.

Agroforestry system types practiced by different categories of farmers in Himachal Pradesh were appraised for their biological yield levels and net returns accrued. Large variations were observed in both the parameters. Hence, to have an estimate of variation and to judge the resilience of an agroforestry system type coefficients of variation for each were calculated. In sub-temperate sub-humid mid hills at Solan (Fig 5.2, 5.3) five agroforestry system types viz: AS, AHS, ASH, PS, PHS and PSH were common among different groups of farmers. Coefficients of variation for biological yield as well as net returns were maximum in case of PSH and ASH system types, respectively.

The above two agroforestry system types (PSH and ASH) were the least resilient tested in terms of biological yield as well as net returns. However, it may not always prove similar. Each parameter can yield different results. The studies conducted involving lowland and upland agroforestry systems in Balh valley of Himachal Pradesh explains this clearly. Among the three commonly practiced agroforestry system types viz., AS, ASH and PS in low lands coefficient of variation for biological yield was maximum with ASH but for net returns with AS (Fig 5.4, 5.5). Similarly in uplands four agroforestry system types (AS, ASH, PS and PSH) were common amongst farmers groups. Coefficient of variation for biological yield was higher with AS system however, for net returns it was higher with ASH system (Fig 5.6, 5.7) type.

AGROFORESTRY IN MITIGATING CLIMATE CHANGE

Agroforestry practices can reduce atmospheric concentrations of CO_2 by increasing carbon storage in biomass, decreasing emissions at source and by modifying agricultural practices to increase the quantity of carbon stored in soil organic matter. Besides agricultural lands, wastelands including degraded lands could be used for the production of biofuel crops such as willow, bamboos and wattles etc. in short rotation coppice system. These crops also known as energy crops, can be used to substitute for fossil fuels and thus to decrease emissions at source. Preserving and enhancing the land carbon sink, in both trees and soils will be the main focus of agroforestry management strategies. This will depend on the type of the agroforestry system put in place as well as temporal arrangement of the system components. To prove its mitigation potential some questions are needed to be answered: i) whether agroforestry system can increase the carbon sequestration and yet meet food and fodder needs; ii) extent to which wood from agroforestry can substitute the wood harvested from traditional forests for fuel and small timber purposes thereby helping to conserve the forest carbon pool; iii) opportunities for carbon trading.

In Himachal Pradesh, as per the land capability classification potential lands for introducing trees are class II, III and IV lands where intensive agriculture is least feasible due to undulating topography, medium to steep and long slopes and lack of irrigation water, etc. Government policies are also conducive to transform the barren hill slopes, pastures and large rainfed agricultural tracts into agroforestry. Undoubtedly, this will have a big impact on the flux and medium to long-term storage of C in biomass as well as soils. Soil and forest degradation as a result of change in landuse from forest to agriculture, horticulture, industries, hydropower projects, urban settlement, etc. have been major causes of C loss in Himachal Pradesh contributing to CO_2 accumulation in the atmosphere. Hence, while integrating agroforestry technologies with other landuse systems to improve the productivity of different production systems it will also provide opportunities to create and enhance existing C sinks. In one of the study carbon sequestration potential of silvipastoral and agrihortisilviculture systems was evaluated to be very high with respect to other systems considering natural grassland systems as control. Total carbon stocks (plant+soil up to 40 cm depth) for agroforestry systems *viz.*, hortipastoral, silvipastoral, agrisilviculture, agrihorticulture, agrihortisilviculture and natural grasslands in sub temperate sub humid mid hills region were 7.86, 153.71, 62.70, 57.04, 85.5 and 79.40 t ha^{-1}, respectively.

SOCIAL AND ECONOMIC GAINS OF HILL AGROFORESTRY

Farmers in Himachal hills and mountains rarely practice monoculture systems. Leaving apart upland paddy fields in valley regions trees find an integral place all over the agricultural landscapes. Occurrence of trees on agricultural fields while supplement fodder, fruit, fuel and timber needs of the farmer, they also ensure optimal use of available lands, family labour and moneyfory resources. The income of the farmers especially marginal, small and medium in rainfed areas (about 80 %) can be increased from the sale of wood, fruits and other tree products. However, crop yield due to tree shade, release of allelochemicals, competition for nutrient, moisture and land, itself is reduced. Improved plant and canopy management practices viz., pollarding, lopping, coppicing, hedging, etc can however reduce the adverse influence of trees on associated crops yield. *Grewia optiva* the most preferred and nutritious fodder tree in low and mid hills does not allow any crop to grow underneath its canopy. Pollarding *Grewia* trees at 1 m or 1.5 m height and planting on the

Table 5.1. Average size, lopping time and fodder yield of important fodder tree species Found In Sub-tropical Low Hill and Sub-temperate Mid Hill zones of H.P.

S. No.	Name	Height (m)	Canopy diameter (m)	Main lopping period	Mean leaf yield of an adult tree (Kg green matter)
1	*Grewia optiva*	7	8	Oct-Feb	15-20
2	*Bauhinia variegata*	12	9	Oct-Feb	12-15
3	*Robinia pseudoacacia*	12	8	Sept-Oct	10-15
4	*Dendrocalamus hamiltonii*	12	15	Jan-Feb	30-40
5	*Terminalia arjuna*	30	20	February	40-50
6	*Quercus incana*	12	14	Feb-April	8-10
7	*Albizia chinensis*	12	10	Dec-Jan	10-15
8	*Morus alba*	10	7	Sep-Oct	8-12
9	*Celtis australis*	16	12	i) Nov-Feb ii) June	15-20
10	*Acer oblongum*	5	2.87	Nov-Dec	12-15
11	*Bauhinia retusa*	7	3.56	Oct-Jan	30-35
12	*Mallotus philippinensis*	7	2.95	Dec-Feb	20-25
13	*Albizia stipulata*	7	3.97	Oct-Dec (Nov)	-
14	*Ougenia oozeinensis*	6	-	Sep-Nov (Oct)	-

Table 5.2. Effect of tree management practices on leaf, branch, grass biomass and their total in single and mixed tree species based silvipastoral system.

*Management practice	Leaf biomass	Branch biomass (q ha^{-1})	Grass biomass (q ha^{-1})	Total biomass (leaf+branch +grass) (q ha^{-1})
T1	11.40 (39.34)	17.54 (60.61)	97.00	125.94
T2	7.99 (19.14)	33.76 (80.86)	47.83	89.58

Contd...

Table 5.2. Contd...

T3	8.53 (26.31)	23.90 (73.69)	71.69	104.13
T4	10.72 (20.87)	40.64 (79.13)	48.28	99.64
T5	8.31 (29.74)	19.65 (70.26)	54.11	82.06
T6	7.85 (21.36)	28.90 (78.64)	84.45	121.21
T7	4.09 (22.14)	14.38 (77.86)	75.21	93.67
SE (d)	1.48	4.02	9.95	10.07
CD0.05	3.05	8.27	21.46	20.71

*T1: *Grewia optiva* pollarded at 4 m height + grass

T2: *Leucaena leucocephala* pollarded at 1 m height + grass

T3: *Grewia optiva* pollarded at 4 m + *Leucaena leucocephala* pollarded at 1m height + grass

T4: *Leucaena leucocephala* pollarded at 2 m height + grass

T5: *Grewia optiva* pollarded at 4 m + *Leucaena leucocephala* pollarded at 2 m height + grass

T6: *Morus alba* pollarded at 4 m height + grass

T7: *Morus alba* pollarded at 4 m + *Leucaena leucocephala* pollarded at 1 m height + grass

Table 5.3. Aboveground biomass and productivity of *Eucalyptus tereticornis* at different densities (7 years old)

Attribute	Density (trees/ha)		
	D1 (27,778)	D2 (12,346)	D3 (6,944)
Biomass (ton ha^{-1})	127.0	70.85	49.49
Productivity (ton ha^{-1} yr^{-1})	18.15	10.12	7.07

Table 5.4. status of operational land holdings among different categories of farmers.

Category	Percentage	
	Farmers	Land area(ha)
Marginal (upto 1ha)	63.80	21.30
Small (1-2 ha)	20.00	23.30
Semi-medium (2-4 ha)	11.30	25.50
Medium (4-10 ha)	4.30	20.30
Large (10-20ha)	00.60	09.60

Table 5.5. Occurrence of agroforestry systems in subtropical submontane low hills at Balh valley (874 m.sl) in Mandi district of Himachal Pradesh.

Agroforestry system	Farmers category*					
	Uplands			Lowlands		
	Marginal (<1 ha)	Small (1-2ha)	Medium (2-5 ha)	Marginal (<1 ha)	Small (1-2ha)	Medium (2-5 ha)
Agrisilviculture	√	√	X	√	√	X
Agrisilvihorticulture	√	√	√	√	√	√
Pastoralsilviculture	√	√	√	√	√	√
Pastoralsilvihorticulture	√	X	X	√	√	√
Pastoralhorticulture	√	X	X	X	X	X
Hortipastoral	X	√	X	X	X	X
Pastoralhortisilviculture	X	X	X	√	X	X

√ = system present **X = system absent**

* There was no farm household in large categories

Table 5.6. Occurrence of agroforestry systems in subtemperate subhumid mid hills region at Nauni (1275 m.sl) in Solan district of Himachal Pradesh.

Agroforestry system	Farmers category			
	Marginal (<1 ha)	Small (1-2 ha)	Medium (2-5 ha)	Large (> 5 ha)
Agrisilviculture	√	√	X	X
Agrihorticulture	√	X	X	X
Agrihortisilviculture	√	√	√	X
Agrisilvihorticulture	√	√	X	√
Pastoralsilviculture	√	√	√	X
Pastoralsilvihorticulture	√	√	X	√
Pastoralhortisilviculture	√	√	X	X

√ = system present X = system absent

Table 5.7. Economic analysis of planted agroforestry developed in H.P

Agroforestry system	Functional Units	Productivity (q ha^{-1})	Total costs (Rs. ha^{-1})	Net returns (Rs. ha^{-1})
Agrihortisilviculture	Plum-Mulberry-Pomegranate-Soybean	101	3449	14203
Agrihortisilviculture	Plum-Mulberry-Soybean	146	17097	23915
Agrihortisilviculture	Pomegranate-Mulberry-Soybean	99	7127	9408
Agrihorticulture	Plum-Soybean	158	26100	43654
Agrihorticulture	Pomegranate-Soybean	37	6159	5556
Agrihorticulture	Peach-Maize-Lentil	185	16521	31942
Agrihorticulture	Peach-Soybean-Wheat	175	15977	30195
Agrisilviculture	Mulberry-Soybean	189	8094	17374
Agrisilviculture	Mulberry-Maize-Lentil	228	14248	29039
Agrisilviculture	Mulberry-Soybean-Wheat	282	13689	25140
Agrisilvihorticulture	Mulberry-Peach-Maize-Lentil	207	17222	31554
Agrihortisilviculture	Mulberry-Peach-Soybean-Wheat	196	16678	28810
Hortisilvimedicinal	Peach+Morus+ Tulsi		21357	118820
Hortisilvimedicinal	Peach+Morus+ Mucuna		21307	109980
Hortisilvimedicinal	Peach+Grewia+ Tulsi		21307	118551
Hortisilvimedicinal	Peach+Grewia+ Mucuna		21307	111901
Agriculture	Maize-Lentil	174	13230	27913
Agriculture	Soybean-Wheat	165	12656	2431

Annexure - 5.1. Important Fodder, Fuel wood, Fruit, Timber and Medicine yielding tree and shrub species in different agro climatic regions of H.P.

1. Sub-tropical Sub-montane and low hills

S. No.	Botanical Name	Common Name	Tree/ Shrub	Altitude (ft)	Uses
1.	*Dodonea viscosa*	Mehndu	S	Upto 4500	Hdg,Fo,Cop
2.	*Acacia catechu*	Khair	T	Upto 3000	Med,Fo,Cop

Contd...

Annexure 5.1. Contd...

3.	*Albizia chinesis*	Ohi, siris	T	Upto 4000	Fo Flv,Cop
4.	*Grewia optiva*	Bihul	T	Upto 6000	Fo.Fw.Fib.Agi,Cop
5.	*Celtis australis*	Khirak	T	Upto 8000	Fo.Fw,.Ti.Agi,Cop
6.	*Morus spp.*	Mulberry	T	Upto 9000	Fo.Fw.Agi, Fr,Cop
7.	*Pistacia integerrima*	Kakkar	T	1500 to 8000	Med,Fw,Agi,Fo,Cop.
8.	*Bauhinia variegata*	Kachnar	T	Upto 4000	Fo, Veg, F,Cop
9.	*Mangifera indica*	Mango	T	Upto 3000	Fw, Ti, Frn, Fr,Cop
10.	*Carissa opaca*	Karaunda	S	Upto 5000	Fw, Wfr, Cop
11.	*Ficus palmata*	Fig	T	Upto 5000	Fw, Fo, Wfr, Fr, Cop.
12.	*Pinus roxburghii*	Chir pine	T	1500-7000	Fw, Ti, Frn, Fr,Cop.
13.	*Zizyphus spp.*	Ber	S	Upto 7000	Fod,Fw,Fr,Wfr,Cop.
14.	*Pyrus pashia*	Kainth	T	3000-8000	FW, Wfr, Cop.
15.	*Dendrocalamus strictus*	Bainj	T	Upto 3500	Fw, Fod, Ti, Cop.
16.	*Bambusa sp.*	Bainji	T	Upto-5000	Fw, Fo, Cop.
17.	*Dalbergia sissoo*	Shisham	T	Upto 4000	Ti, Fw. Frn, Cop.
18.	*Melia azedarach*	Darek	T	UPto- 5000	Agi, Fw, Cop.
19.	*Flacourtia spp.*	Kangu	T	Upto-3000	Fw, Agi, Cop.
20.	*Vitex negundo*	Bana	S	Upto-5000	Fw, Med, Cop.
21.	*Aegle marmelos*	Bel	T	Upto- 4000	Med, Frn, Fw, Cop.
22.	*Eucalyptus spp.*	Safeda	T	Upto-4500	Fw, Ti, Med, Cop.
23.	*Shorea robusta*	Sal	T	Upto 4000	Ti, Frn, Fw, Cop.
24.	*Anogeissus latifolia*	Chalil	T	Upto 4000	Ti, Fw, Cop.
25.	*Cassia fistula*	Amaltash	T	Upto- 4000	Med, Fw, Cop
26.	*Leucaena leucocephala*	Leucaena	T	Upto-4000	Sw, Fw, Cop.
27.	*Salix tetrasperma*	Beuns	T	Below 4000	Sw, Fw, Cop.
28.	*Mallotus philipinensis*	Kaml	T	Upto 4000	Fo, Fw, Cop.
29.	*Emblica officinalis*	Amla	T	Upto 4500	Med, Fw, Frn, Cop.

Contd...

Annexure 5.1. Contd...

30	*Azadirachta indica*	Neem	T	Upto-2000	Fw, Fo, Frn, Cop.
31.	*Syzygium cumini*	Jamun	T	Upto-000	Fw,Fo,WFr,Ti,Cop.
32.	*Toona ciliata*	Toon	T	3000-8000	Ti, Fw, Frn, Agi, Cop.
33	*Sapindus mukorosii*	Ritha	T	Upto 5000	Med, Fw, Cop.
34.	*Berberis spp.*	Kashmal	S	Upto 11000	Med, Fw, Cop.
35.	*Terminalia tomentosa*	Sain	T	Upto 4000	Med, Fw, Cop.
36.	*Terminalia bellerica*	Bhera	T	Upto 2000	T_i, M_{ed}
37.	*Terminalia chebula*	Harad	T	Upto 2000	T_i, M_{ed}
38.	*Santalum album*	Chandan	T	Upto 2000	T_i, M_{ed}
39.	*Woodfordia floribunda*	Dhau	S	Upto-5000	Agi, Fw, Med, Cop.
40.	*Olea cuspidata*	Kahu	T	Upto-6000	Frn, Agi, Fw, Cop.
41.	*Xanthoxylum alatum*	Timru	S	Upto 6000	Fw, Med, Agi, Cop.

2. Sub-temperate Sub-humid mid hill zone

S. No.	Botanical Name	Common Name	Tree/ Shrub	Alt. (ft)	Uses
1.	*Quercus leucotrichophora*	Ban	T	5000-8000	Fo, Fw,Cop.
2.	*Robinia pseudoacacia*	Robinia	T	2800-10500	Fo, Fw,Cop
3.	*Pistacia integerrima*	Kakkar	T	3000-5000	Fo, Fw,Ti,Cop.
4.	*Toona ciliata*	Toon	T	3000-8000	Ti,Fw, Frn, Agi,Cop.
5.	*Grewia optiva*	Bihul	T	Upto 6000	Fo,Fw.Fib.Agi,Cop.
6.	*Celtis australis*	Khirak	T	Upto 8000	Fo.Fw.Ti.Agi,Cop.
7.	*Albizia chinensis*	Ohi, Siris	T	Upto 4000	Fw.Fo,Cop.
8.	*Ficus palmata*	Fig	T	Upto 5000	Fw, Fo, WFrCop.
9.	*Bauhinia variegata*	Kachnar	T	Upto 4000	Fo, Veg, Fw,Cop.
10.	*Pyrus pashia*	Kainth	T	3000-800	Fw.WFr,Cop.
11.	*Prunus paddus*	Paja	T	3000-6000	Fw,Cop.
12.	*Vitex negundo*	Bana	S	Upto 5000	Fw.MedCop.
13.	*Salix tetrasperma*	Beunus	T	Below 3000	Sw.Fw,Cop.

Contd...

Annexure 5.1. Contd...

14.	*Acacia catechu*	Khair	T	Upto 3000	Med Fw,Cop.
15.	*Pinus roxburghii*	Chir pine	T	1500-7000	Ti,Fw,Re,Pc,Cop.
16.	*Melia azedarach*	Darek	T	Upto 5000	Agi, Fw,Cop.
17.	*Populus spp.*	Poplar	T	Upto 10000	Pc,Fw,Cop.
18.	*Eucalyptus spp.*	Safeda	T	Upto 4500	Fw,Ti,Cop.
19.	*Prunus armeniaca*	Wildapricot	T	Upto 9000	Fw.WFr.Fr,Cop.
20.	*Myrica nagi*	Kaphal	T	Below 6000	Fw, WFr. Fr,Cop.
21	*Morus indica*	Mulberry	T	Upto 5000	Fo. Fw. WFrCop.
22.	*Indigofera pulchella*	Kathi	S	Upto 5000	Fw.Med,Cop.
23.	*Berberis spp.*	Kashmal	S	Upto 11000	Med,Fw, WFr,Cop.
24.	*Machilus duthici*		T	4000-9000	Fw, Fod,Cop.
25.	*Prinsapia utilis*	Baikhal	S	4000-8000	Fw, Med., Oil yielding, Cop.
26.	*Prunus persica*	Peach	S	4000-8000	Fw,Cop.
27.	*Prunus saliciana*	Plum	T		Fr, Cop.
28.	*P. communis*	Plum	T		Fr, Cop.
29.	*Desmodium triflorum*		S	Upto 4000	Fw,Cop.
30.	*Xanthoxylum alatum*	Timru	S	Upto 6000	Fw,Cop.
31.	*Olea cuspidata*	Kahu	T	Upto 6000	Frn, Agi, Fw.Cop.
32.	*Woodfordia floribunda*	Dhau	S	Upto 5000	Agi, Fw,MedCop.

3. Wet temperate High hill zone

Shimla district (except Rampur Teshil), parts of Kullu, solan, Chamba, Mandi, Kangra and Sirmour, height above mean sea level is above 1800m.

S. No.	Botanical Name	Common Name	Tree/ Shrub	Alt. (ft)	Uses
1.	*Quercus leucotrichophora*	Ban Oak	T	5000-8000	Fw, Fo,Cop.
2.	*Alnus nitida*	Kunis	T	3000-9000	Fw, NFT,Cop.
3.	*Quercus dilatata*	Mohru	T	6000-9000	Fo,Fw,Cop.
4.	*Q. semicarpifolia*	Kharsu	T	8500-12000	Fw,Cop.

Contd...

Annexure 5.1. Contd...

5.	*Celtis australis*	Toon	T	3000-8000	Ti,Fw,Cop.
6.	*Cedrus deodara*	Deodar	T	3500-12000	Ti,Med,Fw,Cop.
7.	*Pinus wallichiana*	Kail	T	6000-10000	Ti, Re,Fw,Cop.
8.	*Indigofera pulchella*	Kathi	S	Upto 5000	Med,Fw,Cop.
9.	*Prunus armeniaca*	Wildapricot	T	-	Fw.WFr.Fr,Cop.
10.	*Prunus paddum*	Paja	T	3000-6000	Fw,Cop.
11.	*Berberis spp.*	Kashmal	S	Upto 11000	Med,Fw,Cop.
12.	*Abies pindrow*	Fir/Rai	T	8000-10000	Ti,Pc,Fw,Cop.
13.	*Picea smithiana*	Pruce/Tosh	T	6000-11000	Ti, Pc, Fw,Cop.
14.	*Robinia pseudoacacia*	Robinia	T	4800-10500	Fo, Fw,Cop.
15.	*Salix spp.*	Beuns	T	7000-10000	Sw.FwCop.
16.	*Aesculus indica*	Bankhor	T	5000-10000	Ti,Fw,Med,Cop.
17.	*Juglans regia*	Akhrot	T	7000-10000	Fr,Med,Ti,Fw, Cop.
18.	*Ficus palmata*	Fig	T	Upto 5000	Fw,Fo,WFr,Cop.
19.	*Prunus persica*	Bahmi	T	Upto 6000	Fw,Cop.
20.	*Corylus colurna*	Bhutia	T	5500-10000	Fo.Fw,Cop.
21	*Ulmus wallichiana*	Elm	T	3500-10000	Fo,Fw,Cop.
22.	*Desmodeium sp.*		S	Upto 4000	Fw,Cop.
23.	*Morus serrata*	Kimu	T	4000-9000	Fo,Fw,Cop.
24.	*Malus domestica*	Apple	T	5000-9000	Fr, Fw,Cop.
25.	*Pinus roxburghii*	Chir pine	T	1500-7000	Ti,Fw,Re,Pc,Cop.

4. Dry temperate high hill zone

S. No.	Botanical Name	Common Name	Tree/ Shrub	Alt. (ft)	Uses
1.	*Alnus* sp.	Alder/Kunis	T	3000-9000	Fw, NFT, Cop.
2	*Salix* sp.	Beunus	T	7000-10000	Sw, Fw, Fo, Cop.
3	*Populus* sp.	Poplar	T	Upto 10000	Pc, Fw, Fo, Cop.
4	*Robinia pseudoacacia*	Robinia	T	2800-10500	Fo, Fw, Cop.
5	*Pinus wallichiana*	Kail	T	6000-10000	Ti, Fw, Pc, Cop.

Contd...

Annexure 5.1. Contd...

6	*Pinus gerardiana*	Chilgoza	T	7000-11000	Fr, Pc, Cop.
7	*Pinus roxburghii*	Chir pine	T	1500-7000	Ti, Fw, Pc, Cop.
8	*Cedrus deodara*	Deodar	T	3500-12000	Ti, Med, Fw, Cop.
9	*Quercus* spp.	Dhek/Oak	T	5000-12000	Fo, Fw, Cop.
10	*Prunus armeniaca*	Apricot	T	5000-10000	Fw, WFr, Cop.
11	*Juglans regia*	Walnut	T	7000-10000	Frn,Med, Ti, Fw,Cop.
12	*Celtis australis*	Khirak	T	4000-8000	Ti, Agi, Fo,Fw,Cop.
13	*Morus serrata*	Mulberry/ Kimu	T	4000-9000	Fo, Sw, Fw,Cop.
14	*Aesculus indica*	Bankhor	T	5000-10000	Fw,Med,Cop.
15	*Prunus persica*	Wild peach	T	Upto 6000	Fw, WFr,Cop.
16	*Betula* spp.	Birch/Bhoj Patra	T	7000-14000	Fw,Cop.

Key:

Ti = Tree, timber (in column of uses)

Shr = Shrub

Cop = Coppicer

Hdg = Hedges

Med = Medicinal

Fw = Fuel wood

Fo = Fodder

Fib = Fiber

Agi = Agricultural implements

Veg = Vegetables

WFr = Wild fruit

Frn = Furniture

Re = Resin

Fr = Fruits

Sw = Sport wood

Pc = Packing cases (fruit and vegetables)

NFT = Nitrogen Fixing Tree

Annexure - 5.2. Functional Major components of dominant agroforestry system in different agro climatic regions of H.P.

i) Agrisilviculture

Agroclimatic zone	Functional components						
	Agriculture				Trees		
	Cereals/pulses		Vegetables		Timber	Fodder	Non timber trees of economic importance
	Kharif	Rabi	Kharif	Rabi			
Sub-tropical sub-montane low hill zone	Maize, Mash, Soybean, Pigeon pea	Wheat, Gram, Lentil, Mustard	Tomato, bean	Peas, Cabbage, Cauliflower and Potato	*Toona ciliata, Melia azedarach, Albizia lebbeck, A. chinensis, Ulmus villosa, olea ferruginea Annogeissus latifolia*	*Grewia optiva, Morus laevigata, Bauhinia variegata, Annogeissus-latifolia, olea ferruginea* Bamboo species	*Acacia catechu, Santalum album, Pinus roxburghii, Terminalia spp. Emblica officinalis, Sapindus mukorosii etc.*
Sub-temperate sub-humid mid hill zone	Maize, Soybean, Mash, Soyabean	Wheat, Gram, Lentil	Tomato, Zinger, Bean	Peas, Cabbage, Cauliflower	*Toona ciliata, Albizia chinensis, Salix tetrasperma, olea ferruginea.*	*Celtis australis, Grewia optiva,* Morus alba, Bauhinia variegata, Bamboo, *Salix tetrasperma, olea ferruginea, etc.*	*Acacia catechu, Quercus leucotrichophora, Pinus roxburghii, etc.*
Wet temperate high hill zone	Maize, Millets	Wheat, Barley Buckwheat, Beans, Potato	Tomato, Peas	Cabbage, Cauliflower	Deodar, Kail, Fir, horse chestnut, Spruce, *Ulmus wallichiana*	*Robinia pseudoacacia, Morus serrata,* Ulmus wallichiana Oak species, *Salix tetrasperma*	*Quercus spp, Prunus americana*
Dry temperate high hill zone	-	Wheat, Potato, Barley, Millets	Potato, Peas	-	Populus, Salix, Deodar, Kail	*Robinia pseudoacacia, Salix spp.*	*Pinus gerardiana* and Juniper's species

ii) Agrihorticulture system

Agroclimatic zone	Functional components				
	Agriculture				Trees
	Cereals/pulses		Vegetables		Fruit trees
	Kharif	Rabi	Kharif	Rabi	
Sub-tropical sub-montane low hill zone	Maize, Mash, Soybean, Pigeon pea	Wheat, Gram, Lentil, Mustard	Tomato, bean	Peas, Cabbage, Potato	Mango, Citrus, Guava, Papaya and Litchi.
Sub-temperate sub-humid mid hill zone	Maize, Soybean, Mash, Soyabean	Wheat, Gram, Lentil	Tomato, Zinger, Bean	Peas, Cabbage, Cauliflower	Peach, plum, Apricot and Pear.
Wet temperate high hill zone	Maize, Millets	Wheat, Barley Buckwheat, Beans, Potato	Tomato, Peas	Cabbage, Cauliflower	Pear, Apple, Apricot, Kiwi and Persimon.
Dry temperate high hill zone	-	Wheat, Potato, Barley, Millets	Potato, Peas	-	Apple, Apricot and Grape

iii) Agrihortisilviculture system

Agroclimatic zone	Functional components							
	Agriculture				Trees			
	Cereals/pulses		Vegetables		Timber	fodder	Non timber trees	Fruit trees
	Kharif	Rabi	Kharif	Rabi				
Sub-tropical sub- montane low hill zone	Maize, Mash, Soybean, Pigeon pea	Wheat, Gram, Lentil, Mustard	Tomato, bean	Peas, Cabbage, Potato	Toona ciliata, Melia azedarach, *Anogeissus latifolia* and *Albizia lebbeck and Terminalia spp.*	*Grewia optiva, Morus laevigata, Bauhinia variegata, Anogeissus latifolia,* Bamboo spp.	*Acacia catechu, Santalum album, Sapindus mukorosii*	Mango, guava, Litchi, Papaya, Aonla
Sub-temperate sub-humid mid hill zone	Maize, Soybean, Mash, Soyabean	Wheat, Gram, Lentil	Tomato, Zinger, Bean	Peas, Cabbage, Cauliflower	*Toona ciliata, Albizia chinensis, Salix tetrasperma*	*Celtis australis, Grewia optiva,* Morus, Bauhinia, Bamboo, Salix	*Acacia catechu*	Plum, Pear, Apricot, Peach
Wet temperate high hill zone	Maize, Millets	Wheat, Barley Buckwheat, Beans, Potato	Tomato, Peas	Cabbage, Cauliflower	Deodar, Fir, Spruce, Oak	Robinia, *Morus serrata,* Oak, Salix	Oak, *Prunus americana*	Apple, Apricot, Persimon
Dry temperate high hill zone	-	Wheat, Potato, Barley, Millets	Potato, Peas	-	Populus, Salix, Deodar, Kail	Robinia, *Salix* spp.	Junipers and *Pinus wallichiana*	Apple, Apricot, Grape,

iv) Silvipastoral system

Agroclimatic zone	Timber trees	Non–timber trees of economic importance	Fodder trees	Grasses
Sub tropical, Submontane low hills	*Dalbergia sissoo*, *Acacia catechu*, Bamboo, *Toona ciliata* Poplar	*Acacia catechu*, *Santalum album*	*Grewia optiva*, *Morus laevigata*, *Bauhinia variegata*, Bamboo spp.	Setaria, Panicum
Sub-temperate subhumid mid hills	*Delbergia sissoo*, *Albizia spp*, *Pinus roxburghii*, Poplar and Eucalyptus species	*Acacia catechu*, *Olea ferruginea*, *Myrica nagi*	*Grewia optiva*, *Morus laevigata*, *Bauhinia variegata*, Bamboo spp.	Setaria, Panicum
Wet temperate high hills	*Pinus wallichiana*, *Pinus roxburghii*, *Aesculus indica*, *Cedrus deodara* and *Ulmus wallichiana*, etc.	*Prunus armenica* Hill bamboo	*Robinia pseudocacia*, *Morus serrata*, *Quercus* spp. *Ulmus wallichiana*	Eragrostis, Apluda, Chrysopogon
Dry temperate high hills	Poplar species, Salix species Junipers spp.	*Pinus gerardiana* Juniper species	*Robinia pseudoacacia*, *Salix* spp.	Poa, Eragrostis

Fig 5.1. Agro climatic zones of Himachal Pradesh

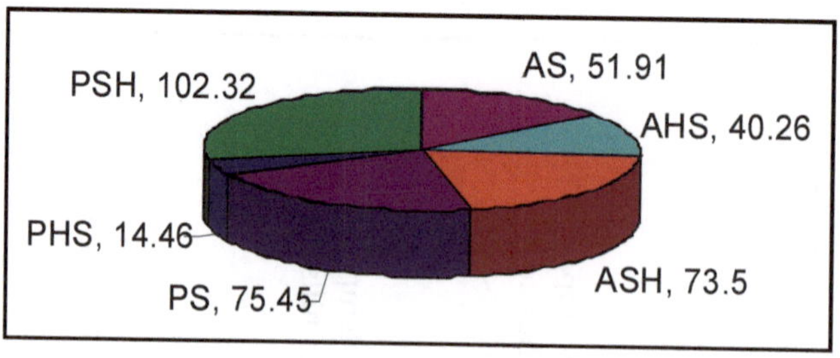

Fig. 5.2. Co-efficient of variation for biological yield (%) of AFS between different categories of farmers (Nauni, Solan H.P)

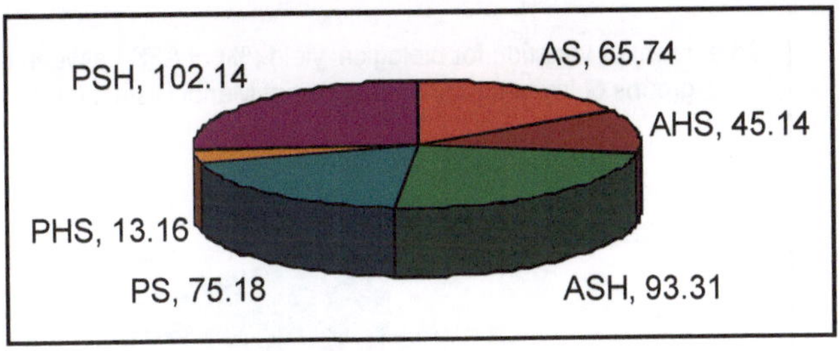

Fig. 5.3. Co-efficient of variation for net returns (%) of AFS between different categories of Farmers (Nauni, Solan H.P.)

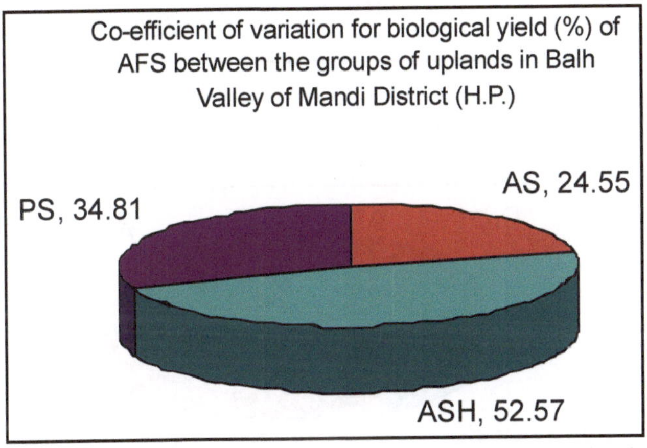

Fig. 5.4. Biological yield of Agroforestry Systems in up lands of Balh Valley

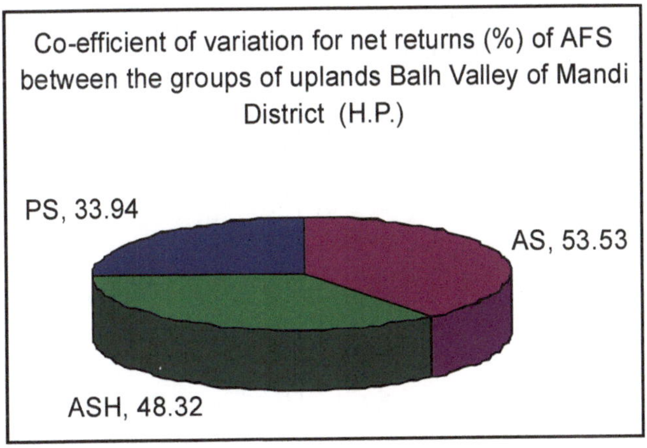

Fig. 5.5. Variation in net-returns of Agroforestry systems in Balh Valley

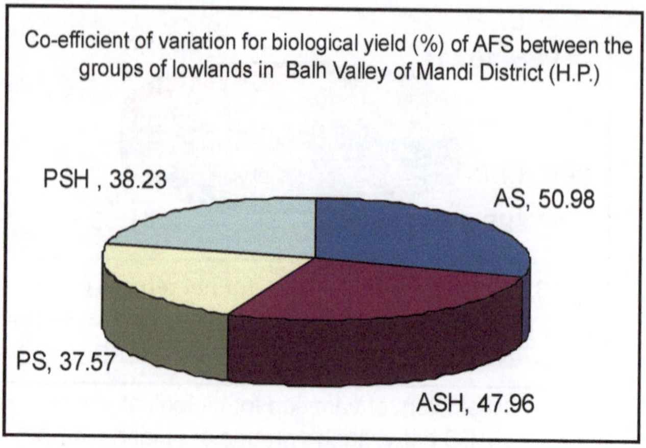

Fig. 5.6. Biological yield of Agroforestry Systems in low lands of Balh Valley

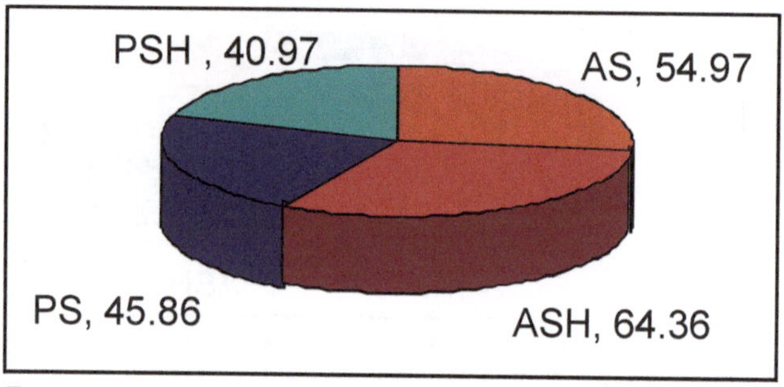

Fig. 5.7. Co-efficient of variation for net returns (%) of AFS between different categories of farmers of lowlands Balh Valley of Mandi District (H.P.)

Photo. 5.1 Agrisilviculture system (wheat + *Grewia*)

Photo 5.2. Silvipastoral system (*Setaria* + *Leucaena*)

bunds at closer spacing (3m apart) facilitates the annual crops growth and enhances yield levels. Tree-crop-soil interaction research during the last two decades have established that integrated management of economic trees and annual crops results in increased financial returns from an agroforestry system adopted (Table 5.7). Hortisilvimedicinal systems integrating Peach, Morus/ Grewia, Tulsi/Mucuna have good potential to enhance farmer's net return substantially.

REFERENCES

Christersson, L and Verma, K S. (2003). Short Rotation Forestry- an alternative Silviculture. *In*: Short Rotation Forestry for Industrial and Rural Development. Pp. 21-27. *Eds.* K.S. Verma, D.K. Khurana and L. Christersson, ISTS, Nauni, Solan, India.

Kachru, S R. (1997). Diagnostic survey and productivity appraisal of agroforestry systems in sub-temperate and sub humid region of H.P. *M.Sc thesis*, Deptt. of Silviculture and Agroforestry, Dr. Y.S. Parmar University of Horticulture, Nauni, Solan (Himachal Pradesh), India.

Lehria, S K. (2004). Biomass production and nutrient dynamics in silvipastoral systems of mid Himalaya. *Ph.D. thesis*, Deptt. of Silviculture and Agroforestry, Dr. Y.S. Parmar University of Horticulture, Nauni, Solan (Himachal Pradesh), India.

Lodoen, D. (2000). Path to prosperity through Agroforestry: ICRAF'S corporate strategy 2001-2010. pp. 1-43.

Nayak, B K. (1996). Studies on biomass productivity and nutrient content in *Eucalyptus teriticornis, Leucaena leucocephala* and *Melia azedarach* under high density plantation. *M.Sc thesis*, Deptt. of Silviculture and Agroforestry, Dr. Y.S. Parmar University of Horticulture, Nauni, Solan (Himachal Pradesh), India.

Sharma, A K. (1989). Studies on growth performance and site amelioration under high density plantation. *M.Sc thesis*, Deptt. of Silviculture and Agroforestry, Dr. Y.S. Parmar University of Horticulture, Nauni, Solan (Himachal Pradesh), India.

Upadhyaya, K. (1998). Diagnostic survey of agroforestry systems in Balh valley of Himachal Pradesh. *M.Sc thesis*, Deptt. of Silviculture and Agroforestry, Dr. Y.S. Parmar University of Horticulture, Nauni, Solan (Himachal Pradesh), India.

Verma, K. S., Mishra, V. K., Kaushal, R., Sharma, J. K. (2003). Temporal change in the soil properties under high density energy plantation of *Acacia mollissima* and *Eucalyptus hybrid*. *In*: Short Rotation Forestry for Industrial and Rural Development. pp99-106. *Eds.* K.S. Verma, D.K. Khuranrand L. Christerson, ISTS, Nauni, Solan, India.

Yadava, J. N. (1997). Biomass productivity and nutrient content of *Morus alba* and *Leucaena leucocephala* based silvipastoral system. *M.Sc thesis*, Deptt. of Silviculture and Agroforestry, Dr. Y.S. Parmar University of Horticulture, Nauni, Solan (Himachal Pradesh), India.

6

Agroforestry in Uttaranchal

SALIL TEWARI, RAJESH KAUSHAL AND R. PUROHIT
Agroforestry Section, G.B. Pant University of Agriculture and Technology, Pantnagar

ABSTRACT

The paper introduces the biophysical environment and landuse pattern of newly formed state. On the basis of ownership of land the scope of adopting agroforestry had been discussed. Existing agroforestry systems under different agroclimatic zones of Uttaranchal are provided with examples. Tree improvement program and system development research carried out under AICRP on Agroforestry for Terai zone of the state has also been discuused in the paper. Constraints in adopting and carrying out research in agroforestry along with future strategies to extend agroforestry in Uttaranchal are also given in the paper.

INTRODUCTION

Uttaranchal, formed on November 9, 2000, is the 27th state in India. The state is spread over 53,483 km^2 of land, which extends between 77^0 34′ and 81^0 02′ E longitude and between 28^0 43′ to 31^0 27′ N latitude. It covers 1.67 per cent of the country's total area and is home to 84.7 lakhs of people as per the provisional census report 2001. The entire state forms part of the Central Himalayas. The state is strategically located and forms part of the Northern boundary of the country sharing its borders with Nepal and China. The entire region, except the southern most foothills belt, is mountainous with a series of ridges and valleys. Starting with an altitude of about 240 m in the *tarai* on the southern fringes, the terrains rise to 7817 m in the north with in a short width span of about 160 km. It touches Tibet in the north, Himachal in the west and northwest, gangetic plains of Uttar Pradesh in the South and Nepal in the east. Uttaranchal has two administrative divisons, the western half known as Garhwal and the the eastern region known by the name of Kumaon.

Garhwal divison consists of seven districts viz. Dehradun, Uttarkashi, Pauri, Tehri, Chamoli, Rudraprayag and Haridwar, while, Nainital, Pithoragarh, Champawat, Almora, Udham Singh Nagar, and Bageshwar forms the six districts of Kumaon region. Eleven districts viz. Almora, Bageshwar, Pithoragarh, Nanital, Rudrprayag, Cahmpawat, Chamoli, Tehri, Pauri, Uttarkashi and Dehradun are situated in hill region having a total area of about 49100 square km while, other two district viz. Udham Singh Nagar and Haridwar form the rest of area. The map of the state is given in Fig. 6.1.

BIOPHYSICAL ENVIRONMENT

Most of the area in the state is in the form of steep slopes, narrow terraces with inward slopes and valley lands. The summer and winter monsoon systems of Asia mainly govern the weather and climate of the region. The winter rains are brought by western disturbance and the summer rains by summer monsoon winds. At times cloudbursts cause unusually heavy rainfall, often leading to disastrous flash floods. During rainfall the region is prone to landslides and floods. There are also wide variations in temperature, due to variations in altitude. Maximum temperature reaches upto 36°C during summer and the minimum fall to 0 °C or even below during winter. Summer is extremely hot in valleys and pool frosts are common during winter.

The soil types of the state can be broadly classified as red loam, brown forest soil, podosol and meadow soil. The soil cover is generally very thin. Soils are generally sandy-to-sandy loam and are mostly shallow, stony and dry on southern aspects. In areas where shales and limestone occurs, the soils are very clayey. The soils in tarai region (US Nagar and Haridwar) are very fertile and support a large number of crops. On the other hand, soils of the hilly region are prone to constant erosion due to steep slopes making it less fertile. The soil classification as per the soil types are given in Table 6.1.

LAND USE PATTERN

Subsistence agriculture is the prime source of livelihood for more than 70 per cent of the population in Uttaranchal. Agriculture in the state is very complex and is interlinked with animal husbandry and forestry to form a production system. A majority of the area of state is under forests and wastelands thus leaving behind only a small amount of land (about 12 %) for cultivation. Out of the total landholdings, 70 per cent of landholdings are less than 1 ha in size and form about 27 per cent of total cultivated area. While about 26 per cent of landholdings are between 1 to 4 ha in size covering about 51 per cent of total cultivated area. Only a little more than 3 per cent of total land holdings are above 4 ha in size constituting about 22 per cent of total cultivated area. Due to the large number of small landholdings, the input cost per unit of the output is quite high for hill agriculture as compared to the plains. On an average only 0.8 ha land is available per house hold of 5-6 family members about 85 per cent of which is rainfed (Negi and Joshi, 2000). The land use pattern of the state is given in Table 6.2.

Major crops of the state include paddy, wheat, mandua, jhingora, urd, soybean, maize. Chillies and ginger are the main kharif cash crops but they account for very less net sown area. In rabi, the predominant crop is wheat which is sometime intersown with barley and/or gram, lentil. Potato is the major rabi cash crop grown in small irrigated pockets. The cropping intensity of the state is 163.79, which is higher than the country average. Fruit cultivation is not practiced in a systematic way, though scattered orchards of apples, apricot, and peaches in cool northern aspects and citrus, mango, litchi in hot southern aspects and valleys are common occurrence. Forests constitute the most important natural resources in Uttaranchal. The forests have direct economic, ecological as well as social bearing on the local people of the state. They are important source of fuelwood, fodder, timber and other major and minor forest produce. Out of the total forest area, 69 per cent is under the control of the Forest Department, 18 per cent under civil soyam, 12 per cent being panchyat forests and 1 per cent under private ownership.

SCOPE OF ADOPTING AGROFORESTRY SYSTEMS

For identifying and extending suitable agroforestry systems/practices, three major land uses and their ownership may be broadly recognized:

1. Agricultural/privately owned land

Several agroforestry systems such as agrisilvicultural, agrisilvipastoral, hortisilvicultural, hortisilvipastoral and some other systems like sericulture, apiculture, etc. are being practiced in the state in varying spatial and temporal distributions. The risers and drain channels for inwardly slopping terraces are being utilized for raising fodder grasses and legumes and trees for fuel wood and fodder. All these agroforestry systems not only provide diverse products but also help in terrace stabilization, moisture conservation, soil improvement and reduction of biotic pressures on existing forest resources and drudgery.

2. Community/Civil Soyam land

The communities in the state have access to grasslands and other non-agricultural lands (including civil/soyam/panchayat grazing lands and poorly stocked forest areas), which is more than 30 per cent of total area of the state. Trees are found scattered on these lands, which suffer heavy pressure from the surrounding villagers. These lands are generally used for grazing cattle by local people or by nomadic graziers. Cut and carry method of grasses and fodder is also common practice on these lands. The productivity of these lands is alarmingly low. However, there is ample scope of improvement of these lands by adopting improved agroforestry systems viz., silvipastoral systems, energy plantations and multiple use vegetation cover, vegetative check dams and stream bank plantations. These systems, if properly managed can improve biological productivity of these lands by 30-60 times or even more and can cater socio-economic needs of the people besides ensuring ecostability and reduce biotic pressure on existing forest resources in the region, This category of lands can be developed using participatory approach.

3. Forest lands

The recorded forest area in the state is 3.429 million hectare, which is about 64 per cent of total geographical area. However, actual forest area is only 1.539 million ha (which includes, dense forest of 0.236 million ha, medium 0.810 million ha and poor 0.493 million ha). Miscellaneous forest and without tree cover are 0.218 and 1.670 million ha, respectively. This clearly indicates the poor forest resource base as dense forests occur only in 4.5 per cent of total geographical area, and seriously need immediate attention. The forests in the state are heavily tapped for fuelwood, fodder, and collecting other minor forest produces and are heavily degraded. The degraded forest areas may be developed in such a way as to support income generating activities/ small scale industries by the fringe communities for their economic diversification. Agroforestry systems like integrated multipurpose trees, medicinal herbs/shrubs plantations, silvipastoral and energy plantations can be developed based on the principles of Community Forest Management. Well-stocked forests can be exploited for herbs, shrubs, grasses, bamboos and other non-timber forest products supportive to income generating activities without impairing the objective of their scientific management.

EXISTING AGROFORESTRY SYSTEMS IN DIFFERENT AGROCLIMATIC ZONES

Agroforestry in real sense is the back bone of cultivation system of the Uttranchal. The state has been divided into four agroclimatic zones. Agroforestry has its own role to play in all these zones. There are structural as well as functional changes in the agroforestry systems in different zones

1. Tarai and bhabar zone (up to 1000 m)

This zone consists of plains, tarai and bhabar areas. Out of these, tarai zone is intensively cultivated because of fertile soil and plenty of irrigation water. Agroforestry systems in tarai zone are well developed and are of commercial nature. In the last two decades, poplar and *Eucalyptus* based agroforestry systems have played a major role in augmenting the economy of the farmers in tarai region and are common amongst the farmers. With intensive management of these fast growing species, the farmers are getting up to Rs. 40,000 to 45,000 per acre per year against approximately Rs. 12,000 to 15,000 from rice-wheat rotation. Many agricultural crops viz. wheat, black gram, rape seed, lentil, coriander, tomato cabbage, sugarbeet, sweet potato, potato, spinach, brinjal, fenugreek, knol- khol, turnip, strawberry, pine apple, onion, lettuce, celary, garlic, etc can be profitably grown with poplar up to 8 year old plantation. Some other crops like pea, chilly, berseem, chickpea can be grown with poplar up to 4 years. Few crops like sugarcane, maize, sorghum, soybean and mentha can suitably be grown only up to two years (Table 6.3).

Taungya system is also one of the important agroforestry system of this region. Since 1962, big farmers have taken taungya system to establish poplar in tarai region. The following combinations are followed:

Trees	Associated agricultural crops and yield (kg/ha)
Shorea robusta, Tectona grandis, Acacia catechu, Dalbergia sissoo, Eucalyptus spp. *Populus* spp.	Maize (946), rice (862), soybean, wheat (380), Chick pea (447), rape seed (392)

Recently, Eucalyptus based agroforestry systems are increasingly being adopted by the farmers due to slash in the poplar prices. These trees are mostly grown on boundries and bunds of farms. These trees not only provide additional revenue but also act as windbreaks.

In this zone, agrihorticulture system are also well developed. Fruit trees viz. *Mangifera indica* (Mango) and *Litchi chinensis (*litchi) are most commonly grown by the farmers. *Psidium guava* (Guvava), *Prunus persica* (Peach), *Pyrus communis* (pear), *Artocarpus integerifolia* (jack fruit), *Carica papaya* (papaya), *Citrus sp* (lemon, kagzi lime, orange, chakotra, galgal ,mosumbi etc) are also of common occurrence on farmers field. Wheat, maize, soybean, lentil, sugarcane, mustard, rainy season vegetables, winter vegetables, leguminous crops like bengal crop, barseem, peas, black gram and green gram are usually grown as intercrops with these fruit trees in initial years. Sometimes, filler trees like peach, papaya, plum, guava etc are also grown in between mango trees. However, these filler trees are generally uprooted after 8- 10 years before any competition set in.

Besides these well managed agroforestry systems, sometimes marginal farmers also retain some of the trees like *Dalbergia sissoo* (shisham), *Leucaena leucocephala* (Subabul), toon (*Toona ciliata*), dek (*Melia azedarach*), *Anthocephalus kadamba* (Kadam), *Holoptelea integerifolia* (kanju), *Adina cardifolia* (haldu) on their farm for meeting the need of fuelwood, fooder and small timber.

2. Sub-tropical Zone (1000-1500 m)

Agroforestry systems in this zone are traditional and are less developed. In this zone, agrisilviculture, agrihorticulture, silvipastoral and agrisilvipastoral systems are generally found. The agriculture crops are grown as per the season, rotation and needs of the farmers. Landholdings of the farmers in this zone are small and highly fragmented. Fodder trees are grown on field bunds and homesteads generally for meeting the farmers own requirement without any geometric arrangement. Bhimal (*Grewia optiva)*, khirak (*Celtis australis*), kachnar (*Bauhinia variegata*) and kimar (*Morus serrata*) are most important tree species of this zone. These species are generally harvested for the fodder and fuelwood after 5-7 years of planting. Among these trees, *Grewia optiva* provides excellent fodder in winters when no other green fodder is available. While, kimar and khirak are lopped in summer to rainy season. There is no regular management of all the above tree species. However, with efficient management practices like pollarding, yield of fodder and fuelwood can be certainly increased. In addition to above trees, shisham (*Dalbergia sissoo*), khair (*Acacia catechu*), dhak (*Butea monosperma*), siris (*Albizia lebbek*), simal (*Bombax ceiba*), bamboo (*Dendrocalamus strictus*), subabul (*Leucaena leucocephala*), tuni (*Toona ciliata*), bakain (*Melia azedarach*), harar/behra (*Terminalia* spp.), Timla (*Ficus roxburghii)*, Mangiri (*Ficus palamta)* etc. are also retained by the farmers (Bhatt & Rawat, 1993; Singh and Gautam, 2004; Singh *et.al.*, 1995; Singh, 2003).

Fruit trees viz., apricot (*Prunus armeniaca*), plum (*Prunus saliciana*) and peach (*Prunus persica)* are also grown by medium scale farmers in this zone. About 250 trees/ha are generally planted. These trees commence the production after five years of plantation. Intercropping of beans and peas are taken for two to three years. However, low chilling varieties of peach can be intercropped for upto six years.

In this zone many new silvipastoral systems have been developed after years of research. These technologies are well suitable and highly profitable. Some of the important recommended combinations are: *Leucaena leucocephala*

+ *Chrysopogon fulvus / Eulaliopsis binata, Dalbergia sissoo* + Bhabar grass, *Albizia lebbeck* + Bhabar grass or other grasses, *Grewia optiva* + fodder grass in steep slopes etc (Naryan *et.al.*, 1998; Dadhwal, 1999; Dadhwal and Tomar, 2000; Singh *et al.*, 1987b, 1988, 1990b, 1991, 1986, 1987a, 1990a). The yield of some important fuelwood/ fodder trees in these silvipastoral system is given in table 6.4 and 6.5

3. Cool Temperate Zone (1500 to 2400 m)

Silvipastoral, agrisilviculture and agriihorticulture systems are mainly prevalent in this zone. Shallow soil depth, low water availability, poor fertility, soil erosion, absence of soil microbial mass, sloppiness and heavy demographic pressure are the main constraint of this zone. Fruit trees viz. apricot, apple, walnut etc are planted as orchards or on agriculture fields. Fruit trees are generally retained at linear spacing of 5-8m in the rows. The major practice in the field is line planting on the terrace risers. These trees are intercropped for five to six years with beans or peas. It is reported that intercropping a young apple orchard with grasses and legumes under rained conditions for one year can give yields of green grass varying from 24,000 to 31,000 kg/ha and those of legumes varying from 24,000 to 26,700 kg/ha

In agrisilviculture system, forest trees viz. *Quercus* spp (banj, moru and kharsu), *Aesculus indica* (pangar), *Robinia, Alnus,* are retained by the farmers for meeting the basic needs of fuelwood, fodder and small timber. Generally, the trees are not planted in any systematic manner and no attention is given to spacing to get full density per unit area. These primitive practices are of not economic importance. However, using appropriate technologies, agroforestry can contribute to sustained production of crops and livestock, on these lands.

In this zone, silvipastoral systems are more commonly found where lopping of trees and grazing of grasses by nomadic graziers is a common practice. People have the right to graze their animals and collect the fodder, fuelwood and timber for their needs. The forests and civil soyam land in this zone suffer from high grazing pressure. Scattered trees are generally found on these lands. However, fodder tree species are now being taken under Joint forest management programme for improvement of these lands. Some of the important trees, shrubs and grasses (Bana *et.al.*, 1996; Pal, 2004) found in this zone are given in Table 6.6.

4. Dry temperate zone

This zone is mainly dominated by the grasses viz., *Danthnia cachemyriana, Koeleria cristata, Calamagrostis emodensis, Festuca lucida, Brachypodium sylvaticum, Trisetum spicatum, Andropogon tristis, Phleum alpinum, Agroptron semicostatum, Agrostis munroana, Deschampsia caespitosa, Helictotrichon virescens* and *Deyeuxia pulchella*. In addition shrubs and. trees like *Hippophae salicifolia, H tibetiana* (chharma/chuk), wild apricot, *Betula utilis* are also found. The tree species are heavily lopped for fodder and fuelwood.

The various agroforestry systems prevalent in Uttaranchal in various agro climatic zones along with their components is given in Table 6.7

WORK CARRIED OUT BY G.B. PANT UNIVERSITY OF AGRICULTURE AND TECHNOLOGY, PANTNAGAR

The work on different aspects of agroforestry at G.B. Pant University of Agriculture and Technology, Pantnagar was started from April, 1983 under All India Coordinated Research Project on Agroforestry. Poplar and shisham are the priority species allotted to the University. In 2004, College of Forestry, Hill Campus Ranichauri was also included as sub center under AICRP with *Grewia optiva* and *Morus alba* as priority species.

At Pantnagar, about 100 tree species have so far been collected and are being evaluated for their survival, establishment and growth (Anon, 1992; Anon, 1999; Purohit *et.al.*, 2005). Depending upon their suitability these species are being taken up for the future research. The work carried out so far (Table 6.8) revealed the performance of some important species in the order *A. cadamba* > *P. deltoides* > *E. hybrid* > *C. equisitifolia* > *E .citridora* > *A. lebbek* > *A. fraxinifolius*> *B. ceiba* > *M. azedarach*, after 8 years of planting. Species like *D. sissoo, A.catechu, M. alba, T. arjuna, T. grandis, T. ciliata* and *T. nudiflora* were also found to suitable under *tarai* condition. Species like *G. robusta, Mangifera, Lagerstromia, P. roxburghii, P. pinnata, T. chebula* and *T. bellerica* were not found suitable for short rotation period of 8 years. After 14 and 20 years of planting the trend almost remain the same and the performance was in the order *P. deltoides* > *A. fraxinifolius* > *A. cadamba* > *A. lebbek* > *E. hybrid* > *C.equisitifolia* > *E .citridora* > *A.lebbek* (Anon, 2002)

The work on the priority species *viz.*, Poplar and shisham is being carried out under the foolowing heads:

Populus deltoides

The work on Poplar is being done under the following heads:

- Poplar improvement
- Evaluation of selected poplar clones
- Optimization of management practices and planting density

Poplar improvement work is being carried out at Pantnagar on various aspects viz. Germplasm collection, provenance testing under agroforestry, development of new clones and their testing and clonal identification based on morphological and molecular markers. In addition, clone Pant Poplar-5 has been developed by the University and recommended to the farmers of the Tarai area. As compared to national checks (G-3, G-48 and D-121), PP-5 has higher productivity and tolerance to blight and stem borer.

Germplasm Collection of Poplar

A total of 167 diverse clones derived from different agencies and University own variants have been well maintained in germplasm block at a distance of 75 cm between the rows and 60 cm between the plants. This germplasm is serving good purpose for supply of pure material for experimentation and commercially to the poplar growers.

Provenance testing of Poplar clones

Fourty nine clones of poplar are being evaluated for their growth performance under agroforestry. The details of the clones being evaluated are given in Table 6.9. The grand mean, range, heritability, genetic advance, phenotypic and genotypic coefficient of correlation are given in Table 6.10.

In another experiment six promising clones of poplar in tarai region (Pant Poplar 5, L 49, Udai, G 48, L 52 and L 188) are being evaluated for their growth performance in agroforestry systems with wheat (PBW 343) and direct seeded rice (Pant Dhan-12). Among all the clones under study, Udai excelled in height and diameter growth performance over G 48. The other clones viz., L 188, L 49 and Pant Poplar 5 were statistically at par with udai (Table 6.11). Grain yield of wheat and rice however were not influenced due to different clones (Anon, 2004).

In another experiment in similar agroecological conditions, growth of Poplar planted at five different tree spacing/ density viz. 5 m x 10 m (200 trees/ ha), 5 m x 8 m (250 trees/ ha), 5 m x 6 m (333 trees/ ha), 5 m x 4 m (500 trees/ ha), 5 m x 2 m (1000 trees/ ha) was evaluated under agroforestry. It was observed that the maximum (14.98 cm) mean DBH of poplar trees was attained with lowest tree density (200 trees/ ha) which decreased with the increased density, producing significantly smaller (13.48 cm) mean DBH with maximum (1000 trees/ ha) tree density (Table 6.12). The height growth followed the reverse but exactly the similar trend as of diameter growth and tallest (13.27 m) mean tree height was observed with maximum (1000 trees/ ha) density, which was significantly more than mean height under other densities except 500 trees/ ha. Average tree height did not differ significantly with density from 200 to 500 trees/ ha. The maximum grain and straw yields were recorded in control (without tree) in both wheat and rice which decreased linearly with increase in tree density (Anon, 2004).

Management practices like pruning helps in better growth of tree species. In addition the practice also solves the problem of fuelwood scarcity besides allowing more solar radiation to the understorey crops. In Poplar based agroforestry system, effect of pruning was studied on growth of Poplar and below ground crop viz., toria and rice. Five different pruning treatments viz., Un pruned trees, 25 per cent pruning, 50 per cent pruning, 75 per cent pruning and pruning for equidistant branching pattern were superimposed on the Poplar. The trees were regular pruned in the month of January. The results revealed that pruning intensities/ pattern did not influence the tree height growth significantly. However, the tree pruning up to 75 per cent of tree height showed adverse effect on tree growth (DBH and height both) but reduction was significant in diameter growth only (Table 6.13). The Fuel wood quantity obtained as a result of pruning treatment varied substantially and maximum (17.52 q/ ha) quantity of air dried fuel wood was yielded from pruning to 75 per cent tree height. The pruning intensities however, failed to influence the yield of rice significantly. In case of toria, the grain yields with 25, 50 and 75 per cent pruning of tree height and for no pruning and pruning for equidistant branching pattern did not differ significantly (Anon, 2004).

Dalbergia sissoo

Keeping in view, the importance of Shisham (*Dabergia sissoo* Roxb.) as a potential tree, research experiments on shisham are being carried out on the various aspects viz,. germplasm collection, provenance testing under agroforestry, development of new clones and their testing and clonal identification on basis of morphological and molecular marker.

Germplasm block of shisham

The experimental material (PS1 to PS107) consisted of 107 diverse genotypes collected from plus trees of shisham from different parts of India viz., U.P., Uttaranchal, Bihar, Himachal Pradesh, Haryana, Punjab, Jharkhand and Nepal. The selection of plus trees was made on the basis of superiority of plus trees for height, diameter at breast height, straightness, bark thickness, health etc.

Provenance testing of shisham

The seeds collected from selected plus trees were planted in Randomized Block Design with three replications. Experiment was planted at a distance of 3.0, 1.5 m during March 1993, 1994, 1995, 1996, 1997 and 1998 from PS1 to PS15 (Set I), PS16 to PS30 (Set II), PS31 to PS60 (Set III), PS61 to PS75 (Set IV), PS76 to PS90 (Set V) and PS91 to PS107 (Set VI), respectively.

On the basis of height and diameter PS-5, PS-6, PS-1, PS-13 and PS-15 were identified as better performing genotypes for set I. For II set PS-20, PS-23, PS-26, PS-16 and PS-29 were selected as better genotypes on the basis of mean height and diameter. Genotypes PS-46, PS-47, PS-59 and PS-60 were selected for III set. On the basis of outstanding height and diameter, genotypes PS-75 and PS-74 were selected in group IV. Genotype PS-76 out performed in set V. In the VI set PS-99 and PS-107 were selected on the basis of high mean height and diameter (Anon, 2004). There were considerable differences in the genotypic and phenotypic coefficient of variability for crown length, crown width, stem straightness. Significant genotypic differences, wide range of means Coefficient of variability obtained is given in Table 6.14

In another experiment (Table 6.15) effect of tree density on growth performance of shisham and Eucalyptus was studied. The trees were planted at density of 5000, 2875, 1887, 1176, 690, 431, 265 and 167 trees/ ha respectively. In both the tree species tree diameter increased with decreasing tree densities. Same trend was also observed in height in case of *Eucalyptus tereticornis.* The maximum tree height (15.91m) was observed for tree density of 2857 trees/ ha and minimum tree height (14.15 m) was for density of 431 trees/ ha (Anon, 2003).

Salix alba

Four clones of *Salix* viz., *Salix alba I cv. coerulea, Salix alba II cv. coerulea, Salix alba cv. vitellina, Salix babylonica* are being evaluated for their growth performance under agroforestry. The results revealed no significant difference in the growth of various clones (Anon, 2004). Further, grain and straw yield of paddy were also not affected significantly due to different clones (Table 6.16).

In another experiment no significant effect on height and dbh was observed in two different clones and four spacing (Table 6.17).

Jatropha curcas

Seven provenances of *Jatropha curcas* viz Kaladhungi, Haldichaur, Gola, Kheda, Bailparao, Pantnagar 1, Bilaspur, Pantnagar 2 are being evaluated at

4x2 m spacing for the seed and oil yield. The experiment was planted in August 2004. The results revealed wide range of variation for different characters under study (Anon, 2005). Height was maximum in Gola provenance while collar diameter was maximum in Kaladhungi provenance (Table 6.18).

In another experiment (Anon, 2005) on *Jatropha curcas* four spacings viz 4x2, 4x3, 4x4, 4x5m are being evaluated to find the optimum spacing under agroforestry. The results revealed 100 percent survival at 4x4 and 4x5 spacing. Height was maximum at 4x4m spacing while the minimum height was recorded at 4x2m spacing. Collar diameter also revealed similar trend and maximum collar diameter of 20.54 mm was recorded in 4x5m spacing (Table 6.19).

Bamboo

The work on Bamboo is being under coordinated approach. Bamboo Coordinating Centre (BCC) has been established in the University, which is coordinating the bamboo research in 13 different centers *viz.*, Institute of Himalayan Bioresource Technology, Palampur; Allahabad Agriculture University, Allahabad; GB Pant University of Agriculture and Technology, Pantnagar; National Botanical Research Institute, Lucknow; Century Laminating Co. Ltd., Hapur; Arid Forest Research Institute, Jodhpur; Jain Irrigation Ltd, Jalgaon; Bidhan Chander Krishi Vishavidyalya Nadia; Uttarbanga Krishi Vishvavidyalya, Cooch Behar; Institute of Wood Sciences & Technology, Bangalore; Institute of Forest Genetics & Tree Breeding, Coimbatore; Growers Ltd, Salem; Kerala Forest Research Institute, Peechi are part of the coordinated research programme. A common framework for multilocational trial is being followed at all the centres with two local and six common species *viz., Bambusa bambos, B balcooa, B nutans, B tulda, Dendrocalamus asper* and *D hamiltonii*. In addition, trials on nutrient and water management, organic farming in edible bamboo species, spacing, clump management, comparison of growth of Macro vs micro propagated planting material and development of bamboo based cropping systems are also being undertaken at the different centres with different species. All the experiments have been initiated in the year 2005. In addition, a bambusetum is being established. So far, about 30 species of bamboos have been collected from different parts of the country and have been established in the bambusetum. Research trials have also been taken for production of quality planting material which will be distributed to the farmers. Commercial plantation of bamboo has also been established in 40 ha area of the University. The produce from this will be helpful in catering the needs of nearby paper and pulp industries.

In hill campus Ranichauri, growth performance of different tree species of hilly regions are being evaluated. Work is also being done on development of package and practices of various fodder trees. Management practices are also being standardized for lopping and pruning. In addition tree improvement work on *Grewia optiva* and *Morus alba* has also been initiated.

CONSTRAINTS AND FUTURE THRUST

- Majority of the farmers in the state are marginal/submarginal with small and fragmented land holdings. Therefore the priority of the farmers is production of food grain crops, though some trees are retained by the farmers on the bunds and terraces for meeting his own requirement
- The economic condition of the farmers of the state is largely poor which poses the constraint in adopting the new technology. However, such types of problem can be overcome by providing loan and subsidies to the farmers.
- Success of any system is governed by market network, which is lacking for tree produce. A well develop market network is required to win the confidence of the farmers. This can be developed by setting forest based industries
- Production and distribution of quality planting material is one of the key factors, which can decide the future of agroforestry. The success of poplar and eucalyptus in tarai region in last two decades reflects the importance of quality planting material.
- Proper extension is the major key to success of any new developed technology. At present, many new technologies are lying in the arsenals of the researchers, which need proper extension for their success.

In past, majority of the work on agroforestry has been done in *tarai* region and the adoption of agroforestry in this area is very high. However, a very little work has been done on the hills where, most of the systems are traditional. With Uttaranchal now as separate hilly state, more efforts are required to focus on the hilly areas. The immediate need is to:

- Improve productivity of the existing lands using agroforestry interventions
- Efficient conservation of soil and water through agroforestry interventions
- Developing low external input sustainable agricultural technologies

Table 6.1. Soil types of Uttaranchal

Area	Altitude (m above mean sea level)	Soil
Irrigated lower hills	600-1200	Alluvial sandy soil
Rainfed lower hills	600-1200	Residual sandy loam
Mid hill south aspect	1200-1700	Sandy loam
Mid hill north aspect	1200-1700	Brown forest soil
High hills	1700-2500	Red to dark
Very high hills	2500-3500	Red to dark black clay

Table 6.2. Land use pattern in Uttaranchal (1997-98)

S. No.	Category	Area (ha)	Percent
1.	Reported area	55,93,342	100.00
2.	Forest area	35,01,295	62.60
3.	Cultivable barren land	2,99,114	5.35
4.	Other land additional to cultivable land	1,64,238	2.94
5.	Barren land	3,22,480	5.76
6.	Pasture	2,27,324	4.06
7.	Orchards and other vegetables	2,19,134	3.92
8.	Present fallow land	11,102	0.20
9.	Other fallow land	67,007	1.20
10.	Grass sown area	12,39,647	22.16
11.	Net sown area	7,81,298	13.97
12.	Net sown area more than one time	4, 58, 349	8.19

Table 6.3. Important Poplar based combination in *Tarai* region of Uttranchal

S. No.	Crops	Age of poplar plantation (year)							
		1	2	3	4	5	6	7	8
1.	Sugarcane	√	√	-	-	X	-	-	-
2.	Wheat	-	-	√	-	-	-	-	-
3.	Paddy	-	X	-	-	-	-	-	-

Contd...

Table 6.3 Contd...

4.	Maize	√	√	-	-	X	-	-	-
5.	Black gram	-	-	-	-	√	-	-	-
6.	Green gram	-	-	-	-	√	-	-	-
7.	Rape seed	-	-	-	-	√	-	-	-
8.	Lentil	-	-	-	-	√	-	-	-
9.	Coriander	-	-	-	-	√	-	-	-
10.	Tomato	-	-	-	-	√	-	-	-
11.	Cabbage	-	-	-	-	√	-	-	-
12.	Sugarbeet	-	-	-	-	√	-	-	-
13.	Pea	√	√	√	√	-	-	X	-
14.	Chilly	√	√	√	√	-	-	X	-
15.	Radish	-	-	-	-	√	-	-	-
16.	Turmeric	-	-	-	-	√	-	-	-
17.	Zinger	-	-	-	-	√	-	-	-
18.	Barseem	√	√	√	√	-	-	X	-
19.	Sorgham	√	√	-	-	-	-	X	-
20.	French bean	-	-	-	-	√	-	-	-
21.	Sweet potato	-	-	-	-	√	-	-	-
22.	Potao	-	-	-	-	√	-	-	-
23.	Chick pea	√	√	√	√	-	-	X	-
24.	Spinach	-	-	-	-	√	-	-	-
25.	Brinjal	-	-	-	-	√	-	-	-
26.	Gram	√	√	√	√	-	-	X	-
27.	Fenugreek	-	-	-	-	√	-	-	-
28.	Knol-khol	-	-	-	-	√	-	-	-
29.	Turnip	-	-	-	-	√	-	-	-
30.	Onion	-	-	-	-	√	-	-	-

Contd...

Table 6.3 Contd...

31.	Garlic	-	-	-	-	√	-	-	-
32.	Soybean	√	√	-	-	-	-	X	-
33.	Mentha	√	√	-	-	-	-	X	-
34.	Lettuce	-	-	-	-	√	-	-	-
35.	Strawberry	-	-	-	-	√	-	-	-
36.	Pineapple	-	-	-	-	√	-	-	-

Table 6.4. Growth, yield of fuelwood, lopped fodder and fiber bark of top feed tree species in association with different grass species

Treatment	Growth parameters		Fuelwood ** (air dry) over bark (kg/ha/yr)	Lopped** Fodder (oven dry) (kg/ha/yr)	Bark fiber** (kg/ha/yr)
Main treatments (tree spp.)	Height (m)	DBH (cm)			
T1 *Albizia lebbek*	5.46	88	500	415	-
T2 *Grewia optiva*	5.24	71	585	208	615
T3 *Bauhinia purpurea*	4.23	81	475	315	-
T4 *Leucaena leucocephala*	3.54	55	298	196	-
Sub- treatments (lopping)					
C1 50% intensity	471	79	480	272	708
C2 75% intensity	453	68	451	296	531
Sub-sub-treatments (grasses)					
G1 *Chrysopogon fulvus*	486	78	579	366	856
G2 *Eulaliopsis binata*	438	70	356	202	379

*10 years growth; ** Av. of 4 years (1986 to 1989)

Table 6.5. Green fodder yield of *Leucaena leucocephala* planted at spacings with and without association with grasses (Av. of 7 years 1983-1989)

Treatment	Yield of *Leucaena leucocephala* (kg/ha)	Yield of grass (kg/ha)	Total yield (kg/ha)
	Main treatment (spacing)		
S1- (1 × 0.5m)= 20000 plants	22270	4911	27181
S2- (2 × 0.5m)= 10000 plants	15223	6844	22067
S3- (4 × 0.5m)= 5000 plants	11692	8748	20440
	Sub-treatments (grasses)		
G0-No grass	16342	-	16342
G1-Hybrid Napier (Var NB5)	13031	12882	25913
G2-Panicum maximum	13361	6872	20233
G3- Chrysopogon fulvus	16076	3299	19375

Table 6.6. Suitable multipurpose plant species for agroforestry in cool temperate zone

Tree species	
1500-2000 m elevation	*Robinia pseudacacia, Grewia optiva, Quercus leucotrichophora, Populus spp., Prunus spp., Cedrus deodara, Melia azedararch, Quercus floribunda, Q. semicarpifolia, Morus serrata, Butryacea* etc.
Above 2000 m elevation	*Acer pictum, Robina pseudacacia, Castanopsis tribuloides, Quercus floribunda, Juglans regia, Salix Spp., Morus serrata, Populus spp., Prunus spp., Celtis australis* etc.
Fodder legumes (Shrubs & herbs)	
1500-2000 m elevation	*Desmodium intortum, D. uncenatum, Medicago sativa, Stylosanthes guyanensis, Pureria thumbergiana, Melilotus spp., Centrosema pubescens, Lotus corniculatus, Vicia villosa, Pureria thumbergiana*
>2000 m elevation	*Trifolium pratense, T. repens, Desmodium taelifolum, Medicago lanticulata, M. truncaluta*
Grasses	
1500-1700 m elevation	*Digetaria decumbense, Panicum coloratum, Chloris gayana, Chrysopogon gryllus, Bracharia decumbens, Pennisetum clandastymum, Setaria sphacelata, Pennisetum orentalis, Pennisatum flaccidium*
1700-2200 m elevation	*Digetaria decumbense, Panicum repens, Setaria anceps, Chrysopogon serrulatus, Pancium cladystymum, Chloris Guyana, Poa annua, Lolium vigidum, Artundinalla nepalensis*
>2200 m elevation	*Lolium multiflorum, Agrostis sp., L. perene, Phleum pratense, Dactylis glomerata, Festuca arundinacea, F. pretensis, Bromis intermis, Poa annua*

Table 6.7. Agroforestry systems found in different agroclimatic zones of Uttaranchal

Zone	Agroforestry system	Functional units
Tarai & Bhabar, sub montane and low hills (upto 1000 m)	Agrisilviculture Agri horticulture	Agriculture : *Triticum aestivum* (wheat), *Zea mays* (maize), *Oryza sativa* (paddy), *Glycine max* (soybean), *Cicer arietinum* (gram), *Lens esculanta* (lentil), *Brassica spp*

Contd...

Table 6.7 Contd...

		(mustard), *Saccharum officinarum* (sugarcane), *Curcuma longa* (turmeric), *Phaseolus mungo* (urd), *P. aureus*mung, *Vigna umbellate* (ricebean), *Cajanus cajan* (arhar), *Pisum* sativum (Pea), *Abelmoschus esculentus* (okra), *Allium sativa* (garlic), *Allium cepa* (onion), *capsicum frutescens* (capsicum), *Vicia faba* (french bean), *Brassica oleracea* (cauliflower), *Brassica capitata* (cabbage) etc Forestry: *Eucalyptus, Populus deltoides* (Poplar), *Anthocephalus kadam*(Kadam), *Dalbergia sissoo* (shisham), *Tectona grandis* (teak), *Toona ciliate* (toona), Acacia catechu (khair), *Holoptelea integerifolia* (kanju), *Adina cardifolia* (haldu) etc Horticulture: *Mangifera indica* (Mango), *Litchi chinensis* (litchi), *Psidium guava* (Guvava), *Artocarpus integerifolia* (jack fruit), *Carica papaya* (papaya), *Emblica officinalis* (Aonla), *Citrus sp* (lemon, kagzi lime, orange, chakotra, galgal ,mosumbi etc)
Sub temperate mid hills (1000-1500)	Agrisilviculture Agri hortisilviculture Silvipastoral	Agriculture : *Oryza sativa* (rice), *Triticum aestivum* (wheat), *Zea mays* (maize), mustard, *Pisum satyvam* (Pea), *Hordeum vulgare* (barley), *Elusine coracana* (mandua), *Paspalum sp.* (jhingora), *Pennisetum typhoiedum* (pearl millet), urd, mung, ricebean, *capsicum frutescens* (capsicum), *Vicia faba* (french bean), *Brassica capitata* (cabbage) Forestry: *Grewia optiva* (bhimal), *Toona ciliata* (toona), *Morus alba* (sehtoot), *Bauhinia variegate* (kachnar), *Celtis australis* (khirak), *Diploknema butyracea* (chura), *Pinus roxburghii* (chir) *Robinia, Corneus oblungum* etc

Contd...

Table 6.7 Contd...

		Horticulture: *Pyrus communis* (pear), *Prunus domestica* (plum), *Prunus armeniaca* (apricot) Grasses: *Apluda mutica, Chrysopogon sp., Heteropogon, Agrostis, Thameda anathera, Arundinella nepalensis, Digetaria decumbense, Panicum coloratum, Chloris gayana, , Bracharia decumbens, Pennisetum clandastymum, Setaria sphacelata, Pennisetum orentalis, Pennisatum flaccidium*
Temperate high hills (1500-2500 m)	Agrisilviculture Hortiagriculture Hortisilviculture Silvipastoral	Agriculture: *Triticum aestivum* (wheat), *Zea mays* (maize), *Hordeum vulgare* (barley), *Elusine coracana* (mandua), *Paspalum sp* (Jhingora), *Pennisetum typhoiedum* (pearl millet), *Pisum satyvam* (Pea), *Vicia faba* (French bean), cabbage, *Amaranthus candatusi* (ramdana), *Solanum tuberosum* (potato), ricebean, *fagio pirum* (buckwheat) Horticulture: *Pyrus malus* (apple), *Juglans regia* (walnut), *Pyrus communis* (pear), *Prunus armeniaca* (apricot), *Prunus amygdalus* (almond), *Pistachio,* Forest: *Aesculus indica* (pangar), *Alnus nitida, Betula, Salix, Quercus leucotrichophora* (Banj), *Q. semicarpifolia*(kharsu), *Q. diltata* (moru), *Fraxinus micrantha, Rhododendron arboretum* (Buransh), *Populus ciliata(Poplar, Taxus baccata* (thuner), *Abies pindrow* (raga), *Cedrus deodar* (deodar), *Myrica esculanta*(kafal) Grasses: *Aragrostis* spp, *Bromis intermis Festuca arundinacea, Dactylus glomerata, Festuca, Kobretia, Cymbopogon distans, Chrysopogon royleanus, Danthonia cachemyriana, Digetaria decumbense, Festuca arundinacea, F. pretensis Kobretia, Panicum repens, Setaria anceps, Chrysopogon*

Contd...

Table 6.7 Contd...

		serrulatus, Lolium multiflorum Pancium cladystymum, Chloris gayana, Poa annua, Lolium vigidum, Arundinalla nepalensis, Phleum pratense,, Poa annua
Dry temperate zone (>2500m)	Agrisilviculture Hortisilviculture Silvipastoral	Agriculture: *Triticum aestivum* (wheat), *Hordeum vulgare* (barley), amaranthus, *Solanum tuberosum* (potato), *Fagop- yrum* (buckwheat) Horticulture: *Pyrus malus* (apple), *Prunus armeniaca* (apricot), almond, Pistachio Forest: *Hippophae, Betula, Salix etc* Grasses: *Danthnia cachemyriana, Koeleria cristata, Calamagrostis emodensis, Festuca lucida, Brachypodium sylvaticum, Trisetum spicatum, Andropogon tristis, Phleum alpinum, Agroptron semicostatum, Agrostis munroana, Deschampsia caespitosa caespitosa, Helictotrichon virescens and Deyeuxia pulchella*

Table 6.8. Performance and suitability index of different agroforestry tree species after 8 and 14 years of planting in tarai region

S. No	Species	Survival		Height increase		DBH increase		Relative score (%)		Performance
		8 years	14 years	8 years	14 years	8 years	14 years	8 years	14 years	
1	*Acacia catechu*	100	100	52.0	62.1	56.7	61.8	70.7	74.7	S
2	*Acrocarpus fraxnifolius*	100	100	66.9	91.1	73.7	90.4	81.4	93.8	MS
3	*Albiza lebbeck*	100	100	66.3	65.2	80.5	82.4	83.6	82.5	MS
4	*Anthocephalus cadamba*	100	100	87.4	79.1	107.9	90	100	89.7	MS
5	*Bahuinia variegata*	100	100	46.9	37.5	57.4	52.8	69.2	63.4	S
6	*Bombax ceiba*	100	100	59.4	52.7	81.2	69.2	81.5	74.0	MS
7	*Cassurina equisetifolia*	95	95	100	84.7	64.3	63.2	87.8	81.0	MS
8	*Dalbergia sissoo*	90	90	61.1	52.6	66.8	68.4	73.8	70.3	S
9	*Emblica officinalis*	100	100	38.3	40.6	45.5	44.5	62.2	61.7	US

Contd...

Table 6.8. Contd...

10	*Eucalyptus citridora*	95	95	100	84.3	60.3	61.5	86.4	80.2	MS
11	*Eucalyptus hybrid*	95	95	104	82.0	64.3	6.9	89.1	61.3	MS
12	*Grevellia robusta*	80	80	17.7	42.6	13.7	38.6	37.7	53.8	US
13	*Lagerstromia flosreginae*	100	100	36.0	38.5	46.6	51.1	61.8	63.2	US
14	*Mangifera indica*	90	90	28.0	29.9	35.7	37.8	52.1	52.6	US
15	*Melia azaderarch*	100	100	62.9	52.8	73.3	60.7	80.0	71.2	MS
16	*Morus alba*	100	100	48.6	39.4	74.0	64.2	75.4	67.9	S
17	*Pinus roxburghii*	80	80	17.1	29.0	28.2	34.6	42.4	47.9	US
18	*Pongamia pinnata*	95	95	33.7	40.9	41.9	48.1	57.8	61.3	US
19	*Populus deltoides*	100	100	90.9	100	100	100	98.5	100	MS
20	*Pterocapus marsupium*	95	95	38.9	62.5	42.2	66.5	59.7	74.7	US
21	*Syzigium cuminii*	100	100	40.6	40.9	54.2	50.4	65.9	63.8	US
22	*Tectona grandis*	100	100	58.9	61.0	56.7	55.6	73.0	72.2	S
23	*Terminalia arjuna*	100	100	44.0	53.3	63.2	63.2	70.1	72.2	S
24	*Terminalia bellerica*	100	100	45.1	69.0	45.9	62.9	64.7	77.3	US
25	*Terminalia chebula*	0	100	55.4	57.9	32.0	60.6	29.6	39.5	US
26	*Toona cilliata*	100	100	55.4	59.2	75.5	64.2	78.2	74.5	S
27	*Trewia nudiflora*	100	100	53.1	57.4	79.4	72.3	78.7	76.5	S

S-suitable, MS-most suitable, US-unsuitable

Table 6.9. List of poplar clones in the germplasm

S. No.	Clone	Source	S. No.	Clone	Source
1	S7C1	USA	26	L-49/82	Lalkuan Selection
2	S7C2	USA	27	L-200/84	Lalkuan Selection
3	S7C4	USA	28	L-62/84	Lalkuan Selection
4	S7C8	USA	29	WSL-45	WIMCO
5	S7C15	USA	30	22-N	FRI, Line
6	S7C20	USA	31	23-N	FRI, Line
7	G-3	Australia	32	25-N	FRI, Line
8	G-48	Australia	33	26-N	FRI, Line

Contd...

Table 6.9. Contd...

9	3167	FRI, Line	34	34-N	FRI, Line
10	82-33-3	FRI, Line	35	36-N	FRI, Line
11	82-35-4	FRI, Line	36	37-N	FRI, Line
12	82-26-5	FRI, Line	37	38-N	FRI, Line
13	421-2	FRI, Line	38	39-N	FRI, Line
14	S13C11	USA	39	40-N	FRI, Line
15	S4C2	USA	40	41-N	FRI, Line
16	D-121	USA	41	42-N	FRI, Line
17	ST-124	Germany	42	43-N	FRI, Line
18	2502	FRI, Line	43	63-N	FRI, Line
19	3324	FRI, Line	44	UD-88	FRI, Line
20	111828	FRI, Line	45	RD-01	FRI, Line
21	113520	FRI, Line	46	PP-5	Developed through γ ray mutation (Pantnagar)
22	A-194	FRI, Line	47	PH-1	By crossing *P.deltoides* ×*P ciliata* Pantnagar
23	PIP-123	FRI, Line	48	PH-2	By crossing *P.deltoides* ×*P ciliata* Pantnagar
24	PIP-124	FRI, Line	49	PH-6	By crossing *P. deltoides* ×*P ciliata* Pantnagar
25	L-34/82 Selection	Lalkuan			

Table 6.10. Estimates of heritability and genetic advance (in per cent) for different characters:

S. No.	Characters	Heritability (%)	*Genetic advance (%)	GCV	PCV	Grand mean	Range
1.	Tree height	91.0	50.4	20.7	21.7	12.4	5.76-17.66
2.	Diameter at breast height (cm)	66.8	43.9	17.8	21.8	14.9	9.26-21.33

Contd...

Table 6.10. Contd...

3.	Clear bole length (m)	87.5	24.1	29.4	31.4	4.3	2.43-7.66
4.	Crown length (m)	78.4	32.6	23.1	25.0	8.1	3.03-11.23
5.	Crown width (m)	67.9	10.9	14.2	17.3	4.5	3.40-5.96
6.	Bark thickness (mm)	91.0	49.8	22.6	23.7	11.2	6.33-17.50
7.	Length of leaf blade (cm)	80.2	40.6	15.0	16.8	11.5	10.96-23.26
8.	Length of leaf petiole (cm)	83.7	31.7	19.1	20.9	7.8	5.90-13.90
9.	Total leaf length (cm)	78.9	61.2	14.2	16.0	23.5	18.00-37.17
10.	Leaf width (cm)	66.4	28.0	12.3	15.1	13.8	9.43-20.00
11.	Leaf area index (cm^2)	99.7	40.8	22.6	22.6	87.8	48.79-121.03
12.	Volume over the bark (m^3)	96.8	23.6	22.6	23.0	0.5	0.354-0.867
13.	Volume under the bark (m^3)	96.2	24.4	28.7	29.2	0.4	0.267-0.779
14.	Leaf fall duration (days)	98.6	50.7	21.2	21.3	116.9	91.00-181.22

Table 6.11. Tree growth (height and d.b.h.) and crops (wheat and rice) under promising poplar clones.

Poplar clones	Tree Growth		Wheat crop yield (q/ha)		Rice crop yield (q/ha)	
	Height (m)	d.b.h. (cm)	Grain	Straw	Grain	Straw
Pant Poplar 5	14.0	18.7	23.5	56.2	7.2	25.0
L 49	14.1	18.3	29.1	64.8	7.0	22.6
Udai	14.5	19.6	22.1	53.3	6.5	18.1

Contd...

Table 6.11. Contd...

G 48	12.4	16.8	27.7	50.5	6.9	21.8
L 52	13.0	18.3	22.7	51.5	6.3	25.7
L 188	14.3	18.3	26.2	53.1	7.2	22.1
SEm±	0.3	0.5	1.7	3.9	0.5	1.4
C.D. at 5%	1.0	1.4	NS	NS	NS	4.2

Table 6.12. Tree growth and crop yields as influenced by the tree density in poplar based agroforestry system.

S. No.	Tree density (trees/ha)	Mean tree Growth		Wheat yield (q/ha)		Rice yield (q/ha)	
		DBH (cm)	Height (m)	Grain	Straw	Grain	Straw
1.	0	-	-	37.8	57.1	16.1	58.3
2.	200	15.0	11.2	31.2	46.0	14.8	46.4
3.	250	14.7	11.6	30.8	44.8	13.8	43.9
4.	333	14.7	11.9	28.9	39.8	10.3	34.7
5.	500	14.5	13.0	26.4	31.8	6.6	31.9
6.	1000	13.5	13.3	25.2	30.6	5.0	18.5
SEm±		0.28	0.3	0.3	1.9	1.5	1.7
C.D. at 5%		1.08	1.1	1.0	6.8	5.3	6.1

Table 6.13 . The tree growth (diameter and height growth and yields (Grain and stover/straw) of Toria and Rice crops as influenced by various pruning treatments in poplar.

Treatment	Mean tree Growth			Toria yield (q/ha)		Rice yield (q/ha)	
	DBH (cm)	Height (m)	Fuelwood (q/ha)	Grain	Stover	Grain	Straw
No pruning	9.7	8.3	-	4.3	18.9	22.0	64.3
25 % pruning	9.6	8.3	3.2	6.4	21.9	23.5	64.8

Contd...

Table 6.13. Contd...

50 %pruning	9.6	8.1	9.5	6.5	22.4	32.5	86.8
75 % pruning	8.2	7.7	17.5	6.9	23.3	34.2	92.0
Equidistant pruning	9.5	8.4	6.8	5.2	21.0	24.0	74.8
SEm ±	0.2	0.3	-	0.4	1.6	3.8	12.0
C.D. at (5%)	0.8	NS	-	1.6	5.9	NS	NS

Table 6.14. Range, Mean, coefficients of variation, for different quantitative characters in shisham

Genotypes	Genotypes	Height	DBH	CBL	CL	CW	LA	SS	UH	BT	CC	Volume
Set I	Range	8.0-10.00	0.096-0.155	2.13-4.85	2.26-4.36	2.08-4.20	0.00-20.00	2.08-3.91	2.91-5.66	0.22-0.98	28.16-37.16	0.045-0.150
	gm	9.2	0.1	3.4	3.3	3.0	6.7	3.0	3.9	0.5	31.8	0.097
	cv.	3.0	3.8	6.4	12.0	17.6	38.5	7.7	7.3	1.8	1.7	8.061
SetII	Range	8.30-15.40	0.084-0.217	3.90-8.83	2.50-6.00	1.80-3.83	0.00-8.33	2.16-5.91	4.00-6.83	0.20-0.89	284.00-313.00	0.044-0.444
	gm	10.7	0.1	6.0	3.8	2.5	0.3	4.7	5.5	0.5	299.1	0.1
	cv.	3.6	4.1	4.9	5.5	8.4	20.5	113.3	4.9	6.0	0.2	8.1
Set III	Range	6.91-8.66	0.078-0.143	2.00-4.70	1.83-3.93	1.46-3.30	0.00-20.00	2.08-3.91	2.91-5.66	0.20-0.96	277.33-301.00	0.030-0.115
	gm	8.0	0.1	3.3	2.7	2.2	6.7	2.9	3.9	0.4	291.4	0.1
	cv.	3.7	4.4	6.4	6.7	8.4	25.1	7.4	7.3	6.2	0.4	8.9
Set IV	Range	9.96-17.01	0.098-0.226	4.16-9.03	3.38-7.08	2.70-4.16	0.00-8.33	2.66-5.91	4.00-6.83	0.22-0.92	28.066-44.066	0.062-0.536
	gm	12.7	0.1	6.2	4.7	3.3	0.3	4.7	5.5	0.5	36.4	0.2
	cv.	3.4	2.9	4.1	12.1	14.1	189.7	112.5	4.9	2.5	1.0	6.5
Set V	Range	8.09-14.66	0.103-0.257	2.50-9.86	1.86-5.70	1.50-3.66	0.00-21.66	1.91-9.83	3.16-12.66	0.40-1.30	278.00-316.00	0.06-0.46
	gm	11.3	0.2	6.4	3.6	2.5	6.8	4.3	6.3	0.7	302.1	0.2
	cv.	3.4	3.2	5.9	5.9	9.0	24.6	6.8	6.6	3.3	0.4	2.6
Set VI	Range	9.59-14.85	0.105-0.176	3.48-7.03	2.96-5.40	2.28-3.20	0.00-20.00	2.00-5.16	2.91-7.50	0.308-0.983	31.06-43.90	0.066-0.280
	gm	13	0	5	4	3	6	3	5	1	37	0
	cv.	4	5	4	18	20	26	7	6	3	1	13

DBH= Diameter at Breast Height, CBL= Clear Bole Length, CL= Crown Length CW= Crown Width,l LA= Lean Angle SS= Stem Straightness UH= Unforked Height, BT= Bark Thickness, CC = Chlorophyll Content, Volume = Volume

Table 6.15. Tree height and Diameter as influenced by tree densities in 13 year old *Dalbergia sissoo* and *Eucalyptus terticornis*.

Planting Densities (no./ha)	*Dalbergia sissoo*		*Eucalyptus tereticornis*	
	Tree height (m)	DBH (cm)	Tree height (m)	DBH (cm)
5000	15.60	15.30	12.63	11.30
2857	15.91	16.80	12.95	13.00
1887	14.82	17.20	13.30	13.90
1176	15.40	18.30	14.15	14.60
0690	14.35	18.50	14.80	15.50
0431	14.15	19.00	15.25	16.40
0285	14.50	20.50	16.30	17.40
0167	14.80	21.00	17.10	17.80

Table 6.16. Tree height and DBH of three year old Salix clones.

Clones	Tree height (meter)	DBH (cm)	Grain yield (kg/ha)	Straw yield (kg/ha)
Control	-	-	5022	5966
Salix alba I cv. coerulea	4.83	6.19	4806	5718
Salix alba II cv. coerulea	4.70	6.36	3510	4870
Salix alba cv. vitellina	5.92	9.02	5346	6698
Salix babylonica	5.45	11.38	4104	6052
CD at 5%	NS	NS	NS	NS

Table 6.17. Effect of two year old Salix clones and tree spacing on tree height and d.b.h.

Treatment	Tree height (m)	Tree DBH (cm)
A. Salix clones		
Salix alba cv coerulea	4.8	4.6
Salix alba cv vitellina	5.1	5.2
CD (5%)	NS	NS

Contd...

Table 6.17. Contd...

B. Salix spacing		
4x5 m	5.2	5.0
6x5 m	4.9	4.7
8x5 m	4.9	4.8
10x5 m	4.9	5.0
CD (5%)	NS	NS

Table 6.18. Mean performance and range of variation in different provenances of *Jatropha curcas*

Provenance	Survival	Height (cm)	Collar diameter (mm)
Kaladungi	92.50	49.88±3.35	22.22±2.02
Halduchaur	96.29	40.6 ± 4.9	21.20±3.28
Gola	96.29	53.5±2.69	20.56±1.87
Kheda	100	46.33±10.4	21.26±2.00
Bailparao	100	47.55±4.14	21.22±0.93
Pantnagar 1	100	41.0±1.0	19.1±1.41
Bilaspur	100	36.67±6.69	17.85±2.94
Pantnagar 2	100	41.8±2.50	14.6±1.07

Table 6.19. Effect of different spacing on growth of *Jatropha curcas*

Spacing (m)	Survival	Height (cm)	Collar diameter (mm)
4x2	83.33	49.55±3.88	14.56±1.58
4x3	91.66	58.72±4.65	17.91±1.96
4x4	100	65.38±6.53	18.76±0.96
4x5	100	68.25±3.56	20.54±1.33

- Development of package and practices of important agroforestry trees.
- Germplasm collection and genetic improvement of important agroforestry tree species
- Strengthening of agroforestry research and education

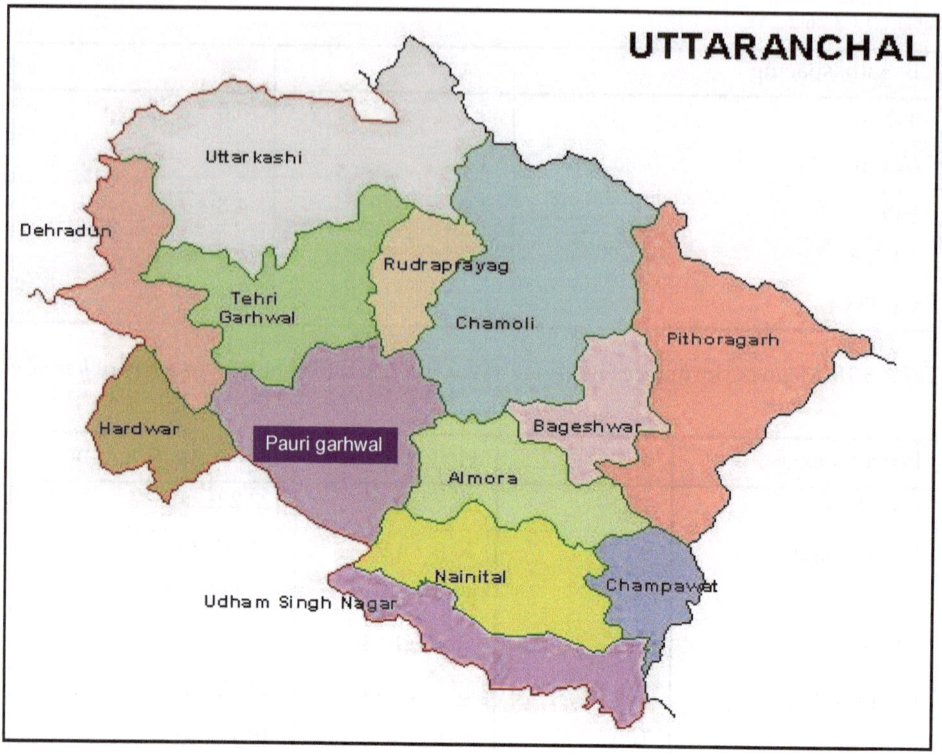

Fig. 6.1. Map of Uttranchal

- Strengthening of market network for agroforestry produce
- Redefining the role of various state forest, horticulture, agriculture, animal husbandry department
- market of agroforestry products should be regulated through Mandi Parishad Act and the Forest Corporations of the State. A provision of buy-back guarantee/minimum support price should be provided to farmers.
- Establishing institutional mechanisms for developing linkages between farmers and industry.
- Institutional finance through NABARD and other banks to needy farmers under low interest rates.
- Change in Government policies so that restrictions on felling, transit and conversion of agroforestry trees from farmers land are removed.

REFERENCES

Anonymous (1992) *Annual Report. AICRP on Agroforestry.* G.B. Pant University of Agriculture & Technology Pantnagar. Uttaranchal.

Anonymous (1999) *Annual Reports. AICRP on Agroforestry.* G.B. Pant University of Agriculture & Technology Pantnagar. Uttaranchal.

Anonymous (2002) *Annual Report. AICRP on Agroforestry.* G.B. Pant University of Agriculture & Technology Pantnagar. Uttaranchal.

Anonymous (2003) *Annual Report. AICRP on Agroforestry.* G.B. Pant University of Agriculture & Technology Pantnagar. Uttaranchal.

Anonymous (2004) *Annual Report. AICRP on Agroforestry.* G.B. Pant University of Agriculture & Technology Pantnagar. Uttaranchal.

Anonymous (2005) *Annual Report. AICRP on Agroforestry.* G.B. Pant University of Agriculture & Technology Pantnagar. Uttaranchal.

Bana, O.P.S.; Singh,V.; Singh, A.K. (1996). forage production in U.P. Hills Status, Constraints and Development Strategies In proceeding of subregional *Working Group on Temperate Asia Pasture and Fodder Working Group (FAO) Meeting,* March 9-14, Deharadun.

Bhatt, A.B. and Rawat, N. (1993). Fodder Resources of Garhwal: A search for Non-conventional Fodder Species. *In* Rajwar, G.S. ed. Garhwal Hemalaya Ecology and Enviornment. New Delhi: Ashish Publishing House 237-239.

Dadhwal KS (1999) Arresting environmental degradation of limestone mined lands from north-west Himalaya through agroforestry. *In* : K.R.Solanki, A.K.Bisarria and A.K.Handa (Eds.), pp.95-101 *Short Course on Sustainable rehabilitation of degraded lands through agroforestry,* National Research Centre for Agroforestry, Jhansi.

Dadhwal, K.S. and Tomar, V.P.S. (2000). Management of tree roots in agroforestry land use from marginal lands of north-west Himalaya. *In* : *International Conference on Managing Natural Resources for Sustainable Agricultural Production in the 21st Century,* New Delhi, India, Vol.3 (Voluntary paper), pp.1107-1109.

Malik, K P. S. and Prakash, O. (1996). Changes in some physical and chemical properties of soil in Agri-silvicultural system with some multipurpose tree species in Tarai region of Uttar Pradesh, *In.* Kohli, R K and Arya, K S (eds) *Resource Inventory Techniques to support Agroforestry and Environment Activities.* IUFRO/DNAES, Chandigarh, India pp 249-254.

Narain P., Singh, R. and Joshie, P. (1988) Studies on nutrient and water budget under different land use systems. *Annual Report- Central Soil & Water Conservation Research & Training Institute,* Dehradun, India, pp.55-57.

Narain P, Singh RK, Sindhwal NS, and Joshi, P (1998) Agroforestry for soil and water conservation in the western Himalayan valley region of India. *Agroforestry Systems.* 39: 191-203.

Pal M.S. (2004) Mountain farming systems. Attributes of livestock. *In*: Singh, V. and Gautum, P.L. (Eds.) *Livestock production systems for sustainable food security and livestock in Mountain Area.* pp. .35-39.

Purohit, R, Naithani; Tewari, S.K., Kaushal, R and Gaur, Neeta (2005). growth behavior of selected medicinal tree species in Tarai region of Uttaranchal . Page 23: *Abstract National Consultative Workshop on Medicinal and Aromatic Plants.* June 25-27, keld at Pantnagar.

Sing, V. Sharma, R.J.; and Kumar, A. (1995). Perspectives on the Utilization of Forest Fodders in the Mountaion. *In* Singh, V.; Sharma, R.J. and Kumar, A. (eds) *Ecological Carnage in the Himalaya*. IBD. Deharadun.

Singh H., M.K. Vishwasnatham, P Joshie and M.L. Khybri. (1987b). Economic utilization of bouldery riverbed land for fodder, fibre and fuel. *Annual Report, CSWCRTI,* pp. 48-49.

Singh, H., Vishwasnatham, M.K., Joshie, P and Khybri, M.L.. (1990b). Economic utilization of bouldery riverbed land for fodder, fibre and fuel. *Annual Report, ibid* 1989-90, pp. 41-42.

Singh, H., Vishwasnatham, M.K., Joshie, P and Khybri, M.L.. (1991). Economic utilization of bouldery riverbed land for fodder, fibre and fuel. *Annual Report ibid,* 1990-91, pp. 23-24.

Singh, H., Vishwasnatham, M.K., Joshie, P and Bhardwaj, S.P.. (1986). Effect of spacing and grass combination on fodder yield of subabul in bouldery class V-VII land under rainfed conditions. *Annual Report, ibid.,* pp. 45-46.

Singh, H., Vishwasnatham, M.K, Joshie, P and Bhardwaj, S.P.. (1987a). Effect of spacing and grass combination on fodder yield of subabul in bouldery class V-VII land under rainfed conditions. *Annual Report, ibid.,* pp. 49-50.

Singh H., Vishwasnatham, M.K., Joshie, P and Bhardwaj, S.P.. (1988). Effect of spacing and grass combination on fodder yield of subabul in bouldery class V-VII land under rainfed conditions. *Annual Report, ibid.,* pp. 37-38.

Singh H., Vishwasnatham, M.K., Joshie, P and Bhardwaj, S.P. (1990a). Effect of spacing and grass combination on fodder yield of subabul in bouldery class V-VII land under rainfed conditions. *Annual Report, ibid* 1989-90. pp. 45-46.

Singh, V. (2003). Agroforestry in the Development of sustainable Mountain Agriculture.*In Recent Advances in Agroforestry.* G.B. Pant University of Agriculture and Technology. Pantnagar: Uttaranchal

Singh, V. and Gautum, P.L. (2004) *Livestock production systems for sustainable food security and livestock in Mountain Areas*. G.B. Pant University & technology Pantnagar Uttaranchal.

Singh, V.; Sharma, R.J.; and Kumar, A. (1995). Perspectives on the Utilization of Forest Fodders in the Mountains. In Singh, V.; Sharma, R.J..; and Kumar, A. (eds.) *Ecological Carnage in the Himalaya*. Deharadun: IBD.

• • •

7

Promising Agroforestry Practices in Punjab

H. S. SARALCH, BALJIT SINGH AND S. K. CHAUHAN
Department of Forestry & Natural Resources
Punjab Agricultural University, Ludhiana-141 004

ABSTRACT

Punjab is the richest state in the country in term of per capita income and has earned the name as food bowl of the country by putting 84-85 per cent of its geographical area under highly intensive, technical and mechanical agriculture with cropping intensity as high as 185 per cent and contributing more than one third of rice and half of its wheat production to the central pool of food reserve. The food grain production in the state has increased approximately twelve times in a span of about 50 years after independence from 20 lac tones in 1950-51 to 234.89 lakh tonnes in 2002-2003. Achieving this high productivity without caring for natural resources has resulted in a considerable loss in the inherent production potential and deterioration of soil health. As a result, man is getting serious warning signals in the form of lowering of water table, drought, high rate of environmental pollution, extremes of climatic parameters, etc. Agriculture in the state has become too intensive, wasteful and reckless and the radical changes are required to maintain the balance. Therefore, there is a strong thinking to diversify farming in the state to maintain the sustainability of the whole system. The state government is stressing hard to save the natural resources through the diversification in traditional crop rotation and adopt resource-conserving measures.

With the introduction of fast growing multipurpose tree species in Punjab during sixties, agroforestry has been recognized as a sustainable system of high potential to boost the state's meager forest resources and to check the further degradation of natural resources. Highly productive block planting of poplar intercropped with wheat, sugarcane, turmeric, oats, berseem, mustard, medicinal plants, seasonal flowers, vegetables, etc. has become popular in irrigated tracts of the state. Boundary planting of eucalypts in the wheat- paddy cropping system and scattered trees of *kikar, khair,* shisham, dek, etc. on cultivated lands under rainfed conditions, especially in the sub-montane foot hill zone, play an important role in soil conservation and boosting the socio-economic status of the farmers. These tree species have inherent potential of increasing the productivity of problematic lands of the state. Therefore, the diversion of farmlands towards agroforestry will not only increase tree cover but also achieve the desired objective of National Forest Policy-1988 and conserve the precious natural resources like ground water, soil, biodiversity, etc. and prove to be a viable alternative sustainable land use system, which is renewable and eco-friendly as compared to sole agricultural crops in the years to come. The profitability of tree plantations on farmers' fields will further improve with the implementation of trading of carbon credits under Kyoto Protocol.

INTRODUCTION

Rapidly changing economic and social needs of the society have resulted in putting ever-increasing pressure on our natural resources to "do more with less". The unabated deforestation brought about by increasing population over the years has swelled up the demand for wood coupled with the problems of soil erosion, floods, land slides, silting of reservoirs, drought, air, water and noise pollution, etc. Because of the increasing emphasis on the conservation of natural forests, protection of ecologically sensitive areas and ban on green felling by Hon'ble Supreme Court of India from conventional forest areas, the supply of the industrial timber from forest reserves has gradually declined or negligible over the years, which has resulted into the widening of gap between demand and supply of forest produce. To minimize these problems, an insight into the aforesaid situations clearly revealed that there is an urgent need to evolve some strategies, which should focus on sustainable agriculture and forest development. Therefore, our priorities should be focused on exploring new land management systems as an alternative.

Agroforestry, as dynamic, ecologically based, natural resource management system that integrates trees on agricultural landscape, diversify and sustain production for increased social, economic and environmental benefits for land users at all levels, seems to hold promising potential. This practice owes capabilities not only to provide multiple outputs of tree/crop based products to meet the needs of local people and industries, but also facilitates the amelioration of environment. For millennia, farmers have nurtured trees on their farming landscape for many benefits they provide. Neither the concept nor the practice is new, but the term has been transformed from its ancient practice into a youthful science with the application of scientific technology and is yielding significant results.

Agroforestry, the word coined about two and half decades back, has made its place in the developed and the developing countries of the world. The contributions of research on agroforestry are many which include; the identification and propagation of the right trees for right environments, management practices for optimizing tree and crop productivity, improved soil fertility, better watershed management and increased tree diversity in the agricultural landscape. Research over the last 20 years has confirmed that agroforestry can be biologically more productive, more profitable and more sustainable than agriculture and forestry monocultures. More than a decade of research and pilot projects on this discipline, clearly show that agroforestry constitutes a very important part of rural development and the planting of trees even on small scale farm holders can reduce poverty, improve food security and protect the natural environment.

STATUS OF FORESTRY, AGRICULTURE AND AGROFORESTRY IN PUNJAB

Punjab has a vast expanse of flat alluvial land, the Shiwalik hills in the north have fluvial carvings and deposits of three rivers namely; the Ravi, Beas and Satluj which, flow through central Punjab and sand dunes in the southwest. It is predominantly an agrarian state and has taken big strides forwarded in the fields of agriculture, dairy, poultry, machinery, industry, etc. during the post-independence period. The state is known for its prosperity, apart from total production, the productivity per unit area per unit time compared as best in the world in the field of agriculture and worst in the field of forestry. The success story of Punjab agriculture during the last 35 years has no parallel in the history of mankind but, it has been achieved at the cost of a colossal damage to our natural resources including forests. Therefore, an attempt has been made to highlight the scenario of forestry, agricultural and the need of agroforestry in the state.

A) Forestry

Stating the scenario of forestry in Punjab, the figures are very discouraging. Punjab is having a total geographical area of 50,362 km^2 (1.5 per cent of country's geographical area). The human population of the state is 24.29 million, 2.4 per cent of the country's population, of which 66.1 per cent is rural and 33.9 per cent is urban. The average population density of the state is 482 persons per km^2. The total forest cover of the state as per 2001 assessment (Forest Survey of India, 2001) is 2432 km^2 and this constitutes 4.83 per cent of the geographical area of the state and 0.4 per cent of country's forest cover (Table 7.1) against 20 per cent required as per the National Forest Policy of 1988. Furthermore, the forest area in Punjab is neither equitably distributed nor uniformly stocked and bulk of the forest area is in the sub-montane range of Shivalik hills, which is prone to illicit felling, lopping for fuel wood and fodder and unregulated grazing.

Unfortunately, despite the best efforts by various agencies, there has not been any significant increase in the forest area of the state. There is no spare land to increase the forest cover and practically, it is not possible to divert as such the fertile agriculture land to forestry in view of prevailing socio-economic and agro-climatic conditions favourable for agriculture. The only option, which finds some rays of hope, is agroforestry i.e. integration of tree, crop and animal component simultaneously or sequentially on the same land management unit.

To popularize tree plantations on private lands, Planning Commission has already recommended to Govt. of India for imposing heavy import duty on various forest based products, which can otherwise be produced by the farmers of our country. The State government should encourage the establishment of wood-based industries and act as a facilitator in ensuring buy back arrangement between the tree growers and wood based industries.

B) Agriculture

The Punjab state has brought the country from its being food deficit state to its present status of food grains exporting country. The land of three rivers has ninety five per cent of agricultural land of the state under irrigation through various irrigation sources (Table 7.2). It is one the richest states of the country in terms of per capita income having 84-85 per cent of its geographical area under highly intensive, technical and mechanical agriculture with cropping intensity as high as 185 per cent (Table 7.3). The main agricultural crops, which are being grown in the state are; wheat and rice and some other crops like sugarcane, cotton, maize, sorghum, oats, berseem, barley, gram, groundnut, rapeseed, mustard, soybean, sunflower, medicinal plants, seasonal flowers, vegetables, fruits, etc. But, wheat and rice are the main crops contributing more than 51 per cent in wheat and 33 per cent in rice share to the central pool of food reserves since 1980-81 (Table 7.4). But, intensive use of groundwater, chemical fertilizers, fungicides and pesticides and burning of crop residues have resulted in deterioration of soil health, lowering of water table and high rate of environmental pollution, which is the matter of great concern for the time to come.

In Punjab, degradation of land is mainly of two types i.e. visible and invisible. Visible degradation includes soil erosion (by water and wind) and development of salinity, alkalinity and water logging, while invisible degradation is due to the reduction in productivity caused by lowering of water table, adverse chemical, physical and biological changes. Invisible degradation is relatively more widespread and serious problem than visible degradation. A major cause of land degradation is its over-exploitation as the per capita availability of cultivable land is almost less than 0.2 ha, thus, reflecting an enormous pressure on our finite land resources resulting in various types of degradation in different zones of the state (Table 7.5).

i) Question of sustainability

Punjab at this stage is under serious question with respect to the sustainability of the existing production systems. The facts like:

- Excessive pumping of ground water as more than 70 per cent of the irrigation water of rice crop in Punjab is met through tube wells as a result, water table in the central Punjab is receding at an average rate of 42 cm per annum and if ground water withdrawal continues at this rate without any recharge, by the year 2010, no place in the central Punjab will be left with water table less than 17 meters deep.
- Use of higher doses of fertilizers and pesticides. The use of nutrients through application of chemical fertilizers is highest in the state i.e. 367 kg/ha, which has developed deficiencies and toxicity of many macro and micro- nutrients.
- Presence of chemical residues in the drinking water like selenium and other toxic elements, which render it unfit for drinking purposes.
- Total factor productivity on rice crop in the state of Punjab is on the decline. The compound growth rate of rice yield, which was 1.74 per cent in the period 1979-80 to 1989-90, has declined to 0.09 per cent during 1989-90 to 2001-02 (Johl, 2005).
- Higher rate of environmental pollution through burning of paddy and wheat stubbles in the fields, which creates heavy smoke pollution and leads to serious breathing and other related problems to the human beings.

Thus, it is the huge contribution of food grains to the central stocks by this tiny-big state, which has exhausted its soil, water resources and clean environment. It has put the Punjab agriculture sector on crossroads and the state, is now left licking its wounds. Therefore, the state is seized of the situation and has to re-prioritize its future strategies (Aulakh, 2005).

ii) Need for diversification of agriculture

Presently, agriculture in the Punjab state has been one of the targets of criticism with regards to the sustainability of production systems. Therefore, there is a strong thinking to diversify farming in the state to maintain the sustainability and the state government is stressing hard to save the natural resources through the diversification in traditional crop rotation and adopt resource conserving measures through less demanding crops. With the acceptance of Johl Committee report on diversification of agriculture in state of Punjab by the Planning Commission, the farmers may adopt other alternative crops like cotton, pulses, oilseeds, tree farming, etc. having very less production and area under these crops on larger scale, other than wheat and rice.

C) *Agroforestry*

i) Concept

The concept of agroforestry to Punjab (i.e. practice of growing agricultural crops under scattered trees on farm lands) is as old as our civilization and does not seem to have changed for centuries. Though, this venture as a business enterprise on commercial scale is not only new to Punjab, but to the whole country. In recent times, an increasing awareness has been created across the country regarding the multiple utilities of the trees. Since, Punjab is basically an agricultural state having 84.80 per cent area under agriculture, 6.10 per cent forest area, 8.30 per cent area under miscellaneous uses (not available for cultivation), 0.52 per cent fallow lands and 0.28 per cent other un-culturable lands excluding fallow lands (Fig.7.1). Therefore, to make agroforestry discipline a success, the farmers of the state must be taken into confidence as the state is having less than one per cent area under fallow lands and bulk of the land is with the farmers.

ii) Need

Punjab is endowed with less forested area, fertile land resource and abundant water, whereas, its earth resource (mines, fossil fuel, etc.) are negligible. Agriculture is the mainstay of the state. The success story of Punjab agriculture during the last 35 years has no parallel match in the history of agriculture, but it has been achieved at the cost of a colossal damage to natural resources. The scientists, intellectuals and public at large have expressed serious concern that both land and water resources have been overexploited, while their conservation has been slow and inadequate. So the need of the hour is to check the further depletion of these precious natural resources, as we can not take the soil granted, because it is not an inert substance, but is fragile entity of unrivalled complexity. Keeping in view the above problems, agroforestry, seems to be one of the viable alternative land use system for maximum productivity on sustainable basis.

iii) Present status

There is a tremendous change in the composition and intensity of tree species on farmer's fields since the inception of green revolution. Before the green revolution, the important tree species which were grown on the farmer's fields in different agroclimatic zones were; *kikar, neem, shisham, lasoora, toot, siris, dhak, ber, mango, dek, khair, beul, bahera, jand,* etc. These tree species were included in the farms for small timber, fuel wood, fodder and shade. During the consolidation of land holding, almost all the trees were felled and removed because of the fear of transfer of ownership. But, the real transformation towards agroforestry occurred with the introduction of two fast growing exotic tree species (*Eucalyptus tereticornis* and *Populus deltoides*) in Punjab during 1960's and 1970's. The success of these two important exotics on the farmers' field as well as non- conventional forestry areas in the state has revolutionized and replicated the concept very fast. As a result, the state is having 3.2 per cent tree cover of the state's total geographical area other than the conventional forests and approaching to parallel figures of forest cover of the state, which is 4.83 per cent. Agroforestry has become one of the popular land use system in the recent past in the state. Presently, farmers grow these two commercial tree species on a large scale in addition to their traditional species like *Acacia nilotica, Melia azedarach, Dalbergia sissoo, Azadirachta indica,* etc. on their farming landscape.

iv) Scope

Major portion of wood now produced in the state, comes from tree plantations established on the farmers fields. Therefore, there is a possibility of expanding the scope of agroforestry on private lands which include, in its fold, environmental goods and services like carbon sequestration, replacement or substitution of fossil fuels, bio-diversity conservation, soil and water conservation and eco-tourism in addition to the conventional forestry/ agricultural products in which the participants directly share the economic values attached to these goods and services. The important potential tree species which can be commercially exploited in the different agroclimatic zones of the state are; *eucalypts, poplar, dek, gamari, toot, kikar, subabul, shisham, sagwan, arjun, bahera, harar, pahari kikar,* etc. and tree borne oilseeds (TBO's) like *jatropha, neem, karanja, mahua,* etc under different agroforestry models. The diversion of farmlands toward agroforestry will not only increase the forest cover of the state, which is less than 5 per cent and conservation of precious natural resources, but also prove to be an alternative sustainable economically viable land use system as compared to monocultures.

AGROFORESTRY PRACTICES IN PUNJAB

1) Traditional Agroforestry

The traditional agroforestry has been in practice in the state since ages. Farmers retain local trees like *dek, kikar, dhak, mango, shisham, siris, jand,* etc. and local bushes for their domestic use. But, the traditional agroforestry was greatly affected by the consolidation of land holdings after independence, availability of irrigation facilities and dominance of wheat and paddy cultivation as a result of green revolution. Consequently, the primitive extensive agriculture dominated by silvipastoral systems was replaced by highly mechanized irrigated intensive farming. The tree density on farm lands went on diminishing. The Punjab state has been divided into three agroclimatic zones on the basis of soil and water management programmes (sub-montane foot hill zone, central zone and semi-arid zone) and agroforestry has its own role to play in all the agroclimatic regions. The important traditional agroforestry systems prevalent in these zones are described as:

A) Sub-montane foot hill zone (Kandi Area)

The zone is located in the foothills of the Shiwalik range and extends from Derabassi block of Patiala district to Dhar block of Gurdaspur district, covering Ropar, Nawanshahar, Fatehgarh Sahib, Hoshiarpur and Gurdaspur districts. In these areas, soil erosion due to flash floods is very common. It covers 9.5 per cent of the total area of the state. The lands are uneven, undulating and heavily eroded. Terracing and leveling of undulating land is necessary to make some headway in developmental activities. The annual rainfall varies from 1000-1400 mm of which, 80 per cent is received during the monsoon season. The land holdings are small and fragmented. The resource poor farmers supplement their low farm income by integrating large number of trees, animal husbandry along with their rainfed crops.

Some of the common trees retained by the farmers are: khair (*Acacia catechu*), kikar (*Acacia nilotica*), phulai (*Acacia modesta*), dhak (*Butea monosperma*), shisham (*Dalbergia sissoo*), ber (*Zizyphus mauritiana*), toot (*Morus alba*), neem (*Azadirachta indica*), mango (*Mangifera indica*), dhau/chhal (*Anogeissus laitfolia*), kachnar (*Bauhinia variegata*), rajain (*Holoptelea integrifolia*), bel (*Aegle marmelos*), amla (*Emblica officinalis*), etc. These trees provide diversification in agriculture and income and also fodder and fuel wood during the lean periods. Other species which can hold promise in this region are siris (*Albizzia lebbek*), beul (*Grewia optiva*), sohanjana (*Moringa oleifera*), subabul (*Leucaena leucocephala*), toon (*Toona ciliata*), khirak (*Celtis australis*), Dek (*Melia azedarach*), harar (*Terminalia chebula*), bahera (*Terminalia belerica*), etc. A number of shrub and herb species are found in the region, which could be made use off in the agroforestry component. Prominent agroforestry systems of this region are:

i) *Silvipastoral/ tree fodder production :* The foothill farmers keep large number of animal in the rainfed farming systems. The milch cattle require green fodder round the year. But, there is hardly any area earmarked for fodder cultivation. A field survey conducted in the selected watersheds of the Kandi area revealed an overall deficit of fodder to the extent of 40.4 per cent of the requirement. The fodder requirements during the rainy season are partly met out from the crop residues of sorghum, pearl millet and cluster beans. Some of the requirements are also met out from culled out plants from maize fields, weeds and green grass from agricultural fields. The acute shortage of green fodder is, however, experienced during the winter season. In areas close to forests, some of the green fodder requirement is met out by lopping the branches of forest trees like *Butea monosperma, Grewia optiva, Anogeissus laitfolia,* etc.

   The fodder trees like beul (*Grewia optiva*), bahera (*Terminalia belerica*), kachnar (*Bauhinia variegata*) dhak (*Butea monosperma*), dhau/chhal (*Anogeissus latifolia*), khirak (*Celtis australis*) for feeding cattle and jhar ber (*Zizyphus nummularia*), ber (*Zizyphus mauritiana*) for feeding goats, are raised on the field bunds and their branches are lopped during winter to get nutritious leaf fodder. The lopping reduces shade to the companion crops during winter. The green foliages of the trees are rich in nutritive values and when mixed with inferior grade roughages, the overall quality of forage is improved (Randhawa, 1977; Kapoor, 1978).

ii) *Shisham (Dalbergia sissoo) and babool (Acacia nilotica) based agroforestry :* The two important native tree species namely, shisham (*Dalbergia sissoo*) and babool (*Acacia nilotica*) are hardy and growing everywhere in and around the agricultural fields. Because of their ability to fix atmospheric nitrogen, these are commonly found both on arable and non-arable lands of the region. The farmers of this region keep variable number of trees on their farm lands depending upon the availability and quality of land (Dhanda, 2003). The frequency of shisham trees is comparatively low to that of kikar. Under irrigated lands, the number of kikar trees retained on field bunds vary from 5 to 10 per ha. Under terraced agricultural areas, having no irrigation facilities, about 35-50 kikar trees are generally seen on the field bunds of each hectare. But in case of coarse textured undulating lands with uncertain prospects of getting sustainable crops, 50-100 trees per ha are retained on the fields. This fact has also been supported by another study (Anonymous 1991), which has revealed that the relative frequency of kikar trees was 87.6 per cent under rainfed as compared to 20.7 per cent under irrigated conditions.

A study conducted in the village Rail Majra of district Hoshiarpur during 1990-91, has found that out of 1936 trees growing on 50.6 ha land of 58 farmers, 1622 (82.8 %) were of *Acacia nilotica*, 79 (4.8 %) of *Dalbergia sissoo* and remaining of miscellaneous species (Samra *et al.*, 1991). The data indicated that *A. nilotica* is the most preferred tree species in this zone. In some locations, *A. catechu* (khair) and *A. modesta* (phulai) trees are also raised along the field boundaries of gravelly lands under undulating terrain and grasses like; *Saccharum munja* (locally known as sar, sarkanda, munj or kanna), *Saccharum spontaneum* (locally known as kans, kahi or kan) and *Eulaliopsis binnata* (bhabber grass) come up naturally along with the trees. These grasses have been in great demand in this zone for their multiple uses.

iii) *Fruit tree based agroforestry (mango and citrus)* : The district of Hoshiarpur and few pockets of Gurdaspur and Nawanshahar are known for different kinds of citruses and desi mango varieties. Fruits of both the tree crops are preferred for making pickle. During initial years of growth, when tree canopy is not fully developed, seasonal field crops are usually raised in association with young trees. As the trees gain growth, the cultivation of crops becomes difficult, because of dense shade. Since, open grazing is a serious problem in this zone; farmers have been protecting their orchards by raising live-hedges of *Carissa carandas, Agave cantala, Euphorbia helioscopia* and *Ipomea* sp., and *Jatropha curcas* as live posts.

B) Central zone

The central zone comprises of the districts; Patiala, Ludhiana, Jalandhar, Kapurthala and Amritsar. Here, the land is level and there are extensive urbanization, industrialization and agricultural activities. This zone is intensively cultivated because of fertile soil and availability of plenty of irrigation water where, farmers do not prefer to grow trees on the cultivated fields. The major rice-wheat crop rotation of the region has become non-sustainable and much emphasis is being given to diversify this rotation. The traditional tree species which are growing or can be grown with agricultural crops in this zone are toot, shisham, dek, etc. and fruit plants like lemon (*Citrus aurantifolia*), amrud (*Psidium guajava*), ber *(Zizyphus jujuba)*, mango *(Mangifera indica)* and jamun (*Syzygium cumini*). But these traditional tree species have been now dominated by two fast growing exotics i.e. *Eucalyptus tereticornis* and *Populus deltoides.*

C) South-western zone

The south-west zone comprises of the districts; Sangrur, Bathinda, Mansa, Mukatsar, Moga, Faridkot and Ferozepur. Sand dunes are common in the southern most area, adjacent to Rajasthan and are being reclaimed gradually to make the land suitable for agricultural purposes. Sand scraping is also common. Since, the underground water is brackish, the soils are getting relatively sodic as a result, and such areas are going out of cultivations. The forest species found on the field boundaries/scattered in cultivated land in this zone are *khejri, kikar, neem, jand, shisham*, etc. Very few systematic agroforestry systems are available in this zone. Because of the failure of cotton crop, farmers are shifting towards agroforestry, where the water logging and salt problems are common. Agroforestry models based on fruit and forest tree species will find its place in this zone.

Though the traditional agroforestry has undergone various changes depending upon the locality and socio-economic conditions of the farmers, but the native tree species like kikar (*Acacia nilotica*), jand (*Prosopis cineraria*), shisham (*Dalbergia sissoo*), toot (*Morus alba*), neem (*Azadirachta indica*), jamun (*Syzygium cumini*), dek (*Melia azedarach*), mango (*Mangifera indica*), beul (*Grewia optiva*), khair (*Acacia catechu*), etc. have still survived as scattered trees or as boundary plantations in different zones of Punjab because of their multiple uses.

2) Commercial Agroforestry Practices

Punjab farmers are now shifting to commercial tree farming along with their agricultural crops on their farms to meet the demand of industries for pulpwood, plywood, matchwood, packing cases and other diversified needs of the industries. Amongst the agroforestry models adopted, the predominant tree species in organized planting programmes are *Eucalyptus tereticornis* and *Populus deltoides.* The 'miracle tree' i.e *Leucuena leucocephala* has great potential in the Shiwalik hills of Punjab. Dek (*Melia azedarach)* is another potential agroforestry tree species, which finds its suitability under agroforestry.

A) Eucalyptus based agroforestry

This is the main agrisilviculture system, which is dominating in paddy-wheat cropping system mainly under the irrigated environments of the central Punjab. Initially in the early sixties, very few farmers planted eucalypt trees on the farm and field-boundaries and along farm approach paths. After 15 to 20 years, they had earned handsome return (Rs 150-300 per tree, at that time) from these single row trees. This encouraged them and many others joined to plant eucalypt trees on a much wider scale. Around orchards, the farmers were planting two or three rows of tall trees of eucalypts along with mulberry or other trees as windbreak.

The Forest Department played a pivotal role in supplying eucalypt seedlings on a nominal price to the farmers. Because of large scale planting of *Eucalyptus* by farmers during late seventies to mid eighties, the Punjab state had become surplus in eucalypt timber (Dhanda 1989). By the time these tree plantations were ready for felling, there was a slump in prices for farm grown eucalypt timber. The farmers were disappointed and started uprooting their eucalypt plantations, as no large-scale consuming unit like paper or pulp mill was established in the State. But, interestingly, the Punjab state started exporting farm grown eucalypt timber to neighboring Haryana state as poles and pulpwood and as wooden crates for apples to Himachal Pradesh and J & K (Dhanda 1989).

However, eucalypt which was earlier rejected by the farmers because of ecological and economic constraints has got its momentum again among farmers of the state because of better economic returns. Farmers are shifting towards this species from poplar due to present mono-utility and downfall of prices of poplar wood in Punjab, and availability of site specific clones of eucalypt. Presently, the farmers are growing regular agricultural crops in their fields with eucalypt trees on border rows.

B) Poplar based agroforestry

In the mid eighties, many farmers discouraged with *Eucalyptus* experience, adopted another fast growing trees species i.e. poplar on their agricultural landscape. These days, poplar (*Populus deltoides*) farming has become very popular among the farmers of central zone, because of favorable characteristics of this fast growing tree species for integration with the agricultural crops such as winter deciduous, easy decomposable leaf litter, pruning tolerant nature and high economic returns of the species. However, degree of awareness, attitude and efforts are important factors in adopting poplar based agroforestry. Farmers with the good awareness, positive attitude, and potential for making good efforts and to overcome the constraints, have successfully adopted poplar based agroforestry in the central zone. Most of the farmers are cultivating agricultural crops including fodder crops and fruit trees under poplar. The popularity of poplar can be mirrored through the fact that presently, there are approximately 95 plywood industries in Punjab, which are solely operating on poplar wood and their wood requirement is yet not fully met.

Poplar based agroforestry has been accepted by the farmers in the state as an economically and ecologically viable alternative land-use system particularly under complete irrigated conditions (poplar based agroforestry is one of the alternatives to diversify the rice-wheat rotation system as rice can not be grown with the poplar). Poplar grown along with agricultural crops has a very high productivity as compared to pure poplar plantations, because of response of trees to fertilizers, irrigation and other management practices applied to intercrops. The poplar trees planted at a spacing of 5 x 4m are being harvested at 6-7 year rotation, the farmers were earning about on an average of Rs 35,000-40,000 per acre per annum from the poplar only and intercropping as a bonus. The cultivation of all agricultural crops is possible up to 2-3 years of age including sugarcane during both the seasons (kharif and rabi), but after 3 years age, only fodder crops as intercrop can be grown during kharif due to heavy shade effect. However, in rabi season, wheat, gobhi sarson, oats, berseem, etc. are being grown during the whole rotation period of poplar.

Presently, the earning of farmers has declined due to the non- availability of proper marketing structure and exploitation of farmers by the traders and middle men (locally known as *Arhtiyas*). But, it is definite that prices of poplar wood will rise due to short supply of raw material to the industries and farmers are getting aware of the situation with respect to the marketing of their poplar wood.

C) Leucaena based agroforestry

Considering the shortage of fuel wood and fodder especially in Kandi area, *Leucaena leucocephala* has a great potential for meetings these demands and thus can contribute to the sustainability and profitability of local agricultural practices. An innovative silvihortiagriculture practice was experimented upon for the rainfed Shivalik hills. An irrigated class II and plot 0.11 ha (36 x 32 m) was planted with *Leucaena leucocephala* (one peripheral row for fuel wood and fodder (130 plants), and one mid-plot row (34 plants) for organic mulch, 34 plants of lemon (*Citrus aurantifolia* at 3.5 m spacing as second ring after *Leucaena* inside the peripheral ring and 320 plants of papaya (*Carica papaya*) as the middle rows, and turmeric (*Curcuma domestica*) and okra (*Abelmoschus esculentus*) as under storey. The trees and crops received four to five irrigations each of 10 cm per year, whenever there was a moisture stress. The soil and moisture were conserved by mulching with the leaves and twigs of *Leucaena* (5 tonnes ha^{-1}) and other plants. This silvohortiagriculture practice provided an average net income of Rs 17,066 per ha over a period of four years as compared to an average net income of Rs 7,852 cropping alone. The former was 117 per cent more than the latter (Grewal *et al.*, 1992).

The economics of these practices at Chandigarh reported by Grewal *et al.* (1992), indicated that *Leucaena* grown alone at spacing of 375 x 25 cm was significantly less economical than that grown at a spacing of 75 x 25 cm. Among the three agricultural crops, black gram grown alone was the least economical. The net cash returns from the three combinations of *Leucaena* with crops, and the pure *Leucaena* at spacing of 75 m x 25 cm, were comparable and not significantly different.

D) *Melia azedarach* based agroforestry

Melia azedarach based agroforestry system is an other potential agroforestry system which has great acceptability among farmers of the state because of its diversified uses, fast growth, compatibility with agricultural crops (being winter deciduous) and economic returns comparable with eucalypt. Presently, the wood of dek has dominated the furniture market of the Punjab and plantations of this species are coming up on larger areas at a faster rate. *Melia azedarach* hold great promise in diversification of dominated poplar and eucalypt based agroforestry systems in Punjab in the years to come.

E) Miscellaneous

In addition to above mentioned tree species, some other potential tree species like *Jatropha curcas, Anthocephalus kadamba, Gmelina arborea, Tectona grandis* and bamboos hold good promise in the commercialization of agroforestry in the state of Punjab. The farmers have already started planting these tree species on their agricultural fields in the form of boundary and block plantations with the assumption of getting high productivity under such well managed production systems of tree-crop interface.

ROLE OF PUNJAB AGRICULTURAL UNIVERSITY IN THE FIELD OF AGROFORESTRY

The research on agroforestry in the Punjab Agricultural University started with the establishment of the Department of Forestry and Natural Resources under the College of Agriculture in the year 1978. But, research on this important discipline in the Department was actually strengthened with the allocation of a coordinating centre of All India Coordinated Research Project on Agroforestry in 1983. In the span of about 20 years research, Department has evaluated and screened various tree-crop combinations of sugarcane, cereals, fodder crops, pulses, vegetables, flowers and medicinal plants with eucalypts and poplar in the central zone and with kikar in the sub- montane zone of the Punjab. In addition to this, the different management and cultural practices of trees and crops have been standardized for optimizing the

Table 7.1. Forest area and cover in Punjab (in 2001)

Recorded Forest Area (km²)	
Reserved Forest (RF)	44
Protected Forest (PF)	1,112
Un classed Forest (UF)	1,903
Total	3,059
Of the State's Geographic Area	6.1%
Of the Country's Forest Area	0.4%
Forest Cover (km²)	
Dense Forest	1,549
Open Forest	883
Total	2,432
Of the State's Geographic Area	4.8%
Of the Country's Forest Area	0.4%
Tree Cover (km²)	
Culturable Non-Forest Area (CNFA)	45,204
Number of trees per ha of CNFA	15.6
Tree Cover	1,634
Of the State's Geographic Area	3.2%
Of the Culturable Non-Forest Area	3.6%
Forest and Tree Cover (km²)	
Total Forest and Tree Cover	4,066
Of the State's Geographic Area	8.1%
Of the Country's Forest and Tree Cover	0.5%
Per capita Forest and Tree Cover	*0.02 ha (Lowest in the country)*

Source: Forest Survey of India-2001

Table 7.2 : Agricultural Statistics of Punjab since 1995-96

a) Agricultural Production (Lac ton)

	1995-96	1996-97	1997-98	1998-99	1999-2000	2000-01	2001-02	2002-03
Cereals	217.23	214.87	211.05	226.44	251.66	252.79	248.67	234.58
Pulses	0.84	0.78	0.56	0.47	0.41	0.39	0.31	0.31

Contd...

Table 7.2. Contd...

Food grains	218.07	215.65	211.61	226.91	252.07	253.18	248.98	234.89
Oilseeds	1.88	3.36	1.45	1.67	1.04	0.88	0.84	0.86

b) Irrigated Area (2001-02)

Canals	:	1148 thousand hectares
Tube wells	:	2880 thousand hectares
Total irrigated area	:	4035 thousand hectares
Percentage irrigated area	:	95%

c) Farm Machinery (2001-02)

Tractors	:	4, 10,000
Tillers	:	3, 00,000
Combines	:	5,300
Tube wells	:	11, 50,000
Electric	:	8, 59,000
Diesel	:	2, 91,000

d) Animal Husbandry (1997)

Livestock	:	9857 thousand
Poultry birds	:	11457 thousand (2001-2002)
Milk production	:	8190 thousand tones
Per capita availability of milk per day	:	890 grams

Source: Anonymous-2005

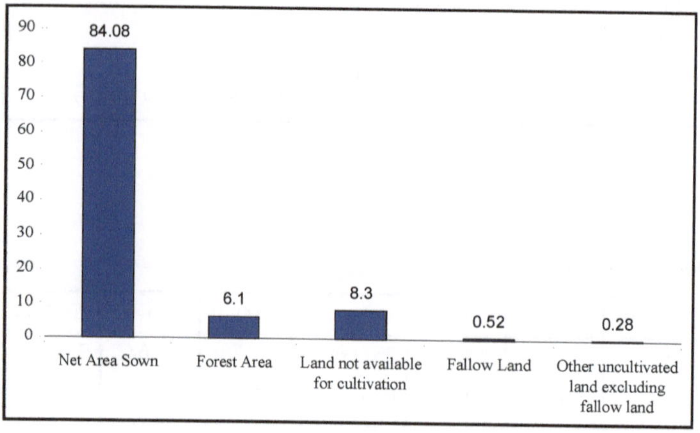

Source : Anonymous 2002.

Fig. 7.1. Land utilization pattern in Punjab 2001-02 (%)

Table 7.3. Land utilization pattern in Punjab since 1970

Year	Geographical area	Total area according to village papers	Forests	Barren and un-culturable land	Land put to non-agricultural use	Total (4 +5)	Culturable based	Permanent pastures and other grazing lands	Land under misc. tree crops and groves not included in net area sown	Total (7+8+9)	Current fellow	Fellow land other than current fellow	Total (11+12)	Net Area sown	Net Area sown as %age to total area	Area sown more than once	Total cropped area (14+16)
0	1	2	3	4	5	6	7	8	9	10	11	12	13	14	15	16	17
1970-71	5036	5031	123	208	416	624	83	5	4	92	139	-	139	4053	81	1625	5678
1980-81	5036	5033	216	98	436	532	41	4	4	49	45	-	45	4191	83	2572	6773
1990-91	5036	5033	222	83	343	436	35	10	12	57	82	A	110	4218	84	3284	7502
1998-99	5036	5033	305	55	326	381	25	7	9	41	38	28	41	4264	85	3681	7945
1999-00	5036	5033	281	46	394	440	20	6	5	31	35	3	38	4243	84	3604	7847
2000-01	5036	5033	279	49	384	443	15	4	5	23	31	3	34	4264	85	3671	7935
2001-02	5036	5033	281	45	409	444	05	3	6	14	25	1	26	4268	85	3673	7941

Cropping Intensity: 185%

Source: Anonymous 2002.

Table 7.4. Contribution of rice and wheat to central pool of food reserve (%) since 1970 and number of operational holdings.

Year	Rice	Wheat
1980-81	45	73
1990-91	41	61
1991-92	46	72
1992-93	42	70.4
1993-94	40	51
1994-95	44	61.4
1995-96	35	59.2
1996-87	35	69
1997-98	42	64
1998-99	39	49
1999-2000	42	55
2000-2001	36	58
2001-2002	33	51
2002-2003	49	52
Number of operational holdings (1997)		
Number (000)	Area (000 ha)	
Marginal (< 1 ha)	144	85
Small (1-2 ha)	192	260
Semi-medium (2-4 ha)	323	863
Medium (4-10 ha)	301	1756
Large (> 10 ha)	78	1168
Total	1038	4132

Source: Anonymous-2005

Table 7.5. Major problems and important traditional tree species of different zones of Punjab.

a) Major zones of Punjab and their problems

Zone/Situation	Sub-montane zone	Central plains	South-western zone
Agro-climatic regions	I & II	III	IV & V
Area (m ha)	0.46	3.16	1.41
Rainfall (mm)	1000-1400	700-1000	<700

Contd...

Table 7.5. Contd...

Major problems	Soil erosion, poor soil fertility, lack of irrigation water	Lowering of water table, nutrient deficiencies	Wind erosion, poor quality underground water, inadequate canal irrigation
b) List of important traditional trees			
Agricultural fields	Kikar, dhak, mango, khair, shisham, beul	Shisham, eucalypts, poplar, toot, dek	Kikar, shisham, jand
Plantation or natural forests	Kikar, dhak, mango, khair, shisham, chhal	Eucalypts, arjun, ber poplar, karer, dek	Phulai, kikar, jand
Home compounds	Dek, kikar, beul, harar toot, bahera, kachnar	Shisham, toot, siris, jamun	Neem, kikar, shisham,
Near ponds, nullahs, rivers	Jamun, mango, bamboo jamoah, khajur, harar	Shisham, siris, toot, khajur, arjun	Shisham, siris, lasoora, neem, kikar
Roadsides, canals, railway lines	Mango, eucalypts, kikar,	Shisham, eucalypts, dek	Kikar, eucalypts, shisham, jand
Community lands like rakhs, birs, worship places	Kikar, khair, phulai, ber, khajur, baur, pipal, mango	Neem, siris, ber, pipal, baur, shisham	Kikar, ber, jand, siris neem

Source : Thind and Dhillon (1994), Sharma and Bir (1978) and Parker (1924)

productivity of different agroforestry models. Based on the findings of the research on agroforestry, farmers of Punjab have adopted and/or adopting the technology developed by the Department.

STRATEGIES FOR PROMOTION OF AGROFORESTRY IN PUNJAB

Agroforestry is an integral rural development approach and to promote this venture among farmers in the state, some essential issues must be addressed with technical descriptions and some hints and advice for the organization of promotion activities:

- Popularization of agroforestry for its diversified sustainable uses.

- Establishment of demonstration blocks/garden/parks of agroforestry to create awareness.
- Availability of quality planting material from certified agencies.
- Involvement of government or corporate sector in regulating marketing of wood and wood based products or minimum support price at par with agricultural crops in the State.
- Loan facilities for establishment of wood based industries.
- Agroforestry know-how-center to link farmers, NGOs, rural co-operatives, government officials and other interested parties.
- Since agroforestry as a science is just two to three decades old, therefore, to promote the research and extension activities being carried out by various institutions/agencies on agroforestry in the state, the financial support from different funding agencies like State Department of ; Horticulture, agriculture, Forest, Industry; Indian Council of Agricultural Research; Indian Council of Forestry Research and Education; Department of Science and Technology; Central Scientific and Industrial Research; NABARD; State Co-operative Banks, etc. is the ultimate need to address important issues related to agroforestry in the state in the future.

REFERENCES

Anonymous (1991). *Annual Report* 1991-92 of Zonal Research Station for Kandi Area, Ballowal Saunkhri, PAU, Ludhiana.

Anonymous (2002). Land utilization in Punjab. *Statistical Abstracts,* Punjab. pp. 158-159.

Anonymous (2005). *Punjab Agricultural University Handbook,* P.A.U. Ludhiana, Punjab. pp. 6-8.

Aulakh, K S (2005). *The report of the Vice Chancellor presented in Annual Convocation* of Punjab Agricultural University Ludhiana, Punjab on April 15, 2005. 24 p.

Dhanda, R. S. (1989). From a deficit state to a surplus state in *Eucalyptus* timber, Punjab (India). *In: Proc. International Conference on Fast Growing Trees and Nitrogen Fixing Trees,* Marburg West Germany. October 8-12, 1989.

Dhanda, R. S. (2003). Agroforestry in irrigated agro-ecosystems in India. *In: Proc. National Symposium on Agroforestry and Sustainable Production,* IGFRI, Jhansi. November 7-9, 2003.

Forest Survey of India (2001). *State of Forest Report.* Ministry of Environment and Forests, Govt. of India, Dehra Dun, 130 p.

Grewal, S. S., Mittal, S. P., Dayal, S. and Agnihotri, Y. (1992). Agroforestry systems for soil and water conservation and sustainable production from foothill areas of north India. *Agroforestry Systems* 17: 183-91.

Johl, S S (2005). *The Convocation Address of College of Agriculture,* Punjab Agricultural University, Ludhiana, Punjab on January 19, 2005. 8 p.

Kapoor, S. K. (1978). Silviculture in Indus Basin. Proceedings of National Symposium on Land and Water Management in Indus Basin (India). *Indian Eco Soc* 11: 358-362.

Parker, R. N. (1924) . "*A forest flora for the Punjab with Hazara and Delhi*". Govt. Printing Press, Lahore, Punjab.

Randhawa, M. S. (1977). Farm Forestry in Punjab. *Indian Fmg* 26 (11) : 75-79.

Samra, J. S., Grewal, S. S., Bansal, R. C., Mittal, S. P., Agnihotri, Y., Goel, P. K., Sastri, G. and Gurmel Singh (1991). *Management plan for environment rehabilitation of degraded Shiwaliks through watershed management* CSWCR&TI, Chandigarh.

Sharma, M. and Bir, S. S. (1978). *Flora of Patiala.* Punjab Univ. Press, Patiala. 257 p.

Thind, H. S. and Dhillon, M. S. (1994). Degraded lands of Punjab and their development through agroforestry. *In: Agroforestry Systems for Degraded Lands,* Vol. 1(Editors: Punjab Singh, P S Pathak and M M Roy) pp 1-12.

• • •

8

Agroforestry Systems and Practices in Haryana

D. P. S. NANDAL AND K. S. BANGARWA
Department of Forestry, CCS Haryana Agricultural University
Hisar - 125 004, Haryana

ABSTRACT

Haryana being predominantly agricultural state had only 2.86 per cent recorded forests at the time of its inception. Agroforestry has been an age old practice in Haryana as in other parts of India. It has existed in various forms through the time immemorial. Farm boundary plantation/natural regeneration and scattered trees in the farmlands were the main agroforestry interventions in Haryana till the seventies. To counter the environmental degradation and to provide fuel, fodder and timber different tree-crop combinations were tried for accessing their compatibility. As a result number of agroforestry systems had been recommended and are discussed in this paper. The paper also include methods of managing the perennial component of agroforestry system. Economics of some of the successful systems has also been discussed.

INTRODUCTION

Haryana state lies in the Indo-gangetic plains of North-west India. It has 44,212 sq.km of land and 90 per cent of this is in plains and the remaining area is in Shiwalik and Aravalli hills. Yamuna plains in the North and sandy plains in the South and West respectively have 50 per cent and 40 per cent of above land. More than 80 percent of the land is under agriculture. Canal and tube well irrigation is available for 65 per cent of arable land. Two crops per annum are a routine practice in Haryana. The climate of Haryana is arid, semi-arid and dry sub-humid. It is sub-tropical, continental and monsoonic type. The range of temperature daily as well as annual is extreme. It varies from 1°C in winter to 47°C in summer. Mean annual average temperature varies from 23°C to 26°C. May and June are the hottest months while

December and January are the coldest ones. The normal annual rainfall ranges from 300 mm in the southwestern parts of Bhiwani, Hisar and Sirsa districts to about 1500 mm in the northeastern hilly tract of Ambala district. The mean daily pan evaporation ranges from 2.4 mm in December and January to 13.4 mm in May and June.

The soils of Haryana have been classified into 5 great groups such as reddish chestnut, tropical arid brown, arid brown, sierozem and desert soils. Major problems are wind and water erosion, salinity and alkalinity, water logging and floods. About 65 per cent of under ground water in the state is brackish/sodic, out of which about 55 per cent can not be used for agriculture. The soils are low in N, low to medium in P and medium to high in K. About 74 per cent soils are deficit in Zn.

Haryana being predominantly agricultural state had only 2.86 per cent recorded forests at the time of its inception. At that time, agroforestry in the state was limited to naturally grown few scattered trees of *Prosopis cineraria, Dalbergia sissoo, Acacia nilotica, Azadirachta indica*. These trees on maturity served as security in drought years and crop failures. With the implementation of social forestry project during the seventies, massive plantations were done on panchyat and farmlands. Due to large-scale plantations of eucalypts in the 1970s and of poplar in 1990s, the forest/tree cover has increased to 7.4 per cent of geographical area of the state (Anonymous, 2001). However, agroforestry in the state received severe blow first after the crash of eucalypts wood prices in 1980s and now due to the rock bottom prices of poplar wood since 2003.

AGROFORESTRY SYSTEMS IN HARYANA

Agroforestry has been an age old practice in Haryana as in other parts of India. It has existed in various forms through the time immemorial (Pathak and Pateria, 1999). Farm boundary plantation/natural regeneration and scattered trees in the farmlands were the main agroforestry interventions in Haryana till the seventies. Organized agroforestry in a scientific way was initiated in early seventies with the start of social forestry programme. Since then various agroforestry practices being followed in Haryana are described below:-

I. Trees on farm boundaries/irrigation channels

This is the most commonly followed practice with all categories of farmers. It was earlier adopted for eucalypts and now a days with poplar. In Yamunanagar district, which is the bowl of poplar production in Haryana,

18.8 per cent of the cultivated area is under boundary plantation of poplar and only 4.8 per cent under block plantation of poplar. Under this practice, trees are planted on field boundaries or on along irrigation channels either on one side or on both sides of the channel. The farmers plant trees at very close spacing and high tree density has been one of the main reasons for failure of eucalypts in Haryana. Now a days also farmers plant poplar on field boundaries or irrigation channels at very close spacing of about 1 m and it results in very poor growth of poplar trees. Poplar or eucalypts planted on field boundaries have been found to affect yield of adjoining crops to varying distances from the tree line depending upon age of trees, tree density, tree species, the associated crop and the management of trees.

At CCS Haryana Agricultural University, Hisar Research Farm (Singh *et al.*, 2003) observed decrease in grain yield of wheat to the extent of 10.2 per cent even at 12-18 m distance from the tree line of N-S planted row of 12 years old eucalypts due to decreased availability of photosynthetically active radiation. At the same research station, Malik and Sharma (1990) have reported 47 and 34 per cent reduction in the yield of mustard and wheat, respectively on both sides of E-W tree line of 3½ years old eucalypts. However, at a farmers field near Hisar, Kumar *et al.* (1998) observed the adverse effect of 8 years old eucalypts (E-W) and 7 years old poplar single row (N-S) boundary plantation only up to 11 m from tree line.

II. Parkland system

It is the oldest practice of agroforestry followed by farmers throughout Haryana. However, with the decrease in land holdings, scattered trees have been left in the fields only in southern Haryana where irrigation facilities are very poor. In other parts of Haryana where canal or tube well irrigation facilities are available practice of naturally grown trees inside the field is rare. However, in southern parts of Haryana particulars in Bhiwani and Mohindergarh districts, the density of scattered trees mainly of *P. cineraria* vary from 5-15 trees/acre at present. Earlier this density used to be very high, however, it is continuously decreasing due to cutting down of these trees during the drought years, introduction of mechanical farming and lopping of *Prosopis cineraria* every year. The presence of scattered trees have both harmful and beneficial effect on the associated crops except that *Prosopis cineraria* has been found to enhance the growth of adjacent crops in parkland system (Arya *et al.*, 1991 b). Puri and Bangarwa (1992) while conducting studies in the semi-arid regions of Haryana to determine the effect of scattered trees of *Azadirachta indica, Prosopis cineraria, Dalbergia sissoo* and *Acacia nilotica* on the yield of irrigated wheat at different distances found no significant effect of *Azadirachta indica* and *Prosopis cineraria* on the yield of wheat, however, wheat yield decreased significantly in the vicinity of *Dalbergia sissoo* and *Acacia nilotica* (Table 8.1).

Prosopis cineraria (in Indian desert known as 'Khejri' or 'Jandi') occurs naturally in the dry and arid regions (south-west Haryana). Khejri is extremely drought tolerant, growing in areas with as little as 75 mm annual rainfall, generally 150-400 mm with dry seasons of eight months or more. Khejri is much valued as a fodder tree. Green leaves along with tender branches provide fodder during November-December. Leaves contain 13.8 per cent crude protein, 20 per cent crude fibre and 18 per cent calcium. Pods also provide a good fodder, containing a dry sweet pulp. Pods are eaten as a vegetable in human diet. Green pods called 'sangar' are boiled and dried for human consumption. The flowers during March-April are valuable for honey production. The bark can be used in leather tanning and yields an edible gum. Bark and flowers are used for medicine purpose. The wood of khejri is used for agricultural implements, house buildings, posts, carts, tools, handles and boat frames. Khejri is the excellent source of firewood for cooking and domestic heating. Khejri is considered as best species for soil improvement and sand dune stabilization. As the khejri tree has deep tap root system, therefore, trees are not expected to compete for moisture or nutrients with crops grown close to trunk.

Maximum yields of fodder are obtained when trees are pollarded on a three-year rotation. Now-a-days farmers lop their trees during winter every year and seed production is very-very low during the year of lopping. Complete lopping of trees annually and consumption of green pods causes acute shortage of seed. Shortage of seed is a major limiting factor for regeneration of this wonder tree. Therefore, farmers are advised to follow a three-year rotation of lopping instead of annual lopping for getting maximum fodder from khejri trees along with maximum seed production for regeneration. For this purpose, farmers are required to make three groups of trees by involving $1/3^{rd}$ number of trees in each group and accordingly trees are to be marked with group number. The trees with group number one, two and three should be lopped completely in November-December of 1st year, 2nd year and 3rd year, respectively. Besides, use of green pods (sangar) should be discouraged. This pattern of lopping will be useful for obtaining maximum fodder along with sufficient amount of seed production for regeneration purpose.

Mahaneem (*Ailanthus excelsa*) is the best tree species for agroforestry plantation in arid and semi-arid region of Haryana. It is also planted on the southwest boundries for protection of agricultural crops from the adverse effect of hot winds. It is an excellent source of nutritive fodder.

III. Agrisilviculture system

Growing of agricultural crops in combination with block plantation of trees mainly poplar is generally prevalent among the farmers of Kurukshetra and Yamunanagar districts of Northern Haryana. In these districts also only farmers having more than 10 acres of land go for block plantation of poplar or in some cases of eucalypts. On the basis of continuous field experience farmers of these districts have adopted a standard cropping sequence under agroforestry. Table 8.2 presents the cropping sequence adopted by the farmers.

Among both the rainy and winter season crops, wheat crop is cultivated throughout the rotation of poplar sometimes from third year onward and in some cases from first year itself. However, some of the agronomic practices to be followed in cultivation of wheat with poplar are different than monocropping of wheat. Working of the field after application of pre-sowing irrigation is very important factor to achieve good germination of wheat under poplar. The field should be ploughed down when it is slightly wet to ensure enough moisture for germination of wheat. Even a single day delay in ploughing and sowing of wheat can result in very poor germination of wheat. Therefore, one should not assume that in November poplar stop growing and do not extract soil moisture. Wheat in poplar is sown early from last week of October to first week of November so that germination of wheat and shedding of poplar leaves do not coincide. Eating of germinating seeds by squirrel is also a problem in some areas. At CCS Haryana Agricultural University, Hisar, while testing 15 wheat varieties for their performance under different shade levels of poplar it was observed that seeds of *Durum*-wheat variety WH-896 are not eaten by squirrel. Moreover, parrots which are a major pest of agroforestry at the time of wheat maturity also do not damage panicles of this variety probably due to the presence of long awns and taste of its grains. The grain yield of this variety was also at par with WH-542, PBW-343 and WH-711 varieties of wheat found suitable for cultivation with poplar.

How much fertilizer should be applied to wheat intercropped with poplar is also very important? Field studies (Table 8.3) at CCS Haryana Agricultural University, Hisar have revealed that after 4 years of poplar plantation at 5x4 m when the yield of wheat is about one third of the sole wheat crop, the doses of fertilizers should be reduced to half, however, under wider poplar spacing (10 x 2.5 m) where the yield is reduced to half of the monocrop of wheat full doses of fertilizers should be applied.

Wheat seed rate studies conducted in the same plantation revealed that when the shade level in poplar increases to the extent that it allows only 40-

50 per cent of the PAR to reach the ground, the wheat seed rate should be increased by 25 per cent to compensate for reduced tillering under shade.

Eucalyptus, the second most important agroforestry tree species of Haryana has only a few fields under block plantations in Northern parts of Haryana. The main reason for limited plantation of eucalypts in agroforestry is its more adverse effect on the yield of associated crops during both the crop seasons. Whereas good crop yields are realized in poplar based system during the winter season. The performance of wheat, potato, mustard, lentil and berseem was compared under two years old eucalypts planted at 6x2 m spacing. The data (Table 8.4) showed maximum yield reduction in mustard (82.4 %) followed by potato (80.8 %), wheat (63.2 %), berseem (52.1 %) and lentil (16.3 %).

Minimal decrease in Lentil yield might be ascribed to non-lodging of the crop under eucalypts due to reduced plant height.

The second reason for poor ranking of eucalypts by the farmers is its high water requirement thereby lowering the water table. Field studies in progress at CCS Haryana Agricultural University, Hisar have established that in the upper soil surface (0-60 cm) water content before each irrigation in wheat sown under eucalypts was two to three times less than the control. Surprisingly, the water content beyond 80 cm soil depth was higher under eucalypts than open field. It might be attributed to continued upward movement of water through capillary movement on account of continuous transpiration by eucalypts. Similarly, Malik and Sharma (1990) at the same research station have also reported that in 3½ years old *Eucalyptus tereticornis* bund plantation eucalypts was found to extract 5 times more water than mustard from 0-15 cm soil profile. Jeet Ram (2004) has also studied effects of two 18 years old eucalypts hybrid plantations on the shallow ground water table during April to September 2004 at Dhob-Bhali research plot located in Rohtak district of Haryana by installing 50 observation wells inside and outside the plantations. He has reported deeper ground water table underneath the eucalypts plantation. The average draw down of groundwater table underneath the plantation was 1.01 m below the average natural ground water table level (4.15 m).

Except for poplar and eucalypts no other timber or fuelwood tree species is planted in block in agrisilviculture system in Haryana.

IV. Hortiagricultural system

This system is followed by majority of the farmers in Sirsa district in western part of Haryana. Under this system Kinnow (Kinnow hybrid of *Citrus deliciosa* x *C. nobilis*) is planted at 5 x 5 m or 6 x 6 m spacing and crops like cotton, moong bean, sorghum (F) and cowpea (F) during the rainy season and wheat, mustard and gram during the winter season are raised in the interspacing during the first four years of plantation. Thereafter when Kinnow attains sufficient growth only fodder crops of sorghum, maize and cowpea are grown in the interspaces during the rainy season. Field trials on crop-tree compatibility conducted by university scientists at the farmers field in Sirsa district have shown cowpea as the most compatible crop with Kinnow throughout its life cycle. Cotton crop was found to reduce branching of Kinnow after four years. Some additions/deletions are carried out on the tractor for inter culture operations in Kinnow orchard. To avoid damage to the body of the tractor and the driver by the side branches of kinnow, silencer of the tractor is removed and specially built flat iron rods are fixed on both sides of the tractor body to avoid scratches by the braches on the tractor body and the tractor driver.

MANAGEMENT OF TREES FOR AGROFORESTRY

Model tree for agroforestry

Conceptually, a model tree for agroforestry system should be characterized by fast vertical growth, small crown, few branches with smaller angle, loose canopy, self pruning habit, straight and clean bole with aggressive apical dominance, higher proportion of main stem, deep root system and deciduous nature.

Justification

Fast growing trees get quickly established and the returns also come comparatively earlier due to reduction in harvest age. The trees with small crown having a few narrow angled branches attain a compact form. Such trees require little horizontal space and hence more trees per unit area can be accommodated. Consequently, returns per unit area are more. Loose canopy permits light penetration. Under such trees, crop can grow better without much shading effect and no interference from the trees. Trees with self-pruning habit do not need any pruning which otherwise costs. Also, the trees with the self-pruning habit produce longer clean bole and thus have larger proportion of main stem. Straight and clean bole provides quality wood and such bole is easy to handle in processing (Faulkner, 1969; Shelbourne, 1969). Straight growing trees occupy less horizontal space and do not interfere much with the growing crops under the agroforestry systems. Sheikh (1988) also proposed that straight growing trees of *Prosopis juliflora* should be used in plantation under arid and semi-arid areas. In the forest trees, most of the economic product comes from the main stem. Obviously, its higher proportion is desired in the model trees. The trees with smaller crown occupy mainly the vertical space and less horizontal space thus again helping in accommodating more trees per unit area. Root system of the trees for agroforestry must be deep. It is all the more important in the dry areas where there is always moisture stress for the growing crops. The trees with shallow root system like *Acacia* (Toky and Bisht, 1992) compete for moisture and nutrients. It has been practically observed that in the dry areas the crops invariably fail due to the effect of *Acacia* in the vicinity of these trees. Also, the shallow root system interferes in the usual agricultural operations like ploughing, sowing and hoeing of the annual crops. Thus, deep root system is one of the major requirements of the trees for agroforestry system. Deciduous trees, on one hand, add organic matter to the soil and improve its fertility while make light and air available to the growing crop on the other. This is obvious from the nature of poplar where crops like wheat can be raised very successfully even very close to the tree trunk.

STRATEGY

Due to complex genetic make up of trees, there is always genetic variation among the growing trees both within and between provenances. Khosla (1985) advocated for the possibility of genetic improvement of agroforestry trees. Bangarwa (1993) conducted a survey of *Dalbergia sissoo* trees growing in northern India and discussed the possibilities of genetic improvement of this species through plus tree selection. On the basis of progeny testing of plus trees, Bangarwa (1996) and Singh (1994) reported that there is a definite possibility of developing a *shisham* genetic strain that will resemble very closely to the present proposed model. Bangarwa (1993) reported high narrow sense heritability for the traits like straightness, apical dominance and length of clear bole. Obviously, these traits can be fixed through selection of plus trees. Thus, a tree breeder firstly should look for the plus trees which resemble, in most respects, to the proposed model trees and perform selection and progeny testing. After 2 to 3 generations of progeny testing followed by selection, there is a possibility of desired genetic improvement. When the most desired traits like fast vertical growth, straightness and length of clear bole are not available in a single genotype, the breeder has to employ hybridization between different genotypes having complementary features of the model tree. Alternatively, variability can be induced through usual induced mutation breeding.

QUALITY OF PRODUCE IN AGROFORESTRY

Although reduced yields are recorded in association with tree species, however, the quality of crop produce in agroforestry is generally better than sole crop. The quality of some crops raised in association with different spacing of poplar was assessed at CCS Haryana Agricultural University, Hisar (Nandal and Hooda, 2005). Potato yield and tuber size under poplar were reduced, however, there was no greening of potato tubers as generally observed in sole crop of potato. The colour of potato tubers grown under poplar was very attractive and thus fetched good price in the market. The same observation has also been reported by the farmers of northern Haryana where potato is also one of the crops intercropped with poplar. The fodder quality of berseem, sorghum and cowpea grown with poplar has also been assessed at CCS Haryana Agricultural University, Hisar research centre. The data in Table 8.5 clearly indicate that all the fodder crops grown under poplar were more nutritious than control grown fodder. The intake of fodder grown under poplar was also higher due to less fibre content. It has also been observed that fodder crops in association with trees are not grazed by stray cattle as long as these are available in open adjoining fields.

ECONOMICS OF AGROFORESTRY SYSTEMS

The growing of agricultural crops with commercial timber tree species can be one of the promising and viable options to achieve diversification as well as assured income to the farmer. Reports are available, where diversification of the cropping systems, consisting of commercial timber species, have provided higher net returns in comparison to sole crop (Chauhan, 2000, Jain and Singh, 2000). In southern Haryana where underground water is brackish and canal irrigation is not assured, farmers during the drought years meet their household needs by selling *Prosopis cineraria* trees standing in their fields. Therefore, in such areas, agroforestry provides insurance against crop failure due to drought.

The productivity and financial viability of poplar based agroforestry system of different periods representing different rotation cycles as well as peak and bottom prices of poplar wood of northern Haryana worked out by Nandal *et al.* (2005) is presented in Table 8.6. The financial analysis of the agroforestry systems at these locations realized undiscounted net returns of Rs. 112142, 279080 and 54585 per acre over six, eight and five years, respectively. It works out to Rs.18960, 34885 and 10917 per acre per year. Whereas, net returns on the basis of cost of cultivation comes to Rs. 2869/acre/year from paddy-wheat rotation, generally followed in this region. Benefit cost ratio, internal rate of return and net present worth financial parameters at all the locations also point out that even with low prices of wood since 2003, poplar based agroforestry system is more economical than monocropping of agricultural crops.

Table 8.1. Grain yield (g m^{-2}) of wheat at different distances from tree under four tree species.

Distance (m)	Tree species			
	D. sissoo	*A. nilotica*	*P. cineraria*	*A. indica*
1	296.7	161.2	378.5	379.5
3	366.7	185.0	385.0	390.2
5	400.2	251.0	410.2	420.0
Control (with trees)	416.0	412.0	420.0	413.0
LSD	18.3	22.0	20.6	19.3

Puri and Bangarwa, 1992

Table 8.2. Cropping sequence adopted by the farmers

Year/ Age of tree	Crop	Average (q/acre)	yield	Girth of Poplar (inches)
1st year	Planting of poplar in February at a spacing of 15x15 ft or 15x12 ft (Average of 200 trees /acre + Sugarcane crop planting)	200-250		11.0
2nd year	Ratoon of Sugarcane	150-200		17.0
3rd year	*Kharif* – Jowar *Rabi* – Wheat	80-100 15-16		22.0
4th year	*Kharif*- Fallow/Turmeric/ Jowar *Rabi*-Wheat/berseem	10-15/70-80 11-15/170-180		26.0
5th year	*Kharif*-Fallow/Turmeric/Jowar *Rabi*- Wheat/berseem	10-15/70-80 11-12/170-180		29.5
6th year	*Kharif* = Fallow/Turmeric/Jowar *Rabi* = Wheat/Berseem	10-15/80-90 11-12/170-180		32.0

Table 8.3. Effect of different fertilizer levels on grain yield of wheat at different poplar spacing

Fertilizer	Poplar spacing (m)			
	5 x 4	10 x 2.5	Control	Mean
Control	4.8	7.0	22.8	11.5
50% of recommended dose	10.2	15.5	34.3	20.0
75% of recommended dose	12.0	19.4	38.2	23.2
100% of recommended dose	12.2	24.0	44.0	26.7
125% of recommended dose	12.0	25.9	44.2	27.4
Mean	10.2	18.4	36.7	-
CD at 5% for				
Spacing	1.2			
Fertilizer	1.6			
Spacing x Fertilizer	2.8			

(Anonymous, 1999-2000)

Table 8.4. Comparative yield (q/ha) of different test crops in eucalypts based agrisilviculture system

Crop	Yield		Paired t-test
	Agroforestry	Control	value
Wheat (grain)	14.8	40.2	8.2
Potato (tuber)	25.1	131.1	22.1
Mustard (grain)	3.2	18.2	9.0
Lentil (grain)	4.1	4.9	NS
Berseem (fodder)	87.1	182.1	47.6

Kumar and Nandal (2004)

Table 8.5 Chemical composition (% DM basis) of fodder crops intercropped with poplar

Quality parameters	Sorghum		Cowpea		Berseem	
	Poplar	Control	Poplar	Control	Poplar	Control
Crude protein	7.82	7.24	21.44	21.00	18.68	17.90
Ether extract	2.72	2.16	5.46	4.88	4.12	3.98
Total ash	7.18	7.36	9.48	9.12	10.86	10.60
Crude fibre	26.10	29.44	21.40	23.60	23.72	25.86

Anonymous, 1999-2000

Table 8.6 Cash-flow analysis of poplar based agri-silviculture system

Year of plantation	Total cost (Rs./acre)			Total returns (Rs./acre)			Discounted Cost (Rs./acre)			Discounted Return (Rs./acre)		
	Bapada	Dhanokheri	Gudda	Bapada	Dhanokheri	Gudda	Bapada	Dhanokheri	Gudda	Bapada	Dhanokheri	Gudda
I	36654	24760	23450	29640	22140	20000	33319	22507	21316	26943	20125	18180
II	25474	18280	15340	30780	25740	8000	21042	15099	12671	25424	21261	6608
III	21050	18140	9315	15600	21300	8000	15809	13623	6996	11716	15996	6008
IV	21900	18730	9510	14800	8150	6500	14958	12793	6495	10108	5566	4440
V	20200	19120	20000	13000	47660	89700	12181	11529	12060	7839	28739	54089
VI	40400	14200	-	172000	28000	-	22826	8023	-	97180	15820	-
VII	-	13900	-	-	192000	-	-	7131	-	-	98496	-
VIII	-	13780	-	-	75000	-	-	6435	-	-	35025	-
Total	163678	140910	77615	275820	419990	132200	120135	97140	59538	179210	241028	89325

Nandal *et al.*, 2005

		Bapada	Dhanokheri	Gudda
	Undiscounted net income (Rs.)	112142	279080	54585
	Discounted income			
(i)	Net present worth (Rs.)	59075	143888	29787
(ii)	Internal rate of return (%)	78	190	70
(iii)	Benefit cost ratio			

Financial analysis of clonal eucalypts plantation raised in April, 1999 and harvested in March, 2005 on 3.5 acre of a farmer in Kurukshetra district of Haryana has also been worked out by Haryana Forest Department. According to their analysis, net income of Rs.21,343/acre/year was realized from clonal eucalypts plantation (Anonymous, 2005).

Kinnow based Horti-agricultural system is highly remunerative. In general, average gross income of Rs.50,000 to Rs.1,00,000/acre is obtained from kinnow plantation after 6-7 years of plantation depending upon tree density and management. Some of the progressive farmers even earn more than Rs.1.00 lakh from an acre. The net return from the agricultural crops after 4years of plantation is very less because of reduced yields.

REFERENCES

Anonymous (1999-2000). *Annual Progress Report of the All India Coordinated Research Project on Agroforestry*, Department of Forestry, CCS Haryana Agricultural University, Hisar.

Anonymous (2001). State of Forest Report. FSI. MOEF, GOI, Dehradun.

Anonymous (2005). Plantations of clonal eucalypts-success story. *Forest News* 2: 5-6.

Arya, S., Toky, O.P., Bisht, R.P. and Tomer, R. (1991 b). *Khejri*. Intensive Agriculture. pp.248.

Bangarwa, K.S. (1993). Studies on provenance variation, progeny testing and reproductive biology in *Dalbergia sissoo* Roxb. Ph.D. thesis, CCS Haryana Agricultural University, Hisar, India.

Bangarwa, K.S. (1996). *Sissoo Breeding*. Agriculture and Forestry Information Centre. India. pp. 156.

Chauhan, H.S. (2000). Performance of poplar (*Populus deltoides*) based agroforestry system using aromatic crops. *Indian Journal of Agroforestry* 2: 17-21.

Faulkner, R.E. (1969). Some characters of secondary importance to stem straightness in the breeding of conifers. *In: Second World Consult of Forest Tree Breeding* Washington, D.C. 1969. FO-FTB-69/3/2.

Jain, S.K. and Singh, P. (2000). Economic analysis of industrial agroforestry poplar (*Populus deltoides*) in Uttar Pradesh (India). *Agroforestry Systems* , 22: 255-273.

Jeetram (2004). Agroforestry for the reclamation of water logged areas. Proceedings of National workshop on agroforestry organised by Forest Department, Haryana held at Chandigarh during 22-24 November 2004, pp92-93

Khosla, P.K. (1985). Genetic improvement of agroforestry trees. In: *Agroforestry Systems – A new challenge* (P.K.Khosla and S.Puri, eds.), ISTS Solan, H.P. pp. 151.

Kumar, A., Hooda, M.S. and Dhillon, R.S. (1998). Influence of tree windbreaks on the yield of wheat (*T .aestivum*) in semi-arid regions of India. *Indian Journal of Forestry* 21: 208-213.

Kumar, A. and Nandal, DPS (2004).Performance of winter crops under Eucalyptus tereticornis based agrisilviculture system.Indian J.Agroforestry 6:92-93.

Malik, R.S. and Sharma, S.K. (1990). Moisture extraction and crop yield as a function of distance from a row of *Eucalyptus tereticornis*. *Agroforestry Systems* 12: 187-195.

Nandal, D.P.S. and Hooda, M.S. (2005). Production potential of some agricultural crops under different spacings of poplar. *Indian Journal of Agroforestry* (communicated).

Nandal, D.P.S., Mandal, B.S., Kaushik, J.C. and Chauhan, R.S. (2005). Financial analysis of poplar based agroforestry system - case study. *Indian Journal of Forestry* (communicated).

Pathak, P.S. and Pateria, H.M. (1999). Agroforestry in the Indo-Gangetic plains: an analysis. *Indian Journal of Agroforestry* 1: 15-36.

Puri, S. and Bangarwa, K.S. (1992). Effects of Trees on yield of irrigated wheat crop in semi-arid regions. *Agroforestry Systems* 20: 229-241.

Sheikh, M.I. (1988). Straight trees of *Prosopis juliflora* for desert afforestation. *Pakistan J. For.* 38: 119.

Shelbourne, C.J.A. (1969). Breeding for stem straightness in conifers. In: Proc. Second World Consultation of Forest Tree Breeding. Washington, D.C.: FAO-FO-FTB-69-3/4.

Singh, A., Singh, R. and Pannu, R.K. (2003). Effect of eucalypts plantation on moisture extraction pattern, water use and yield of wheat. *Indian Journal of Agroforestry* 5: 45-49.

Singh, J. (1994). *Progeny testing in Dalbergia sissoo Roxb.* M.Sc. Thesis, CCS HAU, Hisar.

Toky, O.P. and Bisht, R.P. (1992). Observations on the rooting patterns of some agroforestry trees in an arid region of north western India. *Agroforestry Systems* 18: 245.

• • •

9

Agroforestry Practices and Systems in North Gujarat

S.N. JAIMINI AND S.B.S TIKKA
S.D. Agricultural University
Sardarkrushinagar- 385 506, Gujarat

ABSTRACT

The western part of India particularly Gujarat and Rajasthan received erratic and scanty rainfall and have sandy to sandy loam soil. In such climatic and edaphic conditions failure of crop is a regular phenomenon. In such situations a cropping system is required which can protect the associated crops from extremes of climatic condition and/or provide alternative/additional income. Agroforestry has such potentials; as a result lot of research had been done on agroforestry. This paper provides some of the agroforestry system and practices, which are suitable for Gujarat. The hurdles in adoption of agroforestry and its future prospects are also provided.

INTRODUCTION

North and North-West Gujarat fall under scarcity zone, where rainfall is inadequate as well as erratic in nature. Sandy loam soils with low moisture retention capacity pose problems of low and unstable crop production and degradation due to soil erosion. Under these situation, adoption of a high yielding, more remunerative, cost effective, compatible and multipurpose system ie Agroforestry is a dire necessity.

Agroforestry is an age-old land use practice which has been practied by the farmers in various forms in different parts of India. In arid and semi-arid areas of Gujarat and Rajasthan tree species like *Khejri* (*Prosopis cineraria*), Ardu (*Ailanthus excelsa*), Neem (*Azadirchta indica*), Shisham (*Dalbergia sisoo*), Anjan (*Hardwickia binata*), and Ber (*Zizyphys spp*) are grown traditionally as scattered trees, that increases soil organic matter and nitrogen content and increase availability of micro nutrients (Zn, Mn and Cu) and moisture in the surface top soil layer.

The adoption of certain MPTS based agroforestry models will not only help in fulfilling vast demand of fast growing population in a sustainable manner but will also ensure environmental stability and economic security to rural people.

AGROFORESTRY SYSTEMS

1. Agrisilviculture system

Khejri (Proposopis cineraria) based model

Khejri (*Prosopis cineraria*) is the most important tree to combine with agricultural crops in the rainfall zone of 250-400 mm. Its suitability in agroforestry system stems from the fact that a well lopped *Khejri* tree recovers its full crown in period of nine months. In Rajasthan, *Khejri* is lopped in September to facilitate better growth of *kharif* crop or lopped when there is scarcity of fodder during summers. In cases, the fields are to be used for *rabi* crops like wheat and mustard, lopping is continued slowly during summers and fast during monsoon. It is a dominant tree species on alluvial and sandy undulating plains of arid desert. The cultivators prefer to retain *Khejri* tree on their field largely due to their strong conviction of better growth underneath these trees. The higher production from crops growing in the vicinity of *Khejri* may be due to higher fertility of the soil and retention of moisture for longer period under arid condition. Aggarwal *et al.* (1993) studied long term effect of *Prosopis cineraria* on the nutrient status and availability from the underneath desert sandy soil in comparison to adjacent open soil. Their study has shown a significant increase in organic matter, total N,P,S, available N,P,K content and 2-3 times higher dry matter yield of pearlmillet in *Khejri* soil over non-*khejri* soil. Kumar *et al.* (1992) also recorded higher yield from pearlmillet, clusterbean and cowpea in association with *Khejri*. Tiwari (1994) studied various agroforestry systems based on *Prosopis cineraria* and their economic returns. He observed a return of Rs.4500/ ha/yr. more over sole crops.

Jaimini and Tikka (1998) studied *Khejri* based agrisilvi culture system during 1995-96 to 1997-98. They have observed higher income from *khejri* based agrisilvi culture system than the sole raising of crops. The income was maximum when clusterbean was raised as intercrop in between *khejri* rows (Table 9.1). They have also observed higher organic carbon and available N, P and K per ha under agrisilvi culture system of agroforestry as compared to soil with no *Khejri* cover.

Ardu *(Ailanthus excelsa)* based model

Ailanthus excelsa has an excellent market in Rajasthan and North Gujarat where its leaves are sold as fodder for goats, sheep and camel. Its timber has great demand in construction work and in plywood industries. The farmers in villages around Mehsana, Kalol, Nandasan and Gandhinagar have, therefore, planted this species on the boundaries/bounds of their farm. The average, additional income received from planting *Ailanthus excelsa* is Rs.7600/ ha/yr (Tiwari, 1994).

Studies conducted on *ardu* based agroforestry during 1999-00 to 2003-2004 at Sardarkrushinagar have also indicated beneficial effect of ardusa on soil health and economic returns (Table 9.2). The raising of arable leguminous crops *viz.*, cowpea, clusterbean and greengram as sole and as intercrops in between tree (528 trees/ha) rows indicated reduction in production under agrisilviculture system as compared to sole, though there was an improvement in soil fertility status under ardu based agroforestry system. When income obtained from leaf fodder and fuel wood on lopping per ha per year during the period of experimentation was considered, the loss caused in the production of arable crops due to shading effect and competition between crops and woody component i.e. *ardu* was not only compensated by trees, it surpassed the monetary returns obtained from sole cropping of clusterbean and cowpea. Besides, above benefit on an average the trees also added 2933 kg/ha/year litter fall to the soil which improved soil fertility status of soil.

The ardu not only provided nutritious off season fodder for sheep, goat and camel and fuel wood on lopping during its life cycle but also enriched and protected underneath soil.

Neem (*Azadirachta indica*) based model

Neem (*Azadirachta indica*) is planted traditionally by the farmers in most parts of Gujarat and North Gujarat in particular under irrigated and rainfed conditions for their requirement of timber, fire wood, fodder etc. Under arid condition, its leaves provide nutritious fodder for camel, sheep and goats during scarcity period. These characteristics and traditionally planting practices of this species by farmers, particularly in North Gujarat have made it an acceptable agroforestry species. Under agroforestry conditions, this species has been raised as boundary plantations, wind breaks around orchards and shelterbelt in arid areas and also as scattered trees in the agricultural fields.

To study the potential role of neem in dryland agriculture, a trial was conducted during 1999-00 to 2003-2004 at Sardarkrushinagar (Table 9.3). The five years old neem tree plantation spaced at 6 x 6m was intercropped with cowpea, cluster bean and green gram during *kharif* seasons of each year. The crops were also raised sole to make the comparison. The grain and fodder yield (kg/ha) obtained from sole crops were higher than the intercrop which indicated adverse effect of neem on intercrops. The monitory returns obtained from intercrops without considering the income from tree component were low as compared to sole crops. When income obtained from leaf and fuel wood on lopping and neem fruit (*nimboli*) per ha/year during the period of experimentation was considered, the loss caused in the production of arable crops due to shading effect and competition between crops and woody component i.e. neem was not only compensated, it surpassed the monetary returns obtained from sole cropping of clusterbean and cowpea. Besides above benefit, on an average neem tree also added 2176 kg/ha/year of litter fall to the soil, which improved the fertility status of the soil. The results obtained by Jaimini and Tikka (2001) have also shown neem as promising MPTS for agroforestry which not only provided nutritious off season fodder for sheep, goats and camel and fuel wood on lopping but also cash from sale of ripe fruit during its life cycle.

Shisham (*Dalbergia sissoo*) based model

Among local MPTS, *Shisham* (*Dalbergia sissoo*) offers possibilities in supplementing a part of nitrogen requirement of the associated crops through enrichement of site, providing fuelwood, off season fodder and small timber to cater to the basic needs of rural people for their sustenance in dry lands. An experiment was planned to study the role of shisham in agroforestry. The *shisham* plantation made at 6m x 3m (528 tree/ha) in the rainy season of 1996 was used to study its effect on intercrops (cowpea, greengram, clusterbean and sesamum) and soil fertility. The *shisham* plantation was intercropped during *Kharif* season of each year of study i.e. 2001-02 to 2003-04.

The grain yield of intercrops was lower than the sole crops, which indicated the adverse effect of *shisham* on intercrops (Table 9.4). The monetary returns obtained from inter and sole crops without considering the income from tree component in the years 2002-03 to 2003-04 and average of three years were low as compared to sole crops but in the year 2001-02 intercrops with tree gave higher returns. When income obtained from leaf fodder on lopping during the period of experimentation was considered the loss caused in the production of arable crops due to shading effect and competition between crops and woody component i.e. shisham was not only compensated by trees, it surpassed the monetary return from sole cropping. On an average there was 96.3 per cent more income from the *shisham* based agrisilviculture system of agroforestry. However the returns were maximum when green gram was taken as intercrop in between s*hisham* rows. Tiwari (1994) has also reported on an average additional income of Rs.7600/ha/yr from *Dalbergia sissoo* based agroforestry.

Anjan (*Hardwickia binata*) based model

The *anjan* plantation made at 6m x 6m (256 trees/ha) at AICRP on Agroforestry Farm SDAU, Sardar Krushinagar in the rainy season 1996 was used to study its effect on intercrops. The *anjan* plantation was intercropped with clusterbean, cowpea, greengram and sesamum during *kharif* season of each year of study i.e. 2001-02 to 2003-04.

The grain and fodder yield of the arable crops were influenced by *anjan* in all the three years of study except grain yield was higher in intercrops with tree in the year 2001-02 (Table 9.5). The grain yield of intercrops were lower than the sole crops, which indicated the adverse effect of *anjan* on intercrops. The monetary returns obtained from inter and sole crop without considering the income from tree component in the years 2001-02 to 2003-04 and average of three years were low as compared to sole crops. When income obtained from leaf fodder and fuelwood on lopping during the period of experimentation was considered the loss caused in the production of arable crops due to shading effect and competition between crops and woody component i.e. *anjan* was compensated and also, it surpassed the monetary return from sole cropping. On an average there was 53.45 percent more income from the anjan based agrisilviculture system of agroforestry. However the return were maximum when green gram was taken as intercrops in between *anjan* rows.

2. Agrihorticulture system

Ber (*Zizyphus mauritiana* lamk.) based model

Ber (*Zizyphus mauritiana* lamk.) is one of the important arid fruits growing all over India upto an elevation of 1000 m. Trials conducted at S.D.Agricultural University, Sardar Krushinagar have shown that ber plantation is profitable even in arable lands of North Gujarat semi-arid regions having low and erratic rainfall.

Ber trees planted at 10m x 6m in 1986 and budded in the year 1987 were used for this study. Based on the equivalent yield, gross and net realization was worked out, The results indicated that the highest gross income of Rs. 24613/ha was realized with ber + green gram closely followed by ber + sorghum (Rs. 24180/ha) which in turn, recorded 19.64 and 18.20 per cent higher gross realization over ber sole system, respectively (Table 9.6). Further, intercropping in ber recorded 14.80 per cent higher monetary returns over Ber fallow system. This clearly shows the profitability and sustainability of the intercropping system. Taking into consideration the equivalent yield, gross and net realization and sustainability of the system over long period under risk prone dryland agriculture, agri horticultural system as a part of alternate land use system seems to be profitable and sustainable.

3. Silvipastoral system

This system is grass based and in essentially for the areas with degraded lands. In this system the grass is intercropped with trees, which provide fuel and fodder. *Prosopis cineraria, Zyziphus rotendifolia, Acacia senegal* and *Ailanthus excelsa* are some of the suitable trees for the system. The promising grasses suitable for low rainfall conditions are *Cenchus ciliaris, Cenchrus setigerus, Lasiurus sindicus, Chrysopogon fylvus, Stylonthes hamata* and *Dichanthium annulatum.*

The studies on silvipastoral (*Khejri* +*Dhaman* grass) system of agroforestry under rainfed condition conducted during 1997-98 to 2000-2001 have exhibited higher fodder below *Khejri* canopy as compared to in between *Khejri* rows i.e.away from canopy and sole raising of dhaman grass (Table 9.7). There was an increase in organic carbon and available N,P,K under tree canopy. Therefore, it may be concluded that *Khejri* based silvipastoral system is more remunerative. Because of non competitive habit, nitrogen fixing nature, early litter decomposition and best compatibility with crops growing in the vicinity of *Khejri* tree, Paroda and Solanki (1991) very righty called this tree as a wonder tree of the Thar desert.

REASONS FOR SLOW ADOPTION OF AGROFORESTRY

In spite of so many advantages of agroforestry, farmers are not adopting agroforestry on a big way. The probable reasons of slow adoption may be:

- Establishment of trees is poor due to harsh climatic condition viz., high wind velocity, extreme temperature and water stress during summer etc.
- Small land holding (in between 2-3 ha) discouraging plantation in the cultivated area.
- Small grazing area causing over grazing and soil erosion.
- Farmers belief that trees dominate the arable crops and compete for water and nutrients and encourages bird damage to nearby crops.
- Lack of market support for fuel wood and other agroforestry products.
- Forest laws prohibiting cutting of trees on private land.

- Poor socio-economic back ground of farmers.
- Stray grazing, lack of irrigation, non availability of inputs at proper time, poor survival of saplings.
- Lack of genetically improved planting material and improved agroforestry technology.

FUTURE PROSPECTS OF AGROFORESTRY

The statistics of wood demand and supply hint not only at the future fuel wood crises but also at a serious ecological disaster that the country is heading for. We cannot avoid this mishap, unless a massive afforestation programme along with encouraging agroforestry is implemented throughout the country.

Presently no systematic agroforestry except agrihorticulture in few pockets of North Gujarat particularly in Mehsana and Gandhinagar districts is being followed. To meet ever increasing demand for fuel wood, timber and raw material for agro-based industries, there is no way out except adopting agroforestry as a land-use system for sustaining agricultural production and productivity. It is expected that availability of assured irrigation and remunerative prices from agroforestry products in future will encourage cultivation of cash crops, fruits and plantation of trees for timber and raw material for various agro-based industries. For instance farmers are showing keen interest in plantation of trees like teak, *nilgiri*, *ardu*, mangium, *babul*, *neem*, etc. initially on farm boundaries. Later on if satisfied in alleys in the field. Thus, agroforestry has bright future in this region.

Table 9.1. Grain fodder yield and monetary return of different field crops with and without association of Khejri (*Prosopis cineraria)*

Treatment	1995-96			1996-97			Pooled monetary return (Rs/ha)
	Grain yield (kg/ha)	Fodder yield (kg/ha)	Monetary return (Rs/ha)	Grain yield (kg/ha)	Fodder yield (kg/ha)	Fodder yield (kg/ha)	
Khejri + Cowpea	214	1095	2672	643	2071	7564	5236
Khejri + Greengram	90	714	1786	345	1190	5026	3971
Khejri + Clusterbean	702	2047	7608	595	3369	8750	8982
Cowpea sole	71	857	1126	429	905	4871	3150
Greengram sole	42	1095	1282	214	654	2962	2126
Clusterbean sole	847	2976	9579	381	2310	5607	8616
S.Em	-	-	548	-	-	495	688
C.D. at 5%	-	-	1613	-	-	1486	1986

Table 9.2. Grain and fodder yield and expected return of different field crops with and without association of *ardu* (*Ailanthus excelsa*) at Sardar krushinagar.

S. No.	Treatment	2001-2002			2002-2003			2003-2004			Total returns without tree (Rs/ha.)	Total income from the tree produce (Rs/ha.)	Av. Returns from the system
		Grain yield kg/ha.	Fodder yield kg/ha.	Return Rs/ha.	Grain yield kg/ha.	Fodder yield kg/ha.	Return Rs/ha.	Grain yield kg/ha.	Fodder yield kg/ha.	Return Rs/ha.			
1.	*Ardu* + Clusterbean	74	394	888	81	382	1754	132	457	1774	4416	3788	2736
2.	*Ardu* + Cowpea	241	419	2619	241	407	3925	288	675	3250	9794	3788	4527
3.	*Ardu* + Green gram	275	634	4722	272	657	6476	239	701	4828	16026	3788	6605
4.	*Ardu* + Sesamum	126	581	2074	112	583	2746	86	393	2507	7327	3788	3705
5.	Clusterbean sole	75	498	974	82	532	1931	139	480	2877	4782	-	1594
6.	Cowpea sole	303	701	3383	274	556	4528	264	792	3744	11655	-	3565
7.	Green gram sole	328	1252	5880	290	1296	7351	287	829	5800	19031	-	6340
8.	Sesamumm sole	166	745	2725	139	750	3410	90	417	2664	8799	-	2922
	SEm ±			294.2			148.0			170.1			293.0
	C.D. 5%			882.1			443.8			510.2			835.0
	C.V. %			19.7			7.18			10.0			13.8

Table 9.3. Grain and fodder yield and expected return of different field crops with and without association of *neem* (*Azadirachta india*) at Sardarkrushinagar.

S. No.	Treatment	2001-2002			2002-2003			2003-2004			Total returns without tree (Rs/ha.)	Total income from the tree produce (Rs/ha.)	Av. Returns from the system
		Grain yield kg/ha.	Fodder yield kg/ha.	Returns Rs/ha.	Grain yield kg/ha.	Fodder yield kg/ha.	Returns Rs/ha.	Grain yield kg/ha.	Fodder yield kg/ha.	Returns Rs/ha.			
1.	*Neem* + Clusterbean	85	555	1102	86	498	2153	148	509	1993	5248	5456	3568
2.	*Neem* + Cowpea	325	930	3720	298	841	5100	291	908	4168	12988	5486	6158
3.	*Neem* + Green gram	365	1023	6363	312	872	75345	255	736	5136	19034	5804	8278
4.	*Neem* + Sesamum	112	398	1837	111	426	2709	97	400	2950	7496	5317	4271
5.	Clusterbean sole	85	532	1082	92	490	2053	170	578	2281	5416	-	1805
6.	Cowpea sole	405	993	4546	307	826	5224	330	957	4681	14451	-	4817
7.	Green gram sole	569	1331	9780	330	1012	8030	303	902	6130	23940	-	7980
8.	Sesamumm sole	128	599	2185	116	542	2859	106	463	3216	8260	-	2753
	SEm ±			541.9			1344			295.4			507.7
	C.D. 5%			162.9			432			886			1522
	C.V. %			27.60			6.33			16.30			15.49

Table 9.4. Grain and fodder yield and expected return of different field crops with and without association of *shisham (Dalbergia sissoo)* at Sardar krushinagar.

S. No	Treatment	2001-2002			2002-2003			2003-2004			Total returns without tree (Rs/ha.)	Total income from the tree produce (Rs/ha.)	Av. Returns from the system
		Grain yield kg/ha.	Fodder yield kg/ha.	Return Rs/ha.	Grain yield kg/ha.	Fodder yield kg/ha.	Returns Rs/ha.	Grain yield kg/ha.	Fodder yield kg/ha.	Returns Rs/ha.			
1.	*Shisham* + Clusterbean	86	664	1184	81	394	1770	134	449	1789	4743	11012	5252
2.	*Shisham* + Cowpea	450	1076	5038	305	831	5198	290	855	4125	14361	10934	8465
3.	*Shisham* + Green gram	465	854	7872	334	620	7813	241	708	4869	20554	11797	10784
4.	*Shisham* + Sesamum	111	502	1826	96	447	2348	87	415	2651	6825	10859	5895
5.	Clusterbean sole	83	463	1013	83	461	1878	150	508	2008	4899	-	1633
6.	Cowpea sole	294	780	3327	315	590	5177	320	931	4538	13042	-	4347
7.	Green gram sole	426	1205	7418	361	738	8495	270	799	5460	21373	-	7124
8.	Sesamumm sole	107	539	170	99	495	2433	96	455	2935	7128	-	2376
	SEm ±			354.7			143.0			176.5			546.0
	C.D. 5%			1063.0			427.0			529.0			NS
	C.V. %			18.7			6.32			9.70			9.58

Table 9.5. Grain and fodder yield and expected return of different field crops with and without association of *anjan* (*Hardwickia binata*) at Sardarkrushinagar.

S. No.	Treatment	2001-2002			2002-2003			2003-2004			Total returns without tree (Rs/ha.)	Total income from the tree produce (Rs/ha.)	Av. Returns from the system
		Grain yield kg/ha.	Fodder yield kg/ha.	Returns Rs/ha.	Grain yield kg/ha.	Fodder yield kg/ha.	Returns Rs/ha.	Grain yield kg/ha.	Fodder yield kg/ha.	Returns Rs/ha.			
1.	*Anjan* + Clusterbean	82	277	861	84	294	1716	134	459	1799	4376	8461	4279
2.	*Anjan* + Cowpea	339	613	3696	278	485	4539	246	741	3507	11752	8960	6904
3.	*Anjan* + Green gram	298	775	5160	291	652	6884	233	669	4702	16746	9292	8679
4.	*Anjan* + Sesamum	89	350	1459	106	395	2592	89	386	2708	6759	8579	5113
5.	Clusterbean sole	73	261	780	88	322	1818	150	563	2063	4661	-	1553
6.	Cowpea sole	405	870	4448	288	530	4722	275	824	3918	13128	-	4376
7.	Green gram sole	641	1238	10880	318	662	7485	278	775	5585	23950	-	7983
8.	Sesamumm sole	83	468	1372	111	449	2717	98	390	2999	7088	-	2363
	SEm ±			554.6			105.5			148.6			745.0
	C.D. 5%			1662.0			316.0			445.0			NS
	C.V. %			30.10			5.06			8.49			13.2

Table 9.6. Economics of ber based agri-horti system.

Treatment	Gross realization (Rs/ha)	Cost of cultivation (Rs/ha)	Net realization (Rs/ha)
T_1: Ber + Pearl millet	22243	12413	9812
T_2: Ber + Green gram	24613	11514	13099
T_3 : Ber + Sorghum	24180	12294	11886
T_4 : Ber + Laster bean	22141	11622	10519
T_5 : Ber + Fallow	19779	9492	10287

Table 9.7. Fodder yield of dhaman (*Cenchrus ciliaris*) from khejri (*Prosopis cineraria*) based silvipastoral system at Sardar Krushinagar.

S. No.	Treatment	Fodder yield (sundried) Q/ha.				
		1997	1998	1999	2000	Mean
1.	Beneath khejri canopy	62	60	37	39	49.50
2.	Between khejri rows	51	47	30	33	40.25
3.	Sole dhaman grass (without khejri)	45	40	25	25	38.75

REFERENCES

Aggarwal, R.K., Kumar P. and Raina, P. (1993). Nutrient availability from sandy soils underneath *Prosopis cineraria* compared to adjacent open site in arid environment. *Indian For.* 119 (4):321-325.

Jaimini S N and Tikka, S.B.S. (1998). Khejri (*Prosopis cineraria*) based agroforestry system for dryland areas of north-west Gujarat. *Indian J. Forestry* 21(4): 331-332

Jaimini S N and Tikka, S.B.S. (2001). Studies on multipurpose tree species for agroforestry in dryland agriculture *Indian J. Forestry* 24(2): 185- 188.

Kumar, V, Yadav H O and Sharma H C. (1992). Agroforestry the suitable farming system for arid and semiarid region. *Haryana farming* 21(2) : 15-16.

Paroda, R.S. and Solanki, K.R. (1991). Khejri: A wonder tree of the thar desert. *Indian Fmg.* 41(7): 12-14

Tiwari, D.N. (1994). Agroforestry for sustainale development. *In*: Agroforestry systems for degraded lands (Edited Panjab Singh, P.S.Pathak and M.M.Roy). Oxford & IBH publishing Co. pvt. Ltd.

• • •

10

Agroforestry Systems in Arid Regions of India

L.N. HARSH AND J.C. TEWARI
Central Arid Zone Research Institute, Jodhpur - 342 003 (Raj.)

ABSTRACT

Scattered trees/shrubs in agricultural fields or grazing fields are common features of landscape in arid regions of India. These trees were considered as insurance to the people during drought years. In recent past importance of these trees in such areas had been recognized by the scientific world. Lot of research had been made and is continuing to identify and research the tree-crop combination. This paper reviews the tested agroforestry systems in arid region of country with particular reference to Rajasthan. Economic gains due to technology intervention in agroforestry system is also given.

INTRODUCTION

The people in the arid regions of India, since time immemorial has developed location specific agroforestry systems. The scattered trees / shrubs in the agricultural fields or grazing fields are common features of the landscape. Since the beginning of cultivation the farmers deliberately allow the trees /shrubs to grow with the crops. They have screened the trees / shrubs based on their utility. They have considered that these trees are insurance to them during drought years. These trees provided lean period fodder, fruits, vegetable, timber, fuel, fiber etc. as farmers knew that occurrence of drought is common phenomenon in Rajasthan and due to failure of rains the crops failed to grow or their production is abysmally low. In the recent past efforts have been made to identify existing agroforestry systems for different agro-ecological zones (Malhotra, 1984 ; Harsh *et al.*, 1992 ; Saxena, 1997). Efforts have also been made to study the production systems under each agroforestry systems and also to improve the existing systems through technological interventions (Harsh and Tewari, 2003; Tewari *et al.*, 2004; Harsh *et al.*, 2004).

DISTRIBUTION OF HOT ARID REGION

The hot Indian arid zone is spread in 31.7 million hectare of which major part is in northwestern India (28.57 m ha) and 3.13 m ha in south India. The north western arid region occurs between 22°.30″ to 32°.05″ N and from 68°.05″ to 75° 45″E covering western part of Rajasthan (19.6 m ha) Gujarat (6.22 mha) and south western part of Haryana and Punjab (2.75 m ha) (Faroda *et al.*, 1999). The region is bordered by the irrigated Indus plains in the north, Aravalli hills range in the east, and international border with Pakistan in west and black soil region and alluvial plains of Sabarmati in the south.

In Gujarat, whole region of Kutch and partially the district of Jamnagar (80%) Surenderanagar (29%). Junagarh (28%). Banas Kantha (18%), Mehsana (7%) and Rajkot (6%) fall under arid zone. Ten percent of hot arid region is present in Punjab and Haryana. The percentage area of districts in these two states, which come under arid zone, is Bhatinda (88%), Ferozpur (77%), and Sangrur (8%) in Punjab and Hissar (90%) and Mahendragarh (9%) in Haryana. In Andhra Pradesh 7 per cent of the area falls under arid zone. The percentage area covered by arid zone in Anantapur, Kurnool and Cuddapa district is 67, 31 and 9 respectively. Three per cent of hot arid zone falls in Karnataka. The eastern portion of Bellary district (25%), western portion of Raichur (39%) and parts of Dharwar (1%) and Chitradurga district (4%) falls in the arid zone. Very small pocket in west Solarpur district and some part in Dhulia and Satna district falls in arid region covering 1290 km^2. (Velayutham and Bhattacharyya 2000).

CLIMATIC FEATURES OF INDIAN ARID ZONE

Rainfall

The major part of the rainfall is received during the southwest monsoon period, June-September. The contribution of this seasonal rainfall to annual rainfall is very high (91 to 96%) over the whole of Gujarat and South and Central part of western Rajasthan. In north of this region, the percentage contribution of SW monsoon reduces to 78 per cent in Punjab and 80-85 per cent in Haryana and Northwest Rajasthan because of some contribution (6-12 %) by winter rains due to influence of western disturbances. In contrast, the percentage contribution of June-September rainfall to annual rainfall decreases in Maharashtra region (61-63%). It is lowest in Karnataka and Andhra Pradesh (48 to 67%), as these regions receive considerable rainfall (21 to 32%) during the North East monsoon season. This duration of monsoon season is larger in the southern arid zone, where monsoon sets in the first fortnight of June and withdraws in second fortnight of November as compared to arid region of Northwest India, where monsoon sets in first fortnight of July & withdraw in the first fortnight of September (Ramakrishna *et. al.*, 2000).

Another interesting feature of the rainfall is the high coefficient of variability in the annual rainfall which often exceeds 50 per cent in NW Indian Arid Zone and is higher than 70 per cent in the extreme western regions of arid Rajasthan, where the annual rainfall is very low. (Ramakrisna and Rao, 1992). In contrast, the coefficient of variability is about 30 per cent in the arid zone located in the Peninsular India.

Temperature

In arid zone of India there is a grate variation in the diurnal and annual temperatures. During the winter season, the arid regions in NW India experience very cold temperatures, which ranges from 3° to 10° C in Punjab, Haryana and arid Rajasthan. Frost often occurs in the northwest Indian arid zone. Gujarat and Maharashtra on the other hand, experience minimum temperature of the range of 11-15°C while Karnataka and Andhra Pradesh records 16-18°C. Lowest temperature in North Indian arid zone experienced in January, while in South Indian arid zone in December.

Wind speed is very high in Gujarat followed by western Rajasthan, where wind speed is more than 20 kmph during April to July months. This period also coincides with the occurrence of dust storms (Andhi) and during that period wind speed may go to as high as 80 kmph. The mean wind regime, however, is weak over the Punjab and Haryana varying from 2 to 10 kmph, whereas in southern arid zone, wind regime is more than 18 kmph in Anantapur and Kurnool regions of Andhra Pradesh, and Raichur region of Karnataka during June to August.

Potential Evapotranspiration (PE)

In hot arid region of India the annual potential evapotranspiration value is 130 cm in Punjab to 160 cm in Haryana, and to over 200 cm in Rajasthan. During May and June, the PE values are at its peak varying from 18-20 cm in Punjab, 22-23 cm in Haryana, and 21 to 31 cm in Rajasthan. However, in these areas winter PE values ranges between 3-8 cm. The annual PE values in southern arid region ranges between 180 and 185 cm, which are comparable to those over western Rajasthan.

Soil resources

Like climate, a great variability in soil characteristics have been reported within the hot arid region of India. In hyper arid zone of Western Rajasthan, soils are torripsamments associated with dune and sandy waste. The soils of dunes or interdunal plains are deep and excessively drained and are calcareous / noncalcareous in nature. The lower topographic position in the interdunal plains are classified as typic, calciorthids, typic palaorthids and typic salorthids

(Table 10.1). Whereas in Bagar region of Rajasthan, north Gujarat plain and southwestern Punjab, the dominating soils are deep to very deep sandy ones with or without underline calcareous layer. These occupy gently slopping plains with sporadically dotted hummocks. Such soils are present in northern parts of Jodhpur and Pali District, respectively.

In arid regions of south Kutch and north Kathiawar, Peninsula soils are moderately to strongly alkaline and salinity increases with depth. The surface texture is sandy loam to loamy sand with clay content increases from 13 to18 per cent with depth. The soils of the Deccan Plateau arid zone comprise mixed red and black soils. These soils are highly susceptible to erosion and frequent droughts. The consistence of the subsoil is very hard when dry and act as hardpan for root penetration and infiltration of water. In general, the clay content is high.

Vegetation resource and crop association

Since climatic and soil features vary in different arid regions, the natural vegetation and arable crop species also vary enormously. The natural vegetation of hyper arid region comprises only of sparce and sporadic groves of tropical thorn forest with common tree/shrub species like *Acacia senegal, Prosopis cineraria, Calatropis procera, Euphorbia caducifolia, Heloxylon salicornicaun, Calligonum polygonoides, Capparis decidua,* etc (Satyanarayan, 1963). Their densities per hectare are also very poor. These trees are often grown in associated with major grasses viz *Lasiurus sindicus* and *Panicum targidum* and crops like pearl millet and guar.

The vegetation type varies in Bagar region of Rajasthan, North Gujarat Plain and South-Western Punjab Plain of hot arid region. Due to higher rainfall than that of hyper arid region, the common species are - *Prosopis cineraria, Acacia nilotica, Zizyphus nummularia / Z. rotundifolia, Tecomella undulata, Albizia sp. and Dalbergia sissoo* are common. The density of tree/ha is higher than hyper arid region. The common grass species are *Cenchrus ciliaris and C. setigerus.* These trees are commonly grown in association with arable crops like bajra, jowar, til, moong, moth, guar, chawla etc.

In south Kutch and north Kathiawar Peninsula, the vegetation is sparse and classified as southern tropical thorn forest. The common species are: *Acacia nilotica, A. catechu, Azadirachta indica, Zizyphus spp.*

In alkaline soils, common species are *Acacia nilotica, Salvadora oleoidis* etc. *Prosopis juliflora* in recent past invaded in almost all in Kutch region. The common grass species are *C. ciliaris, Dichanthium annulatum* etc. In Central plateau and Deccan plateau, the vegetation is classified as tropical thorn forest with common tree species of *Acacia leucophloea, Acacia nilotica, A. ferruginea,*

Azadirachta indica, Butea monosperma, Madhuca sp., Albizia lebbek, etc. These trees are grown in association with sorghum, guar, pigeon pea and cowpea.

From the above it is very clear that even in hyper arid and arid region of India, though climate, soil and vegetation differs but growing of arable crops/ grass with woody perennials is common practices (Harsh *et. al.*, 1992) (Table 10.2). The reason being that growing of woody perennials provides an insurance to dry land against recurring failure of crops due to droughts. They provide, leaf fodder, fruits, tannins, gum and fuel to farmers during drought years. Though farmers are getting some returns from the existing agroforestry systems, but our efforts are here, to further improve the existing system without any adverse effect on fragile ecosystem.

A) Agroforestry In Hyper-arid Ecosystem

(i) The hyper arid ecosystem is dominated by *Lasurus sindicus* grassland along with scattered shrubs or trees. The economy of the region is livestock based. Even with a low rainfall, the *L.sindicus* resprouts and provides 20-30 q/ha/yr dry fodders. If this grassland is reseeded properly, than 40-45 q/ha/yr dry fodder can be harvested. The nutritive value of grasses can further be increased by cultivating guar in between grass rows. By adopting this grass-legume system, in long run productivity of livestock would be increased.

ii) Dune complex with scrub and grasses. About 58 per cent of western Rajasthan is covered by different types of dunes. They are devoid of vegetation except scattered plants of *Acacia jaquimontii, Calligonum polygonoides* and few other shrubs. These dues are cultivated during Kharif season with Bajra and guar by the farmers. Dunes owned by the Revenue department or Forest department are planted by *Acacia tortilis* for their stabilization only. The growing of crops on dunes becomes uneconomical because of soil erosion and failure of monsoon. Farmers hardly get Rs. 1500-2000 per ha. Similarly growing of trees on dunes does not have economic returns to farmers as its main role is for dune stabilization and gestation period for economic return is very long. Therefore, CAZRI has developed model which can increase the economic returns as well as stopped dune movement. (Harsh, 2002)

iii) *Dune Agroforestry model :* In general for dune stabilization, trees are planted at the spacing of 5x5m from crest to foot hill. But if trees are planted in belt of 5 rows of trees leaving a space of 50 m between the two belts for sowing of grass or crops. This model will provide crops produce as well as soil erosion will be stopped (Kavia and Harsh 1991).

Recently from this model, the economic returns have further enhanced by technological interventions Viz., gum from *A. tortilis* and wood processing for small timber. By injecting ethephon, on an average 400 g gum/tree can be harvested after 12th year of plantation. (Harsh and Mathur, 2002). The comparative economic analysis revealed that growing only bajra on the dunes can give the gross return of Rs. 1600 per ha/year, and growing *A. tortilis* on the dunes for fixation purpose can give the gross return of Rs. 2045 per ha/year keeping 20 years of rotation. But by simple technological intervention, the same system can be much more economical and sustainable, beside, utilizing *Acacia tortilis* as fuel wood, which fetches only Rs. 100 per quintal, if trees are treated with gum inducer chemicals after 15 years of plantation upto the age of felling and if wood treated for seasoning, than from each tree 2 ft^3 wood can be harvested and which can fetch Rs. 200 per ft^3. Hence total returns can be four to five times than pure cultivation of bajra (1:4.6 bajra to *A. tortilis* plantation) (Table 10.3).

B) Agroforestry On Sandy Plain

In low rainfall areas (>150-300 mm per year) common agroforestry practice is growing of bajra or legume crops in association with *A. senegal*. In these regions, the tree density varies from 10-20 trees per ha. Trees are deliberately maintained in the field because *A. senegal* provide seeds (approximately 1 kg/tree) used for vegetable. From the traditional agroforestry farmers are getting gross income of Rs. 1800 per ha. But by adopting CAZRI's gum exudation, techniques (Khan and Harsh, 1994) in *A. senegal*, farmers can get gross income of Rs. 3,000 per ha per year, which is almost double over the traditional system. (Table 10.4) (Harsh *et.al.* 2000). Besides, gum exudation, if green seeds are collected and processed like Heena, Pea, than green seeds can be sold in Rs. 100 per kg, whereas dry seeds are sold @ Rs. 20 per kg.

Another agroforestry system is Khejri based cropping system existed since time memorial. Khejri (*Prosopis cineraria*) green pods are being used as vegetable and green leaves (dried) called loong used as cattle feed (Shankarnarayan, 1984). Trees do not have any competition with crops. Khejri is known as "Kalpvarkhsa" and also declared as State tree of Rajasthan". The economic analysis of traditional Agroforestry revealed that from bajra alone gross income of Rs. 1600 per ha can be obtained but from the Khejri (30 trees per ha) inter cropped with bajra is capable of providing an income of Rs. 4600 ha. Thus it revealed that *P. cineraria* + bajra crop is much remunerative. In the existing system not much value addition has been done so far. But some efforts have been made to improve the Khejri for fast growing.

Besides, Khejri, farmers are also growing *Ziziphus nummularia* and *Z. rotundifolia* inter cropped with bajra or legume crops. *Z. nummularia* and *Z. rotundifolia* are utilized as livestock feed as well as fence material. After the harvest of bajra crops, in the month of October-November, *Ziziphus* sp. are harvested from the ground level and allowed to dry and dry leaves called "Pala" being collected and stored as lean period feed for livestock. This existing Agroforestry system can give the gross return of Rs. 7000-8000 per ha.

CAZRI, Jodhpur has made some efforts to improve this existing system. Some of the *Ziziphus* plants were budded with improved cultivars of Gola and Seb. These budded ber even from the first year started bearing fruits but 2nd year onward their fruit yield increases to 30 kg/tree/yr. If farmers maintain 50 trees per ha besides natural growing *Zizyphus* farmer can earn gross returns of 12,000-15,000 per ha (Faroda and Harsh 1999). Similarly, Sharma (1996) and Ram Newaj *et al.,* (2001) have reported that growing of crops with ber and aonla is much profitable than cultivation of pure crop.

C) Agroforestry On Sandy Plain With Inland Drainage

This zone covers almost 8.0 million ha. Irrigation through well water is available. Therefore, farmers of the regions are growing kharif as well as Rabi crops with *Prosopis cineraria* and *Acacia nilotica vr. Cuppresiformis*. The gross returns are much higher than rainfed sandy plain sub zone. Traditionally farmers are earning gross income of Rs. 15,000-20,000 per ha/year.

But this income can be further enhanced by adapting improved Agroforestry models. *viz. Ailanthus excelsa* (Ardu) based cropping system. Ardu was initially introduced as fodder tree and from mature tree of 15-20 years old about 2-4 q. green leaves can be harvested. The green fodder can be sold @ Rs. 100 per q., but in the recent past its wood is used in plywood industries as well as in matchstick industries. The market value of green wood is around Rs, 75 per q. (Harsh, 2002). Hence if a farmer plant 40 trees per ha on the farm boundary than he can earn Rs. 3000-4000/- per ha per year in addition to his traditional farming system.

Salvadora based agroforestry system is very common where soils are clay loam or sandy loam. The trees have multipurpose uses. It does not have any adverse effects on the companion crops. Camel browses its leaves and the native people eat fruits. Seed contains 50 per cent non-edible oil. Native people maintain 10-15 trees/ ha. The single tree produced 30-80 kg fruits of which hardly 10 per cent fruits are utilized while rest fall on the ground and perish. Khan *et al* (2004) and Harsh *et al.,* (2005) have developed the technologies for utilization of fruits for preparing squash, jam etc. If the fruits are utilized fully than farmers can earn approximately Rs. 13,000 per ha every year whereas without *Salvadora* sp. the returns form the crops will be Rs. 10,500/ha/yr.

Similarly, in Deccan Plateau of hot arid region, *Ceiba pentendra* are planted on the farm boundary. A mature tree yields on an average 80-100 floss which contains the Silk cotton and has a high market value. By planting 50 trees/ha of *Ceiba pentendra*, farmers can earn Rs. 5000 more returns than its traditional system. Itnal (1992) reported that intercropping of *Agave sisalana* with *Leucaena leucocephala* give additional return by the harvest of fiber (215 kg/ha) from Agave.

Mann *et al.*, (2001) and Awasthi *et al* (2001) also reported that adoption of Agroforestry in semi-arid regions become more remunerative than sole cropping. Gajja *et al.*, (1999) studied the economic returns from Silvipastoral, Horti-pasture and annual crops on marginal agricultural lands of Arid Zone of Rajasthan and reported that alternate land use systems are more profitable than arable farming. The return from hortipasture system is highest.

Besides, in the recent past, most of farmers of irrigated area, where water table has gone down (>150′), have shown their interest in plantation forestry. Some people have planted medicinal plans; some have planted biofuel plants, which required less water for irrigation. Few of the farmers in Rajasthan have planted *Simmondsia chinensis* (Jojoba) in 1992 and after 10th year of plantation, farmers have been earning gross income of Rs. 1,25,000-2,00,000 per ha per year. Jojoba requires irrigation from flowering to fruiting stage (November – April) in fortnightly intervals. Jojoba starts bearing seeds after 3-4 years and become fully productive at the age of 10th year when plants bears more than 1.0 kg seed/plant/year (Harsh *et al* 1987). Keeping in view long gestation period, intercropping with legume crops is recommended. The results are presented in table 10.5 (Sandhu 2001).

Silvipastoral Systems : In arid region silvipastoral system has a very important role as whole economy of people rest on the livestock. To maintain the live stock round the year farmers required fodder. Therefore, pasture development along with fodder trees have a unique role. The silvipastoral systems in arid regions have been precisely documented by Yadav *et al.*, 1997; Shankarnarayan *et al.*, 1987; Muthana *et al.*, 1985; Harsh *et al.*, 1992; Sharma and Dutta, 1999; Sharma *et al.*, 2001 and Sharma 2004.

Harsh *et al.*, (1992) have reported that by growing grass *Cenchrus ciliaris* with *Hardwickia binata* and *Colophospermum mopane*, the average carrying capacity can be increased over pure cultivation of grasses besides substantial

Table 10.1 Physical and chemical properties of Typic Torripsamment and typic comborthid.

Depth (cm)	Clay (%)	PH (1:2.5)	EC (dSm^{-1})	$CaCo_3$ > 2 mm (%)	OC (%)	CES (mole p^+) kg^{-1}
Typic Torripsamment						
0-25	5.7	8.0	<0.2	6.0	0.1	2.3
25-50	6.2	8.2	0.2	6.6	0.1	2.2
50-100	6.5	8.2	0.2	8.0	0.03	2.5
Typic comborthid						
0-25	8.2	8.5	0.4	1.4	0.01	2.8
25-50	10.9	8.6	0.5	8.2	0.01	0.8
50-100	10.2	8.5	0.6	10.7	0.30	2.9

Source : Velaynthain and Bhatta Chargya 2000

Table 10.2. Traditional agroforestry systems in arid zone of Rajasthan

Zone	**Tree/shrubs with agricultural crops**
Jaisalmer-Pugal-Girob (below 150 mm isohyte)	*Zizyphus nummularia* ("bordi") grown with agricultural crops The *Prosopis cineraria* ("Khejri") trees are very few.
Barmer-Osian-Bikaner belt	Crops grown in association with *P. cineraria* and *Z. nummularia.*
Jodhpur-Nagaur-Shekawati belt	Crops mainly grown in association with *P. cineraria* and a limited number of *Z. nummularia* plants are available.
Nawalgarh-Dikar	Mainly *P. cineraria* dominate in agricultural fields but a few *Ailanthus excelsa* are also in agricultural fields.
Pali	The dominating trees species are *Acacia nilotica* var. *cupressiformis* and *A. nilotica* var. *radiane*. Both the "Kharif" (summer monsoon) and "Rabi" (winter) crops are grown with these trees.
Arid and semi-arid areas bordering Eastern Ajmer	*Prosopis cineraria* dominates but *Acacia leucophloea* is also present. Some crops are grown in association with these species.
Jalore Districts	In this tract *Salvadora persica* and *S. oleoides* dominated; crops are grown with these trees.

Table 10.3. Economic gains due to technological interventions in different agroforestry systems

Systems	Products (q /ha)		Returns (Rs./ha/yr)	Total
Mono-cropping of Bajra on sand dunes	Bajra grain	2 q/ha @ 400 /q	800	1600
	Straw	4 q @ 100 /q	800	
Grass cultivation	Grass	15 q/ha @ 100 /q	1500	1500

Contd...

Table 10.3 Contd...

Grass cultivation	Grass	15 q/ha @ 100 /q	1500	2250
	Grass seeds	15 kg/ha @ Rs. 50 /kg	750	
Acacia tortilis Plantation trees 5x5 m (400 trees/ha)	Wood after 20 years	100 kg/tree 400 q/ha @ Rs. 100 /q	2000	2100
	Pods	2000 kg/ha @ Rs.1/kg	100	
Intervention of Gum technology	Gum	22.5 kg/ha @ 100 /kg	2250	7350
Wood Processing technologies	Lumber	40 cft @ 100 cft	4000	
	Fuel wood	10 q/ha @ 100 /q	1000	
	Pods	2000 kg/ha @ Rs 1 /kg	100	

Table 10.4. Economic returns from *Acacia senegal* (Kumat) and bajra farming system after 11 years.

Tree/crops	Tree density/ha	Total gum (kg)/ *bajra* q^{-1}/ha)	Fuel wood (q^{-1}/ha)	Total returns from the system (Rs.)	Return per year (Rs.)	B/C ratio
A. senegal	1111 (3x3 m)	480	330	96,000+33,000 = 1,29,000	11,727	1:3.25
	494 (4.5x4.5 m)	340	197	68,000+19,700 = 87,700	7,973	1:2.21
	277 (6x6)	288	138	57,600+3,800 = 71,400	6,490	1:1.80
A. senegal + Bajra	140 (12x6m)	160	75	39,500+24,100 = 63,600	5,782	1:1.60
	Bajra Fodder	27.00 133.00	- -			
Bajra pure	Bajra Fodder	44.00 220.00	- -	17,600+22,000 = 39,600	3,600	-

harvest of fuel wood (Table 10.6). Similarly, when *C. ciliaris* was sown in between 8 year old *H. binata* and compared the yield with pure *C. ciliaris* pasture the results revealed that Silvipastoral system is more remunerative than sole pasture as well as sole tree block (Table 10.7).

It is revealed from the above results that growing of trees in association with the under story crops, increased the production as well as provide

Table 10.5. Gross returns (Rs.) from Jojoba based agroforestry system

Gross income (Rs.)	Years	Jojoba seed yield (kg/ha)	Gram seed yield (q/ha)	Sale of seeds		Return	
				Jojoba Rs./kg	Rs./Ql.	Jojoba	Gram
14400	1992	-	12.00	-	12.00	-	14400
14400	1993	-	12.00	-	12.00	-	14400
14400	1994	-	12.00	-	12.00	-	14400
7500	1995	30	-	250	-	7500	-
30000	1996	130	-	300	-	30000	-
185000	1997	370	-	500	-	185000	-
156000	1998	300	-	500	-	156000	-
75000	1999	600	-	125	-	75000	-
86000	2000	500	-	100 kg @ 300 & 400 kg @ 140/-	-	86000	-
-	1	-	-	-	-	-	-
50000	2	250	-	200	-	50000	-
100000	3	800	-	200	-	160000	-
792700					G. Total	72064.00	

Source : Sandhu 2001.

Table 10.6. Fodder production and carrying capacity in different systems

Years initial planting	Pure pasture		Silvi-pastoral systems			
	Grass fodder yield (q/ha)	Carrying capacity (sheep/ha/yr)	Grass fodder yield (q/ha)	Leaf fodder yield (q/ha)	Total fodder production (q/ha)	Carrying capacity (sheep/ha/yr)
1	10	2.4	8.2	-	8.2	1.9
2	15	3.6	12.3	-	12.3	2.9
3	20	4.8	16.4	-	16.4	3.9
4	25	6.0	20.5	-	20.5	4.9
5	20	4.8	16.4	-	16.4	3.9
6	20	4.8	16.4	-	16.4	3.9
7	15	3.6	12.4	23.1	35.5	8.5
8	10	2.4	8.4	6.6	15.0	3.6
9	10	2.4	8.4	6.6	15.0	3.6
Average of 9 years	16.1	3.9	13.3	4.0	17.3	4.1
Average fuel production (q/ha)	-	-	-	2.6	-	-

economic stability to farmer's. But long-term study of different agroforestry systems revealed that trees have a negative effect on the crops or vis-à-vis crops have negative effect on trees. Harsh and Tewari, (1993) and Chopra

Table 10.7. Fodder production (Average of 5 years) under Silvi-pastoral systems (Hardwickia binata)

Systems	Top feed (q/ha)	Grass production	Total fodder	Fuel wood	Economic returns (Rs. /ha)
Sole tree Block (H. binata)	23.44	-	23.44 @ 100	33.121 @ 50	Rs. 4000/ha (30.7 %)
H. binata + C. ciliaris	17.12	21.83	38.95 @100	22.48 @ 50	Rs. 5019/ha (64.0 %)
Pure *C. ciliaris*	-	30.61	30.61 @ 100	-	Rs. 3061/ha

Table 10.8. Effect of trenching near the established Acacia tortilis on the crop growth and production

Crops	Crop yield parameters	No trenching	Single side trenching	Both side trenching
Moong bean	Height (cm)	70.0	76.0	78.0
	Grain production (q/ha)	3.6	4.0	5.0
	Fodder production (q/ha)	-	-	-
Cluster bean	Height (cm)	76.0	119.0	130.0
	Grain production (q/ha)	1.1	4.1	5.4
	Fodder production (q/ha)	-	-	-
Sorghum (fodder)	Height (cm)	67.0	107.0	117.0
	Grain production (q/ha)	-	-	-
	Fodder production (q/ha)	60.0	96.3	119.3

et al., (1993) have reported three phases of interactions in agroforestry systems and suggested that at the early stage of trees and crop planting, crops / grasses have negative effects on the tree seedling growth and suggested cultivation of low height crops. In the second phase both trees as well as crops/grasses shares natural resources equally without interfering each other. While in the third phase when tree grows bigger, full grown canopy and root system have a negative effect on the under story crops. In that case canopy interfere for light and roots compete for soil moisture and nutrients and thereby reduces the crop production (Table 10.8). Thereby, canopy lopping/pruning and trenching near the root zone suggested to reduce the effect on the crop production.

REFERENCES

Awasthi, M.K. L.D. Costa, Upadhya, S.D. and Bajpai, R. (2001). Krishi Yougya Padti Bhomi Prabadhan. Pp.51-55. In Prakratic Sansadhano ke Prabadhan Mai Krishi Vaniki Ka Yogdhan. (Eds. K.S. Dadhwal and K.R. Solanki) *National Research Center for Agro-forestry*, Jhansi.

Chopra, N.K., Chopra, Nisha and Harsh, L.N. (1993). Effect of intercrop and root trenching in agroforestry system. In: *Afforestation of Arid Lands*, (Eds.) A.P. Dwivedi and G.N. Gupta. *Scientific Publishers*, pp. 379-384.

Faroda, A.S. and Harsh, L.N. 1999. From empty bowl to self-reliance: Success of farmers of Desert Villages. *Environmental News.* June-July pp.14.

Gajja, B.L., Bhati, T.K. Harsh, L.N. and Khan, M.S. (1999). Comparative economics of Silvipasture, Horti-pasture and Annual crops on marginal agricultural lands of arid zone of Rajasthan. *Annals of Arid Zone.* 38 (2):173-180.

Harsh, L.N. (2002). Value additions in Agroforestry Systems for higher economic returns. Pp. 212. Abstract. National Symposium on Grassland and fodder Research in the new Millennium. Oct.30-Nov.1, 2002. *Range Mgt Society of India,* IGFRI, Jhansi.

Harsh, L.N. (2002). Drought mitigation and povwerty allevation through tree plantations in arid regions. DEN News 6 (3 & 4) : 7-12.

Harsh, L.N. and Tewari, J.C. (2003). Agroforestry systems in hot arid regions with special reference to value addition. Pp. 159-170. *In*: Agroforestry potentials and opportunities. Eds P. S. Pathak and Ram Newaj. Agrobios India.

Harsh, L.N., Tewari, J.C. Patwal, D.S. and Meena, G.L. (1987). Package of practices for cultivation of Jojoba (*Simmondsia chinensis*) in arid Zone, CAZRI, Jodhpur Pub. No. 32, pp. 1-19.

Harsh, L.N. and Mathur, N.K. (2002). A new development in tapping exudates gum in *Acacia tortilis*. Pp. 51. Abstract & souvenir of XVII National Carbohydrate Conference Nov.21-22, 2002. *Central Food Technological Research Institute,* Musore, Karnataka.

Harsh, L.N. and Tewari, J.C. (1993). Tree-Crop Interactions in Agroforestry Practices. In :Agroforestry for Rural Needs (Vol. II) (eds) DK Khurana and PK Khosla, ISTS, Solan, pp 535-540.

Harsh, L.N., Khan, H.A. and Meghwal, P.R. (2005). A study on economic potential of Salvadora oleoides based agroforestry systems in Araba area, district Barmer. *Indian Journal of Forestry.* (Communicated).

Harsh, L.N., Khan, H.A. Tewari, J.C. and Bohra, M.D. (2000). Exploitation of Minor Forest Products yielding tree to enhance production in arid and semi-arid regions. *In Environmental Protection* (Eds. Vimlesh Choudhary) *Pointer Publications* Jaipur pp. 250-254.

Harsh, L.N., Bohra, M.D. Khan, H.A. and Tewari, J.C. (2004) Effect of drought on seed quality and germination of *Acacia senegal, Indian J. of Forestry* (2):181-185

Harsh, L.N., Tewari, J.C. Buraman, U. and Sharma, S.K.. (1992). Agroforestry in arid regions. *Indian Farming* 42 (5) : 32-37.

Itnal, C.J. (1992). Agave for marginal and submarginal lands. Pp.93-94. *In* agroforestry-Principles and Practices (Eds. B.S. Nadagoudar, S.L. Madiwalar and S.M. Mutanal). *Forestry Department UAS, Dharwar*, Karnataka.

Kavia, Z.D. and Harsh, L.N. (1991). Proven technology of sand dune stabilization – a step to combat desertification. *In:* Afforestation of Arid Lands (Eds.) A.P. Dwivedi and G.N. Gupta, *Scientific Publishers*. Pp. 79-86.

Khan, H.A. and Harsh, L.N. (1994). Quality of Ethephon induced gum from *Acacia senegal*, Annals of Arid Zone 33 (1): 85.

Khan, H.A., Meghwal, P.R. and Harsh, L.N. (2004) "*Salvadora oleoides* Decne-an underutilized economic plant of the desert. Science Tech.Enterpreneur.12(12);

Malhotra, S.P. (1984). Traditional agroforestry practice in arid zone of Rajasthan. In Agroforestry in Arid and Semi-arid zone of India. (Eds.) K.A. Shankarnarayan, CAZRI, Jodhpur, pp.236-266.

Mann, J.S., Sharma, S.C. and Mehta, R.S. (2001).Ardhsuskh Shestron Mai Krishi Vaniki Padyoti Ki Tulanatmak Adhyan. Pp. 46-50. *In Prakartic Sansandhano ka Prabandhan Mai Krishi Vaniki ka Yogdan*. (Eds.), Ks Dahwal and K.R. Solanki. NRC, Agroforestry Jhansi.

Muthana, K.D., Sharma, S.K. and Harsh, L.N. (1985). Study in silvi-pastoral system in arid zone. *My Forest* 21(3): 230-233.

Ram Newaz, Yadav, R.S. and Sukhei, A.K. (2001). Barani Shetra Mai Aonla Ke Sath Anth : Fasalothpadan. Pp. 86-88.Prakarthic Sansadhano Ke Porbandhan Mai Krishi Vaniki Ka Yogdan. (Eds.) K.S. Dadhwal and K.R. Solanki. NRC Agroforestry, Jhansi.

Ramakrishna, Y.S., Rao, GGNS. Kesava Rao, AVR and Vijaikumar, P. (2000). Weather Resource Management. Pp. 247-356.(Eds.) J.S.P. Yadav and G.B. Singh. Natural Resource Management for Agricultural Production in India. *Indian Society for Soil Science*, New Delhi.

Ramakrishna, Y.S. and Rao, A.S. (1992). Climatic features of the Indira Gandhi Canal Region. In Prospects of Indira Gandhi Canal Project. (Eds.I.P. Abrol and J. Venkateswarlu), ICAR, New Delhi.pp.1-10.

Sandhu, J.S. (2001). A case study for successful cultivation of Jojoba under resource constraints. In proceedings National Seminar on Production, Marketing and Processing of Jojoba. (*Simmondsia chinensis*) AJORP, Jaipur.pp.25-26.

Satyanarayan, Y. (1963). Ecology of the Central Luni Basin, Rajasthan. *Ann. Arid Zone* 2: 82-97.

Saxena, S.K. (1997). Traditional agroforestry systems in Western Rajasthan. *In :Agroforestry for sustained productivity* in Arid Region (Eds.) J.P.Gupta and B.M. Sharma. *Scientific Publichers*, Jodhpur India. pp 21-30.

Shankaranarayan, K.A. Harsh, L.N. and Kathju, S (1987). Agroforestry in the arid zones of India. *Agroforestry Systems* 5: 69-88.

Shankarnarayan, K.A. (1984). Agroforestry in Arid and Semi-Arid zones. Central Arid Zone Research Institute, Publication No. 24. pp. 1-295.

Sharma, N. K. (2004). Vaniki-charagrah (Nutan shrub + Dhaman grass), KAK. Jalore / 2004/pub-4 pp. 1-6. Assistant Director Krishi Anusandhan Kendra, Keshwana, Jalore.

Sharma, N.K., Singh, R.P. Singh, Mangeet and Harsh, L.N. (2001). Silvo-pasture management. Agrotech Publishing Academy Udaipur , pp. 1-140.

Sharma, S.K. and Dutta, B.K. (1999). Top feed tree species performance in Silvi-pastoral system of arid alluvial soil. *In* management of arid ecosystem. (Eds) A.S. Faroda, N.L. Joshi, S. Kathju and Amal Kar. Arid Zone Research Association of India and Scientific Publishers, Jodhpur. Pp. 351-356.

Sharma, S.K. (1996). Ber based agroforestry *Annual Progress Report*, IGFRI, Jhansi. pp.39-40.

Tewari, J.C., Sharma, A.K. Harsh, L.N. and Khan, M.A. (2004). Agroforestry thrives in the Thar desert of India. AP News No. 25, pp. 8-10.

Velayuthani, M. and Bhattacharya T. (2000). Soil Resource Management. Pp 3-129. *In* Natural Resource Management for agricultural production in India. (Eds.) J.S.P. Yadav and G.B. Singh. Indian society of Soil Science, New Delhi.

Yadav, M.S., Singh, Manjeet Sharma, S.K. Tewari, J.C. and Burman, U. (1997). Silvi-pastoral systems in Arid and Semi-arid ecosystems. CAZRI, Jodhpur. pp 1-472.

11

Status and Scope of Agroforestry Development in Eastern Part of Uttar Pradesh

B. S. RANA, A. K. SAXENA AND A. K. S. PARIHAR

Department of Forestry, N. D. University of Agriculture and Technology
Kumarganj-224 229, Faizabad (U.P.)

ABSTRACT

Eastern part of the Uttar Pradesh falls in mid-Gangetic plain commonly called as 'Purvanchal'. It is divided into three ecozones : i) North eastern tarai, ii) Eastern Plains, and iii) Vindhyan. Most of the farmers of the region have subsistence agriculture with small land holding. Although total forest area is 10.69 per cent of the geographical area, the actual well stocked forests (medium and good forests) in the region are confined to 3.51 %. Vindhyan zone indicates higher dependence for fire wood (35 % of the total need) on natural forests. Cow dung is major source used as fuel in Eastern plains (65 % of the fuel requirement). The cropping systems are different in rainfed and irrigated lands. Surveys of different parts of the eastern U. P. indicate that people have a progressive attitude towards agroforestry. However, due to land hunger most of the farmers develop homesteads. Agrihorticulture system is most popular system adopted by large and medium farmers in the region. Agrisilvihorticulture system is adopted by the farmers of Eastern plains and North eastern tarai zones. Agrisilviculture, Silvihorticulture, Agrisilvihorticulture and sivipastoral systems researched by us, are recommended for adoption by the farmers.

INTRODUCTION

Agroforestry has been an age old traditional land use in India. However, in present times it has attracted global attention as it can address land owners' need for improved productivity. Our country accounts for 2.3 per cent of global geographical area and 1.6 per cent of forest to meet the requirement of 16 per cent of the global population. According to Chauhan *et al.* (2002), India has 65 million hectares as classified forest area, of which, however only

60 per cent area is under optimum stocking level. Our forest wealth has witnessed a colossal depletion, as a consequence the contribution of forestry to the nation's GDP is only 1.7 per cent. According to the estimates given by NWSSLUP (Sehgal and Abrol 1994), 57 per cent of the land (187.7 million ha) in the country is potentially exposed to various degradation processes (Singh *et al.* 1994; Bhojvaid *et al.* 1996; Saxena *et al.* 1996; Singh *et al.* 1990; Rana *et al.* 2000; Parihar and Rana, 2001) which continues to increase year after year.

Unabated loss of natural forests area is coupled with acute shortage of fuel and fodder sources in almost every part of the country. Decline in country's natural vegetation has increased the hardships of rural women considerably as they have to cater their household need (Solanki, 2002). In eastern part of the Uttar Pradesh small size of land holding and subsistence agriculture with majority of farmers generally creates hindrance to most of the agroforestry programmes. Since food security being the primary option, diversion of their small arable land to usual forestry work seems impossible, therefore, development of home stead around their home may satisfy their fuel, fodder and fruit requirements to a great extent . Large and medium farmers have a definite scope to go for a large scale agroforestry on their unutilized and under utilized/agriculturally unproductive lands in the region.

The present paper discusses practices of agroforestry prevalent in eastern Uttar Pradesh and provides research based information for scope of agroforestry development in the region.

GENERAL DISTRIBUTION OF THE AREA

Twenty five districts in eastern U.P. (Purvanchal) fall in six revenue divisions : Basti, Gorakhpur, Faizabad, Varanasi, Devipatan and Mirzapur. The geographical area of eastern U.P. (79271 km^2) accounts for about 27 per cent of the U.P. It lies (24^0-27^0 34′ N latitude and 81^{0}13′-84^{0}11′ E longitude) in the Indo-Gangetic Plains (Fig. 11.1). The climate of the entire region ranges from sub-humid tropical to tropical, which is governed by south-western monsoon. Total annual rain fall is between 900 and 1018 mm in different agroclimatic zones (ecozones). The year is broadly divisible into: summer (March-mid-June); rainy (mid June- mid September); and winter (November-February). The eastern Uttar Pradesh is broadly divided into three important ecozones which are as follow:

North Eastern Tarai

This zone is located in north of 'Saryu' river. It has an area of 33,178 km^2, which is about 42 per cent of the eastern U.P (Table 11.1). This comprises northern parts of Bahraich, Gonda, Kabirnagar, Padrauna, Basti, Gorakhpur, Mahrajganj, Sravasti, Balrampur, and Sidharthnagar districts. This zone has 2,39,023 ha area under forest, while usar and wastelands occupy 30,258 ha and unculturable wastelands about 36,822 ha. Total annual rainfall is about 1,018 mm, and the zone indicates frequent floods in several locations of river bed areas during rainy season. Texturally, the soils are mainly loam to sandy loam. Soils of this zone are not as much fertile as those of Western Tarai of U.P. Northern parts of Gorakhpur and Deoria districts have peculiar problem of Bhat soils, which are rich in calcium carbonate.

Eastern Plains

This zone lies between the course of Saryu and Ganga rivers. It has an area of 34,784 km^2, which is about 44 per cent of the eastern U.P. This covers, districts Faizabad, Ambedkar Nagar, Barabanki, Sultanpur, Pratapgarh, Jaunpur, Azamgarh, Ghazipur, Ballia, Mau, Balrampur and parts of Varanasi. Total annual rainfall is around 900 mm. The area has limited irrigation facilities. Soils of this zone form the part of famous Indo-Gangetic plain of U.P. The area has been developed on alluvium transported by rivers Ganga, Ghaghra, Gomti and their tributaries. Apart from alluvial land, diara and usar lands (88,933 ha) are very common. Total area under forest is 94,093 ha, and unculturable wasteland is 71,239 ha.

Vindhyan

Besides Mirzapur and Sonbhadra districts, this zone also covers some parts of Varanasi district. This has an area of 11,310 km^2, which is about 14 per cent of eastern U.P. and 4 per cent of the U.P. state. Besides 5,07,651 ha area under forest, there is 48,642 ha Usar, 52,390 ha under unculturable wastelands and 76,146 ha area is flood prone and water logged. The annual rainfall is less than 900 mm. Irrigation facilities are meager. The soils are sandy, clay loam to silt loam. The area under sandy loam and loam silts, comprise 43 per cent and 27 per cent, respectively.

NATURAL RESOURCES AND FARMING SYSTEM

Forest and Fuel Resources

Although total forest area is 10.69 per cent of the geographical area, however the actual well stocked forests (medium and good forests) in the region are confined to 3.51 per cent. If north eastern tarai is excluded 1.4 per cent area remains under medium and good forests (with crown density> 40 %) in eastern Uttar Pradesh. The distribution of common forest and fruit trees indicate a little change amongst three ecozones. The north eastern tarai zone has luxuriant growth of forest trees viz.,*Tectona grandis, Dalbergia sissoo, Shorea robusta, Terminalia arjuna, Populus deltoides,* etc. Among fruits common species are : *Mangifera indica, Artocarpus heterophyllous, Litchi chinensis,* etc. In Eastern plains, *D. sissoo, Azadirachta indica, T. grandis, T. arjuna* and *Madhuca latifolia* are important forest tree species. Major fruit species are: *M. indica, Psidium guajava, Emblica officinalis, Aegle marmelos, Zizyphus mauritiana,* etc. In Vindhyan zone important forest species are: *Acacia catechu, M. latifolia, Bambusa spp.* Among fruits common are: *E. officinalis, Z. mauritiana, A. marmelos* and *Carissa carandos.*

Cooking fuel sources in each of the three zones of eastern Uttar Pradesh is shown in Table 11.2. It is evident that Vindhyan zone has higher dependence for fire wood (35% of the total need) on natural forests. Apart from forests, wood is also obtained from croplands and homesteads by the people. Thus, the contribution of total fuel wood (forest+cropland+homestead) in three ecozones is in the order: Vindhyan (50%)>North eastern tarai (35%)>Eastern plains (15%). Due to extremely poor forest area in Eastern plains, the expenditure on other fuel sources is also steadily increasing in this zone. Cow dung is major source used as fuel in eastern plains (65% of the fuel requirement) followed by north eastern tarai (50%). The population of tribals is highest in Vindhyan. Since ages, tribal communities have been closely associated with the natural forests. However, due to fast depleting forest resources and population growth in this zone, their hardships have increased to an alarming extent. Apart from increasing unemployment problem, poor management of forests has led to a large number of valued medicinal and other important species on the verge of extinction.

Farming System

The main component of farming system based on land holdings of the farmers are agriculture crops in different ecozones. Our data on cropping systems in three ecozones of the region are summarized in Table 11.3. The cropping systems are different in rainfed and irrigated lands. The total rainfed area in the region accounts for about 40 per cent of the total land. In general, irrigated areas have the major crops: rice, wheat, maize, sugarcane, gram, arhar and oil seeds. In rain-fed areas wheat, gram, maize, oil seeds and arhar are the usual crops.

Human Population

Total population in eastern U. P. is over 500 million, which is distributed in the order: 42.6 per cent in eastern plains, 41.6 per cent in north eastern tarai, and 15.8 per cent in Vindhyan. The majority of people in eastern U.P.

are marginal and small farmers. Based on land holding in different zones, the percentage of different categories of farmers is in the order: small farmers (35-71 %) > medium farmers (27-48 %) > large farmers (2-17 %). The number of small farmers is still on rise due to fragmentation of the land because of increase in human population in this most populous zone of India.

Animal Husbandry

Table 11.4 enlists the number of live-stock and source of fodder in three different ecozones. It is evident from the data that large and medium farmers in each of the ecozones use to rear milch animals viz. buffaloes and cow. However, small farmers prefer rearing cow and goat. The main fodder obtained from the agricultural crops is straw of wheat, paddy, jowar and bajra. Burseem is also fed to the cattle. Tree fodder constitutes green foliage of *M. latifolia*, *A. indica*, etc.

PRACTICE OF AGROFORESTRY

Choice of Species

The agroforestry had its primitive form in the past and practically a very little attention was paid towards improved farming. Important indigenous multipurpose tree species used to be retained on the farm boundaries, which were means of catering the diverse needs of the people. Common species raised on farm boundaries were: *Azadirachta indica, Acacia nilotica, Dalbergia sissoo, Madhuca latifolia*, etc. A significant sign of improvement that has become evident in the recent past is that educated youth have begun to be interested for inclusion of better forest and fruit species in agroforestry systems. Different tree species and crops presently included in agroforestry by the farmers of all the ecozones are listed in Table 11.5. The choice of species is usually governed by the economic out put to generate cash flow and the basic needs of farmers, as also suggested by Pathak and Roy (1992).

Adoption of Agroforestry

The approach of the farmers towards adoption of agroforestry remained confined to large and medium farmers. Based on the surveys of Faizabad district conducted by Mishra *et al* (1998), it was found that farmers who have greater than 3 ha land holding practice agroforestry. To examine the farmers' preference for agroforestry, farmers were categorized into five classes based on their land under agroforestry in five selected blocks of Eastern plain zone (Table 11.6). It is apparent from the data that in Bhiti and Pura blocks the maximum farmers practiced agroforestry on 20-40 per cent of their land. In Masodha, about 40 per cent farmers followed agroforestry on < 20 per cent land. On an average, about 38 per cent farmers on 20-40 per cent land and 24 per cent farmers on 40-60 per cent land practiced agroforestry in Faizabad. The farmers are adopting agroforestry in all the three ecozones depending on socioeconomic, agricultural and ecological problems existing in the region. Mandany (1991) opined that the adoption process of the agroforestry is determined by a complex set of the forces affecting the existing production system. Possibly the farmers are well aware with declining supply of the forest produce from the reserve forests which have alarmingly shrunk due to over exploitation and undeniably poor management in the entire region. Further, because the crop, live-stock and tree components of the agricultural systems are strongly interdependent (Thapa *et al.* 1991), the farmers had to include trees considering all their needs in mind.

AGROFORESTRY SYSTEMS

Through repeated surveys of different parts of the eastern U. P., it appears that farmers have progressive attitude towards agroforestry. The types of agroforestry systems practiced by the farmers on their own land are as following.

1. *Homestead*

Majority of the farmers in eastern U. P. are marginal and small farmers. As discussed earlier the large farmers have highest per capita land holding (>15.0 ha) although their number is very small. Therefore, due to land hunger most of the farmers develop agroforestry around their homes. They grow various vegetables, fruit trees viz., *Artocarpus heterophyllus, Psidium guajava,* banana, *citrus spp.* Among forest species they have a fair choice for *A. indica, Dendrocalamus strictus,* etc.

2. *Agrihorticulture*

This system is most popular system adopted by farmers in the region. The mango is most widely cultivated fruit species in every part of the eastern U. P. Under this system of agroforestry, wheat, arhar and mustard crops are grown by the farmers.

3. *Agrisilviculture*

This system stands second in choice by the farmers of the region. The most common woody perennial (forest tree species) adopted by the people are: *T. grandis* and *D. sissoo* in eastern plains and north eastern tarai zones. *Populus deltoides* is most popular species among farmers in north eastern tarai (particularly Srawasti and Behraich districts) (Photo 11.1). However in vindhyan, *D. sissoo* is most prevalent species. Unfortunately due to large scale mortality of *D. sissoo* stands during recent years many people in each of the three zones have started switching over to *T. grandis* (Photo 11.2). Agriculture crops grown by the farmers are wheat, paddy, mustard, sugarcane, pea, etc.

4. Agrisilvihorticulture

This system is adopted by the farmers of eastern plains and north eastern tarai zones. However, its prevalence is greater in the later zone. There are varied combinations of forest species with fruit species and agriculture crops. Amongst fruit species mango dominates in this system in most of the area surveyed by us. Among forest species common trees are: *D. sissoo, T. grandis,* E. hybrid, *P. deltoides,* etc. In agricultural component, wheat, paddy, pea, mustard, potato, etc. are cultivated under this system based on location specific requirements.

Among all the agroforestry systems, the farmers in the area seem to be keenly interested in practicing only those agroforestry systems which include horticultural component (Photo 11.3). Farmers usually have food security and generate cash income as their primary motive for including fruit trees in agroforestry. In addition, fruit trees are also managed to address issues like soil fertility maintenance and supply of fuel, fodder and timber. Singh and Singh (1990) and Aiyelaagbe (1992) have also discussed the relevance of fruit crops in agroforestry.

RECOMMENDED AGROFORESTRY SYSTEMS

On the basis of the research conducted at this University, following agroforestry systems have been found suitable for the region:

Agrisilviculture

Different agrisilviculture systems were studied by us at Main experiment Station of forestry and at farmers' field. Yield of the grain under four agrisilvicultural systems has been found lower than tree less open field condition (Table 11.7). Further, as the distance from tree base increases the grain yield under these systems also increases. It is because when a crop is grown in association with trees, there is a competition for light, moisture, nutrients, etc., and a positive or negative interaction might be expected to develop between them (Hocking,1986). Light is most dominating amongst other environmental factors (Evans *et al.* 1976). Thus, the reduction in grain yield under agroforestry system is due to lower photon flux than that of the open condition (Chazdon and Fetcher 1984). Among all the four agrisilviculture systems the wheat grain yield did not show any marked difference. However, the per cent reduction in wheat grain yield from the open condition was reflected maximum (31%) by *M. latifolia* based systems. Paddy grain yield showed a clear-cut difference amongst the studied systems. Greater grain yield was indicated by *P. deltoides* (2.84 t ha^{-1}) and *Casuarina equisetifolia* (2.42 t ha^{-1}) based systems. The per cent reduction in yield was maximum for *M. latifolia* based system (31%).

Under *C. equisetifolia* and *M. latifolia* based agrisilviculture systems, initially Usar-1 variety of paddy was cultivated for three consecutive years, there after Usar-1 was replaced by Usar-2 variety for the next 4 years. By the end of 7th year Sarjoo-52 variety was introduced. In case of wheat, Lok-1 variety was cultivated from second year of establishment of the system up to 7th year. There after, NW-1012 is being cultivated. *C. equisetifolia* based agrisilviculture system is shown in Photo 11.4.

We have also compared the results on carbon storage under agrisilviculture system with open agricultural land. The carbon storage in biomass of *C. equisetifolia* based agrisilviculture system has been computed for 5 to 10 year old system on sodic land (Table 11.8). The total carbon storage in biomass (above and below ground components of trees and crops) ranged from 22.1 t C ha^{-1} (5 yr old system) to 37.7 t C ha^{-1} (10 year old system). Under open condition, the agricultural crops (wheat and paddy) was the sole component amounting to the storage of 7.4-10.2 t C ha^{-1}, for the same period. These results indicate that agroforestry system has far greater potential for carbon sequestration than cultivation of agricultural crops in isolation.

Silvihorticulture

Three systems with different combinations of tree and fruit species have been examined for their performance. At 7 year age, the growth of *Eucalyptus tereticornis* has been found higher when grown with *E. officinalis* and *Leucaena leucocephala* as compared to combinations with other species (Table 11.9). Growth of *P. guajava* and *E. officinalis* is better than *Aegle marmelos* or *Grewia subinaequalis*. The total production of the system is higher when the *P. guajava* or *E. officinalis* are grown with *E. tereticornis* and *L. leucocephala* compared to species combination of *A. marmelos* + *E. tereticornis* + *G. subinaequalis*.

Silvipastoral

Under this system certain stress tolerant grasses were raised in association with *D. sissoo*. Under this system *D. sissoo* was planted at a distance of 8 x 4 m and root slips of four grasses were planted between tree rows. The fodder yield of all the grasses increased with increase in age of the system (Table

11.10), which may be because of the improvement in soil condition. Among all the grasses, the yield has been found maximum for *Pennisetum purpureum* and minimum for *Panicum maximum*. Seasonal variation in herbage yield of certain grasses under *D. sissoo* based silvipastoral system has also been examined (Table 11.11). The results indicate that amongst all grasses, *P. purpureum* has been found highest herbage yielder in different seasons. A remarkable seasonal change in case of *P. purpureum* reflecting maximum herbage yield in rainy season has also been evident. However, other grasses do not show marked variation in herbage yield across seasons

The production potential (total net production) of *Brachiaria mutica* planted under mixed tree stand and in open condition has been found greater under tree stand (690 g m^{-2} yr^{-1}) than in open (603 g m^{-2} yr^{-1}) (Saxena *et al.* 1996). The production potential of *B. mutica* on both stands was greater than those of arid grasslands, but was lower than humid and sub-humid grasslands (Umashanker *et al.* 1993). In our case, aboveground net production accounted for about 85-87% of the total net production resembling with dry sub-humid grasslands (Singh and Yadava 1974). On both stands, the production has been found maximum during rainy season and minimum during summer. The greater productivity of *B. mutica* under mixed tree stand than in open is perhaps related to higher soil organic carbon and nitrogen added with congenial microenvironment under mixed tree stand.

Agrisilvihorticulture

In this system certain fruit species *(E. officinalis, P. guajava* and *Z. mauritiana*) are planted in association with *C. equisetifolia* and *E.* hybrid separately. Agricultural crops are grown between rows of tree species. It has been observed that the agricultural crops perform better under *C. equisetifolia* based agrisilvihorticulture than under *E.* hybrid based system. Later on, due to increase in canopy of tree and fruit species, tuber crops viz., colocasia and turmeric are cultivated. The *C. equisetifolia* based system shows tuber yields of 3.0 t ha^{-1} and 1.5 t ha^{-1} for colocassia and turmeric, respectively. The inclusion of fruit trees in this system may fulfill the primary object of the farmers to have food security as well as generating cash income (Aiyelaagbe, 1992).

FUTURE PROSPECTS AND STRATEGIES FOR AGROFORESTRY DEVELOPMENT

Agroforestry in eastern U. P. involves a host of problems for its success. The first and foremost limitation being the edapho-climatic complexity followed by frequent floods and water logging of large area (6217.5 km^2). However, agroforestry has a great scope to mitigate all the problems and regional constrains. The tolerance limits to different environmental factors is different for different tree species, which is an advantage over the traditional system. If tree species are properly chosen, the productivity may be assured even under hostile problem soil and water logged conditions. Unemployment is the great problem in the region, as the existing agricultural practices do not seem to support the masses of unemployed youth. As a result, the great number of the rural population migrates to the near by towns. Since most of the forestry activities are labor intensive, considerable employment opportunities can be generated from agroforestry works. Agroforestry may be a good launch pad for simultaneous as well as long term gains to substantiate socio-economic and environmental values even under conditions of the fragile wastelands of the region. It would give far more cash income to the farmers than practicing pure agriculture crops per unit land. In fact, agroforestry has immense potential to provide the farmer with a number of alternatives of agriculture, forestry and horticultural crops.

The agroforestry system would be capable of meeting the demand of raw materials of several agricultural and forest based industries. Oil yielding tree species viz., *Jatropha curcas* and *Pongamia pinnata* are suitable for cultivation even under sodic and other wasteland including ravines in this part of Uttar Pradesh. Development of oil based cottage industries would help in employment generation in eastern plains and north-eastern plains of the eastern U.P. *A. indica* and *M. latifolia* have been retained in the forestry boundaries from the ancient times due to usable oil and economic consideration. Inclusion of these species in agroforestry may open ways for good economy generation. In Vindhyan zone, majority of people are marginal farmers and tribals. Bidi making cottage industry has immense potential for employment generation. Leaves of *Diospyros melanoxylon* (tendu) are the raw material for this industry. Development of Agroforestry systems based on *Buchnania lanzan* (chiroungi) is also a viable possibility for employment generation. In addition, silvipastoral systems may offer an increased opportunity for socio-economic upliftment of the tribal people. The advantage with silvipastoral management is that people would be motivated towards live-stock farming and the present pressure on our natural forests would be reduced substantially. Bamboo (*Dendrocalamus strictus)* is widely grown near the homesteads in rural area. Being multipurpose species, bamboo based cottage industries may offer a big scope for rural employment in all the ecozones of this region.

To boost the present status of the agroforestry as visualized through a number of surveys (Mishra, *et al.* 1998., Parihar and Rana, 2001) in this region it is pertinent to focus on the following future strategies:

Table 11.1. General information on different ecozones

Location and area	Ecozones (Agroclimatic zones)			Entire region
	Eastern Plains	North eastern tarai	Vindhyan	
Location	Between courses of Saryu and Ganga rivers	North Saryu river	Mid Sonbhadra and part of Varanasi	-
Geographical area (km²)	34784	33178	11310	79272
Major districts covered	Faizabad, Sultanpur, Jaunpur, Barabanki, Ambedkaar Nagar, Azamgarh, Ballia, Ghazipur,Mau, etc.	Behraich, Gonda, Maharajganj, Gorakhpur,Padrauna, Deoria, etc.	Varanasi, Mirjapur, Sonbhadra, Bhadoi, etc	-
Area under crops (%)	68.70	60.40	34.20	60.30
Forest area (%)	2.58	7.48	45.07	10.69 (3.51)*
Area under fruits (%)	2.3	3.40	1.40	2.63
Wasteland area (%)	11.8	11.90	15.70	12.40
Area under common property resources+ Miscellaneous area (%)	14.49	18.10	3.80	14.48

Table 11.2. Sources of cooking fuel in three ecozones of Eastern Uttar Pradesh

Source of fuel	Contribution (%)	Tree species used for fuel
North Eastern tarai		
Wood from forest	5.0	*M. latifolia, A. indica* and *A. nilotica*
Wood from crop lands	20.0	*A. indica* and *A. nilotica*
Wood from house steads	10.0	*A. indica A. nilotica and M. latifolia*
Cow dung	50.0	-
Crop biomass	15.0	-
Eastern Plains		-
Wood from forest	0.2	*M. latifolia* and *M. indica*
Wood from crop lands	5.0	*A. nilotica* and *A. indica*
Wood from house steads	9.8	*A. indica* and *M. indica*

Contd...

Table 11.2. Contd...

Wood from house steads	9.8	*A. indica* and *M. indica*
Cow dung	65.0	--
Crop biomass	20.0	-
Vindhyan		-
Wood from forest	38.0	*M. latifolia* and *A. indica*
Wood from crop lands	5.0	*A. nilotica*
Wood from house steads	7.0	*A. indica* and *M. latifolia*
Cow dung	35.0	-
Crop biomass	15.0	-

Table 11.3. Cropping system in rainfed and irrigated areas of eastern U. P.

Ecozones	Per cent area	Cropping system
North eastern tarai		
Rain-fed	49.7	Rice-Wheat; Maize-Wheat
Irrigated	50.3	Sugarcane-Arhar; Rice-Gram
Eastern plains		
Rain-fed	29.0	Maize-Wheat; Maize-Gram; Maize-Oilseed
Irrigated	71.0	Sugarcane-Arhar; Rice-Wheat
Vindhyan		
Rain-fed	58.0	Arhar; Maize-Wheat
Irrigated	42.0	Sugarcane; Rice-Wheat

Table 11.4. Live stock and source of fodder in eastern Uttar Pradesh

Ecozone/Farmer Type	Animal Species	Number of animals per house hold	Main fodder	Tree fodder
1. Eastern Plains				
Large farmers	Buffalo and Cow	4	Wheat straw	*M. latifolia*
Medium farmers	Buffalo and Cow	6	Paddy straw	*Acacia nilotica*
Small farmer	Cow,goat/Sheep and Bullock	10	Bajra, Jowar, and Barseem	
2. North eastern tarai				
Large farmers	Buffalo, Cow and Bullock	6	Wheat straw	*M. latifolia*
Medium farmers	Buffalo, Cow, Bullock	8	Paddy straw	*A. nilotica*
Small farmer	Cow,goat/Sheep and Buffalo	10	Bajra, maizeJowar, and Barseem	
3. Vindhyan				
Large farmers	Buffalo and Cow	3	Wheat straw	*Azadirachta indica*
Medium farmers	Buffalo, Cow and Bullock	5	Paddy straw	*M. latifolia*
Small farmer	Sheep, Goat/Sheep, Bullock and cow	15	Bajra,Maize, chari and Barseem	

Table 11.5. Woody plants and crops grown in three ecozones of eastern U. P.

Ecozone/ Farmer type	Average area per farmer (ha)	Forest tree	Fruit tree	Agricultural crop
1 Eastern Plains				
Small farmers	0.43	*D.sissoo, A. indica, Eucalyptus* hybrid	*M. latifolia, M. indica*	Rice,,wheat, mustard, inseeds
Medium farmers	3.24	*D.sissoo, A. indica, Bambusa spp.*	*M. latifoliaE. officinalis*	Rice, wheat, maize, sugarcane, gram, mustard, linseed
Large farmer	15.4	*D.sissoo, A. indica*	*E.officinalis, Psidiom guajava.*	Rice, wheat, maize, sugarcane, gram, mustard, linseed, arhar
2 North eastern tarai				
Small farmers	0.49	*D.sissoo, A. indica*	*M. indica, Litchi chinensis*	Rice, -wheat , mustard
Medium farmers	3.27	*T. grandis, Terminalia arjuna, D.sissoo, A.indica Bambusa spp.*	*,M. latifolia, E. officinalis*	Rice, wheat, Maize, sugarcane, gram, arhar, mustard
Large farmer	15.43	*D.sissoo, A. indica*	*E. officinalis, P. guajava.*	Rice, wheat, Maize, sugarcane, gram, arhar, mustard
3 Vindhyan				
Small farmers	0.42	*D.sissoo, A. indica, Bambusa spp., A. nilotica*	*M. latifolia, E. officinalis*	Rice, wheat, mustard, linseed
Medium farmers	3.80	*M. latifolia, Bambusa spp.*	*E. officinalis, Zizyplus mauriatiana*	Rice, wheat, maize, sugarcane, gram, arhar, mustard
Large farmer	19.68	*A. nilotica, Bambusa spp.*	*M. latifolia, , E. officinalis, Z. mauriatiana*	Rice, wheat, oilseeds, Maize, sugarcane, gram, arhar

Table 11.6. Percentage farmers with land under agroforestry in certain blocks of Faizabad districts

Block	Land under Agroforestry				
	<20	20-40	40-60	60-80	> 80
Sohawal	30	30	30	0	10
Bhiti	0	60	20	0	20
Milkipur	0	40	40	20	0
Masodha	40	20	20	10	10
Poora	0	40	10	30	20
Average	14	38	24	12	12

Table 11.7. Wheat and paddy grain yield in four agrisilviculture systems.

Distance from tree base (m)	*C. equisetifolia* (13-yr)		*M. latifolia* (13- yr)		*D. sissoo* (5-yr)*		*P. deltoides* (5-yr)*	
	Wheat (NW 1012)	Paddy (Sarjoo-52)	Wheat (NW 1012)	Paddy (Sarjoo-52)	Wheat (NW 1012)	Paddy (Sarjoo-52)	Wheat (NW 1012)	Paddy (Sarjoo-52)
0.50	1.42	1.87	1.41	0.88	2.20	1.58	1.42	2.78
1.5	2.42	1.94	2.42	1.95	2.59	1.94	1.94	2.83
2.50	2.64	2.81	2.73	2.71	2.72	2.07	3.42	2.86
3.50	3.46	3.06	3.15	3.08	2.79	2.65	3.43	2.90
Average	2.48	2.42	2.43	2.15	2.58	2.06	2.55	2.84
Open field	3.43	3.11	3.50	3.13	2.79	2.71	3.46	2.93
CD (5%)	0.861	0.092	0.450	0.141	0.235	0.327	0.321	0.367

*Distances from the tree base were 0.5, 1.0, 1.5 and 2.0 m

Table 11.8. Carbon storage (t C ha^{-1}) in vegetation pool of *Casuarina equisetifolia* based agri-silviculture system and open field.

	Age (yrs) of the system					
	5	6	7	8	9	10
(A) Agrisilviculture system						
i. Tree component	14.2	21.6	24.6	24.8	26.3	27.1
ii. Agriculture component						
Wheat	3.1	3.1	3.1	3.5	3.5	3.8
Paddy	4.8	5.0	5.4	6.1	6.6	6.8
Total carbon storage	22.1	29.7	33.1	34.4	36.4	37.7

Contd...

Table 11.8. Contd...

(B) Open area						
Wheat	2.9	3.1	3.6	3.7	3.6	4.0
Paddy	4.5	4.9	4.5	5.3	5.8	6.2
Total carbon storage	7.4	8.0	8.1	9.0	9.4	10.2

Table 11.9. Species combination, density, growth and production of certain 7-yr old silvihorticulture systems (Singh, 1996)

S. NO.	Tree Species combination	Density (plants ha^{-1})	Stem dbh (cm)	Production (t ha^{-1})
1.	*Psidium guajava* +*Eucalyptus tereticornis* +*Leucaena leucocephala*	144	16.0	2.5*
		100	15.9	10.5**
		80	12.3	3.0***
2.	*Emblica officinalis* +*Eucalyptus tereticomis* + *Leucaena leucocephala*	144	16.1	3.6*
		100	18.1	9.0**
		80	12.8	3.1***
3.	*Aegle marmelos*+*Eucalyptus tereticornis* +*Grewia subinaequalis*	144	14.8	2.2*
		100	14.2	7.8**
		80	9.6	0.1*

* fuel ** wood *** fodder

Table 11.10. Green fodder yield of certain grasses in *D. sissoo* based silvipastoral system for five years (Rana *et al.* 2000)

Species	Fodder yield (t ha^{-1})					
	1992	1993	1994	1995	1996	Average
Pennisetum purpureum	12.10	43.94	50.44	55.10	55.52	43.42
Cenchrus ciliaris	8.72	22.99	32.99	45.00	46.63	31.37
Brachiaria mutica	7.25	38.10	41.04	46.60	47.63	36.12
Panicum maximum	3.70	13.97	23.72	25.23	31.13	19.55
CD (1%)	0.264	0.571	0.187	0.477	0.496	-

Table 11.11. Tree growth and grass yield under *D. sissoo* based silvipastoral system.

Species	Tree height (m)	Collar dia. (cm)	Green herbage yield of the grass (t ha^{-1})				Total herbage yield on dry weight basis (t ha^{-1})
			Summer (May)	Rainy (Sept.)	Winter (Jan.)	Total	
Pennisetum purpureum	4.26	3.42	17.31	19.63	16.52	53.46	26.2
Brachiaria mutica	3.87	3.37	13.2	13.74	14.52	41.42	16.15
Panicum maximum	4.05	3.23	10.41	11.14	11.43	32.98	14.51
CD (5%)	0.2167	0.0967	2.1693	2.0471	1.8210	2.0124	

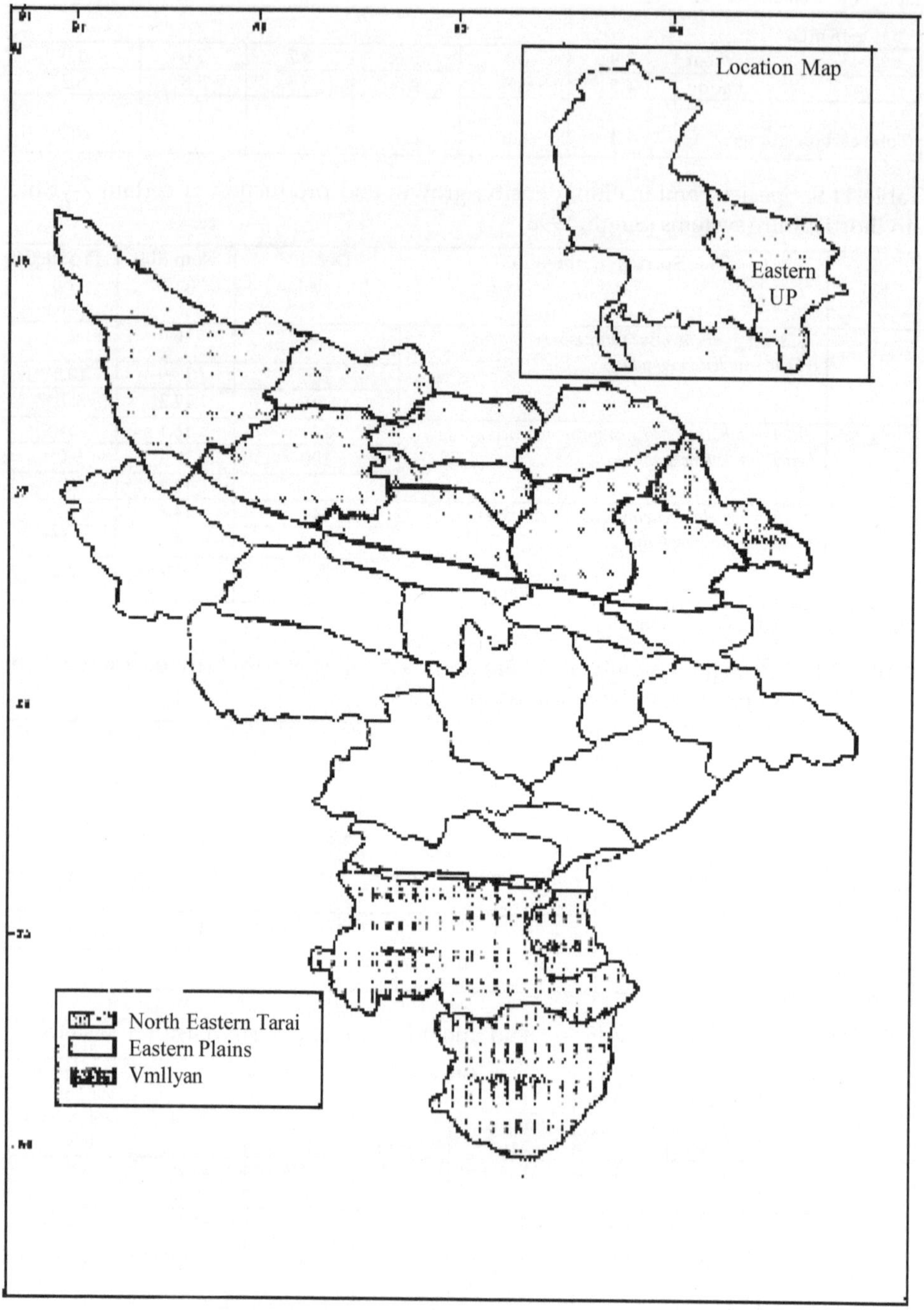

Fig. 11.1. Map sowing three ecozones of eastern part of Uttar Pradesh

Photo 11.1. Populus deltoides + sugarcane based agroforestry systems

Photo 11.2. *Tectona grandis* based agrisilviculture system at farmer's field (mustard is an agricultural component)

Photo 11.3. *D. sissoo+M. indica* based agrisilvihorticulture system developed by a farmer in North eastern tarai zone. *Pisum sativum* is agricultural crop

Photo 11.4. A five year old *C. equisetifolia* based Agrisilviculture system.

- Genetic improvement of indigenous and exotic species (i.e. *D. sissoo, M. latifolia, C. equisetifolia, P. pinnata* and *A. indica*) so that appropriate planting materials may go to the farmers in near future.
- The live-stock quality is generally poor which needs proper attention by the policy makers and planners. Large sized unproductive animals need to be replaced by small sized more productive live-stock. Emphasis on rural pasture land for developing silvipastoral systems should be given by government and other agencies for good livestock farming particularly in tribal areas of Vindhyan zone.
- Large scale plantation of oil bearing tree species viz,. *J. curcas, M. latifolia, A. indica* and *P. pinnata* should be encouraged as the people of the area have a definite attachment with these species since antiquity.
- Sustainable land use management for sodic land, diara land, ravines and other wastelands needs to be done in the region. Development of Agroforestry systems for Vindhyan zone, where *D. strictus, D. melanoxylon, B. lanzan* are important species.
- Identifying disease resistant trees of *D. sissoo* from high tree mortality areas of north eastern tarai zone. The preventive and remedial measures would be possible through an integrated multi-disciplinary approach.
- Identifying bottlenecks of past agroforestry programmes to pave the way for making it a people's programme to generate self financing system in rural areas. This would help in exploring adequate employment opportunities to the rural people through wasteland management in view of promoting agroforestry based by-products and securing ecologically sustainable development in the region.

Acknowledgements

The authors gratefully acknowledge the kind support and facilities provided by Dr. B.P. Singh, Head, Department of Forestry. This study was funded by I.C.A.R., New Delhi under All India Coordinated Research Project on Agroforestry.

References

Aiyelaagbe, L. O. O., (1992). Fruit crops: Are they relevant for agroforestry development? *Agroforestry Today* 4(3):12-13.

Bhojvaid, P. P., Timmer, V. R. and Singh, G.. (1996). Reclaiming sodic soil for wheat production by *Prosopis julifliora* plantation in India. *Agroforestry Systems* 34: 139-150.

Chauhan, S. K., Gill S. S., Chauhan, Rajni and Khajuria, H. N. (2002). *Production Technology and management of Agroforestry Models.* Department of Forestry and natural Resources, PAU, Ludhiana. 334p.

Chazdon, R. L. and Fetcher, N. (1984). Light environment of tropical forests. Pages 27-36 In: E. Medina, H. A. Moony and C. Vazques- Yanes (eds.) *Physiological Ecology of Plants in the Wet Tropics.* Dr W. Junk, Holand.

Evans, G. C., Bainbridge, R. and Rackhan, O. (1976). *Light as an Ecological Factor.* Blackwells, Oxford

Hocking, D., 1986. Some principles of crop complementarity in agroforestry systems. Pages 39-42. *In*: P. K. Khosla, S. Puri and D. K. Khurana (eds.) *Agroforestry Systems- A New Challenge.* ISTS, Solan.

Mandany, M. H. (1991). Living fences: Somali farmers adopt an agroforestry technology. *Agroforestry Today* 3(1):4-7.

Mishra, M, Saxena, A. K. and Singh, B. P. (1998). Agroforestry systems and their yield potential. Pages 339-351. In: D. C. Pandey (ed.) *Managing Agriculture For a better tomorrow.* M. D. Publications PVT Ltd., New delhi

Parihar, A. K. S. and Rana, B. S. (2001). *Impact Analysis Report of All India Coodnated Research Project on agroforestry.* Submitted to the I. C. A. R., New Delhi. 30*p*.Pathak, P.S. and M. M. Roy 1992. Fodder trees in agroforestry: their selection and management. *Range Management and Agroforestry.* 13(1): 63-87.

Rana, B. S., Parihar, A. K. S. and Singh, B. P. (2000). Herbage yield of four grasses in sodic soil under silvipastoral production system. *Range Management and Agroforestry.* 21(2): 213-217.

Saxena, A. K., Rana, B. S., Rao, O. P. and Singh, B. P. (1996). Seasonal variation in biomass and primary productivity of paragrass (*Brachiaria mutica)* under a mixed tree stand and in an adjacent open area in northern India. *Agroforestry systems* 33:75-85.

Sehgal, J. L. and Abrol, I. P. (1994). Soil Degradation in India: Status and Impact. Oxford an IBH Publishing Co.,Pvt. Ltd., New delhi, India .

Singh G., Abrol, I. P. Cheema, S. S. and Gill, H. S. (1990). Agroforestry on salt-affected soils of India. Pages 96-115. *In*: R. K. Singh, A. K. Saxena and I. S. Singh (eds.) *Agroforestry: Present Status And Scope foe Future Development in Farming System.* Raghuvanshi Printers, Faizabad.

Singh, I.S. (1996). *Agroforestry for Salt Affected Lands.* Bulletin, NDUAT, Kumarganj, Faizabad. 23*p*.

Singh, I. S. and Singh, R. K. (1990). Fruits as a components of agroforestry. . In: R. K. Singh, A. K. saxena and I. S. Singh(eds.) *Agroforestry: Present Status And Scope For Future development in Farming systems.* N. D. University of agriculture and technology, Kumarganj, Faizabad.

Singh, J. S. and Yadava, P. S. (1974). Seasonal variation in composition, plant biomass and net primary productivity of a tropical grasslandat Kurukshetra, India. *Ecological Monograph* 44: 351-376.

Singh, P., Pathak, P. S. and Roy, M. M. (1994). *Agroforestry Systems for Sustainable Land Use.* Oxford and IBH Publishing Co. Pvt. Ltd., New delhi. 283 *p.*

Solanki, K. R. (2002). Agroforestry_ Oppurtunities and challenges. Pages 1-6. In. *Production Technology and management of Agroforestry Models.* Department of Forestry and natural Resources, PAU, Ludhiana.

Thapa, B., Joshi, L., Sherpa, S. L. Karki, I. B., Kusle, R. K. Jha, Y. N., Mainali M. P. (1991). The planting and live-stock don't mix.....or do they? *Agroforestry Today* 3(2) : 11-19.

Umashanker, Pandey, H. N. and Tripathi, R. S. (1993). Phytomass dyanamics and primary productivity in humid grasslands along altitudinal and rainfall gradients. *Acta Oecologica* 14:197-209.

• • •

12

Agroforestry in Eastern Uttar Pradesh

S. B. LAL
Dean, College of Forestry, Allahabad Agricultural Institute
Deemed University, Allahabad - 211007, (U.P.)

ABSTRACT

Eastern Uttar Pradesh has 27 districts and lowest per capita net domestic output then other regions of the state. Rainfall is the only source of water. Acute shortage of fuel, fodder and timber makes agroforestry an important proposition for this region. The paper presents different agroforestry models developed by Allahabad Agriculture Institute for the region. Economics of the prominent agroforestry systems has been worked out. Emerging innovative agroforestry practices like aqua farming, honey bee, sericulture are discussed. Recommended species and agroforestry models for utilizing wastelands of the region are also provided.

INTRODUCTION

Eastern Uttar Pradesh covers an area of 87,840 sq. km. ranging from 24^0 - $28^0 20'$ N latitude and 81^0 - 85^0 E longitude (map of Eastern U.P.) with population of 632 lakhs (Census, 2001). It has 27 districts and the lowest per capital net domestic output than other regions of U.P. It has 6 districts which are under developed and 15 districts are industrially backward. The contour of 100m roughly marks the western limit of Eastern Uttar Pradesh. It is mostly filled with deep alluvial deposits by the Ganga and tributaries. The Micro-regional variations are highlighted by the Bangar (the old alluvium) and the Khedar (new alluvium) formations. It enjoys tropical humid climate (rainfall varying from 120 cm to 140 cm). This region is plaughed with recurring floods and droughts. In 1983 Eastern Uttar Pradesh suffered a loss of the order of Rs. 319.12 crores because of the floods; in the same region, drought due to long dry spells left 3.820 lakhs ha. of sown area to wither. The menancing of Ganga - Ghaghara, Doab and recurring drought of trans Yamuna tract of

Allahabad district, parts of Mirzapur and Varansi districts and big chunks of usar and unculturable wasteland (2.93 lakh ha) sap agricultural potential of these areas. Hardly 9.54 percent of total geographical area is under forests. Their paucity accenulates the problems of floods and soil erosion. Most Ganga plain is devoid of mineral wealth excepting kankar in Bhanagar tracts, but southern uplands are rich in minerals like limestone, dolomite, silica, sands, marble, building stone, magnesite and coal; organic matter is rich but nitrogen is deficient. Despite of vast water resources percentage of irrigated to gross cropped area is only 40.36 per cent that led cropping intensity only 145.20 and net irrigated to net sown area is 54.22 per cent. It has depressed economy where unemployment and underemployment is rampant, thus out migration is a typical phenomena.

Eastern districts of Uttar Pradesh have vast potential to increase production in agriculture, horticulture, dairy and fishery products as the Gangetic plain provides fertile lands with adequate water resources. With high pressure of population, frequent recurrence of floods and droughts and low productivities in agriculture, the eastern districts of Uttar Pradesh are regarded as low-income area requiring special attention on bits of development. Backwardness affecting not only living conditions but also economy of the state.

AGRICULTURE AND SOCIO - ECONOMIC CONDITIONS

Agriculture is the fate of Purvanchal accompanied by live stock. This region falls much below then the national average of the Relative Development Index. During agriculture seasons it leads to scarcity of labour due to out migration (Singh, 2001). Agriculture production per hectare is highest in Deoria (Rs. 5404/ha.) and least in Bahraich (Rs. 2434/ha.) as compared to average U.P. figure of Rs. 4502/ha. and country's average of Rs. 3576/ha. Average size of holdings in eastern U.P. is very small as compared to the state and national average. Population Density is 500 persons /km^2 which is above the states average. Irrigation area is 40 per cent of the gross area and Irrigation Intensity is below 12 per cent in 12 districts. Except Mirzapur and Sonebhada the percentage of area under forest is less than national average. Diversification of rural economy has many dimensions and ramifications. However, in the farm sector, income and employment potential could be explored by diversifying agriculture through forestry, apiculture, sericulture, mushroom, dying, mixed farming with live stock, goat rearing and fish culture (Singh and Rai, 2001) which can be achieved by:

1) Creation of woodlots in village community land, revenue wastelands and panchayat land,

2) Planting of trees on the road, canal and rail sides;
3) Afforestation of degraded forest in the vicinity of habitation and
4) Planting of trees on and around agriculture fields, dwelling compounds and on private marginal lands, constituting farm forestry and agroforestry.

Eastern UP's economy squarely hinges upon agriculture but there is excessive pressure on agricultural resources as its per capita net sown area is barely 0.15ha and its agricultural holdings are mostly small and marginal (90.82 %) (Census of India, 1981). Its agricultural economy should be boosted up particularly by providing incentives to small and marginal farmers by way of timely supply of inputs like fertilizers, high yielding varities of seeds, irrigation and credit facilities. Among agriculture crops paddy is dominant, other crops are maize, pulses and other oil seeds. Loss of agricultural resources must be improved by giving high priority to paddy, maize, pulses, and oilseeds cultivation. Farm forestry, social forestry and agroforestry programmes should be encouraged and floods and droughts must be reduced through master plan.

Rainfall, the only source of water for agriculture is scanty in quantity and erratic in distribution. The vicious combination of all these factors has made dryland agriculture a risky proposition leading to lack of investment in rainfed areas and low or non-adoption of technologies (Singh, 2002). It is felt that major part of eastern Uttar Pradesh is under dryland with prolonged drought and hence needs considerable management through agroforestry systems to rehabilitate the eroded lands prone to degradation.

AGROFORESTRY AND AGRARIAN DYNAMISM

Agrarian dynamism refers to several attributes of farmers in a region, it refers to their ability to perceive, intercept and respond to new economic opportunities in agriculture in the context of uncertainty. Adoption of fast growing trees as cash crops appeared to be more prominent in the regions where agriculture was more progressive. In U.P. the difference between tree growing in the agriculturally more stagnant east and rest areas is reflected in the extent of involvement of farmers, number of trees planted, survival, site and purpose of planting. Agroforestry is subsistence oriented in eastern U.P. this is reflected from the primary data of a village Ali of Eastern U.P. (Table 12.1)

AGROFORESTY SYSTEMS

Agroforestry aims at using the land for multiple uses for raising trees, crops, fruits and fodders. It helps to meet the basic requirements of people with regards to fuel, fodder, timber and food. There is an acute shortage of fuel, in eastern Utter Pradesh, people burn cow dung for cooking food, which could otherwise be used for manuring fields. The maximum temperature goes more than 42°C while in winter it may drop down to about 5°C. Traditionally trees are planted in agriculture fields usually on the boundaries of fields and also around homes. Fruit trees are also, commonly planted in which agricultural crops are also raised for a few years. The main species of trees planted are:

Forest trees species

Acacia nilotica, Albizia lebbek, Artocarpus heterophyllus, Dalbergia sissoo, Dendrocalamus strictus ,Eucalyptus sp., Leucaena leucocephala, Moringa oleifera, Syzigium cumini, Tectona grandis, Bixa orellana, Terminalia arjuna, Dalbergia spp., Populus spp., Dendrocalamus hamiltonii, Bambusa balcooa, Pongamia pinnata, Azadirachta indica, Madhuca latifolia, Jatropha curcas (small tree or shrub), *Bombax ceiba, Cassia siamea.*

Fruit trees species

Mangifera indica, Emblica officinalis, Psidium guajava, Punica granatum, Syzygium cuminii, Artocarpus spp., Anona squamosa, Citrus spp., Aegle marmelos, Zizyphus spp.

Agroforestry models

The agroforestry model adoption includes the strategy of relooking and upgrading of knowledge about mixed cropping, intercropping, catch cropping, mixed farming and multi strata cropping. Allahabad Agricultural Institute - Deemed University, Allahabad has already carried out plantations through agroforestry models under agrisilviculture, silvihorticulture, silvipastoral, agrisilvihortipastoral, agrihorticultural, silvihorticultural systems under different districts of eastern Uttar Pradesh (Table 12.2). This provides comprehensive knowledge for successful establishment through desirable tree crop interaction (Anonymous, 2005a).

ECONOMICS OF PROMINENT AGROFORESTRY SYSTEMS

A) Babul based agroforestry system

Acacianilotica has shown successful combination with agricultural crops up to 4-5 years without any adverse effect on agriculture crops. It is used in boundary plantation of the farm land. Its spacing of 3x3m maintained on one acre will require 450 plants. Expenditure on planting of seedlings and its management up to 20 years is Rs. 8394. Timely thinning is necessary for good growth of babul. After 20 years income of Rs. 22,54,427 is obtained in sole babul plantation. While planting across the field boundary in spacing of 4 m can give an additional income of Rs. 3,62,340 per acre after 20 years. However, with paddy / wheat within the field it will provide income of Rs. 22,61,632 / acre after 20 years (Anonymous, 2005c).

B) Dalbergia Sissoo based agroforestry system

It is considered to be one of the best tree of India having its versatile use. It has become an age old practice to plant it on the boundary of agricultural fields as it acts as wind break. In agroforestry it is inter cropped with maize, mustard, chickpea, wheat, pea, sugarcane, cotton, etc.

In pure plantation up to 5-10 years fellings are carried out to be used for fuel wood purpose. After 18 years tree yields 34 cubic feet wood i.e. @ Rs. 125/feet, it yields Rs. 5000 per tree. After 25 years a tree yields 72 cubic feet of wood, which costs Rs. 10,000 Hence, after 25 years from sissoo plantation net return is Rs. 71,37,544/acre.

If sissoo is planted on boundary with 65 trees for an acre and paddy and wheat crops are taken the yield of crop gradually decreases which reaches to 20 per cent upto 6th year. By far, at 25 years a farmer through agroforestry gets net benefit of Rs. 7,59,340 per acre. While within the field through sissoo planting after 25 years in a wheat / rice farming system the additional income is Rs. 78,60,291 per acre (Anonymous, 2005d).

C) Poplar based agroforestry system

A spacing of 3 x 3 m i.e 85 trees per acre is suitable for planting across the boundary of the agriculture fields. Upto 1 to 3 yrs any sort of crops can be cultivated as inter crop. During winter, crops like wheat, jowar, berseem, mustard, seasonal vegetables are preferred due to leaf shedding (IFFDC, 2004). During succeeding years, it gives lesser yields. However, with spreading of crown shade loving plant can be introduced. In wheat/ paddy system it provides additional income of Rs. 1,13,635 per acre. But, in sugarcane based agroforestry the additional net return after 6 year will be Rs. 1,72,194 per acre. Upto 6 years, it gets ready to be cut, but areas having low fertility it is ready to be harvested only after 8 years (Anonymous, 2005 e) (Photo 12.3).

D) Teak based agroforestry system

In a pure plantation of teak after 40 years a return of Rs. 4,11,47,816 can be obtained; With sugarcane teak can be grown on bunds, for this every acre will have 65 trees which will provide a return of Rs. 10,71,623 per acre. When teak is grown on the boundary of the field and in field sugarcane is taken the system will provide a total return of Rs. 4,47,12,936 per acre (Anonymous, 2005f).

E) Eucalyptus based agroforestry system

Based on climate, Eucalyptus can be felled at 6-8 years. With 2 m spacing between trees one acre will have 100-200 trees on bunds. By planting Eucalyptus in one acre of land with agriculture (rice/wheat), after 8 years, it will provide additional net benefit of Rs. 69,175. However, with sugarcane it will provide additional benefit of Rs. 76,790 (Anonymous, 2005g).

EMERGING INNOVATIVE AGROFORESTRY PRACTICES

Honey bee farming

In case of cross pollinated crops were yield is limited by number of pollinators, it has been demonstrated that use of honey bee as pollinator can double and even triple the yield. One person can manage 30 honey bee colonies (Singh and Rai, 2001).

Sericulture

It has tremendous scope. Successful operation of sericulture farm in Gohakhpur distrct is testimony for its suitability to this region.

Dairy Farming

The demand for milk and milk products can be fulfilled through dairy farming. Mixed farming with crop- livestock integration can meet the above. Several research workers have shown the profitability of mixed farming systems over arable farming systems particularly under small land holding situation (Faroda, *et al.*; 1978; Tomar, *et al.*,1982; Singh *et al.*,1984; Singh *et al.*, 1986; Singh *et al.*, 1990).

Goat Farming

Goat is considered poor man's cow and plays important role in the economy of small, marginal and landless farmers. Landless laborers keep these animals for grazing on community land, roadside, canal side, drains and riverside. Since, the area of eastern UP has good consumption of meat, it is more profitable and viable to rear the goats for about 8 months and then to slaughter them.

Aqua Farming

It is now effective to utilize water resources to set up the production of protein rich food. For this, fish culture is a very lucrative business. A farmer can produce 2000-3000 kg fish/ha./year by improved aqua-culture practices. If a farmer can do watch and ward of the fish pond himself, income can be increased by another Rs. 5,000/- per hectare per year. Aquaculture in waterlogged and swampy saline areas not only save the money required for drainage purpose but also give additional income to farmers of about Rs. 30,000-45,000 per hectare (Kumar *et al.*, 1994).

Mixed Farming with Sheep+Goat+Crop farming System

On small and marginal farm of the eastern Uttar Pradesh remarkable success of this system has been observed. The reasons for adoption of this system are:

1. Uncertainty of weather and price changes causing uncertainty in expected net return.
2. Uncertainty of availability of capital, credit, input, etc
3. The psychological propensity of farmers to adopt technical innovations gradually through time, and
4. Uncertainty about the continued profitability of crop enterprises.

AGROFORESTRY FOR WASTELANDS

Vast track of the eastern U.P. consists of waste lands. These lands lie without use through out the year. Some grasses and shrubs come up during the rainy season. These are however destroyed by herds of cows, goats, sheeps etc,. It is surprising to see that maximum number of degraded lands are seen in areas nearest to early civilization. The banks of river Yamuna, which was once the most important trading route, contain most destroyed ravinous tracts. The banks of the holy river Ganga consists of vast streches of Khadar lands that are devoid of vegetation. Seasonally flooded alluvial lands unsuitable for short statured herbaceous annual but may facilitate very high yields of genera such as *Populus, Syzigium, Tamarind, Acacias* etc (Chaturvedi, 1985).

As per the study made by the Department of Forest, U.P. the following trees can be intermixed with each other to reclaim problem of soils.

Problem Soils	Plant Species for plantation
Usar Soil:	
Sandy Soil	In between two consucative rows of poplar, a row or *Termanialia arjuna, Haplophragma adenophyllum, pongamia pinnata, syzigium cumini, comment,* poplar clones G-3, G-48, S7C4 and S7C20 have shown good results.
Clayey Soil	Boundry trench with ridge on inside is effective trench should be about 1.2 m wide at the top. Species- *Agave americana, Euphorbia royleana, Opluntia dillentii, Parkinsonia acculeata, Pithecolobium dulce, Prosopis juliflora, Zizyphus oxyphilla, etc.*
Saline Soil	*Prosopis juliflora, Acacia nilotica, Pongamina pinnata, Terminalia. arjuna, Acacia leucopholaea, Albizia procera, Dalbergia sissoo, Acacia auriculiforms, Cassia siamea, Cordia dichotoma, Syzigum cumini, Haplophragma adenopyllum, Pithecolobium dulce.*
Ravine soils	
Trees and shrubs	*Acacia leucopholaea, Prosopis spicigera, Azadirachta indica, Holoptelea integrifolia, Acacia catechu, Balanites aegyptica, Flacourtia indica, Capparis decidua, Carrisa carandus, Zizyphus numularia laria, Dichrostachys cineria, Salvadora persica.*
Grasses	*Demostachya bipinnata, Apluda mutica, Bothriochloa pertusa*

CONCLUSION

Excessive pressure on agriculture in eastern Uttar Pradesh can be reduced by diversification of the land use/management system The land resources which have not been put in service due to degraded status should be reclaimed and further deterioration be checked by using an integrated land management system. Agroforestry has come up as an important and economical component in land use technologies, which must be incorporated in land management systems in productive and unproductive lands. Further, the agroforestry models those has shown success need to be incorporated in the uncultivable areas of eastern Uttar Pradesh.

ACKNOWLEDGEMENTS

The author of this article would like to thank Dr. Ashwani Kumar Sharma, C.C.F., Allahabad for detail information of forest in the Eastern U.P. The technical help of Dr. A. P. Pandey, Head, Agricultural Economics, A.A.I.-D.U., Allahabad is duly acknowledged.

Table 12.1. Basic data of Ali village situated in Eastern Uttar Pradesh

Items	Unit
Net Cultivated areas	284 ha.
Intesity of cultivation	130 %
Number of cultivators	193
Value of annual crop produced/ha.	Rs.5210
Value of movable assets/cultivator	Rs. 13570
Value of movable assets/ha.	Rs. 9226
Main crops	paddy, sorghum, pulses and wheat
Cultivators with more than 2.5 ha.	17
Cultivated land holdings more than 2.5 ha as a net cultivated land	18 %
Percentage of crop out put marketed	27 %
Cost of chemical fertilizer/ha. of cultivated land	Rs.345
No of trees on farm lands (including homestead lands)	4011
Share of Eucalyptus	82 %

Table 12.2. Agroforestry system established at different places of Eastern Uttar Pradesh:

Places	Agroforestry systems	Components (perennial + annual crops)	
Allahabad	Agrisilvihortifishery system	Arhar + aonla + teak + fish (Photo 12.1)	
	Agrisilviculture	Wheat + poplar	
		Paddy + Eucalyptus	
		Kharif	Rabi
		Bamboo+bajara	Bamboo+wheat, bamboo+teak+ wheat
		Bamboo+paddy	Bamboo+mustard
		Bamboo+sweet potato	Bamboo+arhar

Contd...

Table 12.2. Contd...

	Agrisilvihorticulture	Wheat + teak + mango	
		Kharif	Rabi
		Bamboo+bajara+ aonla	Bamboo+wheat+ guava
		Bamboo+paddy+ aonla	-
	Silvihorticulture	Teak + aonla	
		Kharif	Rabi
		Bamboo+guava	Bamboo+pea
		Bamboo+aonla	Bamboo+aonla+teak
		Bamboo+mango	Bamboo+mango/ guava
	Agrisilvipastoral	Wheat + guava + guinea grass	
Kaushumbi	Silvihorticultral system	Bamboo + mango	
		Teak + guava	
	Agrihorticulture system	Wheat + mango	
Sant Ravidas Nagar	Silvipastoral system	Sesbania + grass	
	Hortipastoral system	Guava + subabul	
	Silvihorticultural System	Teak + guava	
Pratapgarh	Agrihorticulture system	Jack fruit + arhar	
	Silvihorticulture system	Eucalyptus + aonla	
	Silvihorticulture system	Bamboo + mango	
	Agrisilvihorticultrual System	Arhar + bamboo + guava	
	Agrisilviculture	Bamboo around the field boundary (Photo 12.2)	
Mirzapur	Silvihorticulture system	*L. diversifolia* + sesbania + vegetable	
Sonebhadra	Silvipastoral system	Sesbania + grass (guinea) Grewia / *L.diversifolia* + guinea	

Photo 12.1. A side view of an Aquatic agroforestry system prevalent in Allahabad district., eastern Uttar Pradesh

Photo 12.2. Bamboo plantation across the field boundary as observed in the Pratapgarh areas of eastern Uttar Pradesh

Photo 12.3. An overview of the poplar based agroforestry system inter - cropped with jatropha at the university research site of Allahabad Agricultural Institute–Deemed University, Allahabad

REFERENCES

Anonymous (2005a). Annual Report. Development of Wasteland through Agroforestry system in Uttar Pradesh, Project Report Submitted by Allahabad Agricultural Institute- Deemed University, Allahabad to the Ministry of Rural Development, Govt. of India, New Delhi. pp.1–21.

Anonymous (2005b). Annual Report. Eco-rehabilitation of Degraded Lands and Social Upliftment of Rural People through bamboo Cultivation in District of Allahabad, Uttar Pradesh. Project Report submitted by Allahabad Agricultural Institute-Deemed University, Allahabad the Ministry of Rural Development, Government of India, New Delhi. pp.1–25.

Anonymous , (2005c), Babool (*Acacia nilotica*). Booklet published by forest Department of Uttar Pradesh, Rana Pratap Marg, Lucknow, U.P. pp. 1-15.

Anonymous, (2005d). Sisum (*Dalbergia sisso*). Booklet published by forest Department of Uttar Pradesh, Rana Pratap Marg, Lucknow, U.P. pp. 1-15.

Anonymous, (2005e). Poplar (*Poplus deltoides*). Booklet published by forest Department of Uttar Pradesh, Rana Pratap Marg, Lucknow, U.P. pp. 1-15.

Anonymous, (2005f). Teak (*Tectona grandis*). Booklet published by forest Department of Uttar Pradesh, Rana Pratap Marg, Lucknow, U.P. pp. 1-15.

Anonymous, (2005g). Eucalyptus. Booklet published by Forest Department of Uttar Pradesh, Rana Pratap Marg, Lucknow, U.P. pp. 1-15.

Census of India, (1981). Board of Revenue, Uttar Pradesh.

Chaturvedi, A.N. (1985). Fire wood farming on degraded lands: in the Gangetic plain. U.P. Forest Bulletin No. 50. PP. 1-52.

Faroda, A.S, Yadav, R.S and Pal, R.N. (1978). Comparative economics of specialized dairy Farming, mixed farming and arable farming. *J. Res. Harayana, Agric.Univ*. 8: 234-239.

Kumar Rajender ; Agrawal, M.C.and Garg, S.K. (1994). Economical Utilization of brackish ground water for fish culture and control of water logging and soil salinity. Proceedings on National Seminar on reclamation and management of water logged saline soil at Karnal (Haryana) from 5-8, August 1994. vol. II. pp. 126-146.

IFFDC, (2004). A report of Agroforestry Centre, Saranath, Varanasi of Indian Farm forestry Development Corporation Limited, Jail Road, Ray Bareily.

Singh, H.P. (2002) key note. *In*.. Dhople, A.M. (Ed) Agrotechnology for dryland farming. Scientific Publishers, Jodhpur, P.v.

Singh, P. (2001) Agricultural and infra- structural development in Eastern U.P.: present status and suggestions for further development. In: Singh, P.N. (Ed). Economics backwardness of Eastern Uttar Pradesh: problem, strategies of development.

Proceeding of a symposium held between 14-15th, October 1995 at Udai Pratap Autonomous College, Varanasi (U.P.). pp. 7-19.

Singh, S.N. and Rai, K.N. (2001). Analysis prospects of sustained growth of Eastern Uttar Pradesh. In : Singh, P.N. (Ed). Economic backwardness of Eastern Uttar Pradesh: problems and strategies of development, proceedings of a symposium held between 14-15th, October 1995 held at Udai Pratap Autonomous College, Varanasi (U.P.). pp. 20-27.

Singh, N. : Malik B.S. and Kairon, M.S. (1984). Mixed farming for more income. *Indian Farming* 34: 21-23.

Singh, N,; Malik, B.S. and Kadian , V.S. and Mehra, O.P. (1986). Comparative studies of various farming systems. *Haryana Vet.J.* 1(25): 30-33.

Singh. S.N. ; Singh, K.P.; Narinder Singh; Kadian, V.S. and Dahiya, S.S. (1990). Employment potentialities of different farming systems.

Singh, Ravinder; Singh, B.B. and Singh , J.N. (2001). Income and employment increasing possibilities under crop+sheep+goat farming system on small land and marginal farms in Eastern Uttar Pradesh. In: Singh, P.N. (Ed). Economic backwardness of Eastern Uttar Pradesh:Problem & strategies of development, proceedings of a symposium held between 14-15th, October 1995 at Udai Pratap Autonomous College, Varanasi (U.P.). pp. 28-41.

Swaminathan, M.S. (1985). "Forestry and Food production", World Forestry Congress: Plenary session, Mexico.

Tewari, D.N. (1999). Agroforestry – for increased productivity, sustainability and poverty alleviation. International Book Distributors Publication, Rajpur Road, Dehradun. pp. 84 – 87.

Tomer, S.P.; Sairam, R.K.; Hasika, A.S. and Ganguly, T.K. (1982), Comparative efficiency of dairy and mixed farming systems. *Forage Res.* 1: 93-98.

Tolba, Mostfa, K. (1982) "Rolling back the wastelands", lecture to the Society for Promotion of Wastelands Development, New Delhi, 27, December, 1982.

13

Agroforestry Practices in Chhattisgarh — Scope and Potentials

SUNIL PURI
Indira Gandhi Agricultural University
Raipur - 492 006 (Chhattisgarh)

ABSTRACT

Chhattisgarh is a tribal dominating state and is one of the backward and poor area of India in terms of poverty and productivity. The trees have been the part of agriculture in this region since long. The present paper highlights some of the practices which are being adopted by the farmers from time immoral and efforts had been made to fit them in the present day scientific classification system of agroforestry system with their spatial and temporal sequences. Despite of the fact that trees are integral part of farmlands still adoption of agroforestry as a sustainable land use has some constrains which have been discussed and way out suggested. Areas needed in developing agroforestry in the state have also been discussed at the end of the chapter.

INTRODUCTION

There are enormous physiographic, ecological and geological diversities across and within Chhattisgarh region which has greatly influenced soil, land-use pattern, water resources and evolution of agriculture system. The region is largely rainfed in spite of the efforts to expand irrigation potential. Rainfed farming systems are practiced on more then 80 per cent area, rice is the principal crop of the region accounting for about 88 per cent of Kharif sown area in Chhattisgarh plains, and 68 per cent in northern hills and Bastar Plateau. The productivity of traditional farming system is very low and the adoption of improved rainfed technology has been slow in spite of its superiority over that practiced by farmers. Livestock/animals are integral component of farming systems and provide draught power, fuel (in the form of dung cakes), nutrition and cash income from milk etc. however, the animals are weak and hence less productive.

Plantation of multipurpose tree (MPT) species play an important role in restoring productivity, ecosystem stability and biological diversity to degraded lands. The choice of plantation species is likely to influence both the rate and trajectory of rehabilitation process. In Chhattisgarh region where entisols (red laterite) and inceptisols dominate which have negligible productivity, there is a need to find ways to increase productivity by including woody perennial. This need to begin with knowledge of the status of trees and their roles in different land use system.

There occurs a traditional agroforestry in the region which farmer allow to grow *Acacia nilotica* naturally at irregular spacing on the bunds of paddy fields solely or in combination with *Butea monosperma* and *Terminalia arjuna* etc. *Acacia nilotica* , being a multipurpose and nitrogen fixer species, is highly preferred by farmers and as a result, it is widely distributed in the field. However, their unscientific distribution and management may reduce the sustainability of productivity which in turn may jeopardize the system to meet out the requirement of ever-growing population in the region.

Background information of Chhattisgarh region

The Chhattisgarh region with geographical area of 14.4 million ha. which is 30 per cent of the total geographical area of Madhya Pradesh, is spread over between 17^0 41″ N to 24 45″ N latitude east Madhya Pradesh. The region is divided into 3 agroclimatic zones - Chhattisgarh plains, Northern hills and Bastar plateu. It is one of the backward and poor areas of India in terms of poverty and productivity (Singh and Baghel, 1994).

The general geological make up the region comprises of (i) laterites, alluvium and soil of recent age capping over (ii) horizontally bedded sequence of sedimentary rocks of lower vindhyan and on the top, followed successively by gunderdehi shale, Charmuria limestone, and Chandrapur quartzite sandstone and (iii) Archaean granite, gneisses and metasediments of oldest age (Vijayraghavan 1992,Romani 1992). Rock formations in Balaghat are mostly Pre-cambrians and Archaean crystalline complex of igneous and metamorphics which are mostly granite, gneisses and schists. These formations have very limited primary porosity and permeability to water. Hill ranges isolated hillocks, inter- hill valley, low lying plateaus, piedmonts, and inland valley are the primary landform found in the region.

Mahanadi and Godawari are the watershed basins in the Chhattisgarh plains. Drainage system is shaped by widely divergent types of topography. Mahanadi and its tributaries provide drainage outlets to most of the Chhattisgarh plains and to parts of Bastar and Raigarh districts. The Mahanadi emanates from Sihawa in Raipur, turns to southwest and then to northeast and finally flows majestically eastward into plains of Orissa. Its tributaries Aner, Surthi, Hap, Arpa, Kharun,etc. drain western fringe of hillocks and plains to join Mahanadi. Similarly Hasdo, Mand, Kelo and their tributaries drain northern part of the basin into the Mahanadi. Almost all first, second and higher order tributaries of the river system emanate from within the region and are ,therefore, erosive when young. However, subsequently due to plain topography of the basin floor, the river flows rather leisurely through meandering course and therefore has capacity of a sediment carrying and depositing agent. This drainage peculiarity is felt in outfall region of the Mahanadi in the coast where it has built an impressive deltaic tract.

Gawande *et al.* (1963, 1968), Gawande and Biswas (1967) and Biswas and Gawande (1962) studied soils of Chhattisgarh plains. Variation in most soil properties in the region are closely related to their position on the landscape. Soil properties gradually change down the slope indicating that drainage condition, differential transport of eroded material and translocation and deposition of mobile soil constituents played a major role in development of these soils, moving down the slope, there is an increase in soil depth, water holding capacity, cation exchange capacity, and preponderance of calcium and magnesium ions on exchange sites. Along the same direction, there is gradual change in colour from red to darker gray brown, in texture from sandy loam to clayey, in consistency from non-sticky to very sticky, in mineralogical make up towards decrease in illite and concurrent increase in montmorillonite, and in calcium carbonate from almost none to abundant. The topographical variations, have had marked influence in regulating hydrological conditions and thereby on pedologenic processes responsible for morphological features as well as, physical and chemical properties of soils. The soils so developed have been classified into entisols, inceptisols, alfisols and vertisols that widely differ in their production potential and management requirements.

Tropical dry deciduous forests are found in the north, tropical moist deciduous in the east, and tropical thorn in the west and south. The chief tree species present are : *Butea monosperma, Anogeissus latifolia, Terminalia tomentosa, T. chebula, T. bellerica, Acacia leucophloea, A. catechu, A. nilotica, Pterocarpus santalinus, Hardwickia binata, Shorea robusta, Tectona grandis, Diospyros melanoxylon, Albizzia lebbeck, A. procera, Dalbergia sissoo, Madhuca indica, Tamarindus indica, Pongamia pinnata, Azadirachta indica* and Bamboo.(Champion and Seth,1968)

The key perennial grass species in region are *Sehima nervosum, Sichanthium annulatum, Bothriochloa pertusa, Chrysopagon fulvus, Heteropogon contortus,*

Iseilema laxum, Eremopotgon fovealatus, Themeda triandra, and *Cymbopogon coloratus.* The principal annual grass species are *Eragrostis* and *Aristida.* A large number of herbaceous species are associated with these grass communities. Some of the leguminous herbs and shrubs are *Atylosia scarabaeoides, Cassia tora, Clitoria ternatea* and various species of *Aysicarpus, Crotalaria, Desmodium, Indigofera, Mimosa* and *Tephrosia.*

The cultural impact on the natural vegetation is pronounced in most of the areas. Forests and bush have been strongly influenced by human activities through history, by clearing for cultivation, by collecting wood products and by grazing. In spite of these disturbances, the present forest covert of Chhattisgarh region is 43.21 per cent in the three district, the forest area is 32.24, 34.87 and 8.46 per cent in Raipur, Bilaspur and Durg, respectively.

In the Bori forest of Madhya Pradesh teak is associated with *Terminalia tomentosa, Ougenia oojuinensis, Diospyros melanoxylon, Lagerstroemia parviflora, Anogeissus latifolia, Dendrocalamus strictus,* etc. The associates of teak at different places vary a good deal, according to the nature of the soil and its moisture conditions. Recently *Lantana* and *Loranthus* has become a menace in teak forests; teak defoliator and other have become maladies that affect forest on plantations of teak.

On account of its high commercial importance teak is raised in plantations wherever it has been possible to do so. The technique of teak plantation has been the subject of great interest to foresters. In this connection an effort has been made to find plants which could be used as indicators of conditions suitable for teak. *Bambusa polymorpha* and *Cephalostachyum pergracile* are indicators of good sites for teak and *Dendrocalamus strictus* of dry teak forests. Generally in teak plantations there is hardly any ground flora vegetation and this has been a danger sign for the setting in of erosion and efforts have been made to introduce some suitable species in teak plantations.

AGROFORESTRY, TREES AND TREE MANAGEMENT

The following section contains a presentation of agroforestry arrangements and tree management occurring in Chhattisgarh, as well as some use of forest products.

Reasons for having trees in farmlands

Most of the agricultural lands in the region have trees as a component. The farmers have several reasons for integrating trees with farm production and these are listed below, but the list does not reflect to the farmer's priorities.

- Attractive tree products are easily accessible.
- A farmer can claim ownership of tree growing on his own land, but not trees growing in forest land.
- Benefits from trees are, soil improvement, moisture improvement, shade, windbreak, hosting of beehives and prevention of soil erosion.
- Traditional and religious or cultural beliefs often prevent the farmers from felling certain trees situated on the farmland.
- To leave or plant trees on the farm can help secure land tenure as trees can be regarded as a permanent "crop". Attractive land can be saved for children or future expansion.
- National policy is in favour of tree planting and many projects have tree planting as an objective.

Although the majority of farmers consider positive impact on the surrounding environment, there are also arguments against leaving trees in the farmlands under certain conditions. The main is shade to the growing crop. The soil conditions are not favorable to growing trees in their fields and having their crops grow under the canopies. In years with low precipitation there is competition for water which reduces the yield of the crop. Most of the farmers remove trees from farmland because the trees are obstacles to ploughing and may also attract birds.

Farmer's choice of tree species

The origin of the trees on the agricultural land varies and can be categorized into three groups:

i) Trees which are left by farmers during the first clearing for farms and settlements.

ii) Trees that were self- established after settlement and are left to grow and tended for agroforestry purposes.

iii) Trees which are established and planted by farmers.

The farmers gave several reason why certain trees are left when the land is cleared or tended when they occur naturally. The most common reasons are that the trees provide fruit, medicine, wood and shelter or are important to traditional of cultural beliefs. Most species can be used as firewood and for construction purposes. *Acacia nilotica* and *Terminalia spp.* are generally used by farmers. Traditions and forest laws prevent the cutting of several tree species. The spectrum of naturally growing species in a certain area is determined by the prevailing ecological conditions, but also by the history of local land-use which may be unfavorable to the regeneration of certain species. Trees most commonly found in fields are *Acacia nilotica, Butea monosperma, Terminalia arjuna*, neem, karanj etc. Fruit trees like *Carica papaya*, citrus spp, *Mangifera indica, Psidium guajava* are very common and popular. Planting trees species for firewood and fodder as priority products is rare.

AGROFORESTRY ARRANGEMENTS

General aspects

The agroforestry arrangements in Chhattisgarh are classified according to the basic principle proposed by Nair, 1985. It is easiest to start a classification of agroforestry practices by examining the composition of components in the systems as well as their spatial and temporal arrangements. Further, the functions and applicability of the different arrangement can be discussed.

In Chhattisgarh there are not many different types of agroforestry systems and practices, but prominent ones are agrisilvicultural, silvopastoral and agrisilvopastoral arrangement. The complexity varies from highly developed systems where the economics and environmental interactions between the different components are weaker. Complexity, concerning spatial arrangements and outcomes also varies. The combination of trees and agricultural crops are most pronounced, while the interaction of animals often is more diffuse even though most of the farms had all three components in their land use strategy. Different spatial arrangements will be emphasized below, with some aspect of temporal arrangements in the description of woodlots.

Spatial arrangements

It is possible to distinguish two main types of spatial arrangements for trees in agroforestry systems used in Chhattisgarh. The trees are either mixed more or less haphazardly in farm (very rarely seen as mostly trees are on the periphery of fields) or pastureland, or some type of zonal arrangement is used. In the spatial mixed systems, there are no visible patterns in the arrangements of trees. The trees have stood since the land was cleared, and the way they dispersed over the land today due to natural occurrence and the farmer's preference during clearing. Some trees have become established through self-regeneration at a later date. The tree in these mixed arrangements are usually dispersed. Thus, there is a variation in tree cover between the different fields where this arrangement is used.

In the zoned arrangements, single tree rows or wider arrangements make a more or less regular pattern. Depending on the main purpose of the arrangement, the species preferred will vary, as well as the spacing used. Single tree rows are often planted as boundaries to field and farm. The purpose is mainly productive, and the protective role of making borders and providing protection from wind seems to be secondary. Competition between trees and crops is thought to be less when boundary plantation is used than when the trees are planted in the field. By planting roes of single trees in this arrangement, the farmers can produce timber without setting off agricultural land especially for growing trees. Distances between trees in single row vary from 5 - 25 m. As the survival rate of seedlings is often low, large gap in the rows are common. *Acacia nilotica* is the most common species in this arrangement, but other exotic species like *Eucalyptus spp.* are also planted on boundaries.

Arrangements of the trees in farmland are divided into four main types of agroforestry systems in the region:

- Mix of natural growing trees in farm land
- Mix of natural growing trees in pastureland
- Woodlots
- Homegardens

Both mixed and zoned arrangements can be found in the areas around the homesteads (homegarden) and in woodlots, while sparsely growing trees are usually intercropped on outer farm and pasture land.

Temporal arrangements

In different agroforestry practice woody and non-woody components are integrated in various temporal arrangements. The trees are present all the time while agricultural crops and/or livestock are present either more or less continuously, or just for a period.

Farmland with a mixture of naturally growing trees may be seasonally integrated with annual agricultural crops. Agricultural crops are grown only during the time of year when there is enough soil moisture, and the temporal arrangements can be described as intermittent. Post-harvest grazing in this land also occurs periodically, since this practice only can be performed at the end of the growing season. In pastureland with a mixture of naturally growing trees, livestock is commonly grazed more or less coincidentally. The herd is moved around regularly, and returns to graze in the same area after a certain time. However, most of the pasturelands are in a degraded stage as the number of livestock grazing is far in excess then the carrying capacity of land.

It is common to integrate trees and agricultural crops in woodlots during establishment phase, i.e. during the first 2 - 4 years. After this period the trees compete with the crops which are then excluded. When the trees have reached a height which allows grazing of undergrowth without any risk of injuries to the trees, livestock is often integrated in the land use strategy. The temporal arrangement can be called concomitant since the trees are accompanied by the crops during the initial stage, and later on by the livestock.

In the homegardens trees are always integrated with crops (perennials continuously and annuals seasonally), and more occasionally with livestock. Post-harvest grazing is then used seasonally in connection to the harvest in some types of homegardens, but there are homegardens where special fodder crops, like perennial grasses are grown and they are harvested periodically. The temporal arrangement in a homegarden is said to be interpolated. The large number of species in homegarden make year-round harvest possible, which makes the homegarden very important to the families security.

EMERGING TRENDS OF AGROFORESTRY

There are good possibilities for developing and extending existing agroforestry and also for introducing new systems. An assumption is that an important part of the total production of agricultural products and tree products in the region already derives from some type of agroforestry system. Further studies and collection of information on the existing agroforestry systems will give a deeper understanding and provide a good start in the work of increasing the productivity in the system. The introduction of new systems from other areas with similar conditions is also an option. As the literature shows, agroforestry has the potential to maintain or raise soil fertility without intensive external input and to reduce erosion. Agroforestry can be an inexpensive answer to the problems of erosion and declining soil fertility, which are pronounced in the region. Inexpensive measures are of particular interest to the small-scale farmers, but also to society as a whole.

Tree planting

Some farmers plant trees, doing credit to previous efforts to introduce tree planting to the area. The nurseries established by Forest Department, Universities and the Village nurseries provide the farmers with seedlings. Some farmers (very less in number) have established private, small-scale nurseries with varying production capacities. Apparently tree planting has become both accepted and popular, but the high demand also shows the great need for woody products. Thus, increasing the production capacities of nurseries is a valid effort. Those farmers who have private nurseries are important example to others. By spreading simple nursery techniques in extension, more seedlings will be available. The selling of seedlings may provide an important source of income as it already does to private nursery owners.

Agroforestry in new areas

Existing agroforestry systems, where planting is the main way of regenerating trees, occur particularly in the vicinity of homesteads. Several of these systems give high yield and sustainable, long -term production. These successful systems, such as the crop/banana/tree-systems, can be transferred to other parts having suitable environmental conditions. The systems can, when needed, be modified to fit the prevailing conditions in the new area. Due to lack of awareness and environmental constrains the planting of trees on farmland have been limited. If these problems could be overcome, it would be possible to extend the use of more developed agroforestry systems to fields.

Indigenous trees

The emphasis in tree planting has so far been on planting species with a high productive value, and the protective properties and effects of the trees have often been neglected. It may be assumed that in agroforestry systems where the harvest of woody products has been minor, stabilizing organic content in the soil has been more important. Agroforestry systems with naturally growing trees are common in many parts of the state, and these trees seem to be harvested less often than the exotic species used in other systems. However, it is difficult to give more exact estimates of the conservation effects of these systems. The density of trees varies a lot, and this together with qualities specific to the species, determines how much biomass is produced by trees in the field.

Measures should be taken to restrict people from felling trees. This could be a positive example to other villages. These measures will to some extent guarantee present agroforestry with indigenous trees, but not regeneration and continuation of the systems. Traditional agroforestry where trees have had a main protective function should be supported by raising indigenous trees to make replanting possible. The farmers preferences of indigenous species ought to be further investigated.

Agroforestry with fruit trees

Another way to have trees for production integrated in farmland, but still have a relatively low out-take of tree produced biomass, is to use fruit trees. This gives a diversified production, and the trees still have a protective function. There is a wide scope of undertaking tree plantation of horticultural crops, such as, mango, ber, aonla etc.

Trees for firewood production

The increasing population will probably raise the demands for firewood even more in future, necessitating expanded planting of trees suitable as firewood, management techniques which increase productivity. One option is to plant trees with good coppicing properties on boundaries and to coppice them continuously at appropriate intervals. By choosing such species, less planting has to be done. Production of firewood in the vicinity of the homestead will decrease the time needed to collect fuel-wood. Planting trees around the homestead and harvesting some parts for firewood is already practiced, although the trees are intended to be used mainly for construction purposes.

Constraints to agroforestry

In the agricultural production system, especially on the degraded lands, several factors influence the performance of trees. The literature shows a lot of theoretical and also scientifically proven benefits from trees and agroforestry. Most of the arguments given for integrating trees on farm and pastureland are given as productive and protective functions. However, the importance of the protective functions seemed to vary according to the prevailing environmental conditions on each farm. For example, farmers whose land is level do not consider trees as means of controlling erosion since this is not a problem on their farms. The varying conditions of different areas also reflected the type of arrangement used.

The factors that have influenced current agroforestry practices in Chhattisgarh are to some degree different from those that now hinder the introduction or development of new practices. However, several of these constrain overlap in this regards. Factors that inhibit the introduction or maintenance of agroforestry today are emphasized, and are viewed primarily from the standpoint of the small - scale farmer.

Environmental constraint

There is variation of species in natural vegetation, tree species planted and agriculture crops produced. It has been mentioned earlier how natural conditions like precipitation, temperature, terrain and soils vary in different parts of the Chhattisgarh plains. These conditions may act as constraints to growing certain trees and crops and make some land-use or management practice less suitable. The broad variation of natural conditions makes it important for extension workers to consider where a certain system can be successful.

If the prevailing ecological conditions are not considered, agriculture production and other land-use may actually escalate the destructive effects of natural forces like wind and rain. Over-grazing, faulty land preparation practice and deforestation have caused soil erosion. The interactions of human activities and natural conditions have aggravated the destruction, resulting in decreasing productivity.

Irrigation can compensate for insufficient rainfall when tree seedling are establishing. Later, the tree will be better able to supply its own water. However, another problem may arise. Where trees are integrated with crops in the field, trees and crops may compete for water.

Ploughing

When farmland is cultivated with the help of tractor (although tractor ploughing is negligible), or with Ox – bullock (mostly done) trees may act as obstacles. It is however, possible to plant trees on farmlands when appropriate spacing is used. Strips of trees can be planted far apart. The strips should be planted along the contour to prevent the erosive effect of soil preparation performed along the slope. If tree species with a deep root-system are used the roots can be avoided. During the establishment of the trees it may be necessary to prepare soil around the seedlings with a hand hoe. The appropriateness of tractor ploughing for small-scale farmers and on erosion-prone soils is questionable. In some areas, terraces to prevent soil erosion have been destroyed by farmers ignorant of the importance of terraces. Used correctly, tractors can help build terraces ridges with a minimum of labour input and integrate them into an agroforestry land use.

Grazing animals

Uncontrolled grazing is common reason for not planting trees on farmland outside the homestead. It is difficult to protect young seedlings from cattle during post-harvest grazing. Some farmers also have problems with post-harvest grazing by other peoples animals. The farmers regard fencing with thorny branches increase fire hazards and consider fencing methods to be complicated and expensive.

Many farmers consider grazing on farmland a general problem, not only as a constraint to agroforestry. The trampling causes soil compaction and the grazing makes organic matter disappear from the production system (especially when manure is not spread in the field). The majority of farmers in the region depend on livestock for their subsistence. During the dry-period when natural grasses become scarce, crop residues are an appreciated contribution to fodder. Since post-harvest grazing is such an important part of livestock management, great emphasis has to be given to finding alternatives. Introduction of certain agroforestry practices like the growing of fodder trees can be one part of the solution. The problem of post-harvest grazing can probably not be solved by any single measure. A combination of management together with different kinds of restrictions will most likely be needed. For this the participation and involvement of villagers is essential.

Birds

Several farmers believe that trees can host birds that destroy the agricultural crop. Trees are removed from rice field because of this. Termites threaten the establishment of tree seedlings. Indigenous trees are particularly resistant to termites, as well as some exotic species. Apparently, this problem could be reduced by a proper selection of tree species.

UNCERTAINTY OF AGROFORESTRY

Farmers find difficult to consider long-term benefits from agroforestry. In several areas, woody products are still available in the forest or bush and there are no obvious advantages to planting trees on agricultural land. In other areas where trees must be planted to get these products, the advantages of production may be obvious, but it is more difficult to quantify the gains of using trees as protective means. Many farmers use trees in their arrangements to prevent soil erosion. Nevertheless, there are farmlands who do not have trees and even some farmers who want to remove trees from their land.

Ethnic traditions

Ethnic traditions play a role in how agriculture is practiced. In addition, the extent to which and how trees are integrated in the production system, varies with ethnic affiliations. Theoretically, this can effect how well different types of agroforestry are accepted when that are presented in extension. It can also determine how well the farmer will succeed with the new or improved system. It is difficult to quantify the importance of traditions in how successful tree planting has been among different ethnic group. Husband in the family has the most influence on the decisions about timber trees and even other trees. His opinion about of what species to plant may differ from his wife, giving 'false' impression of the total need and preference for tree species. Species choice also influences what type of arrangements will be used when the trees are integrated with farmland.

FUTURE PROSPECTS AND AREAS NEEDING FURTHER ATTENTION

As described earlier it is evident that traditionally agroforestry is widespread among farmers in Chhattisgarh region. There are constrains too but also opportunities for further use of agroforestry in the region. For future work on agroforestry we would like to make a few suggestions on what to emphasize.

* Develop and investigate systems already in use by farmers

Some very productive and sophisticated agroforestry systems are already in use. Their use can be extended in areas with physical conditions similar to those areas where they are practiced today. Other existing systems can be further developed and diversified to contribute more directly to the farm economy. Trees for fodder and firewood production ought to be introduced to a larger extent.

* The choice of species is an important tool to work with

Many indigenous species are drought resistant and meet special needs that exotic species do not, e.g. provide medicine and wood for special tools, etc. People's preferences for indigenous as well as exotic species should be identified.

* Agroforestry requires a multidisciplinary approach

Forest, agricultural and livestock officers, land-use planners and socio-economists should preferably work together on investigations of and recommendations for agroforestry. Agroforestry extension work can only be successful if agroforestry is treated as a system, with emphasis on the interactions interdependence, rather than the disparity, of its individual components. A multi-disciplinary approach to agroforestry is of prime importance to the success of future extension work in Chhattisgarh.

REFERENCES

Biswas, T.D and Gawande, S.P (1962) Studies on genesis of catenary soils on sedimentary formation in Chhattisgarh basin of Madhya Pradesh: I. Morphology and mechanical composition. *J. Indian Soil Sci.* 10: 233 – 234.

Champion, H.G (1936) A preliminary survey of the forest type of India and Burma. *Indian Forest Research* (NS) 1.

Champion, H.G and Seth, S.K (1968) A Revised Survey of Forest Types of India. Manager of Publications, GOI, Delhi.

Gawande, S.P and Biswas, T.D. (1967) Studies on genesis of catenary soils on sedimentary formation in Chhattisgarh basin of Madhya Pradesh. III Chemical Composition of the soils and their clay fraction. *J Indian Soil Science.* 15: 111-118.

Gawande, S.P; Das, S.C and Biswas, T.D. (1968) Studies on genesis of catenary soils on sedimentary formation in Chhattisgarh basin of Madhya Pradesh. IV. Minerology of the soil clays. *J Indian Soil Science.* 15: 111-118.

Gawande, S.P; Vasudeva Rao A.E. and Biswas, T.D. (1963) Studies on genesis of catenary soils on sedimentary formation in Chhattisgarh basin of Madhya Pradesh. II. Mineral assemblage in fine sand fraction. *J Indian Soil Science.* 11: 355-359.

Romani, S. (1992). Artificial recharge possibilities suited to Karstic Terrain. *In*: Workshop on Artificial Recharge of Ground water. Low-cost Rainwater Management for increasing irrigation and ground water recharge in Chhattisgarh Region. Pp 47 - 60. Organised by Central Ground Water Board, Ministry of Water Resources in collaboration with Indira Gandhi Krishi Viswavidyalaya, Raipur (M.P) 8-9 August, 1992.

Vijayaraghavan, V.S (1992) Chhattisgarh Region – its hydrological and ground water development prospects. *In*: Workshop on Artificial Recharge of Ground water. Low-cost Rainwater Management for increasing irrigation and ground water recharge in Chhattisgarh Region. Pp 1-18. Organised by Central Ground Water Board, Ministry of Water Resources in collaboration with Indira Gandhi Krishi Viswavidyalaya, Raipur (M.P) 8-9 August, 1992.

14

Agroforestry Systems in Madhya Pradesh — Potentialities, Possibilities and Opportunities

S.D. UPADHYAYA AND AASHUTOSH SHARMA
Jawaharlal Nehru Krishi Vishwa Vidyalaya, Jabalpur - 482 004

ABSTRACT

Madhya Pradesh is declared as mega diversity state in India. It also harbours diversity in human population as 1/4th of its population is tribal. These tribal people are practicing agroforestry since time immemorial. The state is divided into 11 agro climatic zones. The paper provides the agroforestry systems and species suitable for these zones. The traditional practices and the innovative agroforestry systems developed are also given. Agroforestry systems that are suitable for problematic soil have been briefly discussed. Efforts of different institutes like TFRI, JNKV and projects like TDET, ORP on Agroforestry, AICRP on AF in developing different system and their recommendations are also provided in this paper.

INTRODUCTION

Environment remained in its pristine purity and congeniality till man stayed in the hunting and food gathering stage. But, as he entered in the pastoral age and domesticated animals, he gradually passed from the nomadic stage to settled cultivation. When, he cleared the first patch of earth and upturned the first sod and lifted the first bucket of water, the long story of his interference with environment started. Clearing trees for farming, disturbing the moisture regimes and furrowing the land endangered the phenomenon of erosion leading on to the problem of reducing the soil fertility. At the same time, the art of Agroforestry emerged, though in the most rudimentary fashion, which in varying form and emphasis has continued till today. Coming to recent times, the increasing pressure on land as well as on forests, rising demand for fuel and fodder and finally the requirement of timber for multifarious purposes have strengthen the need of agroforestry as an alternate land use system especially at natural resource conservation front (Singh, 1999). Pathak (2000) opined that agroforestry is a land management system that optimizes the overall land productivity and reduces resources depletion through the utilization of positive interactions between crops-trees-animals in its temporal and spatial dimensions. In this way, agroforestry operates in a given socio-economic frame work; it is not a static but a dynamic structure. An organized research and development is on the move in tropical country like India to encourage agroforestry as alternate land use (Puri and Nair, 2004)

The continuing decline in the availability of cultivable land, rising energy and land casts together with the mounting demand for fruits, fodder and timber have given thrust to the concept of agroforestry plantation in Madhya Pradesh. It is a very intensive form of cropping, which has high relevance to the food and nutritional security of our ever-increasing population. Modern cropping systems differ in many ways from the traditional systems. They require more capital to establish and must be more precocious and productive to be profitable.

MADHYA PRADESH - THE LAND OF UNIQUE OPPORTUNITIES

The vegetation diversity of Madhya Pradesh is a typical representative of Indian biota as it harbours largest gene pool. Due to wide range of climate and corresponding diversity in the vegetation, the State of MP is declared as Mega Diversity State in India. Bio-geographically the State is placed in the semi-arid zone and agro-ecological region-wise, it comes under Central Plateau and Hill Regions considering the physiography, soil types, climate and length of growing period. The State is endowed with various forest types ranging from dry thorn forests to tropical dry deciduous, tropical moist deciduous, sub tropical and semi-evergreen types, which are the storehouse of vast flora and fauna (bio-diversity). Some of which are under threat and some are at the verge of extinction. States 77 per cent rural population depends on these biodiversity.

After the formation of Chhatisagarh, the geographical area of Madhya Pradesh has reduced to 307.44 lac ha. Out of which 49 per cent is under cultivation. Large tract of the State have tribal population (24 %) with forest (23.7 %) and forest resources. There is an acute shortage of fuel, fodder and small timber. States demand for feed (58 million tones of green fodder annually) and fuel (10 million tones) if not attended to, will continue to degrade lands due to overgrazing/ over exploitation of natural resources. During last one decade, the total forest area has come down to 21 per cent of the total geographical area. The majority of cultivated area is under rainfed (69 %) with enormous unexploited agricultural production potential. Eleven distinct agro climatic zones of the State are endowed with large variations in rain fall (1800 - 1600mm), irrigation potential (3 - 64 %), cropped area (3.5 to 33 %), more than 60 per cent is non cropped; soil types (alluvial) terrain from plain to high mountain, one third of the net sown area is unutilized culturable wastelands (having less than 50 % production). These situations pose threat to crop yields and warrant for an alternate land husbandry and crop diversification for sustainable production. There is a great risk of growing food grains in extremely light textured soils and culturable wastelands of Madhya Pradesh, which occupies 22 per cent of the total geographical area of the State. This particular area could be brought under cultivation of suitable agroforestry tree species. Madhya Pradesh is blessed with climatic variability's and diverse soil types. This provides unique opportunities to the farmers to grow multipurpose trees in marginal and culturable wastelands under various agroforestry systems.

AGROFORESTRY - A CONCEPT OF ALTERNATE LAND USE

Agroforestry a term coined in 1977 denotes an age-old practice of having trees mixed in the agricultural landscape. It designates a holistic approach to land use in which woody perennial are deliberately grown on the same land management in spatial or temporal sequence. There must be significant ecological and economical interactions between woody and non-woody components. The objective of such combinations is to increase, sustain and diversify the production of land thereby to help in reducing economic and environmental risks (Dwivedi & Sharma, 1989).

Agroforestry in Madhya Pradesh can be seen as phases in the development of a productive agro ecosystem. Within this ecological framework, farmers can manipulate and manage their trees to take advantage of the benefits or the processes in ecosystem services or products. Therefore, it is suggested that agroforestry should be reconsidered and popularized as a dynamic, ecologically based, natural resource management system, through the integration of trees in farm and rangeland to diversify and sustain smallholder production for increased social, economic and environmental benefits.

All agroforestry systems consist of at least two of the three major groups of agroforestry components i.e. trees and shrubs (perennial), agricultural crops (annual or biennial) and pasture/livestock. There may be other components also, such as fish, honeybees, silk worms, lac insect etc. Based on Diagnosis and Design approach, the following need based location specific agroforestry system with promising MPTs and fruit species in different agro climatic zones of the States have been recommended (Table 14.1). Solanki *et al.* (1999) also reported some of the many more component combination of agroforestry systems prevalent in Madhya Pradesh.

A) Traditional Agroforestry Practices

Shifting cultivation

Shifting cultivation is an age-old practice, which is prevalent up to some extent in hills of Madhya Pradesh (Tejwani, 1994). Shifting cultivation consists of cultivation of crops on steep slopes of hills and hillocks by clearing the trees, bushes; climber's etc. and only some useful plants are left on the field. All the cut and felled material are left for drying and burnt to make the land ready for sowing field crops by the rainy season. After 2 to 3 years, farmers leave the area and a new site is selected. This system is very labour intensive and requires small quantities of inputs and capital. In this system there is huge loss of soil and runoff. So, shifting cultivation in the State has now been discouraged.

Scattered trees on farmlands

The practice of growing agricultural crops under scattered trees on farmlands is quite old. *Acacia nilotica, Mangifera indica* and *Butea monosperma* are cultivated with the arable crops. The species diversity in these systems is very much related to ecological conditions. With the increase in rainfall, temperature and change in edaphic conditions of the area leads to species diversity and system complexity. In Madhya Pradesh traditionally two-tiered canopy configuration (tree + crop) is found. These trees are grown for shade, fodder, fuel wood, fruit, small timber, vegetables and medicinal uses.

Trees on Farm Boundaries

Agroforestry is now in vogue in some parts of the State. Agriculturist especially in paddy growing areas plant trees on their field bunds, which are about 1-2m wide. Species like *Acacia nilotica, Acacia catechu, Albizia procera, Terminalia arjuna, T. indica, Butea monosperma, Zizyphus zuzuba, Dalbergia sissoo, Eucalyptus* species etc are planted on bunds as paddy can tolerate some shade. In wheat growing areas, tree planting is confined to backyard portion of farmlands, which are otherwise not suitable for growing economical agricultural crops in association. *Eucalyptus* is the most preferred species with wheat crop because its canopy does not cause much shade, as wheat crop is susceptible to shade. This type of agroforestry is being practiced by farmers in M.P. since very long and is not assisted by any agency.

Acacia nilotica, Mangifera indica, Butea monosperma, Dalbergia sissoo, Madhuca latifolia, Bamboos, Mehndi etc. trees are grown in agricultural field bunds or on farm boundaries. Mahua tree (*Madhuca latifolia*) has a special place in tribal culture. Its ritual importance, flowers and liquor have importance in everyday life of Maris tribal people. Besides this, it is customary among the Baiga and Maria tribes to have a *bari* attached to the house and it is this *bari* that is used for the cultivation of vegetables, ginger, maize, banana, fruits and in some places coconut etc. These trees which are often around the *bari*, demarcate one *bari* from another. A very important element of tribal agroforestry approach is that normally no two tribes exploit the same segment of the environment in a given region, and this social division creates different socio-economic patterns. Gonds and Baiga share the same ecology on the Maikal hills but they exploit different part of the environment - Gonds are settled cultivators whereas the Baiga are shifting cultivators (Khera, 1993).

Insect on trees outside forest: Entomoforestry

In Central India farmers rear lac insect on *Butea monosperma* growing on field bunds and near by forest for obtaining lac from tree as a minor forest produce under traditional farming system. Lac cultivation has been a traditional livelihood enterprise of tribals for centuries in the Vindhya Region. Baigas and Koles were the earlier practitioners of this enterprise. They mastered the skill with perfection. Gonds adopted lac cultivation some where in the early part of 19th century. Sohagpur (now Shahdol) and Umaria were the chief production center of lac in India during ancient times (Thomas, 2003). Lac is a resinous secretion of minute lac insect *Kerria lacca*. There are about 113 host trees of lac insect, but commercially it is cultivated on Palas, Ber and Kusum trees growing on field bunds and near by forest. There are two types of lac. Superior quality cultivated on Kusum tree is called Kusumi lac, while Rangeeni lac is cultivated on Palas, Ber, Ghont and Arhar. Availability of sufficient number of long and succulent branches per host tree, timely availability of brood lac and its inoculation on host are crucial components of lac cultivation. Pruning is an essential operation of host tree in lac cultivation. Pruning of branches between last week of February and first week of March results in sprouting of comparatively larger number of shoots. Later, every harvest of lac crop leads to automatic pruning of the tree, which will ultimately help in intercrop yield (Thomas, 2003).

In Mandla and near by forest area, farmers rear silk insect on *Morus alba* plantation. In Morena, Jabalpur, Chhindwara and some adjoining district, farmers are raising honeybees on tree crop combinations including floriculture. They usually go for horticulture crops like Mango, Guava etc. with crop combinations like mustard, sunflower etc with flowers combinations of marigold, rose etc.

B) Innovative Agroforestry Practices

Numerous agroforestry systems have been evolved for different areas of the State. Nair (1985) classified the agroforestry systems based on nature of components, structure, function, socio-economic and ecological factors etc. One, which is very convenient, is based on the components, accordingly following are common agroforestry systems adopted from Solanki (1999), which are prevailing in Madhya Pradesh.

- Agrisilviculture (trees + crops)
- Boundary Plantation (trees on boundary + crops)
- Block Plantation (tree + crops)
- Energy Plantation (tree + crops during initial stage)
- Agrihorticulture (fruit trees + crops)
- Agrisilvihorticulture (tree + fruit trees + crops)
- Hortipasture (fruit trees + pasture/ animals)
- Silvipasture (trees + pasture/ animals)
- Agrimedico system (medicinal trees + medicinal crops/ crops)
- Shelter belts (tree + crops)
- Wind breaks (tree + crops)
- Live fence (shrub under trees on boundary)
- Homestead (multiple combinations of trees, fruit trees, vegetables etc.)
- Entomoforestry (tree + sericulture, tree + lac culture + crop, tree + honey bee + crops)
- Aquaforestry (trees + fishes)

Potential agroforestry technologies have been identified for different agro-ecozones by AICRPs on Agroforestry and many ICAR institutes under the Natural Resource Management Divisions (Solanki *et al* 1999), which can be extended at farmer's field on need based and location specific through Operational Research Projects (ORP). Agroforestry technology demonstration has been taken up by JNKVV, Jabalpur, State Forest Research Institute; Jabalpur and Tropical Forest Research Institute; Jabalpur and many other Krishi Vigyan Kendras of JNKVV, Jabalpur through Technology Development

Extension and Training (TDET) project of Ministry of Rural Development, Deptt. of Land Resources on large areas of degraded community lands and also the farmers field to motivate the farmers. This has promoted the farmers to make best use of class V and VI type of land.

Agrisilviculture system

It is experienced that majority of tribal farmer at Jhabua have no knowledge of plantation but some of the farmers grow the multi purpose trees like Eucalyptus, *Madhuca indica* (mahua), Bamboo etc. without following any recommended practices. The fast growing nitrogen fixing tree (NFT) like *Leucaena leucocephala* can be promoted in these areas. In one of the field trial the growth data for 17 yr old *Leucaena leucocephala* var. 'K-28' and 'K-8' planted in 1979 at 1X1 m spacing at Jabalpur, Madhya Pradesh. Var. 'K-28' was thinned to 3X3 m spacing in 1992, and 'K-8' to 2X2 m in 1988 (based on their relative growth rates). Var. 'K-8' was superior in performance, attaining an average girth of 82.5 cm at 17 yr old (compared with 43.97 cm in var. 'K-8)'; respective biomass production was 638.18 and 219.06 t/ha. (Chadhar *et al*, 1997a).

Agrimedicinal trees system

Agroforestry systems involve deliberately growing medicinal trees and shrubs along with arable crops and/or livestock seeking positive synergism. Traditionally several species of trees having medicinal values e.g. Neem, Babool, Harra, Bahera, Aonla were grown in wastelands. Farmers are growing these trees on farm land for different uses including fodder, fuel, timber, gum, medicine, oil, fruit, bee keeping, green manure etc. Further, these woody perennials also help in soil amelioration. Medicinal plants can also be grouped as suitable to different edaphic factors and for different soil problems like salinity and waterlogging (Table 14.2). In this system, the medicinal trees are primary crops and healing herbs are secondary. Inter cropping can be taken for variable periods. Some of the medicinal herbs require shade, for which close planting trees with good coverage of canopy can be preferred; such systems are known as shaded perennial agroforestry systems, for instance; Aonla with turmeric or Khamer with Chitrak.

Four medicinal plant species (*Rauvolfia serpentina, Curcuma aromatica, Chlorophytum arundinaceum, Curculigo orchioides*) were planted under 5-yr-old plantations of 8 tree species (*Acacia auriculiformis, Eucalyptus hybrid* [*E. tereticornis*], *Albizia lebbeck, Albizia procera, Gmelina arborea, Melia azedarach, Pongamia pinnata, Leucaena leucocephala*) near Bilaspur, in eastern Madhya Pradesh (now in Chhatisgarh), on hard lateritic soil (bhataland). The plantations were spaced at 2X2 m, and the medicinal plants (at about 15 cm tall) were planted in the rainy season at 30 cm spacing, in 15 cm wide strips between the rows. Survival varied considerably between medicinal plant species and tree species. Overall, it was lowest in *R. serpentina,* and highest in *Curculigo orchioides.* The best survival and yields were under *E. tereticornis, Acacia auriculiformis* and *L. leucocephala,* and the poorest under *M. azedarach* and *P. pinnata* (Chadhar and Sharma, 1996).

Silvipastoral system

Most of the farmers of Madhya Pradesh have large areas under fodder trees like *Acacia nilotica, Prosopis juliflora,* L. *leucocephola,* ber on farm boundary along with the legume crops like barseem, karabi, etc. and grasses on wastelands raised for stall feeding of animals.

Silvipasture as an agroforestry practice is specifically designed and managed for the production of trees, tree products, forage and livestock. Silvipasture results when forage crops are deliberately introduced or enhanced in a timber production system, or timber crops are deliberately introduced or enhanced in a forage production system. In silvipastoral system, multipurpose trees are planted with grasses and legumes in an integrated way and their forage is utilize through cut and carry system in early years followed by in situ grazing. The silvipastoral experiments in the country have indicated that combination of multipurpose tree species, shrubs, grasses and legumes can play a vital role in not only improving production of animal fodder but also assured availability of fodder round the year. Silvipastoral studies pursued at IGFRI, Jhansi, since 1970, related to establishment and growth of fodder-cum-fuel trees in association with grasses and also compatible tree-grass-legume combinations resulted in the development of silvipastoral models for dry areas of Bundelkhand region. Grasses such as *Cenchrus ciliaris* and *C. setigerus* were found highly compatible in association with *Acacia tortilis* and *Leucaena leucocephala.* Silvipastoral technology is successful in areas receiving less than 1000 mm rainfall with nine months of dry season.

Aquaforestry

In Madhya Pradesh, farmers are now diverting their some portion of land for rearing fish in artificially created ponds, which is surrounded by different species i.e. eucalyptus, Subabul, Mango, *Dalbergia sissoo, Gliricidia sepium* etc. A well managed integrated farming system including Poultry, goat, cow, buffalo etc. with fish in ponds and fruit trees or timber trees on pond boundary make a balanced system for sustainable income to the resource poor farmers. The leaves of many leguminous trees, viz *Gliricidia sepium,*

Leucaena leucocephala, Moringa oliefera etc. have been found to serve as good fish feed (Solankie *et al.*, 1999).

STATUS OF STRESS SITE MANAGEMENT THROUGH AGROFORESTRY

The lands, which are currently not under proper utilization and have a very poor productivity, are known as wastelands. It can be described as "degraded land" which can be brought under vegetative cover with reasonable effort, and is currently underutilized and is deteriorating for lack of appropriate soil/water management on account of natural causes.

In a scenario of decreasing availability of good lands for agriculture, degradation of soil and water resources, increasing pollution of soil and water resources, increasing pollution hazards and threats to the environment and ecosystem, new approaches in farming systems are needed to meet food, fodder, fiber, firewood and timber targets fixed for the 21st century. The future agricultural technologies must assure sustainability while optimizing productivity. Deliberate inclusion of trees in the existing farming/cropping systems and adoption of agroforestry approach on all kinds of wastelands is one of the viable options to achieve sustainability while optimizing productivity.

In a field trial carried out to evaluate *Acacia auriculiformis, Grevillea robusta, G. pteridifolia* (*G. banksii*), *Pinus kesiya* and *P. caribaea* planted as 8-month-old seedlings on bauxite mine spoil it was observed that *Acacia auriculiformis* showed the greatest survival (73%) at the age of 10 year followed by *G. robusta* (62%) (Chadhar, 1995a). Whereas, Singh *et. al.* (1995) reported that the best performing multipurpose tree species seeds was *Samanea saman* (*Albizia saman*), followed by *Pithecellobium dulce, Pongamia pinnata, Acacia campylacantha, Lagerstroemia parviflora, Albizia lebbeck, Albizia procera* sown in pots in copper mine overburden from Malajkhand, Madhya Pradesh. The other species (*Acacia holosericea, Acacia catechu, Adina cordifolia, Acacia auriculiformis*) did not perform well.

Chinnamani (1994) reported that reclamation of the ravines along the Chambal river in Rajasthan, Uttar Pradesh and Madhya Pradesh in N. India is improving the silvopastoral system in relation to the present state of the degraded pastures (dominated by a tree cover of 0.5-5% including *Acacia leucophloea, A. nilotica, Balanites aegyptica* and *Ziziphus nummularia* and by a grass cover of 3-15% including *Dichanthium annulatum, Sporobolus* sp., *Apluda mutica, Aristida* sp., *Eragrostis* sp., *Cenchrus* sp. and *Desmostachya bipinnata*). Methods of rehabilitating the pastures are; closing areas to grazing and tree felling, improving water drainage, controlling soil erosion and selecting suitable tree and grass species for revegetating the areas.

Agroforestry alternatives for different types of stress sites

Large areas in the tropic are affected by salinity, alkalinity, acidity and water logging. Unscientific land use practices have led to a further increase in the area affected by toxicities and deficiencies. Such degraded land can often be reclaimed by agroforestry while providing poor farmers with some income. Many species of trees can grow well in these problem areas where most agricultural crops cannot grow. Singh and Upadhyaya (1999) advocated alternate farming practices for degraded sites as under.

a) Desert Ravine lands and dunes

In Madhya Pradesh there are about 7924.40 ha of ravines lands and gullied area. A major portion of which are confined to Gwalior, Morena, Bhind, Shajapur, Guna districts. These lands come under land capability classes VI and VII and are far below their economic utilization. Small to medium gullies can be put to silvipasture or hortipasture.

Suitable species

Grasses : *Lasiurus sinduicus, Cenchurus cilliaris, C. setigerus*

Legumes : *Arylosia scarabaeoides, Lablab pupureum, Clitoria ternatea*

Trees : *Acacia tortilis, A. arabica, Prosopis cineraria, P. juliflora, Eucalyptus terminalis, E. camaldulensis, Azadirachta indica, Dichrostachys*

b) Acid Soils

In Madhya Pradesh 162.81 ha acid soils falls in the Bhind and Datia region. The soils affected by acidic condition can also be suitably utilized by growing acid tolerant species.

Suitable species

Grasses : *Ponnisolum polystachyon, P. pedicellatum, Paspalum nutatum, Pennisetum clandestinum* (Kikyo grass)

Legumes : *Centrocema pubescens, Desmodium intortum, Stylosanthes guynensis.*

Trees : *Ficus cumeralius, Albizia chinensis, Morus cerrata, Ulmus repelensis, Sehiama wallichi, Bucklandia pupulunea.*

c) Lands affected by shifting cultivation

Suitable species

Grasses	:	*Cenchurus cilliaris, C. setigerus, Pennisetum, Chrysopogon fulbus.*
Legumes	:	*Clitoria ternatea, Lablab pupureum,*
Trees	:	*Lucaena leucocephala, Hardiwicka binata, A. arabica, Acacia tortilis, Dalbergia sissoo, Albizzia spp, Anogeissus pendula.*

Based on the findings, the following agroforestry practices have been recognized and very well established in the State to fulfill the productive and protective objective of land management.

- *Acacia*-babul in rice growing areas
- Khejri (*Prosopis cineraria*) in arid arable fields
- *Ziziphus* in both arable and grasslands of arid ecosystem
- Fruit trees like mango, guava, citrus in semi-arid ecosystems
- NFT and fodder trees in grasslands as silvipastoral,
- Improved fallows/biomass transfer as trees on culturable wastelands, etc.
- Khamer with *Citronella* grass in poor soils
- Bamboo in eroded sites

AGROFORESTRY MODELS RECOMMENDED BY TROPICAL FOREST RESEARCH INSTITUTE, JABALPUR (MP)

Most promising existing agroforestry systems in Central Narmada Valley and Satpura Plateau agroclimatic regions (Agroforestry division, TFRI 2005; http://www.tfri.res.in/Agroforestry%20 Division.htm) are

1. *Silviagriculture models :* (1) *Dalbergia sissoo* with *Cajanus cajan* (2) Babul with Paddy: The components used are *Acacia nilotica* var. *indica; cupressiformis* with short duration varieties of paddy - JR 75 and JR 353.
2. *Silvihortiagriculture models* : The components include Poplar/khamer, citrus (kinnow, seedless lemon and orange) with wheat/ soybean.
3. *Silviolericulture models :* Nine vegetable crops (brinjal, radish, carrot, tomato, spinach, cluster bean, cowpea, French bean and lady finger) were raised in the interspaces of 5 tree species viz. *Acacia nilotica, Albizia procera, Dalbergia sissoo, Gmelina arborea* and *Tectona grandis.*

4. *Silvipastoral models : Dalbergia sissoo* was found successful with *Pennisetum pedicellatum* (Dinanath grass) and *Panicum maximum* (Guinea grass).

5. *Agri medicinal plant model* : Paddy with buch (*Acorus calamus*)

6. *Silvi medicinal model : Chlorophytum borivillianum* (Safed musli) under the idle space of teak plantation.

7. *Alley cropping* : Cowpea/Maize was raised in between the alleys of *Sesbania sesban* whereas maize was grown with *Casia siamea* and *Leucaena leucocephala* under the alley-cropping practice.

AGROFORESTRY MODELS RECOMMENDED BY Jawaharlal Nehru Krishi Vishwa Vidyalaya, Jabalpur (MP)

Significant findings of AICRP on agroforestry experienced for MP

1. Diagnosis and design survey was conducted in 27 districts under nine agroclimatic zones on watershed basis and ecozone wise report was compiled and documented. It was found that the villagers near to forests are neither eager to grow trees on their farms nor to adopt any agroforestry system as they do not have any paucity of fuel wood. Only those villagers who are away from forests showed interest in agroforestry on wastelands. The most demanded system was horticultural woody perennial based agroforestry.

2. A field trial on evaluation of MPTs was conducted for 10 years. Various growth attributes including allometric equations and volume tables were developed for ten MPTs and their performance was evaluated under rainfed to assess the possible financial returns from these trees in addition to agricultural crops. *Dalbergia sissoo, Acacia nilotica, Albiziz lebbeck* and *Leucaena leucocephala* were found most promising species and gave 78, 68, 40 and 108 t/ha biomass yield, respectively.

3. Experiments on tree crop interaction (TCI) involving various combinations of crops and tree species revealed synergistic effect on trees as well as crop component in terms of crop yield and tree volume (Table 14.3). Sissoo, subabool, babool and teak tree species were found to be suitable for planting under various agroforestry system. On the basis of land equivalent ratio (LER), subabool + mustard (1.69) based cropping system was found more suitable among other combinations of field crops such as wheat, linseed and safflower. Guava based agri-horticulture system gave additional yield of 41.60 q/ha from guava fruits besides 19.76 q/ha paddy yield during kharif season. Multiple cropping

index (MCI) was four in this system as four different crops in the sequence of paddy-potato-spinach-spinach were taken. This viable alternate land use system was found advantageous over monoculture.

Findings of Tree Crop Interaction and MPTs Evaluation:

- Guava based agroforestry system provides a return of Rs. 13,360/-ha^{-1} while the return from sole paddy and sole tree are Rs. 11,120/- ha^{-1} and Rs. 4,160/- ha^{-1} respectively. Multiple cropping index (MCI) was four in this system with four different crops in the sequence of paddy-potato-spinach-spinach.
- Under pomegranate based agroforestry system, soybean cv. JS-90-41 and sesamum cv. TKG-21 recorded significantly higher yield over soybean cv. JS 335 and sesamum cv. RT-46, respectively
- *Acacia nilotica* and *Dalbergia sissoo* recorded 32.56 and 70.52 t ha^{-1} biomass yield in 3 x 3 m planting geometry at the age of 11 years. Garlic and sweet potato grown under the shade of these trees recorded 18.5 and 15.1 per cent higher bulb/tuber weight than when grown under open environment.
- An agroforestry system involving *Acacia nilotica* on raised beds, and paddy in adjoining sunken beds on an alkali black soil proved more sustainable for rehabilitation of such soils. The two species had synergistic effect on reclamation and returns.
- Provenances of *Acacia nilotica* showed significant differences in terms of height and cylindrical volume, the latter being significantly and positively correlated with height and diameter at breast height (dbh).

Among multipurpose tree species of forestry plants, tested for their adaptability and growth attributes for flood prone black soil area of Central Narmada Valley, shisham, aonla and arjun have been found suitable. These plants have a capacity to resist drought conditions in the initial stage of plantation. However, these can be recommended for plantation in wastelands as well as in flood prone areas.

In other agrihorticultural experiment, guava (*Psidium guajava*) performed better than aonla (*Emblica officinalis*). However, the growth attributes were not affected due to presence of crops. The yield of soybean was found higher with *P. guajava* as compared to with *E. officinalis* (Table 14.4).

In a trial of agrisilviculture system the Land Equivalent ratio (LER) was computed to evaluate the economic feasibility of the system. The results revealed synergestic effect on trees as well as crop component in terms of total biomass production. *Leucaena leucocephala* + mustard based cropping system found to be more sustainable having higher LER (1.69) which is very well comparable to *Acacia nilotica* + linseed combination *Acacia* and *Albizia* with safflower and *Dalbergia* with all four crops combination were having almost equal LER (1.65).

Findings and recommendation on traditional Acacia-rice agroforestry system

Nitrogen fixing trees (NFTs) are key constitute in smallholder farming system of the tropics. They are major sources of nutrients (particularly nitrogen) from crop growth in addition to fuel; fodder and small timber yield from woody component. *Acacia* –rice; a traditional Agroforestry practice of Kymore Plateau and Satpura hills has been surveyed. The findings revealed that most of the rice growing area of the zone is rainfed (77.11%) with small and medium land holdings (88.42%) comprises *Acacia* plantation on the bunds under variable densities (70 trees/acre to 11 trees/acre). On the basis of the inventerization it has quantified that 6-8 per cent of bund area is occupied by babul. Tree crop interaction study revealed that the management of Babul tree on bunds (East-west direction) and pruning up to 40 per cent LCR (Live Crown Ratio) at four different intervals (alternate months in rainy and once in winter season) not only provides woody and non woody biomass to the user but also ameliorate the bio-physical environment for the growth of Kharif and Rabi crops. Hence the maintenance of 60 per cent Live Crown Ratio (LCR) can be recommended as appropriate for Wheat and paddy crop growth (Upadhyaya and Nema 2003, 2003 (a) and 2004).

Development of agroforestry models through Technology Development Extension and Training project on wasteland development

One of the important wasteland plantation project entitled "Technology development, extension and training in agroforestry for wasteland development" was sanctioned in 2001 by Ministry of Rural development, Govt. of India for execution in 9 centers of JNKVV with its main coordinating centre at Department of Plant Physiology, JNKVV, Jabalpur.

A remarkable change has been seen in the farmer's attitude regarding wasteland development and technology adoption. They prefer aonla based agroforestry model. Aonla is very much promising species for Madhya Pradesh. Its performance in undulating terrain and muramy wasteland is very encouraging. In agroclimatic zone Kamore plateau and Satpura hills, one of the promising models, which have gained popularity among farmers of the region, is Khamer based agroforestry (Medicinal and timber tree) intercropped with Lemon Grass/ Rosa Grass/ *Tagetus* in agrisilvihorticulture system. It is also proved very successful both in terms of biomass production and also in economic return and profitability.

At Sehore center under TDET project the plantation of various tree species has been done. The most of the species are showing good performance according to their suitability and some are failed to service, even under limited water resources condition the results in general are satisfactory. On the basis of three years observations it is found that some of species are in good condition having better survival. These are Khamer, Neem, Baheda, Bamboo, Ritha, Amaltas, Kapok, Deshi babool, Aonla, Deshi Ber, and Custard apple whereas; others are unsuitable for this region.

At KVK Jhabua, Khair, Guggal and Jatropa based agroforestry models are being developed at farmer's wasteland. At Jhabua, frequent severe drought conditions accompanied by very less topsoil warrants for suitable wild fruit/ timber based agroforestry models as compared to traditional agriculture.

Agro-climatic situation of Chhindwara is very unique with respect to soil and climate, which is most suitable for horticultural crops. Hence, fruit based agroforestry system i.e. santara, aonla etc. is recommended for the region and are also gaining popularity among farmers.

Three year old plantations of neem, karanj, harra and aonla at Shahdol had attained good growth and height in the range of 1.5 m to 2.5 m and well established on very poor soils. These plants even under severe moisture stress survived very well. In general the performance of neem and karanj is best among others. Farmers are now much more interested for fruit tree plantation of improved variety seedlings. Survival percentage of fruit seedlings in farmers field are higher mainly because farmers of this area are showing interest in fruit based agroforestry models (Agrihorticultural system) followed by agrisilvicultural system to fulfill their requirements for fruits, fuel and timber. This region has major constraints of undulating terrain, which causes topsoil erosion and ultimately limits crop productivity. This can be mitigated by adopting suitable fruit tree particularly Aonla, Munga and Bel.

At Rajgarh center, agroforestry models like Isabgol + Aonla, Isabgol + Khamer were developed. Plantations of saplings of different varieties of Mango, Aonla & Orange were done for improving the productivity of farmer's wasteland at different degraded sites of Rajgarh Block and these are performing very well.

The plantation of different fruit trees like Guava Var. Lucknow- 49, Aonla Var. local, Kanchan, Chakaiya and NA-5, Mango Var. Lungra and Dashahari,

Mahua, Emali local, Karonda local and forest tree like Eucalyptus, Khamer and Bamboo have been planted under different agroforestry models at farmers field at Sidhi. The plantations of fruit plants have been done under agrihorticulture system and forest trees are planted under agrisilviculture, silvipastoral and boundary plantation. Guava, Aonla, Mango, Eucalyptus and Khamer based agroforestry system has been developed at farmers field as well as KVK farm in which crop component included Arhar, Urid, Moong, Til and Niger at different locations. In silvipastoral system eucalyptus + jowar, Bamboo + natural grasses and karonda has been planted as boundary plantation.

Soil of the Sagar region varies from clay to sandy loam with predominantly medium black soils. The depth is variable that varies from shallow to deep. Medicinal and aromatic plants like Palmarosha, aloe, satawar, bel, arandi and chandrasur were planted in the institutional land for demonstration to the farmers under different agroforestry models (Table 14.5). Mango, eucalyptus, subabool, gulmohar, Glyricidia, mango, aonla etc were performing well under agroforestry model at ZARS, Sagar with respect to tree height, diameter growth and branching pattern. Medicinal crop chandrasur was studied in detail under interaction with aonla var. chakaiya in agroforestry model and found very promising one.

Agroforestry in wastelands – A case study of Jabalpur district

An exploratory survey was conducted in seven blocks of Jabalpur district of central India. The farmers were selected purposively from all the blocks to know the land holding pattern and type of agroforestry system adopted by the farmers in their farmland. During intensive survey of agroforestry systems in Jabalpur district, surveyor find different type of tree combinations raised by farmers in the culturable wastelands. Compared to large farmers, the small farmers are relatively more diversified. Keeping in view the arrangement of tree components on spatial and temporal and their use, a total of 46 different types of tree combination based agroforestry system classes are identified. They are mainly grouped into six-agroforestry system. Based on the nature of components, six main classes, agrihorticulture, agrisilviculture, hortipastoral, silvipastoral, agrisilvihorticulture and silvihortipastoral systems are identified. The inclination towards the adoption of agrisilvihorticulture system was fairly higher (31 to 54%) by all the category of farmers. Majority of farmers raise guava based agrihorticulture system (56.9 per cent) and hortipastoral system (88.9 per cent), eucalyptus based agrisilviculture system (30.4 per cent), eucalyptus based silvipastoral system (50 per cent), bamboo-guava based agrisilvihorticulture system (27.1 per cent) and eucalyptus-guava based hortisilvipastoral system (33.3 per cent) in there field as compared to other tree based agroforestry systems. Farmers of Jabalpur district (32 per cent) show inclination towards mango for future planting purpose. It is practically axiomatic that small and marginal farmers are more inclined towards diversified farming systems to fulfill their livelihood options. All the categories of farmers in Jabalpur district were engaged in agroforestry activities, which show the inclination towards alternate farming especially in changed agricultural scenario (Sharma, 2005).

Transferable Technologies

(i) Natural vegetation along with contour planted vetiver grass, at 1.0 m vertical interval on 6% slope in clay loam soil is found useful to check soil erosion @ 116.3 ton/ha/ year in semi-arable land with conservation practices suitable for cropping.

(ii) Plantation of vetiver grass in three rows each 30 cm apart is helpful in reclaiming washes upto 30 cm deep and 1.0 m wide gullies having gradient upto 20% in land with severe risk of degradation.

(iii) The wastewater of dairies can be successfully utilized for Guava based agrihorticultural system, which give additional yield of 41.6 q/ha. of guava fruits besides 19.7 q/ha paddy yield. Multiple cropping index of four can be achieved in this system by adopting paddy-potato-spinach-spinach.

(iv) Acacia-paddy model is widely accepted in saline discharge areas. Planting of *Acacia* on raised and paddy on sunken bed to arrest land degradation and improve production.

IMPACT ASSESSMENT

Diagnosis survey and research on agroforestry systems that render sustainability to productivity and ensured efficient use of natural resources especially in rainfed and wasteland areas of fragile ecosystem to fulfill the fuel, fodder and food requirement showed biggest impact of the project in the state. Operational Research Project on Wasteland Development through Agroforestry emphasized for adoption of agrisilviculture and silvihorticultural systems in culturable wastelands. These efforts through AICRP on Agroforestry and KVK(s) have resulted in broad area coverage of tree farming in wastelands of different agro climatic zones of the state. It created high degree of awareness among farmers and NGO(s) towards growing trees under suitable agroforestry models leading to environmental and nutritional security.

Babul, Guava, Aonla, Bamboo, and Eucalyptus based agroforestry has been adopted by the farmers. It played a vital role in utilization of culturable wastelands. Reintroduction of NFTs in the existing farming systems (e.g. rice bunds) is one of the approach needs to be popularized in rice based rainfed agro ecosystem to improve the productivity and sustainability through (i) increased fuel wood biomass production, (ii) efficient nutrient cycling and biological nitrogen fixation, and utilization of unutilized farm bunds.

Need of Intensification of Agroforestry Programmes

Agroforestry as a traditional farming system has played and is still playing a very useful role in meeting the multiple needs viz. food, fodder, fuel, fruits and small timber besides soil conservation and wasteland development in the state. There is a need to design and develop problem-oriented systems/models for different agroclimatic zones of the state for increased productivity, sustainability, environmental security and economic reform to the farmers. An inclusion of Horticultural woody perennials and NFTs seems to be the promising options. Organic matter buildup in poor soils, carbon sequestration, soil conservation, biomass production, supply of quality planting material/ tissue culture raised seedlings, raising of kisan nursery, incorporation of medicinal trees and their post harvest technologies including value addition are some of the priority areas needs to be included in future planning.

CONCLUSION

Agroforestry systems hold promise to provide sustainable land management, by arresting degradation and loss of productivity due to excessive use and lower population supporting capacity. Through diversification of agriculture, it is possible to reduce the risk of crop failure due to uncertain weather and erosion hazards. Such interventions lead to the supply of food, fodder, fuel, and timber in a more sustainable way for the survival of man and animal. When the needs of farmers are satisfied in respect of forest products, the dependence on forests will come down, leading to ecological restoration and conservation of soils and forests.

In this context the implementation of agroforestry systems at farmers field and institutional land had a major impact on elevating fertility status, productivity of land, biomass yield as well as improving water holding capacity of the soil from low to high level in terms of high NPK and organic content of the soil, fruit, fuel wood and fodder production etc. This has also pave way to generate various source of employment with regard to engagement in agroforestry models.

Table 14.1. Different promising MPTs and fruit species for various agroforestry systems in farmer's field under agro-ecological zones of Madhya Pradesh.

	Agro climatic zones	Systems	Promising tree species
I	Chhattisgarh Plain	Agrisilviculture	*Terminalia* spp., Shahtoot, bamboo, khamer
		Agrihorticulture	Mango, Jackfruit, guava, drumstick, seedless lime, custard apple
II	Northern Hills of Chhattisgarh	Agrisilviculture/ Agrientomoforestry (boundary plantation is also included)	Traditional Palash based system, eucalyptus, Khamer, karanj
		Agrihorticulture	Jamun, bel, Mango, guava, jackfruit, drumstick, lime, aonla
III	Kymore Plateau & Satpura Hills	Agrisilviculture (boundary plantation is also included)	Traditional acacia-rice, Palash, eucalyptus, teak, shisham
		Agrihorticulture	Guava, mango, aonla, lime, custard apple, jamun, shahtoot
IV	Central Narmada Valley	Agrisilviculture	Bamboo, khamer, shisham, arjun
		Agrihorticulture	Mango, seedless lime, ber, lime, aonla
V	Vindhya Plateau	Agrisilviculture	Khamer, neem, babul, eucalyptus, safed shirish
		Agrihorticulture	Mango, guava, lime
VI	Gird Region	Agrisilviculture	*Prosopis* spp., Babul, subabul, sissoo, Australian babul
		Agrihorticulture	Lime, ber, custard apple
VII	Bundelkhand Region	Agrisilviculture (boundary plantation is also included)	Subabul, neem, karanj, bamboo, Palash, mahua,

Contd...

Table 14.1. Contd...

	Agrihorticulture	Aonla, drumstick, mango, ber, jamun
	Silvipasture (Although in this region no well-defined silvipasture is present, during monsoon the grasses appear, which are used as pasture)	sissoo, Khejri
VIII Satpura Plateau	Agrisilviculture	Bamboo, teak, khamer, shisham
	Agrihorticulture	Santara, Aonla, shahtoot, imli, karonda
IX Malwa Plateau	Agrisilviculture	Khamer, subabul, eucalyptus, kala shirish,
	Agrihorticulture	lime, drum stick, papaya, santara, karonda, aonla
X Nimar Plains Agri-silviculture	Agrihorticulture	Neem, khamer Banana, santara, papaya, lime, Aonla, ber
XI Jhabua Hills	Agrisilviculture	*Acacia* spp, guggal, neem, eucalyptus
	Agrihorticulture	Aonla, guava, mango

Source: Shrivastava (1995), Solanki *et al.* (1999)

Table 14.2. Suitable medicinal plants that do well on different problematic soils under agroforestry system in Madhya Pradesh

No.	Problematic Soils	Name of the species	
A	Shallow, Rocky soil	Herbs	Punarnava (*Boerhavia diffusa*) Anantmool (*Hemidesmus indicus*), Ratti (*Abrus pricatorius*), Manjistha (*Rubia Cordifolia*), Gwar patha (*Aloe barbadensis, Aloe vera*).

Contd...

Table 14.2. Contd...

		Shrub	Gandh Babool (*Acacia farnisiana*), Jhand (*Prosopis cineraria*), Karonda (*Carissa carandus*) Gurmar (*Gymnema sylvestre*) Marorphali (*Helicteres isora*).
		Trees	Neem (*Azadirachta indica*), Aonla (*Emblica officinalis*), Sitaphal (*Annona squamosa*), Bel (*Aegle marmelos*), *Cordia oblica*, *Holoptellia intigrifolia*, *Wrightia tinctoria*, Malkagni (*Celastrus paniculata*).
B	Sandy soils	Herbs	Khus (*Vetiveria zizanioides*), *Cymbopogon* spp., Isabgol (*Plantago ovata*), Safed musli (*Chlorophytum borivilianum*, *C. tuberosum*), Kali Musli (*Curculigo orchioides*), Asgandh (*Withania somnifera*), Mulethi (*Glycyrrhiza glabra*),Senna (*Cassia angustifolia*), Akarkara (*Spilanthus acmela*), Sadasuhagan (*Catharanthus roseus*).
		Shrub	Chitrak (*Plumbago zeylanica*), Sinduri (*Bixa orellana*)
		Trees	Bel (*Aegle marmelos*), Katha (*Acacia catechu*), Jamun (*Syzygium cumini*), Arjun (*Terminalia arjuna*), Salai (*Boswellia serrata*), Khamer (*Gmelina arborea*), *Glyricidia sapium*.
C	Saline soils	Herbs	Satawar (*Asparagus racemosus*), Sankh puspi (*Evolvulus alsinoides*), Khus (*Vetiveria zizanioides*), *Cymbopogon* spp., Chandrsur (*Lepidium sativum*), Kalongi (*Nigella sativa*)
		Shrubs	Karonda (*Carissa carandus*), Adusa (*Adhatoda vasica*), Nirgundi (*Vitex negundo*), Ber (*Zizyphus* spp.)
		Trees	Neem (*Azadirachta indica*), Aonla (*Emblica officinalis*), Katha (*Acacia catechu*), Arjun (*Terminalia arjuna*), Babool (*Acacia nilotica*), Bahera (*Terminalia bellerica*), Harra (*Terminalia chebula*), Sivnag (*Oroxylum indicum*).
D	Alkaline soils	Herbs	Sankh puspi (*Evolvulus alsinoides*), Khus (*Vetiveria zizanioides*), *Cymbopogon* spp., Jangali Pyaj (*Urginia indica*), Van lahsun (*Allium porrum*), Garlic (*Allium sativum*).

Contd...

Table 14.2. Contd...

		Shrubs	Karonda (*Carissa carandus*), Nirgundi (*Vitex negundo*), Guggal (*Commiphora* spp.)
		Trees	Aonla (*Emblica officinalis*), Arjun (*Terminalia arjuna*), Babool (*Acacia nilotica*), Sissoo (*Dalbergia sissoo*), *Eucalyptus teritecornis*, Siris (*Albizzia procera*), Karanj (*Pongamia pinnata*)
E	Water logged soils	Herbs	Bach (*Acorus calamus*), Brahmi (*Bacopa monniera*), Mandook Parni (*Cintela asiatica*), Brangraj (*Eclipta alba*), *Cyperus* spp. Khus (*Vetiveria zizanioides*), Talmakhana (*Astercantha longifolia*).
		Shrub	*Hedychium spicatum*
		Trees	*Eucalyptus teritecornis*, Arjun (*Terminalia arjuna*), Salai (*Boswellia serrata*)

Source: Upadhyaya *et. al.* (2001)

Table 14.3. Performance of various agroforestry systems at Jabalpur

S. No.	Tree + crop system	Volume of 10 year old trees ($m^3 ha^{-1}$)	Average crop yield (q ha^{-1})	Additional income from trees (Rs. In lacks ha^{-1}/10 yr)
1	Pure soybean	-	9.22	-
2	Eucalyptus + soybean	390.75	4.22	13.67
3	Subabool + soybean	215.75	1.85	7.55
4	Babool + soybean	116.85	4.45	4.08
5	Teak + soybean	17.00	4.37	1.19
6	Pure sesamum	-	3.41	-
7	Eucalyptus + sesamum	390.75	1.63	13.67
8	Subabool + sesamum	215.75	0.84	7.55
9	Babool + sesamum	116.85	1.64	4.08
10	Teak + sesamum	17.00	1.58	1.19
11	Pure niger	-	4.57	-

Contd...

Table 14.3. Contd...

12	Eucalyptus + niger	390.75	3.06	13.67
13	Subabool + niger	215.75	2.13	7.55
14	Babool + niger	116.85	2.85	4.08
15	Teak + niger	17.00	3.01	1.19
16	Pure urid	-	4.57	-
17	Eucalyputs + urid	390.75	3.02	13.67
18	Subabool + urid	215.75	2.06	7.55
19	Babool + urid	116.85	3.14	4.08
20	Teak + urid	17.00	3.08	1.19

Source: AICRP Report (1995-2000)

- Eucalyptus, subabool and teak at 2 x 2 m and babool at 3 x 3 spacing
- Eucalyptus, subabool and babool @ Rs. 3500/m^3 and teak @ Rs. 7000/m^3

Table 14.4. Effect of cropping on the growth of fruit trees and crop yield in agrihorticultural system

Species	Tree growth		Tree + crop		Crop yield (kg/ha)
	Height (cm)	CD (cm)	Height (cm)	CD (cm)	
Psidium guajava + soybean	110.1	4.80	103.2	4.70	102
Emblica officinalis + soybean	38.3	1.81	37.8	1.73	98
Psidium guajava + linseed	123.2	5.30	137.6	5.40	435
Emblica officinalis + linseed	46.1	2.19	45.2	2.28	410

Source: AICRP Report (1987-2000)

Table 14.5. Agroforestry Systems developed at Sagar

S. No.	Agroforestry System developed	Name of Trees	Intercrop
1.	Agrisilvicultural System.	Subabul, Siris and Bakine	Soybean (JS 335)

Contd...

Table 14.5. Contd...

2.	Agrisilvihorticultural System.	Aonla, Anar, Karonda and Subabul	Soybean (JS 335)
3.	Agrihorticultural System.	Mango (Amrapali), Anar, (Ganesh) Guava (L-49) and Aonla	Soybean (JS 335)
4.	Agrisilvicultural System.	Shisham	Soybean (JS 335)
5.	Medicinal crop based Agroforestry System	Aonla (Chakaiya)	Chandrasuar
6.	Between rows of MPT'S	Aonla and Gliricidia Gliricidia and Kassid	Palmarosa, Gwarpatha
7.	Live Fencing on bunds	Karonda, Gataran, Shikakai and Ratanjot	-

Source: TDET annual report (2003-04)

Several scopes have been arisen as future thrust on D&D survey for growing MPTs under agrisilviculture, agrihorticulture and silvipastoral systems at different agroclimatic regions of different crop zones of the State. Further scientific investigations are needed to assess the suitability of these systems under different tree species of different crop zones for its sustainability under different agro climatic conditions. The studies need to be conducted on biological interaction including light, nutrient and water use efficiency.

The traditional land use systems have become very fragile and unsustainable due to unbalanced utilization of natural resources, resulting in large-scale degradation and consequent creation of wastelands. The primary cause of degradation is the demographic pressure on land, leading to loss of vegetative cover through deforestation. Harvesting of timber, collecting firewood, overgrazing and encroachment of forest areas are some of the reasons for deforestation in State. Besides that, ignorance of proper soil conservation measures, non-judicious use of fertilizer and pesticides, faulty irrigation and water management, discharge of industrial effluent, sewage/sludge are also responsible for soil degradation to a considerable extent. The rationale of developing agroforestry in Madhya Pradesh is to obtain a higher total, a more diversified and/or a more sustainable production of the combined tree and other agricultural and/or livestock resources that is possible with other forms of land use. Agroforestry based land use development in Madhya Pradesh will claim sustainability, increased agricultural productivity besides fruit fuel-timber productivity and also helping small scale farmers by generating employment and income.

REFERENCES

AICRP on Agroforestry Report (1987-2000) on background information, significant findings, transferable technologies, impact assessment, Deptt. of forestry, JNKVV, Jabalpur (MP).

AICRP on Agroforestry, QRT Review Report (1995-2000) Deptt. of forestry, JNKVV, Jabalpur (MP).

Chadhar-SK (1995a) Biomass production capacity of *Gliricidia sepium. Vaniki-Sandesh.* 1995, 19(1): 29-31.

Chadhar-SK (1995) Performance of some exotic species on bauxite mine overburdens. *Vaniki-Sandesh.* 19 (4): 1-4.

Chadhar-SK; Mehta-SS; Tiwari-BP; Tiwari-SK (1997). Performance of fourteen provenances of *Gliricidia sepium. Vaniki-Sandesh.* 1997, 21(2): 2-7.

Chadhar-SK; Sharma-MC (1996) Survival and yield of four medicinal plant species grown under tree plantations of bhataland. *Vaniki-Sandesh.* 1996, 20(4): 3-5.

Chadhar-SK; Sharma-MC; Haldkar-A; Tiwari-BP (1997a). Biomass production from plantations of two strains of *Leucaena leucocephala. Vaniki-Sandesh.* 1997, 21(1): 1-3.

Chinnamani-S (1994) Silvipasture in Chambal ravines. *Range-Management-and-Agroforestry.* 1994, 15(1): 79-86.

Dwivedi, A.P. and Sharma, K.K. (1989). Agroforestry: Its potential, paper presented in the seminar on social forestry and agroforestry, F.R.I., Dehradun.

Gupta-BN; Williams-AJ; Banerjee-SK (1995) Impact of thermal power plant emission on vegetation and soil. Proceedings-*of-the-Indian-National-Science-Academy.-Part-B,-Biological-Sciences.* 1995, 61(6): 457-469.

Khera, PD (1993). Agroforestry practices in tribal areas of India. Workshop proceedings "Agroforestry for rural needs" Vol II edited by DK Khurana and PK Khosla.

Nair, P.K.R. (1993). An Introduction to Agrofoestry. Kluwer Academic Publishers. Dordrecht. The Netherlands. 450.

Pathak P.S. (2000) Agroforestry: a tool for arresting land degradation. Indian Farming (Feb 2000), 15-19.

Puri S and Nair, P.K.R. (2004). Agroforestry research for development in India: 25 years of experiences of a national program, *Agroforestry Systems,* 61: 437-452.

Sharma, Aashutosh (2005). Unpublished PhD thesis. "Studies on non-forest wasteland development through agroforestry and its impact on socio-economic status of farmers in Jabalpur district of Madhya Pradesh" Guru Ghasidas University, Bilaspur (CG)

Shrivastava, S.S. (1995). Agroforestry. Central book house publisher, Raipur (CG) pp 384

Singh G.B. (1999) Agroforestry Research in India- Issues and strategies. *Indian J. Agroforestry*, 1 (1): 1-14

Singh, Punjab and Upadhyaya, S.D. (1999). Alternate land use systems for problem soils. *Advances in soil science.*

Singh-AK; Singh-RB; Bhowmik, AK; Chand, S; Banerjee, SK (1995) Relative suitability of different tree species for copper mine overburdens. *Van-Vigyan*. 1995, 33(2): 122-125.

Solanki K.R., Bisariya, A.R. and Handa, A.K. (1999) A Decade of Research (1988-1999). National Research Centre for Agroforestry, Jhansi :1-44.

TDET Annual Report (2003-04). Report submitted to Ministry of Rural Development, DOLR, GOI, New Delhi. pp 26.

Tejwani, R.G. (1994). Agroforestry in India, Oxford & IBH Publishing Co. Pvt. Ltd. New Delhi.

Thomas, Moni (2003) Lac to lakhs, Reviving self reliance, a unique experience of Shahdol, MP.pp20.

Upadhyaya, S.D. and Nema, Sharad (2003). Nitrogen supplement in paddy field through *Acacia nilotica* planted at farm bunds- A sustainable tree based cropping system. *Indian J. Plant Physiology* (special issue) 13 (5):72-75.

Upadhyaya, S.D. and Nema, Sharad (2003 a). Tree crop interaction in *Acacia* – rice based agroforestry on bunds at farmers field. *JNKVV Research Journal*, 37(2): 20-24.

Upadhyaya, S.D. and Nema, Sharad (2004). *Acacia*–rice – A traditional agroforestry for food-fuel and environmental security- An exploratory study. *J. Tropical Forestry* 18 (No. 2&3) pp 23-26.

Upadhyaya, S.D., Tiwari, Gyanendra and Sharma, Aashutosh (2001) "Scope of Medicinal Plants cultivation in wastelands of Madhya Pradesh" Published in the SOUVENIR of National Seminar on "*Commercial Cultivation Processing and Marketing of Medicinal and Aromatic Plants.*" Organized by Department of Plant Physiology, College of Agriculture, JNKVV, Jabalpur-482004 (MP) held during Nov. 27-29, pp.24-31.

15

Agroforestry System in Maharashtra

V. M. ILORKAR
All India Co-ordinated Research Project on Agroforestry
College of Agriculture, Nagpur - 440 012

ABSTRACT

Maharashtra ranks second in terms of area under tree cover, however, still large chunk of land is lying without use or are underutilised because of their degradation status. Such degraded lands can be put to use through agroforestry. Agroforestry is being adopted in this state in one way or the other by raising trees on farm bunds, around cattle sheds etc. since earlier times. New innovative technology has now been developed by all the agricultural universities of state. The paper provides detailed information of different agroforestry systems which have been developed and practised in the state.

INTRODUCTION

Maharashtra is the third largest state of India having the geographical area of 307,713 km^2, which constitute 9.4 per cent of the country's geographical area. The human population of the state is 96.75 million (9.4 % of country's population) of which 57.6 per cent is rural and 42.4 per cent is urban. The average population density is 314 persons per km^2. The tribal population is 9.3 per cent of the state population. The state supports livestock population of 36.4 million, which is 7.7 per cent of the country's livestock population. The state rank second amongst all the state/UTs in terms of area under tree cover and third in term of recorded Forest Area (Anon, 2001).

Maharashtra state is divided into six revenue divisions from administrative point of view. Landuse Classification of the state is given in Table 15.1. Requirement of wood and fuel in Maharashtra is given in Tables 15.2 and 15.3, respectively.

Soil and Climate

Soil of the state are broadly classified as :

1) *Red lateritic soils :* Konkan region
2) *Black cotton soils :* Western Vidarbha, Marathwada and Khandesh
3) *Alluvial soils :* Eastern Vidarbha, Eastern part of Central Vidarbha.

Maharashtra has been divided into 11 different agroclimatic zones. The jurisdiction of different agriculture universities with regards to agroclimatic zone of Maharashtra is given in Table 15.4.

Variety of soils, rainfall, temperature has given rise to different type of cropping system in the state. Availability of irrigation and market has developed orchard system mostly in the western Maharashtra and Konkan region.

Agroforestry interventions are more likely to be accepted in the dryland area as an alternative land use system. The main occupation of farmers in the rural area in dryland area is seasonal agriculture and animal husbandry. Agroforestry is believed as the source of fodder for animal, fuel and small timber.

AGROFORERSTRY SYSTEMS

Farmers are growing the useful trees on their farm traditionally. The estimates have shown that there is 236,441 km^2 culturable non forest area (CNFA) that support about 8269 km^2 of tree cover. This is 2.7 percent of State geographical area. The farmers of the region follow various types of agroforestry practices as enumerated below:

Traditional Farm Forestry

Farmers grow and maintain trees on their farms along the farm road, farm bunds, nalas, cattle sheds etc. The tree species are useful to the farmers as fuel, fodder, wood for farm implements, fruits and timber. Tree species commonly found are *Tectona grandis* , *Gmelina arborea, Terminalia arjuna, Albizia odoratissima, Acacia nilotica, Azadirachta indica, Butea monosperma.* Fruit trees are *Mangifera indica, Zyziphus zujuba, Terminalia bellerica, Syzygium cummni, Annona sqamosa.* Bamboo are found most useful for construction and fencing.

Farm Boundary Plantations

Systematic agroforestry interventions are taking place at lower pace. People are growing trees along their farm boundary to protect their farms from stray cattle, for fuel, small timber, green manuring. Species grown are *Pongamia pinnata, Acacia nilotica, Dalbergia sissso, Azadirachta indica, Leucaena leucocephala, Eucalyptus species, Glyricidia maculata, Acacia catechu. Albizia lebbeck, Albizia procera, Tectona grandis ,Gmelina arborea and bamboo (Dendrocalmus strictus , Bamboosa arundenacia, Bamboosa vulgaris).*

Block Plantations

Systematic monoculture plantations have been raised by large farmers on barren and waste land and to some extent on cultivable land (Table 15.5.).

These plantations are meeting the industrial raw material requirement to a great extent. Paper and pulp industries, ply wood and particleboard industries are encouraging plantations of *Leucaena* and *Eucalyptus.*

Agrisilviculture

It refers to the planting of trees in the existing cropping system. Benefit cost ratio of cultivation of bamboo in agrisilviculture system after 12 years of study with different rate of nitrogen application is given in Table 15.6.

- Bamboo planted at 8 x 6 meter spacing
- *Intercrops* : Soybean, black gram
- Soil type shallow medium.clay loam

Market Price of Bamboos.
1998-99 @ Rs. 9.00/Bamboo.
2000-01 @ Rs. 10.00/Bamboo.
2003-04 @ Rs.11.00/Bamboo.

Few farmers are now convinced to grow the trees in alleys along with their crops especially on dry land and paddy farming systems. Teak (*Tectona grandis)* based dryland farming agrisilviculture system have been released for the cultivators based on 14 years of research. The prominent features are summarised in Table 15.7.

Procurement Prices (Rs.) of Teak:

Particulars	Year	
	1998-99	2003-04
One Pole	60.00	100.00
Timber (cum)	17000	21000

Hortiagriculture

During the growing period of fruit trees, inter crops mostly the food grain (wheat) and vegetables, are grown in orchards. Oranges and vegetable, Banana and vegetables, Sapota and vegetables, Mango and vegetables are very common examples. In Konkan region and western Maharashtra region under the irrigated cropping systems i.e. along with grapes, banana, alphonsa, cashew growing of flowers and vegetables, spices and medicinal plants are found to be more economic systems.

Silvipasture / Hortipasture

These are found naturally. However successful attempts have been made to establish the artificial pasture on barren degraded lands. The grass species associated with trees at different soil types is presented in table 15.8.

Shelter Belts/ Wind Breaks

It is a common practice in Marathwada, western Vidarbha and Khandesh region of the state to maintain some trees along the farm side road, boundaries, village common land, religious places and parks and garden .The densely placed tree rows protect the crops/animal from hot wind currents. Species found suitable are:

Farm wood lots : Farmers in dryland region mainly plant tree species on their farm or near their home . This serve as source of fuel, fodder and small timber. The tree species varies from region to region. Most common are *Acacia nilotica, Terminalia arjuna* and *Leucaena leucocephala.*

Nala, and river bank plantations : For the stabilization of nala (stream), river banks *Casuriana, Eucalyptus, Terminalia arjun* and *Leucaena* have been planted.

Live Hedges/ Fencing : Live hedges and fencing is almost common in all part of the state. Many shrubs and trees have proved useful for live fencing, eg., *Prosopis juliflora, Zyziphus oenoplea, Cesalpenia seperaria, Dodonia viscosa, Jatropha curcas, Lawsomia inermise, Lantana camara, Acacia senegal, Vitex negandu, Ipomea cornis, Jatropha gossipifolia, Bamboo species.* (Ilorkar and Totey ,2001a).

Forest Based Industries.

1) *Cottage industries*
   i) Bamboo crafts and domestic items
   ii) Tussor culture

Table 15.1. Land use classification of Maharashtra.

S. No.	Land use	Land Area (lakh hectare)	% of TGA.
1	TGA (Total Geographical Area)	307.58	100
2	Forest	52.96	17.21
3	Barren and uncultivable land	16.97	5.51
4	Fallow land	13.01	4.22
5	Culturable wasteland	9.03	2.93
6	Permanat Pastures	13.41	4.35
7	Scrub Forest	2.26	0.73
8	Current Fallows	11.89	3.80
9	Other barren land	11.71	3.80
10	Cultivable land	176.36	57.33
11	Irrigated land	52.07	16.93
12	Per Farmer land holding (ha)	2.18	--

Table 15.2. Average Wood requirement (M^3 per capita/year) in Maharashtra.

S. No.	Item	Locality from forest		
		0-5 km	>5 km	Town& cities
1	Building construction	0.944	1.447	0.274
2	Furniture	0.035	0.043	0.030
3	Agril. Implements	0.008	0.009	0.003
4	Total timber	0.987	1.499	0.307
5	Fire wood	0.894	1.177	0.814
6	Bamboo	19	18	4

Table 15.3. Alternative fuel consumption in rural area in Maharashtra (per capita / annum)

S. No.	Item	Unit	Land holding Group (ha)		
			0-1	1-2	>2
1	Cowdung	kg	122.6	114.33	92.41
2	Agril. Waste	kg	82.29	87.73	95.64
3	Kerosene oil	Lit.	6.24	6.51	7.84
4	Charcoal	kg	4.01	7.23	14.65

Table 15.4. Agroclimatic zones of Maharashtra.

S. No	Zone	District/Stations	Rainfall	
			Max	Mini
University Jurisdiction : Mahatma Phule Krishi Vidyapeeth, Rahuri				
1	Western Ghat zone	Igatpuri,Lonawala, Mahabaleshwar, Radhanagari	6690.0	3505.0
2	Sub Montane zone	Kolhapur,Vadgaon-Maval, Karad, Gadhingalaj	1258.8	835.6
3	Plain	Pune, Niphad, Digraj, Pimpal gaon	1256.6	300.5
4	Central Plateu	Jalgaon	534.1	NA
5	Scaricity	Solapur, Dhule, Rahurio, Chas, Jeur, Pdegaon, Pandharpur, Kopergaon, Mohol	783.0	263.5
University Jurisdiction : DR. Panjabrao Deshmukh Krishi Vidyapeeth, Akola				
6	Western Vidarbha	Achalpur,Amravati, Akola, Buldhana, Washim	682.0	550.0
7	Central Vidarbha	Katol, Nagpur, Yeotmal, Kutki,Tharsa	899.0	560.0
8	Eastern Vidarbha	Sakoli, Sindewahi, Nawegaon, Sonapur, Ekarjuna	1393.8	724.1
University Jurisdiction : Marathwada Agriculture University, Parbhani				
10	Central Mahatahtra plateu	Parbhani, Nanded, Latur, Ambejogai, Basmatnagar, Badnapur, Tuljapur, Aurangabad, Jalna	700.5	544.0
University Jurisdiction : DR. Balasaheb Sawant Konkan Krishi Vidyapeeth, Dapoli				
10	North Konkan	Palghar, Karjat, Panvel, Repoli	2805.1	2123.3
11	South Konkan	Dapoli, Muladhe, Phondaghat, Wakwali	3595.6	3164.3

Source: Anon 2005

Table 15.5. Systematic monoculture plantations in farmers fields

Species	Spacing (meter)	Remark
Tectona grandis	2 x 2, 2 x 1, 3 x 3	Highest growth was recorded under the spacing 3 x 3 m
Dalbergia Sissoo	2 x 2, 3 x 2,	Found suitable for partially Waterlogged soils
Eucalyptus sp	1 x 1, 2 x 1, 2 x 2	Grown as a raw material for paper and pulp industry and small timber.
Acacia mangium	2 x 2, 3 x 2, 3 x 3,	Recommended for Konkan region
Khaya senegelensis	2 x 2, 3 x 2, 3 x 3,	Recommended for Konkan region
Casurina equisetifolia	2 x 2, 3 x 2 4 x 4	Mostly grown along west coast of Konkan and foot hills of Western ghat.
Leucaena leucoephala	1 x 1, 2 x 1, 2 x 2	Grown as a raw material for paper and pulp industry and small timber , fodder and green manuering
Bamboo (*D. strictus*)	4 x 4, 5 x 5, 5 x 4	Domestic and industrial purpose

Table 15.6. Cost of Bamboo production, Gross return and Benefit cost ratio over a period of Twelve years under Agrisilviculture system.

Treatment	Bamboo Harvested Per ha*	Cost (A+B) Rs./ha	Discounted Cost Rs./ha	Gross Return Rs./ha	B/C Ratio
N0 (0 Kg N)	19446	70844	134433	215133	1:1.60
N1 (50 Kg N)	23692	81583	155043	261925	1:1.69
N2 (100 Kg N)	23894	84644	160987	264380	1:1.64
N3 (150 Kg N)	24418	87045	167791	270216	1:1.61

* Total Bamboos harvested at age 6th, 8th, 10th years.

Table 15.7. Production of poles and gross monetary return of teak planted as agrisilviculture system.

Thinning Year	Spacing (m)	DBH (cm)	Tree Density /ha	No. of Trees Harvested	Wood Yield/ha		Return/ha (Rs)		
					Timber Volum (cum.)	Pole umber	Timber	Pole	Total
1998-99	8 x 2	9.06	600	300	-	300	-	18000	18000
	12 x 2	8.21	400	200	-	200	-	12000	12000
2003-04	8 x 2	15.81	300	150	7.361	150	154577.3	15000	169577.3
	12 x 2	14.65	200	100	3.794	100	79677.9	10000	89677.9

- Intercrops: Soybean, black gram

Soil type shallow - medium.clay loam

Table 15.8 species combination under Silvipasture system in different soil types

Soil type	Pasture Species	Tree Species
Shallow soils	*Cenchrus ciliaris, Stylo hamata*	*Hardwikia binnata Albizia lebbeck Zizyphus zujuba,*
Shallow –medium soil	*Dicantheu, pauna,* cowpea, sorghum, cluster bean	*Acacia nilotica, Zyziphus zujuba, Annna sqamosa.*
Medium -deep soil	Napier, Cluster bean, cowpea	*Erithina indica Mangifera indica, Embelica officinalis.*

iii) Tendu leaves collection and bidi making

iv) Leaves plate making

v) Firewood collection and sale

vi) Wild mushroom collection

2) Large Industries

i) Papermills

ii) Plywood factories

iii) Toy making

iv) Match stick and match boxes

v) Saw mills and wood work

These forest based industries provide handful employment to the rural masses. These industries also procure wood, bamboo etc. as raw material from government as well as private tree growers and hence there is great potential for growing trees in appropriate agroforestry system (Ilorkar and Totey, 2001b).

SCOPE AND LIMITATIONS

The land use classification shows that about 21 per cent (63 lakh ha) of land may be made available for one or another type of agroforestry system. It is necessary to support more research base to design and develop the agroforestry system at microlevel suitable for all type of edaphoclimatic conditions. Existing agriculture extension machinery should encourage agroforestry adoption.

REFERENCES

Anonymous (2001) *State of Forest report 2001*, Forest Survey of India, (Ministry of Environment and Forest), Dehradun.

Anonymous (2005) *Joint Agresco Report* 2005, Maharashtra Agricultural Universities. At Parbhani, from 8 to 10 Jun 2005.

Ilorkar, V.M. and Totey, N.G. (2001a) Phytosociological studies in Naveagon National Park. *Indian Forester*. 129 (3).

Ilorkar, V.M. and Totey, N.G. (2001b) Leaf litter production and Nutrient Concentration in Natural forest Ecosystem of Navegaon National Park. *Jr. Range Management and Agroforestry*.22 (2).

Section 3

Agroforestry: Systems and Practices in East, North-East and Southern India

16

Agroforestry Systems and Practices Prevailing in Bihar

O.P. CHATURVEDI[1] AND D.K. DAS[2]

[1] National Research Centre for Agroforestry, Jhansi-284003, U.P. India [2] Department of Forestry, Rajendra Agricultural University Pusa, Samastipur - 848125, Bihar

ABSTRACT

Bihar is a state whose economy is solely dependent on agriculture. Though it has deep fertile land of alluvial origin, still agriculture of this state is facing problems due to frequent flooding. Population pressure on forest had led to creation of large extent of degraded lands. Agroforestry seems to be one of the viable option to avert these adverse conditions. Traditionally people used to integrate trees in farm lands in different forms. However, green revolution reduced these practices. The present paper provides a detailed account of prevalent agroforestry practices in Bihar. Results of different agroforestry systems which had been researched under AICRP on Agroforestry in Bihar since 1987 are also given in this paper.

INTRODUCTION

Bihar, the twelth largest state of India is bound on the north by Nepal, south by Jharkhand, east by West Bengal and on the west by Uttar Pradesh. The state lies between 24° 20′10" to 27° 31′15" N latitude and between 83° 19′50" to 88° 17′40"E longitude at an elevation of 53.12 m above the mean sea level and comprises an area of 93595.68 sq.km. The topography of Bihar is practically plain. The whole of the Bihar plain is traversed by a number of fairly large rivers, viz., Ganga, Ghagra, Kosi, Bagmati, Burhi Gandak, Kamla balan etc. which cause intermittent floods and engulf several districts of Bihar each year.

Climate

The climate is sub-tropical humid to sub-humid type. Minimum and maximum temperature varies between 7.1°C in January to 41.8°C in June.

Based on climatic variations, the year is divisible into three main agricultural seasons viz. summer (March to mid June), rainy (mid June to September) and winter (October to February). The state is divided into three agroclimatic plain zones.

1. *North-West Alluvial* : It comprises of Saran, Tirhut and Darbhanga divisions. The average annual rainfall is about 1211 mm.
2. *North-East Alluvial*: This comprises of Saharsa, Purnia and part of Begusarai divisions. The average annual rainfall is about 1405 mm.
3. *South Bihar Alluvial*: This zone is located south of river Ganga and comprises of Patna, Gaya and Bhagalpur divisions. The average annual rainfall is about 1110 mm.

Soils

The soils in Bihar are an alluvial deposit with very good soil depth, high fertility and are of recent origin. There are occasional patches of problematic soils with high amount of calcium carbonate and salinity problems. The soil is light to medium textured except in those districts away from influence of rivers where the soil is medium to heavy. The soils under influence of rivers are mostly calcareous. Average soil pH, organic carbon, N, P_2O_5 and K_2O varied in the state between 6.8-8.3, 0.38-0.55%, 235-290 kg ha^{-1}, 14-25 kg $ha^{-1}$and 115-165 kg ha^{-1}, respectively.

LAND USE

Economy of Bihar is solely dependent on agriculture as it is devoid of big industries for source of revenue. Only 7 per cent area is under forest as against a minimum 33 per cent required for ecological balance. Per capita availability of forest in the state is 0.01 ha which is much lower than the world and Indian average of 0.80 ha and 0.08 ha, respectively. The pressure on existing forest is too high to meet the demand of people for timber, fuelwood, fodder etc. in adequate quantity. Area of waste land is about 6.59 lakh ha categorized under barren and uncultivable land (4.37 lakh ha), cultivable waste land (0.5 1 lakh ha), fallow lands other than current fallows (1.52 lakh ha), grassland/ permanent pastures etc. (0.19 lakh ha) (Table 16.1).

The flood besides causing water stagnation in certain areas washes out field crops causing its shortage. The water stagnation delays the sowing of succeeding crops. Not only crops are damaged but also in flood prone area, stored food grains and seeds are also spoiled. The flood damage in Bihar accounts 20-40 per cent of total damage in the country (Singh, 1995). Like flood, drought also affects several districts of the state regularly. The drought causes the failure of standing kharif crops and affects the possibilities of taking good rabi crops in rainfed areas.

About 2.24 lakh hectare of land in Bihar is salt affected (Table 16.2, Prasad, 1995). They are either saline-alkali or alkali soils. The high water expenditure, faults in soil and water management, excessive seepage from canals, high monsoon rainfall, undulating topography and recurring floods have resulted in a large area of Bihar under waterlogging. Such condition has also rendered quite a large area unproductive on account of secondary salinization (Prasad and Sinha, 1998, 1999; Prasad *et al.* 1999). These salt affected soils are often put to arable farming by our farmers despite the fact that these are not suitable for arable cropping. This has resulted in agriculture being not sustainable because of the fact that crops grown in most of these areas are under rainfed situation and have a poor organic matter and waterholding capacity which results in poor productivity.

Diara and tal lands are flood prone areas. When perennial rivers during monsoon are accompanied by high and continuous rainfall, it causes deluge. Diara and tal land in Bihar occupy nearly 10.25 lakh ha (Table 16.3, Yadav, 2000). These areas remain submerged from mid June to November. However, chaur lands may remain submerged from June to mid January. Agriculture in these areas is risky and so people remain poor.

Such environmental degradations have brought human population under socio-economic crisis and ecological imbalance. Under such situations, it is imperative to minimize the farmers' risk through better utilization of natural resources. There is need to improve land productivity through development of sustainable land use models and farming systems. Agroforestry is viewed as practice to overcome these crises by bringing the wasteland and other unutilized land into productive use.

CROPPING SYSTEMS

The area is suitable for growing a fairly large number of crops. Kharif crops cover approximately 79 per cent of the net cultivated area. Rice, wheat, maize and pulses (redgram, blackgram, greengram, lentil, peas) are the prominent crops followed by sugarcane and oilseeds. Rice occupies approximately 44 per cent, wheat 20 per cent and maize 9 per cent of the gross cropped area. Among oilseeds, rapeseed, mustard and linseed occupy major place. Paddy-wheat and maize-wheat are the most popular rotations under irrigated condition. Major tree species and crop sequences under different agroclimatic zone of Bihar are presented in Table 16.4.

In diara land during pre flood crop season (early June to August) short duration maize and millets are sown with the first rain in Rohini Nakshtra and harvested before the floods, which generally occur, by middle of August. Due to flood there is always problems of harvesting, threshing, drying and storage of the produce during August. During post flood crop season (mid September to mid November) farmers broadcast black gram on wet soil. Toria is also sown in mid September, which matures by early December and then rabi crops are sown on residual moisture. Rabi crops usually suffer from moisture stress at reproductive stage and so yields are low and unstable. Agroforestry has a great potential in upgrading the productivity of such lands.

Degraded Forests and Fuelwood Situation

The total forests area in Bihar is 6.16 lakh ha. of which about 1.5 lakh ha. (25% of the total forest area) is under degraded condition. The condition of degraded forests reflects the pressure for fuelwood and grazing in such areas.

On the basis of the diagnostic survey in north Bihar, Chaturvedi *et al.* (2001) reported that fuelwood crisis, though a common problem among all categories of the farmers, is more acute with the marginal (0-1 ha land) and small farmers (1-2 ha land). On an average 72 per cent of the small and marginal farmers adjusted their fuel crisis by collection from the crop fields, orchards, homesteads and nearby houses. About 3.8 months in a year were reported as average fuel shortage period, which varied from 3.0 months for large to 4.5 months for marginal farmers (Table 16.5 & 16.6).

Fodder Situation

In Bihar the estimated shortage of dry and green fodders increased to 15 million tonnes and 40 million tonnes, respectively, in post green revolution period. The shortage of green fodder is experienced generally in the month of October - November and May-June. Gupta (2001) reported different fodder production systems for unutilized lands to minimize the shortage of the green fodder (Table 16.7). Fodder based cropping systems for the different agroclimatic zones of Bihar are presented in Table 16.8.

The diagnostic survey conducted in seven districts (Samastipur, Vaishali, Madhubani, Darbhanga, Muzaffarpur, Sitamarhi and Begusarai) of Bihar located in agroclimatic Zone-I (North west alluvial plain) revealed rearing of livestock has been the traditional occupation of the inhabitants of the villages (Table 16.9). Number of bullock per household varied between 0 (marginal) to 2.0 (small farmer) whereas each type of farmers had at least 1.0 cow and 0.5 buffalo and sheeps were being reared by only lower income group of farmers (i.e. small and marginal farmers). In order to feed their livestocks, farmers grow some fodder crops such as jowar, barseem, oat, meth, teosenite and maize in their own field. However, the marginal farmers do not grow but gather fodder. During scarcity of crop fodder farmers also use the tree leaves of subabul, kachnar, bamboo, jangali jalebi and Ficus species as fodder.

The land use statistics showed only 4.8 per cent of the total land of villages is under grazing and pasture land including some specific grazing system viz. silvopastoral, agrisilvopastoral, hortipastoral and block plantation (Table 16.10). The area under fodder crops (3.7%) was confined only to the small, medium and large type of farmers who used to grow green fodder with the average production ranging from 480 q/ha (Oat) to 575 q/ha (Jowar).

Sources of fodder were of different types like crop byproducts, green fodder, tree fodder, grasses and weeds (Table 16.11). Keeping in view the percent utilization of different type of fodders, crop byproducts ranked first irrespective of the farmer's category. Large and medium farmers used about 70 per cent of total fodder through crop byproducts for their livestocks, while poor farmers (small and marginal) used about 41% of the total fodder as crop byproducts followed by grasses and weeds. Tree fodders were more prevalent among lower income group of farmers.

Except the marginal farmers, all categories of farmers used to grow crop fodder besides other food crops under rainfed and irrigated conditions during kharif and rabi seasons (Table 16.12). Of the total land holdings about 10 per cent in kharif and 9 per cent in rabi were utilized to grow fodder crops by the higher income group of farmers. Small group of farmers used only 1.1 per cent in rainfed and 2.6 per cent in irrigated conditions during kharif whereas 1.4 per cent in rainfed and 2.4 per cent of total land holdings in irrigated conditions during rabi season to grow crop fodders.

PREVALENT AGROFORESTRY PRACTICES

Over the years, farmers have been integrating tree-crop-animal components for their livelihood. With the passage of time the agricultural crops received priority over trees for sustaining self-sufficiency in food grains due to green revolution. Now tree farming is gaining ground in the state under various afforestation programs. Growing tree in agricultural lands is now becoming popular among the farmers. In return the farmers get the cash crops benefits and also the return in the form of timber, fuel wood, etc. Some of the major agroforestry systems practiced in Bihar are listed in Table 16.13.

(i) Agrihorticultural/Agrihortisilvicultural system

Growing of fruit trees is an important cash-cropping practice in Bihar as it is major supplier of fruits in India. A total of 2.86 million hectares of land in Bihar is under fruit orchards, of which mango, litchi, guava and lime occupy areas of 1.55 m ha., 0.26 m ha, 0.31 m ha and 0.29 m ha, respectively. The fruit trees take a long time to bear fruit and are generally spaced widely. Till they grow and cover the entire land farmers raise cereals, pulses, vegetables, tubers, medicinal plants and fodder crops in the interspaces (Table 16.13). Papaya and dwarf varieties of banana are also intercropped depending upon the edaphic conditions. With increase in canopy cover of trees, tuber crops are preferred. Improvement in soil conditions through various agronomical management required for intercropping helps in improvement of fruit bearing capacity of trees.

Under Agrisilvihorticultural system farmers often put a trench on the boundary of their orchards and ridge of the trench is planted with multipurpose tree species like *Dalbergia sissoo, Wendlandia exserta, Phoenix sylvestris, Borassus flabellifer, Tectona grandis* etc. These trees protect the orchard from the high velocity of wind and desiccating heat and provide fuel, fodder and timber to the farmers. The interspaces in the orchard are utilized by cultivating crops.

(ii) Agrisilvicultural system

In this system the trees are planted at wide spacing and cultivations of crops are done in the inter-spaces till the canopy closes and restrict light reaching the ground. Under this system trees are also planted along the boundaries of agricultural fields. In Bihar *Acacia nilotica, Dalbergia sissoo, Bombax ceiba, Acacia lenticularis, Borassus flabellifer, Phoenix sylvestris* and *Artocarpus heterophyllus* are the important species grown with cereals and pulses. Recently, *Bombax ceiba, Anthocephalus cadmba* and *Populus deltoides* have received greater attention due to their fast growth and many benefits.

(iii) Silvipastoral/hortipastoral system

Naturally occurring grasses are present in interspaces of the horticultural or silvicultural component. Most of the livestock graze freely. Scattered tree species like *Acacia nilotica, Acacia catechu, Zizyphus mauritiana, Dalbergia sissoo* and *Mangifera indica* are often found growing on these grazing land covering a large portion of flood prone and salt affected areas. The grasses associated with these grazing grounds are *Dichanthium annulatum, Chrysopogon fulvus, Heteropogon contortus* and *Cencrus ciliaris*. Lopped prunings from trees are often used for fuel and fodder. The tree also provides shelter to the grazing animals during summer and rainy seasons. Controlled grazing can help in improvement of the productivity of these grazing lands.

(iv) Block planting

Dalbergia sissoo, Tectona grandis, Bombax ceiba, Acacia lenticularis, Gmelina arborea, Populus deltiodes, Dendrocalamus strictus and *Acacia nilotica* are the commonly grown species in the woodlots. These species provide fuel, fodder and timber, apart from helping in conservation and amelioration of soils.

(v) Homegardens

Homegardens contain mixed vegetation with various perennial trees and annual vegetable crops. Selection of the species is determined by the environmental factors and needs of the family. Chaturvedi *et al.* (2001) reported that the average size of the homegarden varied between 0.05 to 0.42 ha. A positive correlation between size of farm and homegarden was found to exist. *Dendrocalamus strictus, Dalbergia sissoo, Mangifera indica, Litchi chinensis, Musa paradisica and Carica papaya* are the common species associated with homegardens. A good number of vegetables and spices are found growing in homegardens.

(vi) Apiculture with trees

In this system various honey (nectar and pollen) producing tree species frequently visited by the honeybee are planted on the boundary or mixed with agricultural crops. The main purpose of this system is the production of honey. In Bihar, the potential rewards of bee keeping are great but apiculture is not achieving their full potential mainly due to shortage of flora. Agroforestry may offer some solutions based on symbiotic relationship between bees, field crops and trees. The increased presence of bees could lead to better pollination of both tree and field crops and thus to greater yields of timber, fruit, crop and honey. Consequently, it would bring farmers increased cash income. Semal, litchi, palas, shisham, karanj, chakundi, imli, siris, subabul, guava, jamun, arjun, kadam, safeda etc. are suitable for apiculture in agroforestry system.

(vii) Aquaforestry

Fishponds are an important and intensive form of production in many parts of Bihar, particularly in Chaurs (low-lying waterlogged areas). Planting trees such as *Dalbergia sissoo, Pongamia pinnata, Tectona grandis, Acacia lenticularis, Bombax ceiba* and *Wendlandia exserta* around fishponds augment the fish diet. The trees are managed by the farmers for production of fuel, fodder, timber and soil conservation.

AGROFORESTRY RESEARCH

Agroforestry research in Bihar was initiated on different aspects with the start of All India Coordinated Research Project on Agroforestry during 1987. Some of the research findings have been listed below:

(i) Agrihorticultural systems

Litchi is a major commercial crop in Bihar. Growing a number of other crops in association with litchi plantation is a widespread practice in all litchi-growing areas. Yield of intercrops under two cropping patterns *viz.*, paddy-wheat-greengram-ginger and maize-toria-greengram-pointed gourd was studied by Chaturvedi and Jha (1998) across an age series of litchi plantations during years 1-9 following planting. Relative light intensity under tree plantations decreased with increasing canopy depth and tree height. There was a marked reduction in yield of all intercrops with increase in age of the plantation. In first cropping pattern paddy, wheat and green gram were cultivated till 6 years of planting followed by cultivation of shade loving ginger crop upto 8 years of planting. In second cropping pattern maize, toria and greengram were cultivated upto 5 years of planting followed by pointed gourd upto 8 years of planting. The benefit : cost ratios of intercrops and fruit after 9th year of planting in cropping patterns I and II were 2.17 and 2.73, respectively. On a nine-year cycle of a litchi orchard, establishment, management and harvest of fruits and intercultural operations of intercrops generate employment opportunities of 130-140 man days ha^{-1} year^{-1}. With development of suitable cropping pattern for agri-horticultural systems, more and more farmers are adopting this diversified technology helpful in generating employment and improving economy.

Jha (2001) reported 30 – 39 per cent yield of tuber and spices crops under heavy shade situation of 80 per cent canopy closure in old litchi orchard as compared to their respective yield in open. Generally at this stage the orchards remained fallow and these additional yield from unutilized land resulted in benefit: cost ratio of 2.19. These yields included an extra yield of tuber and spices crops as well as enhanced yield of litchi fruits by 10-15 per cent due to better management of soil and tree in the orchard.

(ii) Agrisilvicultural system

In two cropping patterns viz., (i) "maize (*Zea mays*)-wheat (*Triticum aestivum*)-turmeric (*Curcuma domestica*)" and (ii) "pigeonpea (*Cajanus cajan*)-turmeric" under *Populus deltoides* G3 plantations at 5 x 4 m spacing Chaturvedi and Pandey (2001) reported very little reduction in yield of maize, wheat and pigeonpea in the first year of poplar plantation. But with increase in age of the plantation, yield decreased and maximum reductions of 64.6 per cent in maize, 41.0 per cent in wheat and 67.0 per cent in pigeonpea were recorded during 4th year. With this drastic reduction in yield, shade loving turmeric crop was cultivated from the 5th year in cropping pattern (i) and 6th year in cropping pattern (ii). The yield reduction of turmeric during 6th year of planting varied between 17.8 to 24.7 per cent in the both the cropping patterns and reached maximum 60 per cent in the ninth year. The overall benefit: cost ratio including poplar wood after harvest at the end of 10th year was 5.13 in "maize-wheat-turmeric" and 6.68 in "pigeonpea-turmeric" cropping patterns.

Annual transfer of litter nutrient to the soil by vegetation was 37.3 to 146.2 N, 5.6 to 17.9 P and 25.0 to 66.3 K kg ha^{-1} year^{-1} in young (three-year-old) and mature (nine-year-old) plantations. Turnover rate and time for different nutrients ranged between 0.86 to 0.99 year^{-1} and 1.01 to 1.16 year, respectively. This study shows that *Populus deltoides* has a potential of cropping under an agroforestry land use system and can provide standing biomass of 90.6 Mg ha^{-1} at a short rotation of 9 years. The disappearance of litter to the tune of 3.71 to 4.13 (young) and 10.55 to 10.74 Mg ha^{-1} (mature) protects the soil health from further deterioration and thus maintains sustainability (Das and Chaturvedi, 2005a & b).

Field trials were conducted in the mid-indo gangetic alluvial plain with *Leucaena leucocephala* to assess the biomass production along an age series and the potential of different pruning regimes to produce forage, firewood and its influence on alley cropped wheat, maize and green gram grain yields (Chaturvedi and Jha, 1994). High forage and dry wood yields were obtained with increasing pruning height and reduced pruning intensity. However, for higher grain production lower pruning height and higher pruning intensity were beneficial. Reduction of 55 and 47 per cent yields of maize and wheat respectively were recorded in the lines adjacent to hedgerows in comparison to the lines lying in the middle.

In another experiment conducted on farmer's field with four types of tree species viz. *Dalbergia sissoo, Wendlandia exserta, Borassus flabellifer* and *Phoenix sylvestris* growing on field bund alongwith two crop rotations "Paddy-wheat" and "Maize-tori-mung", it was found that soil fertility was higher near to tree base while grain yield increased with increase in distance from tree base (Chaturvedi and Das, 2002a). All the fields associated with different trees had sown greater availability of N, P and K in the soil as compared to the field without any association of trees. Grain yields were adversely affected in association of trees in the pattern *P. sylvestris* < *B. flabellifer* < *W. exserta* < *D. sissoo*. The average reduction value of grain yield across all crops and associated trees varied between 16.3 to 37.3 per cent. Besides, these trees provide fuelwood and timber. The juice extracted from *B. flabellifer* and *P. sylvestris* is the primary economic product, which is used for preparation of gur (molasses). Thus, these trees strengthen economic security of poor people in case of disaster and emergencies.

A comparative study of the rooting depth of 4-year-old agroforestry tree species namely, *Populus deltoides* (G_3), *Wendlandia exserta, Dalbergia sissoo* and *Acacia lenticularis* interplanted on boundary of wheat crop field was conducted (Chaturvedi *et al.* 2005). The results confirm the general view that *Wendlandia exserta* has a deep root system, which interferes less with shallow-rooted wheat crops. *Dalbergia sissoo* and *Acacia lenticularis* with more fine and bigger roots density and biomass in top soil (0-20 cm), have potential for efficient soil and water conservations, whereas not much preferred in agroforestry due to more competition with crops for nutrients and moisture from top soil. If competition for these species is to be minimized, tree planting must be combined with more appropriate management practices.

Bamboo as a component of agroforestry system is generally planted in homestead or on the boundary of agricultural lands. Bamboo is a fast growing species, and it attains maturity within a short period. Mature culms are harvested annually. Bamboos are extensively used in building small farmhouses, cattle shed etc. Small baskets and strings are made from bamboos. Large bamboo pipes are used as water conveyers in the farm irrigation/ drainage system. Bamboo leaves serve as excellent winter fodder for the goats. *Bambusa vulgaris* and *Dendrocalamus strictus* are widely planted in Bihar. Study on the effect of *Dendrocalamus strictus* row on the yield of some agricultural crops in Bihar showed that turmeric, ginger and dinanath grass produced 32.4 to 40.5 per cent relative yield under full light. Rice and finger millet gave the lowest yield near bamboo row and their yield increased with increasing distance from bamboo row (Annual report, 2000). Thus ginger, turmeric and dinanath grass can be grown successfully in bamboo shade depending upon grower's choice and economic return.

(iii) Silvipastoral/hortipastoral system

In a study on seasonal variation in ground vegetation under 3-year-old *Bombax ceiba* and *Acacia lenticularis* plantations spaced at 5 x 5 m, it was noted that the above ground live shoot ranged from 312-673 g m^{-2} under Bombax, 185-551 g m^{-2} under Acacia and 118-550 g m^{-2} in open field. Peak value was recorded in September. Standing dead biomass increased in February, thereafter it declined in subsequent months due to transfer into litter. The ground vegetation were represented by *Cynodon dactylon, Panicum repence, Setaria glauca, Eleucene indica, Desmodium cynosurvides* and *Dicanthium annualatum* (Chaturvedi and Das, 2002b).

In another study on seasonal variations in biomass, net primary productivity and turnover of dry matter of herbaceous layer under *Albizia lebbeck* (L.) Benth and *Populus deltoides* G_3 Marsh and in an adjacent open stands showed that all the three stands attained peak values 682, 665 and 512 g m^{-2}, respectively in live shoot biomass in September (Das *et al.*, 2005a). Net aboveground production was 749, 692 and 678 g m^{-2} $year^{-1}$, respectively under *A. lebbeck, P. deltoides* and in open field. The belowground net primary production of herbaceous layer was greatest in *A. lebbeck* (89 g m^{-2} $year^{-1}$) followed by *P. deltoides* (110 g m^{-2} $year^{-1}$) and lowest in the open (67 g m^{-2} $year^{-1}$). Maximum aboveground and belowground net primary productions were recorded in rainy season. Total net primary annual production of herbaceous layer was about 12 per cent and 8 per cent higher in *A. lebbeck* and *P. deltoides* plantations than open stand, respectively. The turnover rates of total plant biomass were maximum in the rainy season and the least in the summer season in all the stands. The system transfer function showed that the production of herbaceous vegetation in all the three stands was aboveground-oriented, accounting for 86 to 91 per cent of annual total net primary production. This study concludes that the herbaceous vegetation under monsoon subtropical condition shows seasonality in biomass production. Under the plantations, herbaceous biomass attains higher net primary production, which can be utilized to sustain secondary production of livestock.

Studies were conducted by Jha and Chaturvedi (1990) to evaluate the seasonal variation in dry matter and nutrient concentration in fodder of two commonly associated tree species in the agroforestry systems viz., *Bauhinia variegata* and *Dendrocalamus strictus*. The growing season between March-June was assessed to be suitable period for harvesting the *D. strictus* leaves to feed the animals. The suitable period for *B. variegata* appeared to be July/ August, when the crude protein was moderate and tannin content was lowest. Among the two tree fodders, D. *strictus* was better than *B. variegata* with greater value of nutrients and crude protein.

In another study, effects of cutting intervals on the forage yield from guinea, napier, dinanath and orchard grasses under a 6-year-old plantation of *Acacia lenticularis* were studied by Das and Chaturvedi, (2005c). Guinea, napier and dinanath produced maximum dry matter and crude protein yield when they were cut at 60 days interval, whereas ochard grass showed maximum dry matter and crude protein yield at 40 days of cutting intervals. Crude protein, nitrogen free extract, ether extract, total ash and insoluble ash showed the increasing trend with increasing cutting intervals, while crude fibre showed reverse trend.

The protein contents of green fodder cultivated in shade under orchards were higher than the fodder grown under open areas (Jha, 2001). The increase in protein content was due to decrease in the level of cellulose and lignin in the fodder produced in the orchard.

(iv) Block planting

Under 12-year-old plantations of 12 tree species when soils were analysed it was found that there was an increase in fertility level of soil over the values recorded before planting. A two fold increase in the organic carbon content was observed in *Albizia procera* (0.64%), *Leucaena leucocephala* (0.63%) and *Acacia lenticularis* (0.61%) plots when compared with the initial value (0.31%). The efficiency of areas in enriching the soil with nitrogen, phosphorus and potassium was more with the leguminous tree species than the non-leguminous one (Das *et al*, 2005b).

Culm recruitment, dry matter dynamics and carbon flux were studied at the ages 3, 4 and 5 years of bamboo [*Bambusa arundinacae* (Retz.) Wild] plantations (Das and Chaturvedi, 2005d). The recruitment to culm population varied between 16 and 21 per cent and shoot mortality from 3-5 per cent per year. Net accumulation of green culms between 3rd and 4th year was 675 and between 4th and 5th year 975 ha^{-1}. The total biomass was 170.8 Mg ha^{-1} in the 3-year-old to 257.3 Mg ha^{-1} in the 5-year-old plantation with an average per cent contribution of culms (65%), branches (21%) and leaves (4%). Total net primary production (NPP) ranged between 44.6 Mg ha^{-1} (4-year-old) and 60.5 Mg ha^{-1} (5-year-old), of which aboveground net production was 41.2 to 55.6 Mg ha^{-1} (between 3-4, and 4-5 years, respectively). Turnover rate and time for litter ranged from 0.77-0.80 year^{-1} and 1.25-1.30 years, respectively. The total carbon storage in standing biomass of 4- and 5-year-old bamboo plantation varied from 83.3 to 103.8 Mg ha^{-1}, of which 68 per cent was distributed in stem, 21 per cent in branch and 9 per cent in belowground biomass. A substantial amount of carbon varying from 2.3 (4-year-old) to 2.8 Mg ha^{-1} year^{-1} (5-year-old) was transferred to soil through litterfall and root.

(v) Phenology

Phenology and reproductive biology of some trees of north Bihar were studied between March, 1999 to February, 2000 (Mandal and Chaturvedi, 2001) (Table 16.14). Vegetative growth, which started with bud burst in February to March, was completed by May in 88.0 per cent tree species. In most of the species, leaf drop and simultaneous leafing occurred during warm-dry period between February and March of the year. About 68.0 per cent of species showed multiple leafing. In deciduous trees, flower (28.0%) and fruit (4.0%) formations occurred in March to February was again ahead by one month in deciduous species, than evergreen tree species.

(vi) Allelopathy

The phytotoxicity of *Leucaena leucocephala* leaf leachate was evaluated using rice and a few tree species as test crops (Chaturvedi and Jha, 1992). The phytotoxity of subabul leaf leachate differed according to the amount of mimosine present in it. The Leucaena leaf exposed to sunlight lost over 33-42 per cent of the total mimosine content. Among the three successive leaf extracts tested, the first extract showed an inhibitory effect while the third extract exhibited a stimulatory effect on rice radical growth. The effect of second successive extract was inconsistent.

Allelopathic effect of three agroforestry tree species *viz. Dalbergia sissoo* Roxb., *Acacia lenticularis* L. and *Bombax ceiba* L. was examined on seed germination and seedling growth of three paddy (*Oryza sativa* L.) varieties (Prabhat, Saket-4 and Kamini) (Mandal *et al.*, 2005a). The aqueous leaf extracts reduced the seed germination and seedling growth of paddy. On the basis of value of per cent reduction in seed germination and Germination Relative Index (GRI), the inhibitory effect of different leaf extracts was maximum in Kamini followed by Prabhat and Saket-4. Leaf extract of *Dalbergia sissoo, Acacia lenticularis* and *Bombax ceiba* showed inhibitory effect on sugar and protein and stimulatory effect on total amino acid contents of paddy seedling. The results indicated that the allelopathic effect of leaf extracts of *Bombax ceiba* on different varieties of paddy was maximum in terms of all physiological and biochemical parameters.

In another experiment, when the same treatments were applied to the wheat seeds, the inhibitory effect of different leaf extracts was maximum in Sangam followed by UP 1109 and RR 21 wheat varieties (Mandal *et al.*, 2003). In this case, allelopathic effect of leaf extracts of *Bombax ceiba* on different varieties of wheat was also maximum in terms of above physiological and biochemical parameters.

Potentiality of different concentrations of aqueous leaf extracts of *Populus deltoides* (G_3) Marsh was investigated for germination and seedling growth in three varieties of greengram (*Vigna radiata* L.) *viz.*, T_{44}, Sona and PC_{16} (Mandal *et al.* 2005b). On the basis of reduction in seed germination, germination relative index (GRI) and vigour index (VI), it was observed that the inhibitory effect of leaf extract was maximum in PC_{16} followed by Sona and T_{44}. The shoot and root length along with its respective dry weight decreased, whereas root/ shoot ratio increased as concentration of leaf extract increased. Leaf extract of *Populus deltoides* also showed inhibitory effect on soluble sugar and protein content, whereas stimulatory effect was recorded on total free amino acids content. The result indicated that allelopathic effect of leaf extract of *P. deltoides* was maximum in case of PC_{16}.

(vii) Rooting pattern

Studies on rooting patterns of 5-year-old important agroforestry tree species indicated that deep rooted species like *Acacia nilotica, Dalbergia sissoo* and *P. dulce* are suited to dry areas and may be mixed with shallow rooted species in plantations or in agroforestry systems (Chaturvedi and Das, 2002c). Since all the species had most of their root biomass in the upper soil layer, the deep rooted species like *A. nilotica* and *D. sissoo* will be preferred for intercropping as they will provide greater anchorage and will have potential to pump nutrients from deeper zone of the soil. However, to reduce root competition with companion crops the lateral roots of the tree have to be pruned. *P. dulce, A. procera, A. nilotica* and *D. sissoo* have high values of the soil binding factor and thus have potential for soil conservation.

(viii) Litter decomposition

Study on decomposition rates of leaf litter of *Eucalyptus tereticornis, Dalbergia sissoo, Sesbania grandiflora* and *Leucaena leucocephala* showed that the species having highest nitrogen percentage had the fastest decomposition rate, and that with lowest nitrogen content decomposed slowly (Das and Chaturvedi, 2003). Incorporation of nutrient-rich *L. leucocephala* leaves which result in more rapid loss of dry matter has important implications for synchronizing nutrient release with crop uptake in low input cropping systems in the tropics where turn over rates, climatic factor such as frequency of rainfall should receive much more consideration when these materials are incorporated into the soil in humid tropic conditions.

In another experiment, rate of decomposition of leaf biomass of six multipurpose tree species viz. *Populus deltoides, Dalbergia sissoo, Bombax ceiba, Acacia auriculiformis, Acacia lenticularis* and *Tectona grandis* were studied by Das *et al.* 2004. Leaf of all the tree species showed higher rate of decomposition during early stages. However, after initial one month of leaf biomass disappearance in case of *Acacia auriculiformis* was the slowest (4.4%). The highest rate of decomposition was observed in the leaf of *A. lenticularis* followed by *T. grandis, P. deltoides, B. ceiba, D. sissoo* and *A. auriculiformis.*

Table 16.1. Land use utilization pattern in Bihar (in lakh ha).

S. No.	Items	Area
1.	Total geographical area	93.60
2.	Forests	6.16
3.	Barren and unculturable land	4.37
4.	Land put to non-agricultural uses	
	(i) Land area	12.71
	(ii) Water area	
	(a) Permanent	2.07
	(b) Temporary	1.38
	Total (i + ii)	16.16
5.	Culturable Wastes land	0.51
6.	Permanent pastures and other grazing lands	0.91
7.	Land under miscellaneous tree crop and groves not included in area sown	2.33
8.	Follow lands other than current follows	1.52
9.	Current fallows	6.32
10.	Total uncultivable land	37.56
11.	Net sown area	56.03
12.	Total cropped area	79.46
13.	Area sown more than once	23.43

Table 16.2. Extent of salt affected area in the various districts of Bihar.

Districts	Salt affected area (ha)
Muzaffarpur	31,600
Vaishali	30,270
Samastipur	42,840
East Champaran	24,310
West Champraran	4,330
Saran	26,570
Siwan	33,750
Gopalganj	30,600
Total :	2,24,270

Source : Prasad (1995).

Table 16.3. River-wise flood prone areas.

S. No.	River	Area (lakh ha)	Affected agroclimatic zones
1.	Ganga	2.40	I, II, III
2.	Gandak	1.40	I
3.	Burhi Gandak	2.25	I
4.	Sone	1.15	III
5.	Kosi	1.50	I, II
6.	Bagmati	0.80	I
7.	Adhwara group	0.75	I
	Total :	10.25	

Source : Yadav (2000).

Table 16.4. Trees and crops under different agroclimatic zones of Bihar.

Important Tree Species	Cropping Systems		
	Kharif	Rabi	Summer
1. North-West Gangetic Plain			
Irrigated	Early paddy	Maize	-
Dalbergia sissoo	Maize	Potato	Moong
Tectona grandis	Maize/early Paddy	Rai	Moong
Swietenia mahogany	Paddy	Wheat	-
Bombax ceiba	Maize	Potato	Vegetables
Wendlandia exserta	Groundnut	Wheat	-
Populus deltoides	Paddy	Wheat	Moong
Syzygium cumini	Ragi	Maize	-
Anthocephalus cadmba	Paddy	Sugarcane	Sugarcane
Artocarpus heterophyllus	Maize	Potato	Onion
Gmelina arboea	Paddy	Berseem	-
Cocos nucifera	Maize	Wheat	Moong

Contd...

Table 16.4. Contd...

Mangnifera indica Rainfed			
Litchi chinensis	Paddy	Gram	-
Borassus flabellifer	Paddy	Wheat	-
Phoenix sylvestris	Paddy	Paira lentil	-
	Paddy	Sugarcane	Sugarcane
	Maize	Sweet Potato	Moong
	Ragi	Sweet Potato	-
	Mazie + turmeric/ Ginger	-	-
	-	Tobacco + Garlic	-
	-	Tobacco	-
2. North - East Gangetic Plain			
Dalbergia sissoo Irrigated	Paddy	Wheat	
Bombax ceiba	Paddy	Paddy	-
Acacia lenticularis	Early Paddy	Potato	-
Accacia nilotica	Jute	Maize	-
Wendlandia exerta	Jute	Rai	Maize
Tectona grandis	Paddy	Potato	Onion
Azadirachta indica	Maize	Tori	Groundnut
Artocarpus heterophyllus	Jute	Wheat	-
Cocos nucifera	Jute	Pulses	-
Borassus flabellifer Rainfed	Jute	Wheat	-
	Paddy	Gram	-
3. South Gangetic Plain	Paddy	Wheat	-
Dalbergia sissoo	Paddy	Wheat	Urd
Bombax ceiba	Paddy	Maize	-
Acacia lenticularis	Paddy	Potato	Moong

Contd...

Table 16.4. Contd...

Albizia lebbeck	Paddy	Onion	Moong
Gmelina arboea	Paddy	Toria	Moong
Terminalia arjuna	Paddy	Oilseed	-
Mangifera indica	Maize	Vegetable	-
Zizyphus mauritana	Maize	Berseem	
Spondias pinnata Rainfed	Maize	G. Nut	
Emblica officinalis	Paddy	Paira Lentil	-
Artocarpus heterophyllus	Maize	Oilseed	-
Borassus flabellifer	Fallow	Gram	-
	Fallow	Lentil	-
	Urd	Wheat	-

Table 16.5. Annual fuel shortage for different categories of farmers

Farmer type	Shortage period	Total No. of months
Marginal	Mid June - October	4.5
Small	July - October	4.0
Medium	July - Mid September	3.5
Large	July - September	3.0
All types	Mid June - October	4.5
Average		3.8

Source : Chaturvedi *et al.*, 2001.

Table 16.6. Fuel shortage adjustment mechanism in different farmer categories

Crisis management	% farmers responded				
	Marginal	Small	Medium	Large	Average
Collection	81 ± 6	71 ± 7	38 ± 4	32 ± 3	72 ± 6
Buying	24 ± 3	38 ± 4	20 ± 4	17 ± 2	21 ± 3
Previous stock	8 ± 2	17 ± 1	33 ± 6	42 ± 5	20 ± 3
Branch cutting and twings	0	17 ± 2	25 ± 5	31 ± 2	15 ± 3

Source : Chaturvedi *et al.*, 2001.

Table 16.7. Fodder production system adopted for management of unutilized land

S. No.	Fodders	Date of sowing	Availability of fodder	Yield (q/ha)
1.	Maize (African Tall) + Meth	Feb. - April	April-July	1000
2.	Bajra (L72 or L74) + Meth	Feb. - April	April-July	1000
3.	Jowar + cowpea	Feb. - April	May-August	1500
4.	Guinea grass	Feb. - May	April-August	500
5.	Maize + Meth	May - June	July-Sept.	1000
6.	Stylo	May - June	Sept.- Nov.	500
7.	Berseem + Mustard	Oct. - Nov.	Dec.- April	1000
8.	Lucern (T 9)	Oct. - Nov.	Dec.- July	1000
9.	Hybrid Napier (NH 21)	Feb. - Sept.	March-Nov.	1000

Source : Gupta, 2001.

Table 16.8. Major fodder production system in different agroclimatic zones of Bihar.

Agroclimatic Zone	Crop rotation
Zone I & II	Jawar/Bajra - Pulses
	Rice/Maize-Berseem
	Maize + Cowpea (fodder) - Wheat
	Maize/Rice - Berseem + Maize
	Rice/Maize - Lucerne
	Rice - Oat
	Napier + Berseem in Rabi
	Dinanath Grass - Berseem
	Rice - Oilseed - Maize + Cowpea
Zone III	Maize/Jowar-Berseem
	Jowar/Maize/Napier-Puses/Oat
	Rice-Berseem/Oat
	Jowar-Pulses-Maize + Cowpea
	Rice - Oat

Table 16.9. Animal husbandry and fodder situation in the villages of Bihar

Farmers Type	Animal species	No. of animals	Main fodder crops	Tree fodder in need
Large	Bullock	0.5 ± 0.08	Jowar	Subabul
(>6 ha)	Cow	6.5 0.38	Berseem	Kachnar
	Buffalo	2.6 0.18	Oat	Bamboo
	Goat	Nil	Meth	
	Sheep	Nil	Teosente	
			Maize	
Medium	Bullock	1.0 ± 0.04	Jowar	Bamboo
(>2-6)	Cow	3.5 ± 0.17	Berseem	Subabul
	Buffalo	1.5 ± 0.08	Oat	Kachnar
	Goat	Nil	Meth	
	Sheep	Nil		
Small	Bullock	2.0 ± 0.14	Jowar	Bamboo
(>1-2)	Cow	1.5 ± 0.10	Oat	Jangali Jalebi
	Buffalo	1.0 ± 0.06	Meth	Kachnar
	Goat	3.5 ± 0.12		Ficus
	Sheep	2.0 ± 0.18		
Marginal (0-1 ha)	Bullock Cow	Nil 1.0 ± 0.08	Don't grow fodder crops but gather fodder	Bamboo Jangali Jalebi
	Buffalo	0.5 ± 0.02		Kachnar
	Goat	5.5 ± 0.32		Ficus
	Sheep	3.2 ± 0.24		

Table 16.10. Average grazing pasture area, system followed and area under fodder crops in Bihar plains.

Grazing & pasture (%)	Grazing system followed	Area under fodder crops (%)	Average green fodder production (q/ha)	
4.8 ± 1.2	Silvipastoral	3.7 ± 0.8	Jowar	575 ± 38
	Agrisilvopastural		Oat	480 ± 38
	Hortipastoral		Berseem	500 ± 72
	Block-planting		Teosinte	550 ± 46
			Meth	250 ± 28

Table 16.11. Source of fodder for animals in Bihar.

Farmers type	Source	% of total fodder used
Large	Crop byproducts	68.6 ± 8.5
	Green fodder	27.9 ± 4.8
	Tree fodder	1.4 ± 0.2
	Grass & Weeds	2.1 ± 0.4
Medium	Crop byproducts	71.4 ± 6.4
	Green fodder	20.7 ± 8.2
	Tree fodder	0.7 ± 0.1
	Grass & Weeds	7.1 ± 1.2
Small	Crop byproducts	44.3 ± 4.5
	Green fodder	16.4 ± 2.8
	Tree fodder	12.1 ± 3.6
	Grass & Weeds	27.1 ± 4.2
Marginal	Crop byproducts	37.9 ± 5.6
	Green fodder	5.0 ± 0.8
	Tree fodder	21.4 ± 4.2
	Grass & Weeds	35.7 ± 3.9

Table 16.12. Seasonal distribution of area under crops in Bihar (% of total land holdings).

Farmer Type	Season	Rainfed	Irrigated
Large	Kharif	10.2 $\pm$ 0.8	10.7 $\pm$ 0.4
	Rabi	12.1 $\pm$ 1.2	9.3 $\pm$ 0.4
Medium	Kharif	10.7 $\pm$ 0.6	9.4 $\pm$ 0.5
	Rabi	6.7 $\pm$ 0.3	6.9 $\pm$ 0.4
Small	Kharif	1.1 $\pm$ 0.1	2.6 $\pm$ 0.2
	Rabi	1.4 $\pm$ 0.2	2.4 $\pm$ 0.2
Marginal	Kharif	Nil	Nil
	Rabi	Nil	Nil

Complete disappearance of leaf biomass in *A. lenticularis, T. grandis, P. deltoides, B. ceiba* and *D. sissoo* was observed in 5th, 7th, 8th, 9th and 10th month of placement in soil, respectively. Only 91.8% weight loss had occurred in *A. auriculiformis* by the end of 10th month. Nitrogen and phosphorous passed through the net immobilization phase in all the species except *A. lenticularis.* However, no immobilization was observed in K in all the species.

Though the adoption of different agroforestry practices at farmers level in Bihar is poor, but the systems adopted reveal the increasing diversification in cropping, resulting in reduced risk from market fluctuation of the dominant crops. Based on the present situation following factors need to be emphasized:

(a) Use the broad-based approach to achieve improvements in large number of conditions directly influencing rural welfare.

(b) Emphasis to put on evaluation of more research and appropriate technologies suited for upliftment of all categories of farmers.

(c) More emphasis is to be laid on extension of evolved appropriate agroforestry technologies to the users.

The programs of agroforestry development will further succeed with more and more village level interaction and cooperation of the farmers with the scientists.

Table 16.13. Species associated with different Agroforestry systems of Bihar.

Agroforestry Systems	Tree/shrub yielding, timber, fuelwood etc.	Fruits trees	Cropping including vegetable and spices	Grasses and fodder crops
1. Agrihortisilvi-cultural	*Dalbergia sissoo*	*Mangifera indica*	*Triticum aestivum*	*Sorghum biccolor*
	Phoenix sylvestris	*Litchi chinensis*	*Lense esculenta*	*Avena sativa*
	Moringa obleifera	*Cartca papaya*	*Trichosanthes argunia*	*Euchlaena mexicana*
	Wendlandia exerta	*Musa paradisiaca*	*Curcuma domestica*	
	Tectona grandis	*Cocos nucifera*	*Sacrharum afflcinarum*	
	Bombax ceiba		*Ipomea batata*	
	Accacia lentacularis		*Oryza sativa*	
2. Agrisilviculture/ Boundary planting	*Dalbergia sissoo*		*Triticum aestivum*	
	Bombax ceiba		*Zea mays*	
	Phoenix sylvestris		*Curcuma domestica*	
	Wendlandia excreta		*Zingiber officinal*	
	Populus deltoides		*Nocotina tobacum*	
	Acacia lenticularis		*Avena sativa*	
	Acacia nilotica		*Capsicum annum*	
	Cocos nucifera			
	Moringa oelifera			
	Artocarpus heterophyllus			
	Syzygium cumini			
3. Silvipastoral	*Acacia nilotica*			*Imperata cylendrica*
	Acacia catechu			*Cynodon dactylon*
	Dalbergia sissoo			*Dicanthium annulatum*
	Phoenix sylvestris			*Digitaria ciliaris*
	Syzygium cumini			*Chrysopogon fulvus*
	Zyzyphus mauritiana			*Heteropogon contortus*

4. Block planting (mixed) or monoculture	*Dalbergia sissoo (mixed and monoculture)*		
	Tectona grandis (mixed and monoculture)		
	Albizia lebbeck (mixed)		
	Populus deltoides (Monoculture)		
	Gmelina arborea (mixed)		
	Bombax ceiba (mixed)		
	Syzygium cumini (mixed)		
	Acacia lenticularis (mixed)		
5. Homegardens	*Dalbergia sissoo*	*Tamarindus indica*	*Curcuma domestica*
	Wendlandia exserta	*Cocos nucifera*	*Zephyronthes officinale*
	Tectona grandis	*Syzygium cumini*	*Solanum tuberosum*
	Acacia lenticularis	*Artocarpus heterophyllus*	*Solanum melongena*
	Dendrocalamus strictus	*Annona squamosa*	*Allium cepa*
	Bombax ceiba	*Aegle mormelos*	*Allium sativum*
		Mangifera indica	
		Litchi chinensis	
		Ficus glomerata	
		Psidium guava	
		Carica papaya	
		Musa paradisiaca	

Table 16.14. Phenological characteristics of some hortisilvicultural trees of Bihar

Tree species	Leaf drop	Leaf flushing	Vegetative growth	Flowering phase	Fruiting phase	Seed maturity phase
Acacia lenticularis	February March	March	April	April June	May June	March April
Albizia lebbeck	February March	March	April May	March	April Dec.	Feb. March
Albizia procera	February March	March	April	March May	April Dec.	Jan. March
Artocarpous heterophyllus	January February	March	March June	March	April June	June August
Azadirachta indica	January February	Feb.	Feb. July	February	May June	July
Bauhinia variegata	February March	April	April June	October November	Dec. April	May June
Bombax ceiba	January February	March April	March July	February	March April	May
Cassia fistula	February March	April	May August	May June	July Feb.	March April
Dalbergia sissoo	February	March	March	March May	April June	Dec. Jan.
Delonix regia	February March	March April	May July	May	June July	Feb.
Emblica officinalis	December January	March	April	April June	May Nov.	Dec. Jan.
Erythrina indica	January February	April	April July	March	April May	June July
Leucaena leucocephala	February March	March	March October	March	April Sept.	Oct. Nov.
Melia azedarach	February	March	March June	April	April Sept.	Oct.
Pongamia pinnata	February March	April	May June	April	May August	March
Populus deltoides	November February	March	March June	-	-	-
Psidium guajava	September October	Feb.	February June	June	June July	August Sept.
Tectona grandis	March April	April	April August	July August	August October	Dec. January
Toona ciliata	January	Feb.	February March	March	April May	June July
Ziziphus mauritiana	January	Feb.	March August	September October	Nov. Dec.	January Feb.

REFERENCES

Annual Report (2000). All India Co-ordinated Research Project on Agroforestry (ICAR), RAU., Pusa, Samastipur.

Chaturvedi, O.P. and Das, D.K. (2002a). Effect of bund trees on soil fertility and yield of crops. *Range Mgmt. & Agroforestry.* 23 (2): 90-94.

Chaturvedi, O.P. and Das, D.K. (2002b). Influence of seasons and tree plantations on biomass and primary productivity of herbaceous vegetation in eastern India. *Journal of Tropical Forestry.* 18 (II & III): 11-22.

Chaturvedi, O.P. and Das, D.K. (2002c). Studies on rooting patterns of 5-year-old important agroforestry tree species in north Bihar, India. *Forest, Trees and Livelihoods.* 12: 329-339.

Chaturvedi, O.P.; Das, D.K. and Jha, A.N. (2001). Household fuel situation and agroforestry in north Bihar plains - A diagnostic study. *Range Mgmt. & Agroforestry.* 22 (1): 94-105.

Chaturvedi, O.P.; Das, D.K. and Kumar, R. (2005). A comparative study of the rooting depth of four agroforestry tree species interplanted on boundary of wheat crop field in Bihar. *International Journal of Ecology and Environmental Sciences.* 31 (1) 49-52.

Chaturvedi, O.P. and Jha, A.N. (1992). Studies on allelopathic potential of an important Agroforestry species. *For. Ecol. Mgmt.* 53: 91-98.

Chaturvedi, O.P. and Jha, A.N. (1994). Production potential of *Leucaena leucocephala* and intercrops grown under different alley management systems in mid Indo-Gangetic plains. *Range Mgmt. & Agroforestry.* 15: 93-103.

Chaturvedi, O.P. and Jha, M.K. (1998). Crop production and economics under *Litchi chinensis* Sonn. plantation across 1 to 9 year age series in north Bihar, India. *International Tree crop Journal.* 9: 159-168.

Chaturvedi, O.P. and Pandey, I.B. (2001). Yield and economics of *Populus deltoides* G_3 Marsh based intercropping system in Eastern India. *Forest, Trees and Livelihoods.* 11: 207-216.

Das, D.K. and Chaturvedi, O.P. (2003). Litter quality effects on decomposition rates of forestry plantations. *Tropical Ecology.* 44 (2): 259-262.

Das, D.K., Chakraborty, A.K., Chaturvedi, O.P. and Kumar, R. (2004). Growth, biomass production and leaf decomposition pattern of some agroforestry trees. *Range Mgmt. & Agroforestry* 25 (1): 30-36.

Das, D.K. and Chaturvedi, O.P. (2005a). Structure and function of *Populus deltoides* agroforestry systems in eastern India: 1.Dry matter dynamics. *Agroforestry Systems* (In press).

Das D.K. and Chaturvedi O.P. (2005b). Structure and function of *Populus deltoides* agroforestry systems in eastern India: 2.Nutrient dynamics. *Agroforestry Systems* (In press).

Das D.K. and Chaturvedi O.P. (2005c). Effect of cutting intervals on the grass production and quality under *Acacia lenticularis* based silvipastoral system. *J. Tropical Forestry* (Submitted).

Das, D.K. & Chaturvedi, O.P. (2005d). Culm recruitment, dry matter dynamics and carbon flux in *Bambusa arundinacae* (Retz.) Wild plantation in eastern India. *Bamboo Research* (Submitted).

Das, D.K.; Chaturvedi, O.P.; Mandal, M.P. and Kumar, R. (2005a). Effect of tree plantations on biomass and primary productivity of herbaceous vegetation in eastern India. *Tropical Ecology* (Submitted).

Das, D.K.; Chaturvedi, O.P.; Mandal, M.P. and Kumar, R. (2005b). Reclamation of degraded soil through the plantation - Litter and fertility changes. *Indian Forester* (In press).

Gupta, B.S. (2001). Fodder production system for management of unutilized lands. *In* Summer Institute on *Agribased Entrepreneurship Development.* V.P. Gupta, P. Kumari and V.K. Shahi (eds.), RAU, Pusa, Samastipur. pp. 194-197.

Jha, A.N. and Chaturvedi, O.P. (1990). Seasonal variation in dry matter and nutrients in the leaves of *Bauhinia variegata.* L. and *Dendrocalamus strictus* Roxb. *Range Mgmt. & Agroforestry* 11: 207-212.

Jha, R.C. (2001). Agrihorticulture Systems. *In* Summer Institute on *Agribased Entrepreneurship Development*, V.P. Gupta, P. Kumari and V.K. Shahi (eds.), RAU, Pusa, Samastipur, pp 198-201.

Mandal, M.P and Chaturvedi, O.P. (2001). Phenology and reproductive biology of some trees of north Bihar. *Journal of Tropical Forestry* 17 (4): 37-40.

Mandal, M.P.; Das, D.K. and Chaturvedi, O.P. (2003). Allelopathic effect of three agroforestry tree species on seed germination and seedling growth of certain wheat varieties. *Indian J. Ecol.* 30 (2): 159-167.

Mandal, M.P.; Das, D.K. and Chaturvedi, O.P. (2005a). Allelopathic effects of agroforestry tree species on seed germination and seedling growth of paddy varieties. *Range Mgmt. and Agroforestry* (In press).

Mandal, M.P.; Das, D.K. and Singh, A.K. (2005b). Allelopathic impact of leaf extract of *Populus deltoides* Marsh on germination and seedling growth of greengram (*Vigna radiata* L.) cultivars. *Allelopathy Journal* (Submitted).

Prasad, R. (1995). Distribution and characterisation of salt affected soils in Bihar and their remedial measures. *In* Refresher Course on *Soil and Crop management for Sustained Productivity*. B. Roy (ed), RAU, Pusa, Samastipur. pp. 130-148.

Prasad, R., Kumar, P., Prasad, S. and Sinha, B.P. (1999). Critical ground water table depths for avoiding secondary soil salinization in Patna district under sone command area. *J. Appl. Biol.* 9 (1): 58-60.

Prasad, R. and Sinha, S.P. (1998). Favourable ground water table depth for development of soil sodicity and unfavourable for sustainable agriculture in Samastipur district. *J. Appl. Biol.* 8 (1): 154-155.

Prasad, R. and Sinha, S.P. (1999). Critical ground water table depth for sustainable reclamation of sodic soil in Muzaffarpur district. *J. Appl. Biol.* 9 (1): 61-63.

Singh, V.P. (1995). Contingency crop planning for adverse situation (drought and flood affected areas). In Refresher Course on *Soil and Crop Management for Sustained Productivity*. B. Roy (ed.), RAU, Pusa, Samastipur. pp. 36-46.

Yadav, C.P. (2000). Crop System Planning for Flood Prone Areas. In Refresher Course on *Agro-technologies in relation to ecological changes. A-Agricultural Development and Planning*. B.N. Verma (ed.), RAU, Pusa, Samastipur. pp. 22-26.

17

Agroforestry Systems and Practices in Jharkhand

M. S. YADAVA[1] & S. M. SULAIMAN QULI[2]
[1] All India Coordinated Research Project on Agroforestry,
[2] Department of Extension & Social Forestry, Faculty of Forestry
Birsa Agricultural University, Ranchi - 6

ABSTRACT

The paper gives a comprehensive detail of the state with regards to its inception, socio-economic, climatic and edaphic conditions. In addition states agroclimatic zones along with their characteristics have been described in the paper. Extent of agriculture crops, horticulture, animal husbandry, fisheries and forestry along with their production and contribution to national pool is provided elaborately. Details of agroforestry systems like silvipastoral and agrisilvipastoral for land with very high erosion and hortisilviculture and hortiagrisilviculture, aquaforestry, entomoforestry recommended for land with moderate to low erosion had been explained in details. Significant results obtained in agroforestry research under AICRP on Agroforestry and Horticulture and Agroforestry Research program have also been presented in this paper. Finally some tips to make agroforestry successful and future strategies have been given.

INTRODUCTION

Jharkhand state evolved as 28th State on 15th November 2000 from the southern part of Bihar, with a total area of 79.5 lakh ha and population density of 274/Km^2 consisting of 28 per cent schedule tribes and 12 per cent schedule caste population. A total of 29.61 per cent of the geographical area of the state consists of forest area with 28.4 per cent actual forest cover (Annon., 2001). Most of the forest area of the state consists of mines, which do not allow any agriculture. The continued large-scale mining operations of Coal and Iron ore are generating serious threats for the area coming under down streams. The areas lying under sharp undulating topography are under threats of soil erosion, which in turn affect the low lying water body's carrying capacity *i.e.* due to continuous siltation, all such anomalies emphasizes the need to improve the water and land use efficiency in order to ensure better storage and retention of the rain water, for improving the productivity of the land.

The state has humid to sub-humid climate with average annual rainfall of 1300 mm and covers Sub-zones: Central & North-eastern Plateau; Western Plateau and South-eastern Plateau of the Zone VII (Eastern Plateau and Hill region) of the country. Major river systems (North Koel, South Koel, Damodar, Shankh, Ajay, Karo, Swanrekha, Barkar, and Kharkai *etc*), are mostly seasonal causing severe flood hazards. Main bottlenecks for the development of state are : about 56.8 per cent of the total state's population is below poverty line; approximately 39 per cent area is wastelands as compared to India's average of 37.42 per cent; agricultural production records 50 per cent deficiency; has lowest per capita availability of food grain of just 230 g as compared to India 532g; has just 22.68 per cent net sown area as against 46.69 per cent of the country; Problem of soil acidity (pH<5.5) in 5 per cent of total geographical area (spreading in Dumka, Jamtara, Ranchi, Gumla, Garhwa and East Singhbhum districts) makes the agriculture economically unviable; Only 8.68 per cent of net sown area is irrigated in Jharkhand as against 37.63 per cent in India.

The topographical features as well as the biophysical challenges collectively generate the risk leading to degradation in the water and land use efficiencies. The entire land forms of Jharkhand state for the purpose of best management requires multiple land use concept implementation, which endorses implementation of Agroforestry on priority.

AGRO CLIMATIC ZONES WITH BROAD CHARACTERISTICS

The state is divided broadly into three agroclimatic zones and these are:

I. Central and North Eastern Plateau

This zone covers an area of 4.14 million ha having human population of 15.32 million and 6.58 per cent irrigated area. Low water retentive capacity of the soil particularly of the uplands. Rainfall is erratic and uneven and usually arrives late and retrieves early. Water harvesting practices for raising *rabi* crops are practically absent as a result drying of tanks and wells by February restricts *rabi* crop production.

II. Western Plateau Zone

It covers an area of 2.52 million ha having human population of 7.62 million and 9.65 per cent irrigated area. Moonsoon arrives late and retrieves early as in case of central and northeastern plateau. Rainfall is erratic and unevenly distributed. Soils have low water retentive capacity and areas lack in soil and water conservation practices.

III. South-Eastern Plateau Zone

This zone is spread over an area of 1.32 million ha having human population of 4.06 million and irrigated area of 4.58 per cent. Uneven distribution of rainfall, low water holding capacity, highly eroded soil with poor fertility are some of the usual characteristic of this zone.

Agriculture Crop Production

Rice is the most important crop of the State. It covers about 70 per cent of total cropped area and over 80 per cent area under cereal crops. Average yield of rice is 12.35 q/ha against India average of 19.68 q/ha. The highest productivity in the year 1999- 2000 was recorded in Deoghar district (15.26 q/ha) followed by Giridih (18.8 q/ha) and Pakur (14.30 q/ha). The lowest yield rate was observed in Garhwa district (6.74 q/ha), Palamu and Latehar had 8.36 q/ha each. Maize is the second important crop of the State occupying nearly 6 per cent cropped area with 15.80 q/ha yield rate against national yield rate of 17.92 q/ha, there is ample scope to extend the area. Wheat is the third important crop which occupies over 3 per cent of the cropped area. The average yield rate was 14.54 q/ha against 27.78 q/ha of India's average. Ragi grows under rainfed condition and covers about 2.45 per cent of cropped area with nearly 9.38 q/ha of productivity. Pulses account for roughly 8 per cent area under food grain crop and contribute about 2.5 per cent of total food grains production. The State has higher yield rate of 8.12 q/ha than the national average of 6.35 q/ha. Area under oilseed is hardly 3 per cent. Its yield rate is 6.94 q/ha against India's average of 8.53 q/ha. There is ample scope to extend area under oilseeds in the State.

Horticultural crop production (1997-2000)

A. Fruits

Horticulture crops covers about 8.6 per cent of the State's cropped area. Mango is the most important fruit both by area and production. This fruit shares about 23.16 per cent area and 22.21 per cent production of total fruits in the State. Banana shares 5.84 per cent area and 13.57 per cent production. The average yield rate of banana is 19.62 tonnes/ha. Santhal Pargana regions covers over 63 per cent area and its production. Of the total fruits, citrus covers over 16.87 per cent area and over 13.62 per cent production. The average yield rate is 10 tonnes/ha against 10.54 tons/ha of India's average. The major contributors of this fruits are Ranchi and Palamau divisions. The average productivity of fruits is 12.38 tons/ha as against 13.96 tons/ha in India. Thus, the productivity of fruit in Jharkhand is about 13per cent lower than India's average.

B. Vegetables

Vegetable covers about 81.55 per cent area and 84.57 per cent production of total horticultural crops. It is 63.14 per cent area and 80 per cent production of the country. With 15.35 tones/ha, Jharkhand has lower productivity of vegetables as compared to India's average of 16.18 tones/ha. About more than half of the total vegetable production is shared by zone I comprising the districts of Santhal Paragna & Hazaribagh. Okra, Brinjal, Cauliflower, Cabbage, Tomato, Onion, Potato, French beans are major vegetable crops.

Animal husbandry and Dairy

The state share's only 2.46 per cent geographic area of the country but maintains 3.68 per cent livestock and 3.31 per cent poultry population. This reveals that the livestock wealth of the state is almost 50 per cent higher in comparison to its geographic area. The percentage share of smaller animals is comparatively higher in Jharkhand as it maintains over 8 per cent pigs and about 5 per cent goats of the country. Although, cross-breeding programme was started from the Second Five Year Plan, 4 decades before, hardly 20 per cent cows are estimated cross-bred and 25 per cent improved poultry. To increase livestock production, the state has gigantic task to execute cross-breeding programme in vigorous way. Among the milch animals, over 65 per cent cows and buffaloes remain either dry or not calved even once. The percentage of milking animals would be hardly 35 per cent. This is a great blow to the animal husbandry practices and causing low milk yield. In spite of an impressive livestock and poultry resources in the state, the production of milk, egg and wool is at low. The state shares hardly 10.03 per cent milk, 1.39 per cent egg and 0.40 per cent wool in total production of the country. The Dairies maintained by the State, hardly handle 63 per cent of their handling capacities. About 37 per cent of milk handling capacities along with capital investment remain unused.

Fisheries

Per capita availability of fishes in Jharkhand during 2001 was 7.4g against minimum requirement of 15g. This availability at national level is 25.15g.

Thus state has deficit supplies in fish production to the extent of 50 per cent. On the other hand, the country produces about 68 per cent more fishes against its minimum requirement. Triennial average (2000-02) production of fishes in the state was about 43987 Metric Tones with average income generated at whole sale prices was about Rs. 93.55 crores. Bokaro district had maximum number of Government ponds and Garhwa had the minimum *Jalkar* area and number of Ponds. Total number of Government ponds/tanks in the state during 2002-03 was 11224 with area of 9,142.58 hectares. Number of settled *Jalkars* were 1441 spread in the area of 1563 hectares.

Forestry & Wild life

Forest covers about 29.27 per cent of total geographical area of the State, out of this, only 14.79 per cent is covered by dense forest, 13.61 per cent by open forest and the rest 0.87 per cent by scrub in the year 2001. The State maintains one national park at Betla and 11 sanctuaries. According to the first wildlife census of Jharkhand, 2002, the State maintains 34 tigers, 164 leopards, 758 elephants, 62 antelopes, 18.08 sloth bears and a number of langoors, monkeys and wild bears etc. The details of the forest is given in table 17.1.

PREVALENT AGROFORESTRY PRACTICES

With the efforts of Scientists (AICRP- Agroforestry) of Birsa Agricultural University and some NGO like SHRUMS (Society of Hill Resource Management School) Palamu, the farmers are motivated to practice Agroforestry in their marginal and wastelands. Some of the agroforestry systems practiced by the farmers are as follows:

Agrisilviculture (using fast growing trees like: Gamhar, Khair, Neem , Karanj, Arjun, Semal, Teak, Sisso *etc.,* either in rows or on the boundaries with the upland crops: paddy, pigeon pea, cow pea, maize, niger, millet *etc.*). **Hortiagrisilvicultutre** (using fruit trees like Aonla, Papaya, Jack fruit, Guava, Mango, Litchi, Banana, Ber, Custard Apple, Bel *etc,* along with the aforesaid combinations). The growing demand for fodder and decrease in the carrying capacity of the pastures have motivated the farmers to plant fodder trees at large scale in the pastures, generating examples of **Silvipastoral system**. The trees used include: Bakain (Melia), Su babool, Gamhar, Sissoo, Ficus (Pakur), Ber, Neem and Bel *etc.* The Lac and Tassar culture on the existing host plants can be recorded as prevalent examples of **Entomoforestry** systems being practiced on crude basis. At few places the farmers have raised trees on bunds of pond, which can be regarded as an example of the **Aquaforestry**. The most prevalent system in the tribal area is the **Homestead gardens** where the farmers/ tribal raise multipurpose trees (Chakiundi, Bakain, Gamhar, Amla, Papaya, Gamhar, Su babool, Ber, Sissoo, Karanj, Teak *etc.*) along with vegetables and rhizome crops (Tomato, Capsicum, Green Chilly, Maize, Cauliflower, Ginger, Potato, Colocassia, Turmeric, Cow pea, Yam, beans *etc.*) to fetch their day to day needs for fruit, fuel , fodder *etc* and the sacred grooves or *Sarna sthals*, where the tribal raise trees of their religious importance for worshipping.

RECOMMENDED AGROFORESTRY SYSTEMS

For proper utilization under Agroforestry the entire area is categorized into following two categories:

I. Land with very high erosion

Such lands are normally located on steep slope and are very fragile, normally having no land use because of their sensitive nature and may or may not have permanent cover. Heavy to moderate erosion (some ravines, large number of gullies and rills present), moisture and nutrient deficient soil, with highly fragile surface located mostly on the upper reaches of the steep areas. For such fallow areas two models exist, namely:

i) *Silvipastoral model :* Quli & Siddiqui (1996) reported the excellence of this model for eco-restoration and poverty alleviation of villagers of Chotanagpur and Santhal Paraganas. This advocates afforestation on the unculturable wastelands in combination with native grasses. Suitable trees as per the region specific needs are planted in rows/random design, with simultaneous planting of grasses (preferably endemic) on the floor, for stabilizing the soil and conservation of moisture.

   The tree planting is preferably done by using nursery-raised seedlings, however, for grasses, planting by broadcast or solid sod method. In the problem areas, having rills and gullies, soil working and catchment's treatment is done to arrest soil erosion. The steep slopes should be planted after digging contour trenches for water conservation. Plants having good economic potentials and which are good soil stabilizer like Doob grass (*Cynodon dactylon*), Khas (*Vetiveria zizanoides*) and Agave (*Agave sislana*) are very good for such problem areas. This model has been evaluated to be very profitable for the farmers. After proper site stabilization direct sowing of *Acacia nilotica, Cassia tora, Acacia catechu, Zyzyphus xylocarpa,etc.* are recommended.

ii) *Agrisilvipastoral models :* This model is a slight modification of the Silvipastoral models in which plantation of some agricultural crops like groundnut, millets, maize, cucumber, pumpkins and tubers *etc.*, which come up well in sandy and well drained soils, are recommended. This model is recommended for the areas, which are comparatively better with less erosion problems. While selecting the annual/seasonal crops, care should be taken to use the local varieties (as per the region specific demand), which are fit for growing on moisture and organic matter deficient soils.

For very sensitive areas falling in Categrory-1 the following specific measures / techniques are recommended:

- Specific rain water harvesting structures like: staggered contour trenches, bench terraces, retaining and breast walls, brush wood dams and rock fill dams should be constructed following the standard engineering norms at appropriate places with plantation of quick growing plants (shrubs, grasses and trees) on the upper and lower ends ;
- The entire slopes should be revegetated using grasses of local varieties by *solid sodding* and some fast growing trees like Subabool (*Leucaena leucocepahala*), Khair *(Acacia catechu), Babool (A. nilotica), etc.,* should be planted by broadcasting the seeds during pre monsoon showers;
- The plantation of Agave (*Agave sislana*) along with Khas grass (*Vitveria zizanoides*), should be used on high risk zones;
- In case of areas with heavy grazing pressures, the impalatable plants like *Calotropis, Lantana camara, Ipomia cornea, Vitex negundu etc.* among shrubs and Eucalyptus, Chakundi (*Senna siamia*), Akashi (*Acacia auriculiformis*) and *Karanj (Derris indica)* should be used;
- The area should be closed against grazing and all other biotic interferences;

II. Land with moderate to low erosion

Sites located on comparatively gentle slopes and are normally used by the villagers for upland farming (basically marginal farming). The nature of such land is fragile and is under threats due to improper land use. These lands have moderate to low erosion (some gullies and rills present), comparatively low moisture and nutrient, located mostly on the moderate to lower slopes. For such watersheds the following types of Agroforestry models are recommended.

i) *Hortisilvicultural model :* This model has long gestation period, but after 4-5 years recurrent income in terms of sustained supply of fruits along with fodder and fuel wood is highly encouraging. Such a model is highly profitable in the ecological perspective and being some sort of permanent features, they have a very far-reaching positive impact in poverty alleviation as well as environmental amelioration, besides prime objective of optimizing the water and use efficiency. The spacing for this model has to be kept quite large ranging from 5 - 10 meters between the plants, as the horticultural trees have large size crown. The forest trees like, Arjun (*Terminalia arjuna*), Karanj (*Derris indica*), Neem (*Azadirachta indica*), Bakain (*Melia azedarach*), Kend (*Diospyros melanoxylon*), Semal (*Bombax cieba*) and Kusum (*Schleichera oleosa*), Moonga (*Moringa oleifera*), Ber (*Zizyphus mauritiana*), Cashew (*Anacardium occidentale*), Mahua (*Madhuca indica*), *etc.*, are recommended for this model. To further optimize the economic potentials of this model Tassar rearing, lac culture, and bee keeping should be encouraged when the respective host trees become ready for such ventures.

ii) *Hortiagrisilvicultural model :* Depending upon the interest and working ability of the farmer this model can be developed by introducing certain shade loving/ tolerant agricultural crops, in the Hortisilvicultural models. Use of the crops like Ginger, Turmeric, Elephant foot, Sweet potato and Arwi/Peckchi (Colocassia sps.) further adds to the profit of the farmers.

iii) *Homestead gardens :* Raising fruit and fodder plants in the vicinity of their hut/houses like neem, mango, papaya, guava, drumstick, banana, ber, jack fruit and jamun *etc.*, along with akashi, chakundi, bakain, teak, sissoo and gamhar, for fodder, fuel and timber are recommended under this system. In deep interior tribal areas yams, colocassia, ginger, cowpea, beans and turmeric are used in the home gardens.

iv) *Aquaforestry model* : In some of low-lying areas, the Aquaforestry model is recommended. The pond banks should be planted with fast growing multipurpose tree species yielding fodder and fruits. Such plants are helpful in providing shade for the pond, establishing the banks and the usufructs fetch quite sufficient amount of the fuel and fodder. The various species like: *Dalbergia sissoo, Acacia nilotica, Dendrocalamus strictus, Musa paradisiaca, Aegle marmelos, Syzygium cumini*, are used on the bank, which enrich the fish habitat by adding biomass in terms of leaf litter as well as the droppings of birds, which add to the micro fauna and ultimately enhance the food availability in the ponds.

v) *Entomoforestry :* For West Singhbhum, Ranchi, Dumka, Gumla, Godda, Chatra and Palamu districts, the farmers should use Palas tree for lac culture and Arjun tree for Tassar culture. The examples of beekeeping using the litchi, karanj, guava, mango, *etc.*, for enhancing their income are also very common in most of the areas. Some of the most common trees recommended are *Terminlia arjuna, Butea monosperma, Mangifera indica, Litchi chinensis, Psidium guajava, Carica papaya, Musa paradisiaca, Derris indica, Artocarpus heterophyllus.*

AGROFORESTRY RESEARCHES

Oraon (1996) studied growth behavior and phenology of twenty economically important trees of this region and developed phonological calendar for ranchi district, which is very useful for designing Agroforestry model in this region. Islam (1998) on the basis of analysis of important factors responsible for motivation in Villages of Palamu district recorded economic gains as most important motivational factor for adoption of Agroforestry of the Society of Hill Resource Management School Project of *Chakriya Vikas Yojna.* Minj (1999) on the basis of socio-economic analysis of Agroforestry done in the villages of Palamu district recorded highest cost benefit ratio of 1: 34, for the Agrisilvihorticultutre model raised on waste and marginal land by the Society of Hill Resource Management School Project of *Chakriya Vikas Yojna.* Oraon (1999) reported land use pattern of village Kumharia in Ranchi district as 55 per cent under Agriculture, 17 per cent fallow, 13 per cent home gardens, 15 per cent Agroforestry (Agrisilvi and Silvipasture).

Agrisilvicultural system

Research on Combination of crop and long rotation trees was conducted for five years at Darisai (BAU) farm in agro-climatic zone IV of south eastern plateau zone. Different agroforestry systems can easily be practiced to utilize about 240 thousand hectare barren, eroded and waste lands in Jharkhand state where farmers like to grow upland crops. Agronomical crops (groundnut, ragi, blackgram, rice) and long rotation trees (maturing in 15-20 years) in combination (*Gmelina arborea, Dalbergia sissoo* and *Acacia auriculiformis*) were laid out with different possible combination in a field study from 1998 to 2002. It revealed (Tables 17.2, 17.3 and 17.4) that crop yields failed in 5th year due to shade of tree canopy. Little light intensity (10 to 15 per cent) was available on ground where annual crops could achieve good germination but failed to perform reproductive growth.

Gmelina arborea planted at 5m x 3 m apart rows performed best in agro-forestry combination with groundnut crop followed by black gram, ragi and upland rice. A general trend of decrease in yield of crops was observed in different tree combination. The trees were benefited by nutrition applied to agricultural crops. The tree volume increased markedly in third year onwards and affected crop yield badly thereafter in 4th and 5th years (on the basis of per cent reduction in yield). Therefore, up to three years the agricultural crops can be grown successfully and in fourth and fifth years, the cultivation of grain crops is not recommended. Shade loving crops should be taken namely ginger, turmeric, colocacea, elephant foot (corm of *Amorphophallus*) etc. as inter crop in agroforestry system. The timber tree *Gmelina arborea* gave lowest volume upto 5th year (11.33 cm ha^{-1}) and *Acacia auriculiformis* gave the highest (135.2 cm ha^{-1}). But considering the choice and timer quality by the farmers Gamhar would produce economical yield of timber in terms of money with the increase in the age of tree. The soil fertility in 2000 (pH 5.2, organic carbon 3.42 g/kg soil, available P_2O_5 and K_2O 19.1 and 153.4 kg/ha, respectively) was found to be improved under agroforestry system in three years with pH 5.4, organic carbon 4.6 g/kg available P_2O_5 25.6 kg/ha and available K_2O 178.7 kg/ha during 2003.

Different agro-forestry systems can easily be practiced to utilize about 240 thousand hectare barren, eroded and waste lands, in Jharkhand state, where farmers like to grow uplands crops. Agronomical crops (groundnut, ragi, blackgram, rice) and short rotation trees (maturing in 10-12 years) *Acacia auriculifomis, Leucaena leucocephala* and *Senna siamea* were laid out in different possible combination in field study from 1998 to 2002. It revealed (Tables 17.5, 17.6 and 17.7) that crops totally failed in 5th year due to dark shade of tree canopy. A little light intensity (9 to 10 %) was available on ground where annual crops could achieve germination but failed to perform their reproductive growth.

Chakundi (*Senna siamea*) planted in 5m x 2m apart rows performed best in Agroforestry system with crop groundnut or ragi followed by black gram and rice successfully up to 3rd year. However, ragi showed minimum reduction percent in grain yield up to fourth year. The tree volume increased markedly after third year, which caused reduction in crop yield. The fuel wood tree species *Senna siamea* produced total volume 70.17 cum ha^{-1} with Rs.0.95 lakh/ha followed by *Acacia auriculiformis* (87.25 cum and Rs.0.79 lakh/ha) up to 4th year. Due to shade, agricultural crops could not be cultivated under dense canopy of fuel wood species therefore, trees must be utilized as a fuel wood in 5th year or shade loving crops can be taken as intercrops like colocassia, turmeric, ginger, elephant foot (corn of *Amorphophyllus*) etc. Under *Senna seamea* based agroforestry system, soil fertility status, (mean value pH 5.8, organic carbon 4.20 g kg^{-1} soil, available P_2O_5 24.5 and K_2O 191.6 kg ha^{-1}) in 2000 was found to be improved in three years with pH 5.87, organic carbon 5.9 g kg^{-1}, P_2O_5 30.6 kg ha^{-1} and available K_2O 247.8 kg ha^{-1}.

Table 17.1. Jharkhand forests at a glance

Recorded Forest Area	
Reserved forest (RF) Preotected Forest (PF) Unclassed Forest (UF) Total	4,387.20 sq.km. 19,184.78 sq.km. 33.49 sq.km. 23,605.47 sq. km.
Forest Cover	
Dense forest (Canopy density :>40% Open Forest (Canopy density : 10-40%) Scrub Forest	11,787 sq. km. 10,850 sq. km. 976 sq. km.

Table 17.2. Grain yield (kg/ha) and volume production of tree in Agrisilvicultural system (long rotation) 1998 to 2002.

S.No.	Trees crop/Field crop	1998	1999	2000	2001	2002
	Gmelina arborea (m^3/ha)	0.413	0.666	3.33	9.99	11.322
1.	Groundnut	994	291	194	-	23.4
2.	Ragi	309	399	265	45	141.7
3.	Blackgram	113	76	82	18	16.3
4.	Rice	309	399	265	45	27.5
	Dalbergia sissoo (m^3/ha)	0.119	0.666	5.328	17.316	40.626
1.	Groundnut	979	207	167	-	28.1
2.	Ragi	345	354	263	81	79.2
3.	Blackgram	95	85	45	18	16.3
4.	Rice	235	859	187	265	13.6
	Acacia auriculiformis (m^3/ha)	0.406	1.998	20.91	67.93	135.19
1.	Groundnut	936	244	142	-	63.5
2.	Ragi	288	397	243	56	96.80
3.	Blackgram	100	70	51	15	41.30
4.	Rice	237	1099	214	358	12.50

Table 17.3. Economics of yields (Rs/ha) obtained from tree-crop (long rotation AF system)

S.No.	Trees crop/Field crop	1998	1999	2000	2001	2002
	Gmelina arborea	638.08	1,028.97	5,144.85	15,434.55	17,492.49
1.	Groundnut	11928	3492	2,328	-	280.8
2.	Ragi	1,081.5	1,396.5	0,927.5	157.5	145.90
3.	Blackgram	1,356	0,912	0,984.0	216	195.6
4.	Rice	1,081.5	1,396.2	0,927.5	157.5	96.25

Contd...

Table 17.3. Contd...

	Dalbergia sisoo	280.24	1568.43	12,547.44	40,779.18	95674.23
1.	Groundnut	11748	2484	2004	-	337.2
2.	Ragi	12075	1239	920.4	283.5	277.2
3.	Blackgram	1140	1020	540	216	195.6
4.	Rice	822.5	3006.5	654.5	927.5	47.60
	Acacia auriculiformis	365.4	1798.2	18,819	61,173	1,21,671
1.	Groundnut	11,232	2,928	1,704	-	762
2.	Ragi	1,008	1,389.5	850.5	196	338.8
3.	Blackgram	1,200	0,840	612	180	531.6
4.	Rice	829.5	3,846.5	749	1253	43.75

Rates: Ground nut = @ Rs 12/kg, Rice = Rs 3.50/kg, Blackgram = Rs 12/kg, Ragi = Rs 3.50/kg
Gmelina aroborea @ Rs.5/kg, *Dalbergia sisoo* @ Rs. 350/kg,
Acacia auriculiformis @ Rs. 3/kg

Table 17.4. Percent reduction in Crop Yields of AgriSilvicultural (Long Rotation) System compared to control

S.No.	Crops	Years				
		1998	1999	2000	2001	2002
1.	Groundnut	23	20	44	F*	96
2.	Ragi	1	28	37	60	86
3.	Blackgram	1	31	33	35	76
4.	Rice	12	19	58	47	98

F* = Crop Failed

Table 17.5. Grain yield (kg/ha) and volume production of tree in Agrisilvicultural system (short rotation) 1998 to 2002.

S.No.	Trees crop/Field crop	1998	1999	2000	2001	2002
	Senna siamea(m³/ha)	0.104	2.863	21.992	70.165	131.59
1.	Groundnut	92.1	334	105	-	*
2.	Ragi	493	569	166	77	*
3.	Blackgram	26	128	63	10	*
4.	Rice	242	342	214	38	*
	Leucaena leucocephala (m³/ha)	0.072	2.548	10.937	26.20	41.58
1.	Groundnut	1012	325	143	-	*
2.	Ragi	468	571	233	72	*
3.	Blackgram	28	154	87	6	*
4.	Rice	185	300	233	33	*

Contd...

Table 17.5. Contd...

	Acacia auriculiformis (m^3/ha)	0.568	7.360	34.370	87.250	178.98
1.	Groundnut	803	294	112	83	*
2.	Ragi	553	585	152	F	*
3.	Blackgram	23	109	55	13	*
4.	Rice	148	400	215	32	*

* Decline in crop yield is due to shade impact after 4th year

Table 17.6. Economics of intercropping yields (Rs/ha) obtained from tree-crop (short rotation AF system)

S.No.	Trees crop/Field crop	1998	1999	2000	2001	2002
	Senna siamea	140	382.86	29,594.70	94,722.75	1,77,646.50
1.	Groundnut	11,052	4008	1260	-	*
2.	Ragi	1725.5	1991.50	581	269.5	*
3.	Blackgram	312	1536	756	120	*
4.	Rice	847	1197	749	133	*
	Leucaena leucocephala	97.20	3439.80	14,764.95	35,370	56673.00
1.	Groundnut	12144	3900	1716	-	*
2.	Ragi	1638	1998.5	815.5	252	*
3.	Blackgram	336	1848	1044	72	*
4.	Rice	647.5	1050	815.5	115.5	*
	Acacia auriculiformis	511.2	6624	30933	78,525	1,61,082.00
1.	Groundnut	9636	3528	1344	996	*
2.	Ragi	1935.5	2047.5	532	F	*
3.	Blackgram	276	1308	660	156	*
4.	Rice	518	1400	7525	112	*

Rates : Rice and Ragi @ Rs. 350/q, Groundnut and Blackgram @ Rs 1200/q,
Acacia auriculiformis, Lucaena leucocephala, Senna simea @ Rs. 300/q
* Decline in crop yield is due to shade impact after 4th year

Table 17.7. Per cent reduction of crop yields in agrisilvicultural (short rotation) system compared to control (BAU, center)

S.No.	Crops	Years				
		1998	1999	2000	2001	2002
1.	Blackgram	4	45	44	71	100
2.	Groundnut	9	43	52	F*	100
3.	Ragi	1	4	60	58	100
4.	Rice	29	63	58	75	100

F* = Crop Failed

Mango based models (Nath *et. al*, 2003) with *Stylosenthes* as intercrop and Gamhar as filler crop performed best in case of sloppy topography without irrigation facility. The fruit crop based Agroforestry system for upland utilization in watershed of Eastern India (Nath *et. al*, 2005) had been researched. It was found that fruit based multistoried models are found to be the best for minimizing the risk of sole cropping in uplands. Litchi based models with French bean, cowpea and horse gram as intercrop and guava as filler crop performed best in case of better soil conditions with supplemental irrigation facility. Aonla based multistory systems using black finger millet, black gram or *dinanath* grass as inter crop was found to be best for coarse soils with gravel, when the planting was done after using good soils in the pits.

REFERENCES

Anonymous, (2001). State of Forest Report. *Forest survey of India,* MoEF, GoI, Dehradun.

Islam, A (1998). People's participation in social forestry program in selected villages of Palamu through *Chakriya Vikas Pranali, M. Sc. Thesis.* Department of Extension & Social Forestry, Faculty of Forestry, Birsa Agricultural University, Ranchi.

Minj, A. V. (1999). Socio-economic upliftment of villages through Agroforestry schemes of *Chakriya Vikas Pranali,* in Palamu, *M. Sc. Thesis* (Unpubl.). Department of Extension & Social Forestry, Faculty of Forestry, Birsa Agricultural University, Ranchi, pp: 185.

Mishra, R. K. (2004). Agricultural statistics of Jharkhand at a glance. ADB, B. A. U. pp:111

Nath, V., Das, B., Roy, M., Dey, P., Kumar, S. and Kumar, M. (2003). Mango based cropping systems for upland of subhumid plateau of Eastern India. *Progressive Horticultutre,* 35(2):142-145.

Nath, V., Das, B., Yadava, M. S. and Singh, R. V. (2005). Fruit based agroforestry systems for upland utilization in watersheds of Eastern India. *J. Res. BAU* (In Press).

Oran, P. R. (1991). Studied on Groth and yield performance of different agroforestry systems in Kumharia Village, Ranchi, Jharkhnad, *M. Sc. Thesis.* Department of Forest Managment, Faculty of Forestry, Birsa Agricultural University, Ranchi.

Oraon, B. C. (1996). Phenological studies on economically important trees in and around Ranchi *M. Sc. Thesis.* Department of Silviculture, Faculty of Forestry, Birsa Agricultural University, Ranchi.

Quli, S. M. S. and Siddiqui, M. H. (1996). Restoration of economic and environmental viability in rural areas of Chotanagpur through Silvi-pastoral techniques. *In: Alternative Energy development & Management.* International Book Distributors, Dehra Dun, pp:201-212.

18

Agroforestry Systems and Practices in West Bengal

PANKAJ PANWAR
Department of Forestry, Uttar Banga Krishi Viswavidyalaya
Pundibari, Cooch Behar - 736 165 (West Bengal)

ABSTRACT

West Bengal is the gateway to North-East and has diverse vegetation and social strata. The paper presents the agroclimatic zones of the state along with their characteristic features. The scenario with regard to agriculture, forest cover and livestock shows that practicing agroforestry is the need of the state. For the sake of convenience the state has been divided into two parts: south Bengal and north Bengal. In the tribal dominated parts of the state home gardens are prevalent. Possible interventions in the present land use along with strategies to increase tree cover under agroforestry has also been discussed.

INTRODUCTION

Cultivation of tree species and agricultural crops in intimate combination had been a practice through out the world at one period or another. Trees were integral and important component of the farming system. However, in the middle ages the concept changed and agriculture, forestry and animal husbandry were separated under a monoculture management system. This in long run led to an incredible reduction in forest cover and ecological imbalance. Scientists and managers once again began to think about changing the land use system and integrating trees with agriculture crops (Panwar *et al.*, 2004). In India, research work on agroforestry was initiated during the late 1970's by Indian Grassland and Fodder Research Institute, Jhansi; Central Soil and Water Conservation Research and Training Institute, Dehradun; Central Arid Zone Research Institute, Jodhpur and ICAR Research Complexes in isolation. National commission on agriculture also emphasized starting of agroforestry education in all agriculture university. The formal research activities related to agroforestry started in a big way with the initiation of All India Coordinated Research Project on Agroforestry (AICRPAF) from 7th Feb., 1983 with 20 centers, of which 12 were in State Agricultural Universities and rest in ICAR Institutes. Later more centers were added. The major emphasis was to have research centers in all the major ecological zones of India. In this venture in the first phase AICRPAF was initiated in West Bengal at Jhargram, Distt. Midnapore under the auspices of the then only Agricultural University in West Bengal - Bidhan Chandra Krishi Viswavidyalaya, Mohanpur representing Humid and Subhumid zone of West Bengal.

AGROCLIMATIC ZONES OF WEST BENGAL/ EDAPHIC ZONES

West Bengal had been divided into six agroclimatic zones (Fig.18.1). The details of these zones are presented in table 18.1:

The state is situated in the eastern part of the country between the Himalayas in the north and Bay of Bengal in the south where mighty river Ganga meets the ocean forming well known mangrove forests in Sundarbans. The total geographic area of the state is 88,752 km^2, which is 2.7 per cent of the country's geographic area. The human population of the state is 80.22 million, constituting 7.8 per cent of the country population of which 72 per cent are rural and 28 per cent are urban. The average population density is 904 persons per km^2. The tribal population is 5.6 per cent of the state's population. Livestock population is 47.09 million, which is 7.5 per cent of the country's livestock population. Total forest and tree cover is 13,957 km^2 which is 15.7 per cent of states geographic area and only 1.8 per cent of the country's forest and tree cover. Per capita availability of forest and tree cover in West Bengal is 0.02 ha (Forest Survey of India, 2001). Out of 18 districts the forest cover of the state is mainly concentrated in Jalpiguri, 24 Pargana south, Darjeeling, Medinipur, Bankura and Parulia comprising about 80 per cent of the total forest cover of the state. About 90 - 95 per cent of the area of the state is under cultivation.

Taking into account the pressure of human and animal population, less forest area coupled with the fact that the forest area is unequally distributed and major chunk of the population is rural which requires fodder, fuel and timber in large quantities; need is to increase the forest cover in non-conventional areas i.e. in farmer's field. This will lead to diversification of farms on one hand and on the other farmer needs will be fulfilled and target of having 33 per cent area under forest will also be achieved.

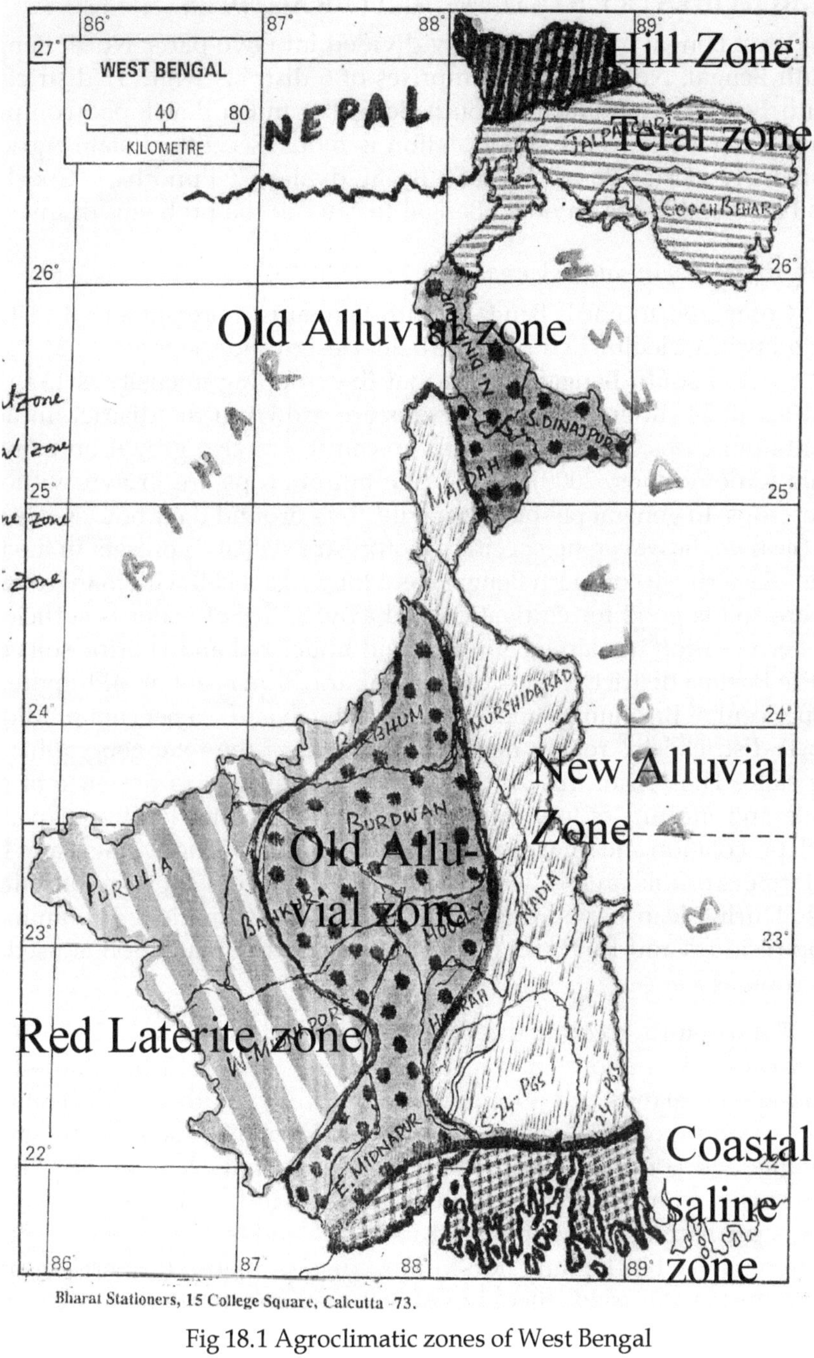

Fig 18.1 Agroclimatic zones of West Bengal

AGROFORESTRY SYSTEMS AND PRACTICES

West Bengal has been broadly divided into two parts: North Bengal and South Bengal. North Bengal comprises of 6 districts while 12 districts are in south Bengal. Economically south Bengal is much developed compared to north Bengal where tribal population is more. Recently a state Agricultural University "Uttar Banga Krishi Viswavidyalaya", Pundibari- Cooch Behar has been established in north Bengal to cater to the problems of this region.

a) Agroforestry in south Bengal

Compared to north Bengal, south Bengal is more advanced in terms of diversity in agriculture crops in the field. Design and Diagnosis of 24-Parganas, a district of south Bengal reveals that the cropping intensity is 137 per cent and about 24 different types of crops are grown in the district. In addition plantation crops, like coconut and arecanut, are also grown on commercial scale (Anonymous, 2000). These plantation crops are grown without any intercrops. In general people grow fruit trees around their houses or scattered on the field, however, no specific agroforestry system is present in the farmers field. Eastern part of south Bengal has a long costal belt with mangrove forests where soil is good for cultivation and availability of water is sufficient. The western portion has large chunk of land under red and lateritic soil covering entire Purulia district, 88 per cent of Bankura, 77 per cent of Midnipore (West), 74 per cent of Birbhum, 34 per cent of Burdwan and 7.5 per cent of Midnapore (East) district. This region makes 28 per cent of the total geographic area of the state. The soil of this zone is difficult to till due to presence of gravels, stone and morum; is low in nutrients; has less water holding capacity and high percolation and uneven distribution of rainfall. The average land holding of farmers of this area is 0.69 ha. Farmers face acute shortage of fodder and fuel. During lean period only poor quality rice straw is fed to animals and crop residues and leaf litter obtained from nearby plantation is used as fuel for cooking.

Taking into account the fuel, fodder and water scarcity, AICRPAF-Jhargram (Midnapore) has recommended that trees be planted in low productive uplands and adoption of agroforestry with suitable combination of trees and crops, following proper management practices. To find out a suitable and compatible combination of trees and crops *Eucalyptus* hybrid was raised in a denuded and degraded upland with 2500, 1250 and 833 plants per hectare maintaining 4x1 m, 4x2 m and 6x2m spacing, respectively. Intercrops like black gram, horsegram, cowpea, maize, turmeric, sesame and groundnut were tried. Black gram, cowpea and turmeric were grown for continuously 4 years and horsegram, maize and sesame for 3 years and ground nut for two years. Average yield obtained, of the intercrops were: blackgram 4.12 q/ha, horsegram 1.42 q/ha, cowpea 130.87 q/ha, maize 49.45 q/ha, turmeric 81q/ha, sesame 1.4 q/ha and groundnut 8.37q/ha. It is concluded that the crops like blackgram, horse gram, cowpea, maize, turmeric, sesame and groundnut can be combined with *Eucalyptus* hybrid maintaining a plant population of 833 - 1250 trees per hectare following normal recommended package of practice in the rainy season (Hansda, 2000). The effect of intercrops on the growth of *Eucalyptus* showed that height of sole plantation was 15.57 m while it was less in all the intercrop combinations except with jute where the tree height was 15.97 m. Similarly the diameter at breast height was also more in jute combination (14.98 cm) compared to sole plantation (14.54 cm).

Under *Tectona grandis* with net plot size of 4 x 11 m, six vegetable and spice crops were grown. Among them maximum yield was obtained in cucumber (15,000 kg/ha) followed by radish (12, 500 kg/ha), ridge gourd, bitter gourd, beans and ground nut. It was observed that growing of flower and vegetable crops with teak has positive effect on the growth of tree, as the height of teak was 5.03 m in sole planting and was higher with all the intercrop combinations. With groundnut it was 7.32 m. The diameter at breast height also showed similar trend being lowest (7.2 cm) in sole plantation and highest with radish as intercrop (10 cm). Agrisilviculture system based on *Acacia auriculiformis* was laid out at a spacing of 6 x 3 m and seven types of intercrops were planted. Maximum yield of 18,666 kg/ha was of sabai (*Eulaliopsis binnata*) followed by stylo (*Stylosanthesis humilis*), rice, horse gram and moong (*Vigna radiata*). The tree-crop combination revealed adverse effect of growing horse gram, moong and rice as intercrop on the height of the tree, while all the rest crops showed positive impact on height growth of tree compared to sole plantation. DBH of trees was found to be less in sabai and rice combination as compared to sole plantation. In another experiment yield of cow pea (1928 kg/ha) was found to be highest which was followed by rice, black gram, horse gram, ground nut and cow pea when they were intercropped with *Gmelina arborea*. Lady finger showed maximum yield of 15,000 kg/ha followed by radish, cow pea, ground nut and moong when they were planted in between the rows of *Embilica officinalis* (Anonymous, 2000).

Eucalyptus and *Acacia auriculiformis* based agrisilvicultural system was studied in farmer field in Midnapore district. The trees were planted in 1997 with 700 plants in eight acres plot. The aman paddy was taken during kharif season. The height of *Eucalyptus* was 8.38 m and *Acacia auriculiformis* was 6.73 m. The grain yield of paddy varied from 12 – 14 q/acre and that of straw 5 -6 q/acre (Table 18.2) (Anonymous, 2002).

Lahari (1980) conducted various intercrop trials with teak in coastal region of West Bengal. He summarized that there is distinct possibility of growing forest crops and agricultural crops in conjugation, which will not only benefit forest plantation but also provide extra food and other raw materials and generate employment. *Corchorous capsularis* (Jute) and cucurbits are not recommended as intercrops as they suppress the growth of forest trees.

Maiti *et al*, (1983) conducted series of experiments on *Leucaena leucocephala* from 1978 to 1982 to evaluate its productivity in West Bengal. At 1.5 m x 1 m spacing the herbage dry weight of *Leucaena* was 54 q/ha in 7 prunings at 60 cm height and the rice and green gram yields were 7.22 and 1.39 q/ha, respectively.

Anonymous, (1986) studied the intercropping of *Vigna munga* (black gram), *Dolichos biflorus* (horse gram), *Vigna sinensis* (cow pea), *Vigna radiata* (green gram) for grain and *Stylonthes quyanensis* (stylo) for green fodder grown under *Eucalyptus* hybrid trees spaced at 4 m x1 m , 4 m x 2 m and 2 m x 2m in West Bengal. The maximum crop yield was observed for cow pea (175.9 kg/ ha) when intercropped with 2 m x 2m spacing of *Eucalyptus*. Stylo recorded the highest dry matter of 1.92q/ha under *Eucalyptus* spaced at 4 x 1 m. Pathak *et al*, (2000) while discussing the agroforestry systems of different AICRP on Agroforestry centers has reported that the prevalent Agroforestry system under rainfed conditions in Jhargram center (south Bengal) is Agrisilviculture with *Butea monosperma*, *Terminalia arjuna* and *Acacia auriculiformis* as prominent trees species with 25 - 30 trees per hectare. In irrigated lands the species grown in Agrisilviculture system is Eucalyptus spp., *Musa paradisiaca*, *Cocos nucifera*, *Areca catechu* with 10 - 15 trees per hectare.

Leucaena leucocephala, *Acacia auriculiformis* and *Dalbergia sissoo* had been tried by Chattopadhyay *et al*, 2003 in south Bengal along with mustard in alley cropping. The experiment conducted also reveled that the system is also beneficial for improving the soil. Maiti *et al*, 2003 had concluded from their studies that elephant foot yam can be profitably cultivated in the alley of *Leucaena* planted at 1.0 m or 0.75 m apart in 4.0 m apart rows and pruned 5 times per annum.

b) Agroforestry in north Bengal

North Bengal is less developed both in terms of agriculture as well as industrialization. The people are poor and land holding is small. The Terai zone includes district Cooch Behar and part of Jalpiaguri where average land holding is less then 0.2 ha and 90 per cent of the population is rural, and more then 47 per cent people belong to SC and ST (Anonymous, 1999). District

Jalpiguri and Darjeeling are the hill districts which have the maximum area under forest. Malda district is famous for mango and large area is under orchards.

The trial on agroforestry in north Bengal goes back to initial stages of agroforestry when Taungya cultivation was brought from Burma in the year 1887. The first teak plantation, in India, with Taungya system was started in North Bengal. Thereafter, no specific development was done with regard to agroforestry in this region. As such, on farmer's field, no agroforestry systems are found in this region. The farmers are still practicing monocropping and are fancy towards growing rice to meet their both end. They take 2- 3 crops of rice in a year. Owing to the low fertility status of soil of this region the production of the crop is less. In pockets people grow potato and vegetables. Some farmers also cultivate Jute but are now discouraged due to its fluctuating price. In some pockets some well to do farmers are raising banana orchards, as it gives immediate returns. Arecanut (*Areca catechu*) is a common species of the area and can be seen in backyard of every house, but not a single tree in the agriculture field. Home-gardens are common having 3- 4 tier comprising of coconut, arecanut, banana, vegetable and flowers. These systems are not only found in villages but also in the main cities of this zone particularly terai zone. Lahari, 1972 while working on effect of teak on intercrops in West Bengal reported that at initial stages agricultural crops may be intercropped with trees while turmeric can be grown in older plantation as the crop is not much affected by shade and provide sufficient income. Experiments with Taungya system in West Bengal showed that yield of teak is not affected at wider spacing and allows additional income through intercropping (Lahari, 1989). In higher altitude of Darjeeling district Large cardamom (*Amomum subulatum*), a native spice crop of Sikkam, is grown in agroforestry system from 600 – 2000 m altitude (Singh and Pradhan, 1989). Among the tree species grown with large cardamom, *Alnus napalensis* is predominant. From tree-cardamom combination about 140 kg of dry capsule yield of the spice alone can be obtained which gives about Rs 8000/ha return with minimum inputs. North Bengal is famous for its tea gardens. The tea gardens can be found from foot hill of Darjeeling from 200 m – 300 m altitude and above. In lower altitude trees are associated with the tea to reduce temperature of the area, through its shade, thus it forms an agroforestry system. The common species found associated with tea in this region are *Albizia lebbeck* and *Albizia procera*. This tea based agroforestry system is the only prominent agroforestry system of north Bengal. However, in higher reaches this system no more remains an agroforestry system as trees are not grown because the temperature is naturally ideal for tea production. Panwar and Dey, 2003 had suggested *Anthocephalus cadamba, Gmelina arborea, Sesbania grandiflora* based agrisilviculture system due to their fast growth and adoption of these species by the people in terai zone of north Bengal. The terai zone had been declared as export potential zone for Pine apple, hence agroforestry systems associating pine apple will also be economically beneficial. This region being humid, homegardens are another system which has potential for this area.

An experiment was conducted to find out effect of integrating banana with arecanut in agroforestry system. Banana and arecanut plants were grown in the year 2003. After first harvest of banana in December, 03 the banana suckers were managed in such a way that they could maintain specific light intensity, varying from 25 - 800 (x 100) lux on the are canut plants growing adjoining to it. The data on growth performance of arecanut was taken for two years and was compared with the plants which were growing in full sunlight. It was found that height, collar girth, number of exposed nodes, leaf length and width, number of leaflets and leaf numbers of arecanut plants growing in 100 x100 lux was highest (Table 18.3). It was also found that as the intensity of light decreased the growth performance of the plants also decreased (Panwar *et al.*, 2006).

A multitier agroforestry system was established in terai zone of West Bengal and after three years of its establishment performance of elephant foot yam was evaluated with six different mulches. Simultaneously, its performance was also evaluated under six different light intensities. It was found that out of the six mulches tried, banana leaves is the best mulch as it gave highest height (71.50 cm), girth (14.26 cm) canopy spread (54.37 cm) and yield (10.32 kg/plot). It was also found that light intensity of 44.1 x 1000 lux is best for elephant foot yam in agroforestry system as it gave maximum yield of 10.32 kg/plot (Panwar *et al.*, 2006).

Mango (*Mangifera indica* var.Amrapalli) trees were planted in the year 1996 with distance of 6 x 6 m and 8 x 8 m in east - west direction. After nine years of planting the average crown spread of trees in east-west direction was 5.013 m and 5.443 m and in north-south direction it was 5.067 m and 57.07 m in 6x6 m and 8x8 m spaced plots, respectively. Elephant Foot Yam (EFY) was planted as intercrop in both the spacings along with control (without trees) in the month of March at two depths (10 cm and 15 cm). It was found that height was maximum in closer spacing (66.07 cm) at 10 cm depth and least in open (46.73 cm) sown at 15 cm depth. Plant girth was maximum (13.37 cm) of plants which were sown at 10 cm depth in 6 x6 m spacing and least (7.43 cm) when the plants were grown in open and at a depth of 15 cm. The maximum yield of 8.60 kg/ plot was obtained when intercrop was sown at a depth of 10 cm in 8 x 8 m spaced mango trees. The least yield of EFY was 6.37 kg/plot when it was grown at a depth of 15 cm in 6 x 6 m spaced mango tree orchard (Panwar and Suresh, 2006).

Banana plants were incorporated in the standard spacing of 5 x 5 m of *Hevea brazilensis* and the performance as well as economics of *Hevea brazilensis* and banana was assessed both in combination as well as pure plantations. It was found that there was no significant effect on the growth of rubber plants due to incorporation of banana as compared to pure plantation. It was also observed that in sole plantation of banana the return per hectare is Rs 98, 933/ ha, however when incorporated under rubber plantation an additional return of Rs 67, 042/ ha was obtained in hortisilviculture system. The quality of the banana fruits was also better in hortisilviculture system as the extent of scaring beetle damage was only 19.52 per cent compared to 93.09 per cent in sole crop of banana (Panwar and Suresh, 2006).

Bamboo species particularly *Bambusa nutans* and *Bambusa balcooa* are most prevalent and every house hold is having at least one to two clump. The clumps are generally maintained on the bunds of the ponds or on the road/ paths. These are generally used for making houses as most of the houses in villages are made of bamboo and locally available wood. These are also an integral part of the homegardens of the area. As such bamboo clumps are not found in or around the farms as it provide shade to the agriculture crops. National Mission on Bamboo Application had recommended that *Bambusa balcooa* can be used as perennial component of agroforestry system in agriculture field as it has compact growing habit. However, regular management of the clump is a must particularly the removal of culms. Trials are going on it the University for assessing growing potential of different crops in bamboo based agroforestry systems.

POSSIBLE AGROFORESTRY INTERVENTIONS

Different agroforestry system integrating medicinal plants and vegetable crops are required to be tested for various climatic conditions and soil types so that ready models are in hand for the extension workers to disseminate it to the farmers of a specific area. Bamboo is central to living of the people of this zone, hence interventions incorporating bamboos in the cropping system will be more economical and will readily be adopted by the people. Multitier agroforestry system can also do well in these areas owing to the tropical and subtropical humid conditions prevailing in this part of the country.

Agroforestry will be useful technology, provided there is support for profitable marketing of its product to encourage the future commitments of

Table 18.1. Agroclimatic zones of West Bengals

1. Hill Zone/ Mountaneous zone	
Districts covered	Darjeeling (sub divisions: kalimpong and Kurseong) excluding Siliguri.
Area	3,115 sq. km which is 3.5 % of the state area
Rainfall	2500 – 3000 mm (mainly during June – September)
Temperature	Max.: 20 – 26^{0}C, Min. : 2 – 15 ^{0}C
Relative Humidity	72 – 95 %
Soil type	Porous, coarsely loam, light textured, high organic matter, shallow in depth
pH	4.8 – 6.0
2. Terai zone	
Districts covered	Siliguri sub division of Darjeeling district, greater part of Coochbehar, Jalpaiguri and Islampur subdivision of north Dinajpur
Area	12,015 sq. km which is 13.5 % of the state area
Rainfall	2100 – 3300 mm (mainly during end of May – first week of October)
Temperature	Max.: up to 32^{0}C, Min. : 7 ^{0}C
Relative Humidity	43 – 90 %
Soil type	Coarse textured, sandy loam to siltyclay sometimes found as mixture of gravels and pebbles
pH	4.7 – 6.5
3. New Alluvial zone	
Districts covered	Entire Nadia, parts of south Dinajpur, Bardhawan, Malda, Murshidabad, Hoogly, 24 Parganas
Area	28,277 sq. km which is 31.4 % of the state area
Rainfall	1400 – 1600 mm (mainly during end of May – first week of October)
Temperature	Max.: up to 33.5^{0}C, Min. : 13.9 ^{0}C
Relative Humidity	36 – 85 %

Contd...

Table 18.1. Contd...

Soil type pH	Soil developed from recently deposited sediments of river Ganga. Soil depth is high and is fertile 6.0 - 7.5
4. Old Alluvial zone	
Districts covered Area Rainfall Temperature Relative Humidity Soil type pH	Western parts of Murshidabad and Hooghly, eastern parts of Birbhum and Bankura, northern part of Howrah and central parts of Bardhaman and Midnapore 18, 358 sq. km which is 20.6 % of the state area 1500 - 2000 mm Max.: up to 33^{0}C, Min. : 11 ^{0}C 38 - 85 % Soil developed from previously deposited alluvium of Vidhya family brought down from the Rajmahal hills of Chotonagpur plateau. Heavier in texture. 5.0 - 7.0
5. Red and Lateritic zone	
Districts covered Area Rainfall Temperature Relative Humidity Soil type pH	Purulia, Bankura, Birbhum, western part of Bardhaman and Midnapore, small pockets of Maldah and south Dinajpur 17,800 sq. km which is 20 % of the state area 1000 - 1300 mm Max.: up to 39^{0}C, Min. : 15.8 ^{0}C 24 - 85 % Ferrugeous, porous, moderate to strong acidic, coarse textured with honey comb structure, low fertility 4.5 - 6.5
6. Coastal Saline zone	
Districts covered Area Rainfall Temperature Relative Humidity Soil type pH	Southern parts of North and South 24 Parganas (Sunderbans) 9630 sq. km which is 10 % of the state area 1400 - 1800 mm Max.: up to 32.5^{0}C, Min. : 18 ^{0}C 65 - 85 % Tidal in orgin, mostly silty clay to heavy clay 6.4 - 7.6

Table 18.2. Effect of trees on the yield of paddy as intercrop in farmers field

Direction	Light Situation	No. of tiller/ hill	No. of panicle/ hill	Panicle length (cm)	No. of grain/ panicle	Yield (q/ha)	
						Grain	Straw
East of rows	Light	11	10	22	178 (12-14)	1R – 36 (05-06)	1R-36
	Shade	09	8	19	161		
West of rows	Light	12	11	18.70	192		
	Shade	10	9	24	181		
South of tree rows	Light	9	7	19	167	Lal swama (10-12)	Lal swama (06-07)
	Shade	7	6	17	146		
North of tree rows	Light	9	8	23	199		
	Shade	7	7	20	182		
Mean	Light	10	8	20.67	184		
	Shade	8	7	20	167		
Open plot		13	11	22	210		

(Anonymous, 2002)

Table 18.3. Growth parameters of four year old arecanut in different light intensities

Light intensity (Lux)	Height (cm)	Leaf number	No. of leaflets	Leaf length (cm)	Leaf width (cm)	Collar girth (mm)	Number of nodes exposed
800x100	278.6	5.3	99.3	128.7	82.7	34.0	3.3
400x100	207.0	5.3	88.7	132.7	93.0	24.0	1.0
100x100	360.0	6.3	104.0	191.7	108.0	40.3	2.3
75x100	293.3	5.7	85.0	150.0	97.7	27.0	2.0
50x100	290.0	6.3	91.3	142.7	92.7	32.7	1.3
25x100	280.0	5.0	90.7	143.3	95.0	27.7	1.3
1200x100 (control)	145.6	3.0	84.0	99.0	68.0	18.3	0.7
SEm	24.2	0.3	5.6	6.7	3.5	2.4	0.5
$CD_{0.05}$	60.8	0.8	14.0	16.8	8.7	6.2	1.4

farmers. Diversification of activities including small scale enterprises based on wood and non-wood forests and tree products can help to increase rural income and thus making agroforestry profitable. Role of NGO's in promoting agroforestry system can also bring about adoptivity of these systems because most of NGO's are small organizations that concentrate on a particular area and have a strong network at the grass root levels. They are able to understand local problems and the methods required to solve them. They may be able to persuade farmers to adopt new techniques because they maintain close ties with them. Strengthening research and extension activities of the state as well as of two State Agricultural Universities will help farmers to adopt Agroforestry and thus increase their income status beside improving climatic conditions.

REFERENCES

Anonymous (1986) *Annual Report AICRP on Agroforestry*, Bidhan Chandra Krishi Viswavidyalaya, Mohanpur, West Bengal.: 64-72.

Anonymous (1999) *Annual Report* of Regional Research Station – Terai Zone, Cooch Behar. Bidhan Chandra Krishi Viswavidyalaya, Mohanpur, West Bengal: 70 p

Anonymous (2000) *Annual Report AICRP on Agroforestry*, Bidhan Chandra Krishi Viswavidyalaya, Mohanpur, West Bengal.

Anonymous (2002) *Annual Report of All India Co-ordinated Research Project on Agroforestry*, NRCAF, Jhansi: 35-36.

Chattopadhyay, A. Bhattacharya, S. Nanda, M.K. and Mukherjee, A.K. (2003). Growth and productivity of different tree species and mustard crop grown under alley cropping system. *In National Symposium on Agroforestry in 21st Century* Feb. 11-14, PAU, Ludhiana: 30-31.

FSI (2001) *State of Forest Report 2001*. Ministry of Environment and Forest, GOI, New Delhi.

Hansda, S.K (2000). Major problems of Lateritic zone and transferable technology. Handout in Group meeting of AICRP of Agroforestry at Dr. Y.S. Parmar University of Horticulture and Forestry, Nauni, Solan.

Lahari, A.K. (1972) Intercropping trials with turmeric in North Bengal. *Indian Forester*, 98 (2):109-115.

Lahari, A.K. (1980) An appraisal of agroforestry experiments in the West Bengal. Proc of 2nd Forestry conference, Dehra Dun, Jan., 1980.

Lahari, A.K. (1989) Taungya based agroforestry trials in West Bengal. *Indian Forester*, 115(3): 127-132.

Maiti, S. Majhi, S.K. and Chatterjee, B.N. (1983) Productivity of subabul (*Leucaena leucocephala*) in the humid sub-tropics under different management systems. *Proc. National Convention on subabul*, 25- 26 May, Bangalore.

Maiti, S. Pal, S. and Mukherjee, A.K. (2003) Response of pruning and densities of Leucaena tree on the biomass productivity of elephant foot yam when grown under alley cropping system. In National Symposium on *Agroforestry in 21st Century* Feb. 11-14, PAU, Ludhiana: 20.

Panwar, Pankaj; Partha Choudhuri; P.S. Medda, C.P. Suresh and R. Chatterjee (2006) Effect of mulch materials and light intensity on performance of elephant foot yam in multitier agroforestry system. *Environment and Ecology* 24S (3A): 930-933.

Panwar, Pankaj; P.S. Medda; C.P. Suresh and P. Chaudhury (2006) Effect of integrating banana at early stages on growth of *Areca catechu* in multitier agroforestry system. *Indian Journal of Agroforestry* 8(1): 8-11.

Panwar, Pankaj and C.P. Suresh (2006a) Growth and production potential of Elephant foot yam under Agrihorticulture Agroforestry system in Terai zone of West Bengal *J. South Indian Horticulture* (submitted).

Panwar, Pankaj and C.P. Suresh (2006b) Supplementary effect of *Hevea brazilensis* on banana in Horti-silviculture agroforestry system (accepted for presentation in National seminar on *Agroforestry for livelihood, security, environment protection and biofuel production*") at NRCAF, Jhansi. 16-18 December, 06.

Panwar, Pankaj and Dey, A.N. (2003).Exploring suitable agroforestry systems for sustainable utilization of Terai zone of West Bengal. *In* : Extended Abstract of *National Symposium on "Agroforestry and sustainable production"* held at IGFRI, Jhansi, Nov 7-9: 95-97.

Panwar, Pankaj; Dey, A.N. and Chakravarty, Sumit (2004). Agroforestry in North-eastern states- its present status and interventions for future *In*: *Agroforestry in 21st century* (Eds: Chauhan, S.K. Gill, S.S. Sharma, S.C. and Chauhan, Rajni) Agrotech Publishing Academy, Udaipur: 49- 54.

Pathak, P.S. Pateria, H.M. and Solanki, K.R. (2000) *Agroforestry systems in India- A Diagnosis and Design Approach.* AICRP on Agroforestry, NRCAF, Jhansi.: 106.

Singh, K.A and Pradhan, I.P. (1989). Annual Progress Report on AICRPAF, ICAR Research Complex for NEH Region, Sikkam Center, Tadong, Gangtok.

19

Agroforestry Systems and Practices in Orissa

A.K.MOHAPATRA, N. BHOLA[1] AND R. K. PATNAIK[1]

AICRP on Agroforestry, [1] Department of Forestry, Orissa University of Agriculture and Technology, Bhubaneswar-751003, Orissa.

ABSTRACT

Orissa is traditionally an agrarian state. It is endowed with vast natural resources, diversified agro-edaphic conditions,rich flora and fauna, wide range of land holdings, various socio-cultural groups and people with multiple economic strata. These factors have led to evolve a variety of agroforestry systems and practices. The traditional agroforestry systems which are prevailing in the state since time immemorial are broadly agrisilvicultural, silvipastoral and agrisilvipastoral systems. The prominent agrisilvicultural systems include podu/jhum cultivation, multipurpose trees on farm lands, multiple species tree gardens and agroforestry for fuel wood production. The important silvipastoral systems are trees on pasture, fodder trees and shrubs on forest land and live fence of fodder trees and shrubs. The very old agrihortisilvicultural system prevailing in the state is homegarden. Interventions of Orissa University of Agriculture and Technology and various Government and Non-Government agencies have resulted in the development of some improved agroforestry practices in recent years. Some of the improved agroforestry practices are *Acacia mangium* based agrisilvicultural practices, Guava based agrisilvicultural practices, coconut based agroforestry practices, Eucalyptus (Clonal) based agrisilvicultural practices, windbreaks, timber tree based agrisilvicultural practices, *Sesbania grandiflora* based agrisilvicultural practice, *Acacia mangium* based silvipastoral system, mangium-guava based silvipastoral practices, improved homegardens, commercial block plantations on farmlands and apisilvicultural practice.

INTRODUCTION

Land is the most valuable natural resources on earth, which has degraded over the years owing to demographic pressure. With growing human and animal population, there has been increase in deforestation and land degradation. By 2020 AD, human and animal population in the country is expected to touch 1400 and 625 million, respectively (Samra *et al.*, 1996). According to an estimate, there is a necessity of 275 million tonnes of food grains, 100 million tonnes of fodder, 235 m cum firewood, 275 m cum timber and 19 m cum industrial raw material by 2015 against our current supplying power of 40, 12 and 4 m cum of firewood, timber and industrial raw material, respectively. The average productivity of our forests is only 0.7 cum ha^{-1}. The ever increasing human and cattle population and the resultant burgeoning requirements of food, firewood, fodder, fiber, timber and fertilizer has to be produced from the never increasing land area. The traditional land use systems have become fragile and unsustainable due to unbalanced utilization of natural resources viz., land, water, biodiversity, etc. Apart from this, lack of awareness on proper soil and water conservation measures, unscientific mining of minerals are responsible for massive land degradation and associated problems. These problems warrant new approaches to farming system research. Association of multipurpose tree species (MPTs) with arable crops in good cultivated lands and development of tree-based land use practices like agrisilvi culture, silvipastoral, agrisilvipastoral etc. in marginal and wastelands have been emphasized at national and international levels. It is stressed that sustainable land management is the key for successful utilization of resources to satisfy growing human needs, while maintaining or enhancing quality of environment and conserving natural resources (Kanwar, 1994). The recent emphasis on agroforestry as sustainable land management systems is one of its consequences.

STATUS OF AGROFORESTRY IN ORISSA

Orissa is traditionally an agrarian state. It possesses diversified agro-edaphic conditions, wide range of land holdings, various socio-cultural groups and people with multiple economic strata. These factors have led to evolve some agroforestry systems in the state since time immemorial, which is briefly discussed below. An annual rainfall range of 1200 to 1500 mm in the state has favored a forest area of 37.33 per cent beside the land under pasture, culturable wasteland, seasonal fallow, barren and unculturable land (26.19 per cent) which can be successfully treated and agroforestry practices can be adopted with economic benefits. The land use pattern in Orissa is presented in the table 19.1. Out of total geographical area of 1,55,71,000 ha, net area under agriculture accounts for only 36.48 per cent.

TRADITIONAL AGROFORESTRY PRACTICES IN THE STATE

Various agroforestry practices are in vogue in different agro-ecological zones of Orissa. A few agroforestry practices have demonstrated their usefulness and considered successful in generating sustainable income to farmers even though at a sub optimal level. The application of agroforestry practices to achieve the optimum level is carried out in a limited scale due to lack of scientific know-how and socio-economic considerations. Agroforestry practices differ with not only the agro climatic zones but also the socio economic condition of the farmers and are never same as those in Eastern Ghats to those in Coastal Plains. Some commonly seen perennial trees in the existing agroforestry practices are given in table 19.2.

1. Agrisilvicultural practices

i) Podu cultivation/Jhum cultivation

Podu cultivation is a very traditional form of agroforestry in tribal districts of Orissa. This is a subsistence form of agroforestry practiced by the tribal communities who are generally poor. This is found in districts of Boudh, Kandhamal, Gajapati, Koraput, Malkangiri, Rayagada, Bolangir, Keonjhar, Mayurbhanj, Kalahandi, Nawarangpur, Sambalpur and Sundargarh. In Oriya Podu means burn. The cultivators cut the bushes and burn them in the month of January-February after harvest of *kharif* crop. They plough the land and incorporate the ash into the soil taking the help of summer rains. Seeds are sown with the onset of south western monsoon during June-July. Crops like maize, ragi, minor millets, arhar, cowpea, horsegram, niger, castor etc are grown. The crops grow along with the trees that develop from coppice shoots. During growing season they prune the branches or clean the bushes. Previously this system of agroforestry was in the form of shifting cultivation in which people were shifting from one patch to another and they were returning to the same site after some years when fertility was regained. However, now a days the shifting cultivation has been turned to settled cultivation due to interventions of different Government and non-government agencies. The condition of shifting cultivators has improved because of settled cultivation. However, the socio-economic condition of the cultivators is still very poor and method of cultivation is deteriorating productivity of land.

ii) Multipurpose trees on farmlands

It is a system in which some multipurpose tree species are widely dispersed in cropped field and form a 2-tier system, one is the tree and other is the crop. Among the trees, *Acacia nilotica, Madhuca indica, Borassus flabellifer, Terminalia arjuna* and *Butea monosperma* are very prominent. *Acacia nilotica* is the most widely grown tree in farm lands along with crop in Orissa followed by *Madhuca indica. Acacia nilotica* is found to grow in most of the districts of Orissa and more prominent in Western Orissa. *Butea monosperma* is also found on farmlands in Western Orissa. *Madhuca indica* is generally grown in Western, North Western and South Western parts of the state. *Borassus flabellifer* is predominant on farmlands in coastal districts. *Terminalia arjuna* is found to grow in relatively moist lands of north and western Orissa. In this agroforestry system, upland crops like ragi, maize, arhar and groundnut are grown along with some upland and medium land rice. In this system, the tree species are present mostly haphazardly.

iii) Multiple species tree gardens

In this agroforestry system, various kinds of tree species are grown like gardens and crops are integrated with them. This system is found in every district of Orissa. The planting arrangements are mostly not organized. The major function of this system is production of food, fodder and wood products for home consumption and sale for cash. Major woody species in this system are: *Tamarindus indica, Cocos nucifera, Psidium guajava, Mangifera indica, Tectona grandis, Bamboos*, lime, etc. The associated herbaceous crops are mostly turmeric, arrowroot, dioscorea , chilli, brinjal, tomato, cabbage, cauliflower, cowpea, okra , medicinal plants, etc.

iv) Agroforestry for fuel wood production

In this system, mostly fast growing fuel wood species are interplanted on or around agricultural lands. This system is found more in coastal districts like Balasore, Bhadrak, Kendrapara, Jajpur, Cuttack, Jagatsinghpur, Puri, Khurda and Ganjam because coastal areas do lack in natural forests. The primary productive role of this system is to produce fire wood and the protective role is to act as fencing, wind break and boundary demarcation. Tree species commonly grown are *Leucaena leucocephla, Gliricidia sepium, Cassia siamea, Samanea saman, Casuarina equisetifolia*, etc.

2. Silvipastoral practices

i) Trees on pasture

In this system, multipurpose trees having fodder value are grown on pasture. Trees are mostly scattered irregularly. This system provides fodder to animals from the grasses grown as well as from tree leaves and fruit pods. The trees also provide shelter to animals from sun and rain. This system is found through out the state. The common tree species associated are *Acacia leucocephala, Acacia nilotica, Ailanthus exelsa, Ficus religiosa, Terminalia arjuna, Leucaena leucocephala, Morus alba, Terminalia tomentosa, Samanea saman*, etc.

ii) Fodder trees and shrubs on forest land

This is a very old system of silvipasture in Orissa. It is a very common practice to allow the domestic animals to graze in nearby forests. However, it is not a recognized system. The grazing and browsing animals damage the forest vegetations considerably. This traditional agroforestry system is existing in large extent since long. The controlled grazing helps in reducing the unwanted plants particularly grasses and weeds whereas overgrazing, adversely affects the growth of desirable species.

iii) Live fence of fodder trees and shrubs

In this system various fodder trees and shrubs are planted as live fence. The plants are grown at closer spacing so that animals cannot pass through it. They are maintained as hedges by regular pruning and the pruned materials are fed to animals. The plant species grown in this system are *Erythrina indica, Gliricidia sepium, Prosopis juliflora, Sesbania grandiflora, Acacia nilotica,* etc. This system is found throughout the state.

3. Agrisilvipastoral practices

i) Home garden

This agroforestry system is very old and found in variable forms in all districts of the state. More intensive forms of this practice are seen commonly in coastal districts like Puri, Jagatsinghpur, Kendrapara, Bhadrak and Balasore. The agro-edaphic condition is favourable to support this intensive agroforestry system. Many species of trees, shrubs, vegetables and herbaceous plants are grown in dense and apparently random arrangements. Most home gardens also support a variety of animals. (Cow, buffalo, bullock, goat, sheep, etc.) and birds (poultry birds, duck). In some cases pigs are also reared. This agroforestry system supplements a variety of basic needs of home such as construction timber, fruits, vegetables, drugs, milk, meat, eggs, fodder, live fence etc. These gardens are highly productive and sustainable. The species diversity and varying production cycles of different components ensure continuous production throughout the year from the home garden system. These gardens are located within the compound of individual houses and are intensively managed by family labour.

The home gardens are characterized by high species diversity and usually 3-4 vertical canopy strata. These consist of an herbaceous layer near ground, a tree layer at the upper level and two intermediate layers. The lower most being at less than 1.0 m in height dominated by different vegetables like okra, chilli, tomato, cabbage, dioscorea, etc. and the second layer is 1.0 –3.0 m height comprising food plants such as banana, papaya and lemon. The third layer is at 5.0 -10.0m height occupied by fruit trees like Guava, drumstick, Custard apple, etc. The upper most layer is tree layer which can be divided into two, consisting of the emergent full grown timber and fruit trees having height more than 20.0m and medium size trees of 10-20 m. In the upper layer, species like *Samanea saman, Bambusa vulgaris, Bambosa tulda, Coconut, Arecanut,* Jack fruit, bael, tamarind, neem, subabul, etc. are grown.

IMPROVED AGROFORESTRY PRACTICES

Research and extension activities on agroforestry by the Orissa University of Agriculture and Technology and other organizations have resulted in development of some improved agroforestry practices in the state. A brief account of such practices is given below.

1. Agrisilviculture system

i) Acacia mangium based agrisilvicultural system

Acacia mangium is found to be a very economic tree species in Orissa particularly in coastal districts. In pure plantations (2m x 2m), one tree is yielding about 5 m^3 timber during a rotation period of 10 years which is quite remunerative. It is compatible with herbaceous crops like turmeric , ragi, sesamum , groundnut, cowpea, colocasia, pine apple and sweet potato. Experiments conducted by AICRP-Agroforestry (OUAT) show that a crop yield of 8.5, 6.9, 6.8, 2.8 q/ha from ragi, sesamum, groundnut and blackgram, respectively is obtained in association with *Acacia mangium*. Sweet potato and colocasia are more remunerative with a yield of 61.5 and 71.2 q/ha, respectively with *A. mangium* at 54 months after planting. The forestry faculty of OUAT including AICRP-Agroforestry is popularizing the plantation of *Acacia mangium* in the state through extension projects. Majority of the farmers of the state are adopting this species.

ii) Guava based agrisilvicultural system

Guava based agrisilvicultural practices seem to be remunerative in the state. Trials conducted by AICRP on Agroforestry, OUAT, Bhubaneswar on such practices gave good result. Tree crops like *Dalbergia sissoo, Acacia mangium* and *Gmelina arborea* are planted at 8m x 2m and guava in the middle and interspace is covered with suitable cereals, oilseeds and tuber crops. Among the intercrops, sweet potato and colocasia yielded significantly more than other cereals and oilseed crops with higher benefit: cost ratio. Sweet potato yield ranged from 53.6 q/ha in Acacia *mangium* to 68.5 q/ha with *D. sissoo* and that of colocasia ranged from 61.7 to 75.4 q/ha under different tree species at 54 months after planting (Annon., 2005).

iii) Coconut based agroforestry system

This practice is very much prevalent in coastal districts of Balasore, Bhadrak, Jajatsinghpur, Kendrapara, Khurda and Puri. Coconut trees are planted at 7.5 x 7.5 m spacing with a population of 175 plants per hectare. The interspaces are covered with fruit crops like banana, papaya, pineapple and vegetables like okra, bottle gourd and drumstick. Tuberose is also grown in the interspaces of coconut. These crops are raised without any yield reduction upto 8 years of planting of coconut. With popularization of paddy straw mushroom, its cultivation is catching up in coconut orchards with more profit in coastal area. Research trials conducted on intercropping of banana, papaya and paddy straw mushroom with coconut under AICRP on palms shows that on an average 25 nuts, 1 to 1.5 kg of drumstick, 20 kg of papaya and 25 kg of banana/plant and 1.3 to 1.5 kg paddy straw mushroom/bed are harvested from high density plantation in coconut orchard against harvesting of 15-16 nuts/plant when coconut are grown as pure crop.

iv) Eucalyptus (clonal) based agrisilvicultural system

The eucalyptus plantations are raised successfully by the people throughout the state. The clonal plantations are becoming more popular because of short rotation. The wood based industries in the state such as J.K. Paper Ltd., BILT and Mangalam Timers Ltd are promoting the eucalyptus clonal plantations. These companies have developed their own clones. These clones are being harvested in 4-5 years rotation. Besides short rotation, the yield and income from the clonal plantation is significantly high. J.K. Paper Ltd., Rayagada (Orissa) has calculated the economics of eucalyptus plantations which is given in table 19.3.

The people are adopting these plantations in marginal and uplands because of good return and assured buy back guarantee given by these wood based industries. Throughout the state, the eucalyptus plantations are being raised. J.K. Paper Ltd is advocating growing crops like lemongrass, ginger, turmeric, etc, with eucalyptus.

v) Wind breaks

In some areas of coastal districts of Kendrapara, Jagatsinghpur, Puri and Balasore, one to three rows of trees and shrubs are planted at 0.5 to 2.5 m apart depending upon the tree species round the farm or garden or crop field. Tall growing tree species are planted in the centre rows and smaller bushy species in the outside rows for higher efficacy of the windbreaks. The tree species used for wind breaks are *Eucalyptus hybrid, Tectona grandis,* Dalbergia *sissoo, Leucaena leucocephala, Acacia nilotica, Cassia siamea, Casuarina equisetifolia, Syzygium cumini* etc. People extract timber and fuel wood from these species.

vi) Timber tree based agrisilvicultural system

Relative performance of cereals, pulse and oilseeds under timber trees such as *Tectona grandis, Dalbergia sissoo, Eucalyptus tereticornis* and *Casuarina equisetifolia* was studied. Paddy (Ghanteswari), ragi (Neelachal), Greengram (Dhauli), Arhar (ICPL-87), Blackgram (Sarala), Cowpea (SEB-2), Groundnut (AK-12-24) and Sesamum (Kalika) were the intercrops. Ragi crop produced 11.0, 10.7, 11.6 and 11.1 q/ha grain under Casurina, Eucalyptus, Sisoo and Teak, respectively with a mean PLER of 0.84, which was the highest among the other intercrops tried. Sissoo was the most compatible tree species causing the least yield reduction followed by *Casurina equisetifolia* (Annual Report of AICRP, Agroforestry, 1995/96). In high altitude areas of Koraput, Rayagada, Boudh, Kandhamal districts, black pepper is grown in association with timber trees like *Grevillea robusta* (Silver oak).The vines of black pepper twines on the tree species and grows. Farmers get an average yield of 3-4 kg of black pepper /plant after 5th years of planting.

viii) Sesbania grandiflora based agroforestry system

Sesbania grandiflora (Agasti) is a very fast growing leguminous tree. The leaves, flowers and pods are rich in crude protein. In southern Orissa, particularly in Ganjam district, people grow this plant as a support for betel vine cultivation. They raise the plants by sowing seeds at a closer spacing and allow the betel vine to grow on it after the sesbania gets established. This is a very sustainable system for betelvine cultivation. Sesbania helps in maintaining fertility of land through its quick leaf decomposition and nitrogen fixation through nodules.

Sesbania leaves are rich in crude protein and because of plentiful availability of sesbania leaves goat farming in Ganjam is done on large scale. The fast growth of this species and high combustibility of its wood make it good fuel wood tree. The flowers are also consumed by people.

2. Silvipastoral system

i) Acacia mangium based silvipastoral system

In the plantation of *Acacia mangium*, fodder grasses like Guinea, Hybrid Napier (NB-21) and Thin Napier are very much compatible. Research under AICRP on Agroforestry at Bhubaneswar showed that Thin Napier was the highest yielder with 351.2 q/ha of green fodder followed by Guinea (253.80 q/ha). Hybrid Napier recorded the least fodder yield with *A. mangium* (182.4 q/ha). The benefit: cost ratio of *A. mangium* + Thin Napier silvipastoral practice was found to be 3.12 (Annon., 2005).

ii) Acacia mangium - Guava based silvihortipastoral system

In this system *Acacia mangium* is planted at 6 m x 6 m interval and horticultural component guava in the middle at 6m x 6m interval. Fodder grasses like Guinea and Stylo are found to be more remunerative under this silvihortipastoral practices. Research Results under AICRP Agroforestry at Bhubaneswar centre reveals that on an average Guinea grass produced 124.6 q/ha of green fodder as compared to stylo which could yield only 98.0 q/ha under same management conditions from 2 cuttings. This practice is well prevalent in coastal districts of Orissa (Annon., 2003).

iii) Acacia auriculiformis based silvipastoral practice

This practices is common in eastern ghat region of Orissa i.e Koraput, Rayagada, Boudh and Kandhamal district. Maize, ricebean (*Vigna umbellata*), cowpea are grown as fodder crop with pure energy plantation of *Acacia auriculiformis* planted at 2m x 2m interval. Research results of AICRP on Agroforestry at Semiliguda, Koraput centre shows that ricebean and maize both produced on an average 68.7 and 58.0 q/ha of green fodder under 1/3rd lopping and 56.7 and 56.0 q/ha of green fodder under unlopped condition when grown in association with *Acacia auriculiformis* (Annon., 1992).

iv) Leucaena based silvipastoral practices

In Koraput district of Orissa, bund plantation of *Leucaena leucocephala* at 1 m x 1m is a practice as it serves the dual purpose of fencing and meet fodder needs of the cattle and goat population of tribal communities. Fodder crops such as oat (*Avena sativa*) and maize are commonly grown with Leucaena. This silvipastoral practice is very much productive. On an average maize and oat produces 70.0 and 380.0 q/ha green fodder, respectively in association with *Leucaena*.

3. Agrisilvipastoral system

i) Improved home gardens

The productivity of home gardens have been improved in recent years in Orissa through better management of different components, introduction of improved varieties and better management practices. People are putting different components in a more planned way in comparison to traditional home gardens. Optimum spacing is being provided to perennial crops. People have now access to get better varieties of plants from the nurseries of Government, Agricultural Universities or genuine private agencies. High yielding varieties of fruits, vegetables, medicinal plants are grown and improved varieties of poultry, ducks are reared by the people. The fast growing timber species like teak, mangium, eucalyptus, bamboo, etc are being grown in homegardens. Different components are better managed. People are applying manure, fertilizer, plant protection chemicals and practicing tending and cultural operations.

4. Commercial block plantations on farm lands

The commercial block plantations of different species like *Tectona grandis, Eucalyptus hybrid, Casuarina equisetifolia, Acacia mangium,* Acacia *auriculiformis* , Bamboo, Rubber are being raised by progressive farmers in certain part of their farm lands. These plantations are being raised as pure crop. Close to these plantations, people are cultivating their normal agricultural crop. Block plantations of *Tectona grandis* and Eucalyptus are most popular in this state. Block plantations of *Casuarina equisetifolia, Acacia mangium , Acacia auriculiformis* and Bamboo are generally practiced in coastal districts of the state. Rubber plantations have been raised successfully in the districts of Dhenkanal, Gajapati and Mayurbhanj.

5. Apisilvicultural system

In recent years, apiculture (raring of honey bees) in association with tree plantations has gained popularity. Trees with high nectar content in flowers like *Eucalyptus, Gliricedia, Grevillea,* Gambhari, Subabul, *Albizia* etc. are suitable for apiculture. The AICRP-Honeybee, OUAT, Bhubaneswar is promoting this practice. Tree having different flowering time in the year help in honey production throughout the year. Tree provide ambient condition to honeybees and helps in maintaining their productivity. Apisilviculture is a good avenue to generate additional farm income in rural areas of the state. Even landless persons can be encouraged to adopt apiculture in an intensive manner in the regions where wide extent of bee-forage plants have been raised.

CONCLUSION

The above agroforestry practices will increase the farm productivity in terms of timber, fuel, food and fodder and minimize dependency on dwindling forest resource. As agroforestry is regarded as a set of land use alternatives and can be practiced in all categories of land with scientific bio-resource management, higher productivity from integrating the tree crops with agricultural and/or fodder crops, animals is achievable without affecting the fertility status of the soil. It is an eco-friendly approach, which is to be followed for environmental upgradation and hither economic benefit to the farmers.

Table 19.1. Area under different land use in Orissa

S. No.	Land use	Area in 000 ha	Percentage to total geographical area
1	Agriculture	5680	36.48
2	Forest	5813	37.33
3	Barren and unsustainable land	843	5.41
4	Land put to non-agriculture use	999	6.42
5	Permanent pasture and grazing land	443	2.84
6	Land under Misc. tree crop and growers	482	3.10
7	Culturable waste land	392	2.52
8	Fallow land including current fallow	919	5.90
	Total	15571	100.0

Source: Statistical Abstracts of Orissa, 2005

Table 19.2. Important trees and fruit trees in different land situations in Orissa

S. No.	Tree component		Fruit trees	
	Scientific name	Local Name	Scientific name	Local name
Farm lands				
1.	*Samanea saman*	Bada Chakunda	*Cocus nucifera*	Coconut
2	*Dendrocalamus strictus* *Bambosa vulgaris* *Bambosa bambos* *Bambosa nutans*	Salia baunsa Sunderkani baunsa Kanta baunsa Badia baunsa	*Mangifera indica*	Mango
3	*Tectona grandis*	Teak	*Artocarpus integrifolia*	Jackfruit
4	*Gmelina arborea*	Gambhari	*Psidium guajava*	Guava
5	*Anogeisus acuminata*	Phasi	*Areca catechu*	Beetle nut
6	*Erythrina indicaa*	Paladhua	*Anona squamosa*	Custard apple
7	*Casuarina equisetifolia*	Jhaun	*Emblica officinalis*	Aonla
8	*Albizzia lebbeck*	Siris	*Ziziphus mauritiana*	Ber
9	Ougeinia oojeinensis	Bandhana	*Aegle marmelos*	Bael
10.	Anogeissus latifolia	Dhaura	*Borassus flabellifer*	Tala

Paddy field bunds

S. No.	Tree component		Fruit trees	
	Scientific name	Local Name	Scientific name	Local name
1.	*Acacia nilotica*	Babul	*Tamarindus indica*	Tamarind
2	*Casuarina equisetifolia*	Jhaun	*Borassus flabellifer*	Tala
3	*Albizzia lebbeck*	Siris	*Madhuca latifolia*	Mahua
4	*Eucalyptus hybrid*	Eucalyptus		
5	*Acacia leucophloea*	Gohira		
6	*Azadirachta indica*	Neem		

Culturable waste/marginal land

S. No.	Tree component		Fruit trees	
	Scientific name	Local Name	Scientific name	Local name
1.	*Schleichera oleosa*	Kusum	*Sizygium cumini*	Jamun
2	*Calophyllum inophyllum*	Polanga	*Diospyrus melanoxylon*	Kendu
3	*Pongamia pinnata*	Karanja	*Madhuca latifolia*	Mahua
4	*Gmelina arborea*	Gambhari	*Anacardium occidentale*	Cashew nut
5	*Casuarina equisetifolia*	Jhaun		

Table 19.3. Per hectare economics of eucalyptus (seed route vs. clonal plantation)

S. No.	Particulars	Unit	Seed route plantation	Clonal Plantation
	First rotation			
1	Cost of raising and maintenance	Rs.	16000	16000
2	Cost of plant material	Rs.	1250	15000
3	Total cost	Rs.	17250	31000
4	Estimated yield/Hectare	Tonnes	60	120
5	Minimum rate/tonne at farm gate	Rs.	1000	1000
6	Total returns	Rs.	60000	120000
7	Net returns in first rotation	Rs.	42750	89000
	Second rotation			
1	Cost of raising and maintenance	Rs.	12500	12500
2	Cost of plant material	Rs.	Nil	Nil
3	Total cost	Rs.	12500	12500
4	Estimated yield/Hectare	Tonne	60	120
5	Minimum rate/tonne at farm gate	Rs.	1000	1000
6	Total returns	Rs.	60000	120000
7	Net returns in second rotation		47500	107500

Photo 19.1. Arrowroot Intercropped in *Acacia mangium* Alleys

Photo 19.2. Stylosanthes hamata In Hortisilvipastoral System. (Guava+ Gmelina +Stylo).

REFERENCES

Anonymous (1992) Annual Report, All India Coordinated Research Project on Agroforestry, Semiliguda Centre, Orissa.

Anonymous (2003) Annual Report, AICRP on Agroforestry, Bhubaneswar Centre.

Anonymous (2005) Annual Report, AICRP on Agroforestry, Bhubaneswar Centre.

Bene, J.G., Beall, H.W. and Cote, A. (1977). Tree, food and people Ottawa, IDRC.

Kanwar, J.S. (1994). Management of soil and water resources for sustainable agriculture and environment. *In*: Diamond Jubilee Symposium, ISSS, New Delhi, 22: 15-25.

Samra, J.S., Narain, P. and Dadhwal, K.S. (1996). Rehabilitation of degraded forests through integrated land use planning with soil and water conservation measures. *Indian J. Soil Cons.,* 24(1): 10-18.

20

Agroforestry Systems and Practices Prevailing in Assam

A.A.AHMED AND D.N. HAZARIKA

Horticultural Research Station, Assam Agricultural University Kahikuchi, Azara, Guwahati-781017

ABSTRACT

Assam has been divided into six agroclimatic zones on the basis of soil, climate, terrain and vegetation. It comprises of Brahmaputra plain zone to Hill zone. The paper deals with the traditional agroforestry systems prevailing in all these agroclimatic zones along with the research activities which are going on in Assam. Salient research achievements with respect to tree-crop compatibility are given at the end of the chapter. Impact of agroforestry research on farmers of Assam in terms of adoption is also given.

INTRODUCTION

Assam, one of the eastern states of India, lies between 24^0 N and $28^0 18'$ N latitudes and $89^0 4'$ E and 96^0 E longitudes. The state is surrounded by Bhutan and Arunachal Pradesh in the north, Nagaland, Manipur and a part of Arunachal Pradesh in the east, Mizoram, Tripura and Meghalaya in the south and West Bengal in the west. The state is endowed with abundant fertile land and water sources with a total geographical area of 78,523 km^2 with population density of 318 per square kilometer.

The climate of Assam is humid subtropical with humid summer and cool dry winter. The average annual rainfall is about 2000 mm with the heaviest fall during summer from June to August (Table 20.1). The pre-monsoon rainfall occurs from March to May receiving 25 per cent of the annual rainfall. But its intensity and distribution is very erratic and unpredictable. Among the districts of Assam, Cachar and Lakhimpur receive the highest rainfall (3000 mm) while the Karbi Anglong receives the lowest (1152 mm). Periodic dry spell generally occurs during winter from November to March. The minimum and maximum temperature ranges from 19^0 C to 35° C in July and 5^0 C to 25^0 C in January. The temperature starts declining from November and becomes lowest in January. The average sunshine hours are less than 6 hours due to continuous overcast sky. The radiation during winter from November to February is not up to the desired extent due to foggy weather.

The texture of the soils of Assam differs considerably depending on the agro-climatic and physiographic units. The major portions of the soils of Assam belong to Inceptisols (49.3%), followed by Entisols (32.3%), Alfisols (12.3%) and Utisols (6.1%). The soils of Assam are acidic in nature having pH range from 4.2 to 5.8. However, the soils of flood plains have slightly high pH (6.0 - 6.5) and the soils of hill districts have higher pH ranging from 6.0 - 7.1. This may be due to the existence of limestone in the isolated areas of the hills in Assam.

LAND USE PATTERN

Assam has a total geographical area of 78,52,300 hectares and out of it, 25.6 per cent are under forest cover and 31.3 per cent are not available for cultivation. The total cropped area is 3984.07 thousand hectares in the state (Table 20.2). Thus, the total net-cropped area is 35.26 per cent of the total geographical area.

PHYSIOGRAPHY

The state has three distinct physiographic units - the plain, the plateau and the hills. The Brahmaputra and Barak Valleys accounts for 80.8 per cent of total geographical area and these two valleys are the major zones for agricultural development in the plains. The plateau covering the districts of Karbi Anglong and North Cachar Hills separates both the valleys. The rivers Brahmaputra and Barak have 40 and 7 numbers of tributaries, respectively. The Brahmaputra valley covers an area of about 5.6 million hectares with gradual slope from east to west. The altitude at Sadiya (east) is 134 m while it is 35 m at Dhubri (west) from the mean sea level. The Brahmaputra valley is an alluvial plain surrounded by hills and interspersed with small hillocks, uplands and swampy lowlands subject to annual flooding

The Barak valley covers an area of 0.7 million hectares comprising undulating topography with small hillocks and swamps. The two hill districts of Assam constitute an area of 1.5 million hectares. The maximum altitude is 930 m in the northern side of the zone.

AGRO-CLIMATIC ZONES

Assam is divided into six agro-climatic zones on the basis of soil, climate, terrain and vegetation. The Brahmaputra valley consists of four agro-climatic zones while the Barak valley consists of one agro-climatic zone only. Both the hill districts comprise the Hills zone. All the zones have zonal research centers to conduct location specific, need based and problem oriented research for each zone.

North Bank Plains zone

The zone consists of Dhemaji, Lakhimpur, Sonitpur, Darrang and Udalguri districts. Rice, rapeseed and mustard, pulses and jute are the dominant crops occupying 75, 9, 4 and 4 per cent of the total cropped area, respectively. The forest covers an area of 23 per cent of the total area. Fruits and vegetables are also grown in some selected pockets. Establishment and maintenance of homestead is a pre-requisite of each household of this zone wherein agroforestry systems are managed to meet the household needs of the family. Cultivation of bamboo, trees for timber and fuel wood, fruit trees, areca nut with vegetables, spices are mostly grown in homestead garden. This zone is characterized by monocropping of sali rice in addition to the deep-water rice in the districts of Lakhimpur.

Upper Brahmaputra Valley zone

Tinsukia, Dibrugarh, Sibsagar, Jorhat and Golaghat are the districts under this zone. Rice, rapeseed and mustard, sugarcane and pulses are the major crops of the zone. The zone comprises 60 per cent of the tea growing areas of the state. Tinsukia and Jorhat districts accounts for mandarin oranges and citronella oil production. The forest covers an area of 30 per cent of the total area. Monocropping of rice is the dominant cropping pattern in the zone. The cropping intensity is 107.1 per cent.

Central Brahmaputra Valley zone

The two districts - Nagaon and Morigaon are under this zone. Rice and jute are the most important crops covering 70.95 and 7.98 per cent of the total cropped area. Pulses, oilseeds and vegetables are also the major crops. The forest area in this zone is the lowest in the state. Like other zones, rice is the most dominant crop but double cropping of jute and rice is the important feature of cropping pattern in this zone. Cropping intensity is 125 per cent.

Lower Brahmaputra Valley zone

This zone is the largest among all the zones in terms of area and comprises of 11 districts namely, Kamrup, Nalbari, Barpeta, Bongaigaon, Kokrajhar,

Buxer, Goalpara and Dhubri. Rice, jute, wheat, pulses, potato, sugarcane, oilseeds are the major crops of this zone. Wheat is extensively grown in the riverine tracts. Small millets and buckwheat cover a sizeable area in Goalpara district. Other crops such as linseed, coriander, tobacco, sweet potato, chillies and cassava are also grown in the zone. Coconut and areca nut production potentiality are maximum in this zone. Cashew nut has been cultivated in Mancachar area in Dhubri district bordering Bangladesh and has tremendous scope for expansion of area in this zone. Kharif crops account for 80.5 per cent of the total cropped area. Cropping intensity is 131.0 per cent. In this zone, 25 per cent of the total area is under forest.

Barak Valley zone

The zone is in the southern part of Assam comprising Cachar, Hailakandi and Karimganj districts. The entire central alluvial tract is a monocropped area, rice occupying 93 per cent of the net-cropped area. Tea gardens are in the hill slopes. Area under forest is also high (39 %). Sugarcane is grown in hillocks. This zone recorded the highest productivity of rice in the state. There is a sizeable area bordering Manipur where pineapple is cultivated extensively. Cropping intensity is 134.0 per cent.

Hill zone

Karbi Anglong and North Cachar Hills districts of Assam comprise the Hill zone. Rapeseed and mustard, maize and sugarcane are the predominant crops of this zone. Out of the total cropped area, rice occupies 70 per cent, rapeseed and mustard 8.6 per cent, maize 8 per cent and sugarcane 3 per cent. Yam, cassava, sweet potato and other tuber crops play an important role in the cropping pattern in this zone. The hills are rich sources of different species of bamboo and other horticultural and plantation crops. Rubber and coffee have already been identified as thrust crops for this region and have been grown extensively in the hill slopes. More than 40 per cent area is under forest. Cropping intensity is not very high (105 %).

Jhum cultivation is the predominant form of agriculture in the hill slopes with some wet rice cultivation in the terraced fields and the mountain valleys.

AGROFORESTRY PRACTICES

Farmers of Assam are less aware of the latest system of land management for rural development through different agroforestry practices. However, some traditional agroforestry practices exist in the entire state.

In the alluvial tracts, most of the lands are subjected to the cultivation of rice, jute, sugarcane in summer and potato, wheat, oilseed and pulses in winter.

Recently, 20 per cent area in Central Brahmaputra valley zone and Lower Brahmaputra valley zone is diverted to boro rice cultivation in winter. Of course, rice has been identified as the major crop for all the zones of Assam. In some high lands, people grow bamboo for dwelling, making fences and household items and for selling to the paper manufacturing industries. They also grow date palm (*Phoenix dactylifera*), coconut (*Cocos nucifera*), areca nut (*Areca catechu*), banana, papaya, pineapple along with the periphery of fish pond. Some farmers grow sissoo (*Dalbergia sissoo*), kadamba (*Anthocephalus cadamba*), neem (*Azadirachta indica*) and arhar (*Cajanus cajan*) along the boundary of the agricultural crops. In some cases, farmers retain few trees within the cultivated field. They use to cut the branches of the trees to avoid shade on the agricultural crops and to meet their fuel requirement. the homestead garden, popularly known as 'bari', farmers grow various fruit trees like mango, jackfruit, litchi, guava, banana, citrus particularly lime and lemon, coconut, areca nut, drum stick. The timbers like sissoo, teak, gomari (*Gmelina arborea*), kadamba, neem are also grown in bari. The minor fruits having medicinal importance like bael, carambola, aonla, *Terminalia chebula, Garcinia cowa* are found in most of the baris. Pine apple, spices like turmeric, ginger and chilli, grasses are grown in the inter spaces of fruit crops as well as timber plants while colocassia is cultivated in the adjoining water disposal drains. Growing of betelvine (*Piper betel*) and black pepper (*Piper nigrum*) with areca nut and coconut as standard is a common practice.

The farmers of all categories i.e. small, medium and large rear animals like cow, bullock, buffalo and goat to meet their requirements. Common grazing land is not available for which the animals have to depend on stall-feeding. The edible crop residues, grasses, legumes and leaves of various trees contribute the roughages for the animals. Some of the fodder trees used by the farmers are *Artocarpus heterophyllus, Artocarpus chaplasha, Ficus hispida, Ficus gibbosa, Ficus religiosa, streblus asper, Anthocephalus cadamba, Leucaena leucocephala, Albizzia sp. Terminalia tomentosa* and bamboo species.

Firewood is the source of fuel and is obtained from the trees growing in home stead garden, farm land and nearby forest. In addition, *Sterculia villosa, Lagerstroemia speciosa, Bischofia javanica, Trewia nudiflora, Terminalia arjuna, Albizzia procera, Albizzia lebbeck, Dysoxylum binectariferum, Albizzia lucida, Toona ciliata, Dalbergia sissoo, Cassia fistula, Salix sp, Tectona grandis, Dilenia indica, Azadirachta indica, Delonix regia* are also grown to meet their demand for fuel, timber and agricultural implements and also as wind barriers. *Bombax ceiba* is grown as commercial enterprise. Approximately 10 per cent farmers grow trees as monoculture, 75 per cent grow trees and agricultural crops concurrently on their holding. Alley cropping is also adopted in traditional way. Very little care is extended to the naturally regenerated trees.

Integration of trees such as *Anthocephalus cadamba, Albizzia lucida, Azadirachta indica* on the border of agricultural land is a common practice for firewood. Even crops like *Cajanus cajan, Sesbania aculeata* and *Tephrocia candida* are raised for fuel and manuring. The main constraints in cultivating the crops are small size of land holding, inadequate water supply during winter, excess rain water in rainy season and sometimes uncertain flood during summer, non availability of animal power and workers, lack of adoption of suitable scientific technology and improper fertility management.

Some of the traditional agroforestry systems prevailing in Assam are as follows:

- *Agrihorticultural* : Castor/sesamum/arhar/turmeric/ginger - Guava/ citrus/ coconut/areca nut
- *Hortisilvipastoral* : Jackfruit/cadamba - Guinea/napier/local grasses
- *Silvipastoral* : *Bombax ceiba*/cadamba - Thatch/Kahua (*Saccharum spontaneum*)
- *Silvihorticultural* : Areca nut/*Erythrina indica* - Betel vine/black pepper/ pineapple
- *Hortihorticultural* : Jackfruit/guava/citrus - Banana/pineapple/papaya

AGROFORESTRY RESEARCH

Researches on agroforestry is mainly concentrated on the following objectives:

- To find out the most suitable combination of different components of agroforestry
- To work out the best arrangement of the components keeping in view the productivity and sustainability of the system
- To standardize the harvesting schedule keeping in view the production of all components
- To find out the economic feasibility of the system compared to monoculture

The agrisilvicultural and agrihorticultural systems on rainfed and irrigated cropland are summarized in the table 20.3.

Anthocephalus cadamba in rice field bund and *Dalbergia sissoo* in the boundary of jute field are the most profitable agroforestry systems identified in croplands.

- With regards to supply of fodder to animal components, the silvipastoral systems are identified for different category of farmers (Table 20.4).
- On the basis of the study on different agroforestry practices in farmers field, the different systems (Table 20.5) are identified that are either profitable or popular among farmers.

SALIENT RESEARCH ACHIEVEMENTS

1. Works initiated on edible bamboo and monopodial bamboo in addition cultural practices of bamboo in horticultural garden.
2. *Gmelina arborea* based agroforestry - yield of turmeric as intercrop in *Gmelina arborea* is good up to 5 years.
3. In coconut based hortiagripastoral system, fodder yield of sorghum, maize and oat was excellent to meet the fodder requirement during winter season.
4. In agrisilviculture system, crop rotation of sesamum-niger was feasible in the interspaces of *Acacia mangium.*
5. During the early growth of *Artocarpus heterophyllus* in hortiagriculture system, the growth and yield of sesamum-niger in interspaces proved success.
6. *Trewia nudiflora, Syzygium cumini, Lagerstroemia flos-reginae* could withstand water stagnated condition. Twenty MPTS, 10 indigenous fruit trees and 5 fodder species were evaluated for different agroforestry system. Among the MPTS, *Acacia auriculiformis* (16.20 m height), *Gmelina arborea* (12.97 m), *Bombax ceiba* (12.90 m), *Albizzia lebbeck* (12.21 m), *Tectona grandis* (12.50 m), *Dalbergia sissoo* (12.33 m) and *Leucana leucocephala* (10.0 m) shoed faster growth in comparison to other species in 7th year of plantation. Out of 10 indigenous fruit trees, *Terminalia chebula, Terminalia belerica* and *Amoora wallichii* showed faster growth of more than 8.0 m height in 5th year of plantation. Hybrid napier, guinea, seteria and broom grasses were grown profitability in the interspaces of coconut orchard. Competitive zone of coconut and grasses was also identified.
7. Turmeric, ginger and pineapple were identified as the most profitable intercrop combinations with arecanut (*Areca catechu*).
8. Betel vine and black pepper as climber on arecanut and coconut proved to be successful multistoried crops.

Table 20.1. Rainfall pattern in Assam

Season	Months	Rainfall (mm)	Characteristics
Pre-monsoon	Mar-May	549.6 (25%)	Low and erratic rainfall with dry spell
Monsoon	June-Sept	1430.0 (65%)	Heavy rainfall and flood
Post-monsoon	Oct-Nov	152.3 (7%)	Low and irregular rainfall
Winter	Dec-Feb	65.9 (3%)	Virtually dry, cool
	Total	2197.7 (100%)	

Table 20.2. Land use pattern in Assam

Uses	Area ('000 ha)
Total geographical area	7852.3
Forest area	2013.3
Land not available for cultivation	2455.6
Other cultivated land	456.2
Fallow land	156.96
Net area sown	2768.87
Total cropped area	3984.07
Area sown more than once	1215.20

Table 20.3. Agroforestry systems in croplands

Condition	Area (%)	Cropping system	No. of trees/ha	Pre-dominant trees
Rainfed	90.0	Rice, wheat, vegetables	100	Cadamba, neem, fruit trees
Irrigated	10.0	Jute, oil seed, pulse	150	Sissoo, Velkol, (*Trewia nudiflora*) Soira

Table 20.4. Silvipastoral systems

Farm size	Animal species	No. of animal	Fodder crops	Fodder trees
Large farmer	Cattle, buffalo, goat	10	*Lathyrus sativus, Pennisetum sp, Vigna sinensis*	*Artocarpus heterophyllus Bauhinia variagata, Cocos nucifera*
Medium farmer	Cattle, buffalo, goat	8	*Vigna unguilata*	*Albizzia sp, Bauhinia variegata*

Table 20.5. Agroforestry systems observed in farmers' field

Agroforestry system	No. of trees/ha	Tree species	Crop species
Rainfed			
Agrihorticultural	177	Coconut, areca nut, guava, citrus	Turmeric, ginger, arhar, sesamum
Silvihortipastoral	200	Kadamba, jackfruit	Napier and guinea
Silvipastoral	100	Simalu (*Bombax ceiba*)	Thatch, kahua
Irrigated			
Silviagrosystem	100	kadamba, neem	Rice - oilseeds

IMPACT OF AGROFORESTRY RESEARCH ON FARMERS

1. Large category farmers (>5 ha) adopted cultivation of *Gmelina arborea, Tectona grandis, Anthocephalus cadamba* and bamboo.
2. Thirty per cent medium and small farmers adopted fruit trees along with bamboo as one of the enterprises in the border of arable crops.
3. Twenty per cent of landless farmers having less than 0.2 hectares of land have adopted areca nut with betel vine and black pepper, vegetables in their homestead gardens.
4. *Trewia nudiflora, Bombax ceiba and Lagerstroemia flos-reginae* have gained popularity for soil conservation especially in river bunds.
5. Forty per cent farmers in the foothill region has adopted turmeric, ginger in areca nut and mandarin orange orchard.
6. Although coconut is non-traditional crop of Assam, its area has been increased recently. Ten per cent of farmers have already started cultivation of grasses and shade loving vegetables and black pepper in the interspaces of coconut.
7. The technology generated in the agroforestry project in relation to comprehensive package of high quality nursery stock has been accepted by the growers. Limited number of farmers adopted the technology commercially.

REFERENCES

Agroforestry Newsletter (1998) pp 9-11.

Annual report, AICRP on Agroforestry, Horticultural Research Station, Kahikuchi.

Impact analysis report, AICRP on Agroforestry, Horticultural Research Station, Kahikuchi.

21

Agroforestry Systems and Practices Prevailing in Meghalaya

J. M. S. TOMAR[1], K. UPADHAYA, O. P. TRIPATHI AND H.N. PANDEY[2]

[1] Agroforestry Division, ICAR Research Complex for NEH Region Barapani, Meghalaya

[2] Department of Botany, North-Eastern Hill University, Shillong - 793 022

ABSTRACT

In the state of Meghalaya, forests are being degraded by age-old practice of shifting agriculture by native tribes and other human activities. Agroforestry has a great potential both from the point of view of restoration and maintenance of soil fertility and to increase agriculture production. The traditional system of agriculture in the state is primitive, labour intensive and dependent on rainfall. The proportion of space occupied by the tree and crop components in the traditional system largely depends upon factors like soil types, plant species and crop arrangements. The common traditional systems practiced in the state may be grouped under agrisilviculture, agrihorticulture, silvihorticulture, pastoralsilviculture and pastoralhorticulture. Agroforestry systems can play an important role in soil conservation and improvement in moisture and nutrient levels through efficient cycling of nutrients, which in turn would enhance agriculture production. The analysis of different agroforestry systems in the state of Meghalaya suggests that further improvement of the existing systems may yield high productivity and make them more sustainable.

INTRODUCTION

Agroforestry is an ideal scientific approach to tackle the problems of degraded lands and to bring about eco-restoration and maintenance of soil resources. The significance of tree based land use systems in general and agroforestry in particular, has been realized in improving the economy of farmers having small land holdings (Chauhan and Dhyani 1990, Beer *et al* 1990, Fisher 1990, Dhyani and Tripathi 1999). The fertility of soils can be improved under the influence of tree cover, which checks erosion, contributes to accumulation of organic matter content, adds nitrogen through symbiotic N-fixation by leguminous or actinorrhizal trees and continuously replenishes the nutrients through effective recycling mechanisms. It offers a practical means of achieving greater outputs and at the same time maintains soil fertility that ultimately helps in increasing productivity of agricultural crops and trees.

Indigenous tribes inhabiting in the state of Meghalaya in northeast India traditionally have developed farming systems in which trees and crops are grown together to meet the household needs on sustainable basis. Scientific innovations of the traditional agroforestry systems have certainly proved to be sustainable, more productive and profitable to the farmers. However, there is still need of systematic study of traditional agroforestry systems, their economics and sustainability.

Meghalaya being a part of the northeastern Indian bio-geographic zone, it constitutes the junction meeting place of paleo-arctic, Indo-Malayan and Indo-Chinese bio-geographic realms. In Meghalaya and northeast India as a whole, forestland has been degraded by age-old practice of shifting agriculture by native tribes and other human activities. Agroforestry has a great potential both from the point of view of restoration and maintenance of soil fertility and to increase agriculture production. About 85 per cent of the population in the state of Meghalaya are dependent on agriculture. The traditional system of agriculture in the state is primitive, labour intensive and dependent on rainfall. The major problems that are adversely affecting agricultural production of the state are the undulating topography, erratic rainfall, heavy soil erosion and continued land degradation. Under such circumstances, agroforestry systems can play an important role in soil conservation and improvement in moisture and nutrient levels through efficient cycling of nutrients, which in turn would enhance agricultural production.

Geology

The state of Meghalaya lies between latitude 25^0 02' and 26^0 07' N, and longitude 89^0 49' and 92^0 50' E with an area of 22, 429 square km. The topography of the state is variable and the elevation ranges from 60 m to 2040 m asl. The central and northern part of Meghalaya plateau is made essentially of highly metamorphosed crystalline rocks of Pre-Cambrian origin, which are often referred to as "Archaean Gnessic Complex" (Anon 1974, Murthy *et al.* 1976). The Dawki fault and Brahamputra lineament mark the southern and northern boundaries, respectively. Rai (1986) has reported that the west massif is bordered by Rajmahal-Garo lineament and eastern boundary is a NE-SW lineament separating the massif from the sediments of Bengal-Assam shelf. The southern parts of the Shillong plateau have experienced tremendous tectonic impacts in the past giving rise to the steep scarp which today stands aloft facing the Sylhet plain of Bangladesh.

Soil

The soils of Meghalaya have been broadly grouped into four categories *viz.,* (i) red loamy soil, (ii) lateritic soil, (iii) red and yellow soil and (iv) alluvial soil. The red loam soil is found in the central part of the Garo Hills and on the uplands of central and eastern Meghalaya. The lateritic soil occurs in the form of a broad belt extending from west to east in the northern part of the state. Red and yellow soil extends from west to east in the southern part. Alluvial soil occurs all along the northern, western and southern fringes of the state (Tripathi *et al* 1996). The soils are acidic (pH 3 - 6.5) in nature. Total Kjeldahl nitrogen content is low (*ca.* 0.15%) at degraded sites and fairly high (*ca.* 0.94%) in the undisturbed forests. Soils are deficient in available phosphorus (15 mg g^{-1}) and medium in available potassium (13 mg g^{-1}), and rich in organic carbon (1.62 %). Soils are not suitable for intensive cultivation due to their poor base (35%) saturation (Singh 1996).

Climate

The climate of the state is monsoonic. The climatic variables like temperature, rainfall and humidity vary widely from place to place in the state due to wide variation in topography. Based on the atmospheric conditions, the year may be divided into summer (March to May), rainy (May to September), autumn (October to November) and winter (December to February) seasons. Three fourth of the total annual rainfall is received during rainy season. Morning fog and frost, and dry weather are the characteristic features of winter season.

The annual rainfall varies widely in the state ranging from 1,600 mm to 11,463 mm. The Ri-bhoi district receives the lowest rainfall, while the northern part of Jaintia hills, Khasi hills and whole Garo hills receive between 1,600 and 4,000 mm rainfall annually where altitude ranges between 600 m to 1400 m asl. At higher altitude in southern Khasi hills annual rainfall varies widely from 4,000 mm to 11,000 mm in different areas. The region receives rain both from northeast and southwest monsoons. The mean temperature ranges between 5^0C and 28^0C.

Vegetation

Meghalaya comprising of the Khasi, Jaintia and Garo hills is endowed with rich natural vegetation, which varies greatly according to the altitude of different areas; ranging from the tropical, sub-tropical to warm temperate types of forest. The state harbours about 3128 species of flowering plants including 1237 endemic species (Khan *et al* 1997). The flora of the state is therefore extremely rich. The rich floristic diversity of the state is threatened by shifting cultivation locally called *jhum,* over exploitation of plants of economic importance, encroachment in forested areas for the developmental activities and settled agriculture.

TRADITIONAL AGROFORESTRY SYSTEMS

Over the years, farmers have integrated trees + crop components for their sustenance. Hill farming system has a variety of constraints compared to large and more commercially oriented farmers. Minimizing risk in food production, lack of cash for purchase of farm inputs, non-availability of labour for intensive farming and basic day-to-day household necessities such as small timber and fuel wood etc. are some of the typical constraints in hill farming systems in Meghalaya. The farmers of the state being aware of these problems have been growing crops along with the trees on hill slopes for time immemorial. Such practice is followed in almost all parts of the state with little variation here and there depending on their requirements and climatic and physiographic features of the area. The diversity of soil type, altitude and climate of the state thus provides ample opportunity of growing a variety of crops ranging from cereals to fruits of temperate and tropical climate.

Several traditional farming systems practiced in northeast India are more or less sustainable and maintain soil fertility over the years. These farming system are reasonably high yielding and socially acceptable as these have been evolved by ingenuity and skill of the local farmers. The proportion of space occupied by the tree and crop components in the traditional system largely depends upon factors like soil types, plant species and crop arrangements.

The survey reveals that agroforestry practices are popular in different agroclimatic zones in Meghalaya (Table 21.1). It was found that farmers cultivate large number of crops with trees in different mixtures on the same unit of land. These traditional systems practiced in the state may be grouped under agrisilviculture (tree + crop), agrihorticulture (fruit tree + crop), silvihorticulture (tree + fruit), pastoralsilviculture (tree + grasses) and pastoralhorticulture (fruit + grasses). Agrisilviculture systems are mostly confined to lower altitude (<700m asl) in the state, where paddy, ginger, colocasia, chilies, sweet potato are grown with *Schima, Acacia, Citrus* and *Artocarpus* tree species. At mid altitudinal range (800-1500 m asl), fruit trees are planted at regular spacing and timber trees are left on the bunds. Paddy, ginger, turmeric, maize and pumpkin are grown in large area in the inter -

spaces of tree species like *Michelia* sp., *Pinus kesiya, Schima wallichii, Erythrina* sp. Grasses like *Thysanolaena maxima* (broom) and *Setaria* are also grown at this altitude. Khasi pine is the most dominant tree species at the higher altitude range (>1400m asl), and agricultural crops like paddy, ginger and maize are grown in its association. Leafy and rhizomatous type crops such as mustard leaf, cabbage, colocasia, ginger, turmeric are grown in the inter spaces of horticultural trees such as *Pyrus, Psidium* and *Citrus* species. Chauhan and Dhyani (1989) have recommended various economically important tree species which are suitable for various agroclimatic zone of Meghalaya (Table 21.2).

Prominent traditional cropping systems

Chauhan and Dhyani (1989) have carried out diagnostic survey of the existing agroforestry system of Meghalaya and identified systems of high production potential which may be encouraged for settled cultivation on hill slopes (Table 21.3).

Replenishing soil fertility

The tribes of the state have evolved various methods to maintain the fertility of soils after crop harvest. Among them application of FarmYard Manure (FYM) is the most important and commonly practiced method through out the state in all the agroclimatic zones. FYM is used for both the crops grown in irrigated agriculture as well as rainfed conditions. Cow dung is the most commonly used manure. The dung is dried in the sun and spread in the field after ploughing and before crop sowing. For crops like maize, beans, cow dung are applied in the pits. To enrich the soil of the agroforestry systems crop residues are often dried and burned and at some places it is left for natural decomposition for slow release of nutrients.

At high altitude the farmers are dependent on poor quality organic resources such as needle of *Pinus kesiya* that form the dominant forest type in this zone. Pine needle decomposes at a slower rate than many of the broad-leaved species. The degradation of primary broad-leaved forest due to shifting cultivation and its replacement by pine forest has created a situation of shortage of rich organic resource of broad leaved trees. This is one of the reasons of short jhum cycle. Mixed cropping with leguminous crops, crop rotations and burning of crop residues in the field are some of the important practices adopted by the farmers for maintaining fertility of the soils.

IMPROVEMENT OF EXISTING AGROFORESTRY SYSTEMS

The analysis of different agroforestry systems in the state of Meghalaya suggests that further improvement of the existing systems may yield high productivity and make them more sustainable. Alley cropping can be introduced on steep hill slopes (>40% slope) to conserve soil and increase production. Hedgerows of broom grass (*Thysanolaena maxima*), on contour bunds along with plantation of nitrogen fixing trees like *Alnus nepalensis, Albizia lebbek, Parkia roxburghii* at a distance of 5 m can be adopted and the alleys can be utilized for growing food crops. Below 800m arecanut could be introduced across the slope at 3m x 3m spacing. Generally no inter-crop is taken with Khasi mandarin. Some crops like maize, beans, peas etc. may be grown with mandarin up to 3rd year, thereafter ginger, turmeric and pineapple could be introduced, which may grow under partial shade without adversely affecting the mandarin. Traditional varieties of different crops having low production potential should be replaced with improved varieties applying recommended doses of FYM, lime and fertilizers etc. to increase production per unit area.

Based on the research work conducted in three improved agroforestry systems (Khasi mandarin, Alder and Khasi pine) at the ICAR research complex for NEH region, Pandey (2004) reported maximum monetary benefit in ginger plus Khasi mandarin based system. Ginger also performed fairly satisfactory with Alder and Khasi pine based systems. Paddy-khasi pine system was most uneconomical (Table 21.4). Improvement in the traditional agroforestry systems could help the farmers to increase crop productivity in the hilly areas.

CONCLUSIONS

In recent years there has been considerable emphasis on agroforestry all over the world since it aims at simultaneous production of food and wood and at the same time minimize the risk of crop failure during drought years. Appropriate agroforestry systems improve soil physical properties, maintains soil organic matter content and promote nutrient cycling. However, suitable agroforestry models for different agroclimatic zones of India are yet to be evolved and tested.

In the north - eastern hill region of Indian forest ecosystems have been undergoing fast degradation during the last few decades owing to deforestation caused by shifting cultivation and other human disturbances. Therefore, there is a need for developing sustainable food production systems in the hills of northeastern region. Indigenous tribes of this region have been growing crops and trees together in different combinations since time immemorial. In such traditional systems, a great deal of variability in crop diversity, crop composition and crop rotation is observed depending on the area and needs of the farmers. Such diversity is primarily because of differences in topography, climate and soil conditions. The needs of the farmers also influence the composition of mixed cropping. A number of multipurpose tree species are grown with cereals, vegetables and other cash crops. Recent studies have reported that soil fertility under tree is improved by increased input of organic matter through litter (Dhyani 1997, Pandey 2004).

Table 21.1. Traditional agroforestry systems/ practices in different agroclimatic zones of Meghalaya

Agro-climatic zone/ distribution	Trees	Crops	Farmers preference
Temperate sub-alpine zone (1500-2000 m asl)	*Pinus kesiya, Schima wallichii, Bambusa pallida, Psidium guojova, Citrus* sp., *Pyrus communis, Prunus* sp.	Paddy, Ginger, Colocasia, Chilli, Dioscoria, Pumpkin, Sweet potato Maize, Bottle-gourd, Betel vine, Pineapple	*Pinus* + ginger, cabbage, Cauliflower and Sweet potato *Schima/Pinus* + ginger, Sweet potato and bean, *Prunus* + cabbage, maize, mustred leaf
Sub-tropical hill and plain zone (800-1500 m asl)	*Citrus reticulata,* *Michelia oblonga* *Pinus kesiya* *Schima wallichii* Banana *Erythrina indica* *Bambusa pellida*	Paddy, Ginger, Maize, Turmeric Colocasia, Chilli, Dioscoria, Pumpkin, Sweet potato, Broom grass, *Setaria*, Lady's finger, Sweet potato,	*Citrus* + maize, bean, cauliflower; *Schima* + pineapple, maize, ginger
Mild-tropical hill and plain zone (200-800)	*Artocarpus hetrophyllus, Litchi chinensis, Psidium guojava, Banana and Areca*	Paddy, Maize, Cabbage, Cauliflower, Mustard leaf, Potato	*Artocarpus*/banana + pine apple; Arecanut + beetle vine, pineapple, black pepper

Table 21.2. Recommended important trees for different agroclimatic zone of Meghalaya

Agro-climatic zone	Timber wood	Fruits	Multipurpose
Temperate sub-alpine zone	*Pinus kesiya, Cryptomeria japonica, Kydia calycina, Michelia champaka, Betula alnoides*	Pear, Plum, Peach, Assam lemon, Appricot	*Exbucklandia populnea, Alnus nepalensis, Bauhinia purpurea, Erythrina indica, Myrica esculenta*
Sub-tropical hill and plain zone	*Pinus kesiya, Vitex* spp., *Michelia champaka, Schima wallichii, Mesua ferrea*	Guava, Cherry, Khasi mandarin, Pears, Assam lemon, Banana, Papaya, Amla	*Alnus nepalensis, Acacia auriculiformis, Erythrina indica, Albizzia* spp., *Ficus* spp., *Morus alba, Indigofera* spp., *Babusa pallida, Cinnamomum tamala, Myrica esculenta*
Mild-tropical hill and plain zone	*Schima wallichii, Tetramelis nudiflora, Terminalia myriocarpa, Albizzia* spp., *Shorea robusta, Tectona grandis*	Coconut, Jackfruit, Arecanut, Banana, Litchi, Guava, Cashewnut, Palm	*Ficus* spp., *Casia fistula, Grewia* spp., *Bauhinia* spp., *Albizzia* lebbeck, *Cinnamomum tamala, Dendrocalanus hamiltonii, Bambusa tulda, Leucaena leucocephala, Havea brasilensis*

Table 21.3. Estimated annual returns of different components of certain important agroforestry systems from Meghalaya (after Chauhan and Dhyani, 1989)

Tree components and density ha^{-1}		Crop components and density ha^{-1}		Yield ((t/ha)		Net income (tree + crop) components (Rs/ha)
				Tree	Crop	
Arecanut	1000	Pineapple	40,000	1.8	20	24,000
Mandarin	400	Pineapple	40,000	4.8	20	29,000
Arecanut	1000	Blackpepper	1,000	1.8	4.8	36,500
Jack fruit	250	Pineapple	40,000	12.8	20	20,000
Arecanut	1000	Betel vine	1,000	1.8	0.8	18,000
Mandarin	400	Ginger/turmeric	40,000	4.8	4	25,500
Arecanut	1,000	Betelvine + pineapple	1,000 + 20,000	1.8	0.8 +10	24,250
Arecanut	1000	Blackpepper+ pineapple	1,000 +20,000	1.8	1+10	42,750

Table 21.4. Estimated monetary benefits of different crops in association with three developed agroforestry systems.

Tree components	Density ha^{-1}	Crop components	Density m^{-2}	Yield (qt/ha)	Net income (Rs/ha)
Alder	416	Maize	6	35	13,500
		Rice	10	18.6	7,300
		Ginger	4	6.8	73,600
Khasi Mandarin	800	Maize	6	56	26,100
		Rice	10	34	19,700
		Ginger	4	80	88,000
Khasi pine	416	Maize	6	17.8	3,000
		Rice	10	5	(-) 3,500
		Ginger	4	54	56,800

Source: Pandey (2004)

The declining productivity in the traditional method of cultivation *i.e.,* shifting cultivation with time and its negative impact on the environment is a major issues of concern (Nair 1985). The traditional practice of agroforestry in the state includes selected woody perennials, which diversify the production base, enhance productivity and add to the sustainability of the systems (Chauhan *et al.* 1993). The crop yield under different agroforestry systems is affected by competition between trees and crop plants for water and nutrients and at the same time reduction in light intensity due to over head tree canopy may also cause reduction in crop growth. Besides shading effect, the similarity in fine root distribution pattern of trees and annual crops in the soil profile may result in over lapping of root activity zones of the two components leading to sharpened competition between them (Scroth and Lehman 1995). It seems that the farmers are aware of these facts; therefore, tree pruning is done in most of the agroforestry systems before crop sowing. In traditional agroforestry systems selection of tree species is most important, and therefore, the farmers prefer deep rooted species to reduce root competition between trees and crops for nutrients and water.

ACKNOWLEDGEMENTS

The authors are thankful to the Director, ICAR Research Complex NEH Region, Barapani and to Indian Council of Agricultural Research, New Delhi for financial assistance in the form of a research project (sanction letter No. 5(8)/98-SW&DF).

REFERENCES

Anon (1974). Geology and mineral resources of the states of India. GSI, Miscellaneous publication. No. 30, pt. III, pp. 124.

Beer, J., Bonnemann, A., Chavez, W., Fassbender, H.W., Imbach, AC and Martel, I. (1990). Modelling agroforestry syatem of Cacao with Cordia alliodora on Erythrina poeppigiana in Costa Rica. V. Productivity indices, organic material, models and sustainability over ten years. *Agroforestry Systems*, 12: 229-249.

Chauhan, D.S. and Dhyani S.K. (1989). Production potential of promising Agro-forestry Practices in Meghalaya. *Indian Journal of Hill Farming* 2: 31-37.

Chauhan, D.S. and Dhyani, S.K. (1990). Traditional agroforestry practices in northeastern Himalayan region of India. *Indian Jorunal of Dry. Agr. Res. & Dev.*, 4 :73-81.

Chauhan, D.S., Dhyani, S.K. and Desai, A.R. (1993). Productivity potential of grasses in association with Alnus nepalensis and pineapple under silvihortipastoral system on agroforestry in Meghalaya. *Indian Jorunal of Dry. Agr. Res. & Dev.*, 8 : 60-64.

Dhyani, S.K. (1997). Analysis of the effect of agroforestry systems on productivity and soil characteristics. *Ph.D. Thesis*, NEHU, Shillong.

Dhyani, S.K. and Tripathi, R.S. (1999). Tree growth and crop yield under agriculture practices in northeastern India. *Agroforestry Systems*, 44: 1-12.

Fisher, R.F. (1990). Amelioration of soils by trees. *In*: Sustained productivity of forest soils. (S.P. Gassel *et al* eds.) pp. 290-300. Proc. N. Am. For. Soil Conf. 7th Vancouver, B.C., Canada.

Khan, M.L., Shaily, M and Bawa, K.S. (1997) Effectiveness of the protected area network in biodiversity conservation, a case study of Meghalaya state. *Biodiversity and Conservation* 6: 853-868.

Leakey, R. (1996). Definition of agroforestry revisited. *Agroforestry Today*, 8: 5-7.

Murthy, M. N. V., Mazumdar, S.K. and Bhaumic, N. (1976). Significance of tectonic trends in the geological evolution of the Meghalaya upland since the Pre-Cambrian. GSI, Miscellaneous publication. No. 23, pp. 471-484.

Nair, P.K.R. (1985). Classification of agroforestry system. *Agroforestry Systems*, 3: 97-128.

Pandey, H.N. (2004). Ecological analysis of selected agroforestry systems in Meghalaya. *Final Technical Report*. North Eastern Hill University, Shillong.

Rai, R.K. (1986). Evidences of rejuvenation of the Deccan foreland, India with particular reference to the Meghalaya Plateau. *In*: International Geomorphology (Gardiner, V. ed.), Part-II. John Wiley & Sons Ltd. Pp. 255-266.

Scroth, G and Lehman, J. (1995). Contrasting effects of roots and mulch from three agroforestry tree species on yields of alley cropped maize. *Agriculture Ecosystems and Environments*, 54:89-101.

Singh, S. (1996). *Technical report on delineation of geo-ecological zones in Meghalaya*. Report No. GE/UGC/SR (1)/SS. Department of Geography, North-Eastern Hill University, Shillong.

Tripathi, R.S., Tiwari, B.K. and Pandey, H.N. (1996). State of Environment of Meghalaya. North Eastern Hill University, Shillong.

22

Agroforestry Systems and Practices Prevailing in Mizoram

U. K. SAHOO
Department of Forestry, Mizoram University
Post Box: 190, Aizawl - 796 001 (Mizoram)

ABSTRACT

Mizoram is predominantly a hilly state where people have been practicing agroforestry in a variety of forms from traditional shifting (jhum) cultivation to the most scientific-induced agroforestry models. However, 'jhum' still remains the main form of land use even though it is known to cause severe soil, water and biodiversity erosion and low crop productivity. The induced agroforestry practices carried out in Miozram have the ability to replace shifting cultivation but these systems again are not properly designed to tackle various biophysical, socio-cultural and economic need of the people, resulting in their poor acceptability. The paper reviews various agroforestry systems prevailing in Mizoram and details some ongoing agroforestry research trials in the State.

INTRODUCTION

Agroforestry is the collective name for land use systems where woody perennials and other crops and animals are deliberately used on the same unit of land in some temporal and spatial arrangement for optimum crop and land productivity. This practice although has been in existence in Mizoram like other places, since time immemorial, a scientific look to the system, however, has received attention in recent years. The agroforestry research has got a boost up after the establishment of ICAR-NEH region, however, the pace of research in the state has been very slow compared to other sister states due to its relative seclusion with rest of N.E states and separation from the Indian mainland by Bangladesh, leaving it connected by a narrow strip of land, to which mountains, hills, plateaus, plains and river system adds an element of complexity for communication, survey and research. Besides, nearly 80 per cent of the land is above 30 per cent slope, further a major chunk of the land is under steep to very steep slope hills limiting cultivation or not suitable for agricultural crops. The area affected annually by 'jhum' (local name for shifting or slash and burn agriculture) is over 4000 km^2 causing the land almost devoid of tree growth (except some bamboo and scrub growth) due to repeated burning, thus posing a threat to the growing timber needs of the people. The growing stock of state owned forest is a mere 35.3cu.m per ha (as per Forest Survey of India, 1989 assessment) which again is not sufficient to meet the ever increasing demand of timber, food, fuel-wood, fodder and other non-timber forest product requirement. Agroforestry, has been slowly developing and accepted by farmers not only to yield sustainable production of 3F (food, fuel-wood, fodder) but also land productivity, and generating adequate employment opportunities and sustained income for them by creating permanent assets in the form of tree-crop orchard. Physiographically, the state encompasses a wide variation in altitude, climate, soil and rainfall resulting in varying cropping pattern, cultivation type and tree-crop combinations. An attempt has been made in this paper to reviews various agroforestry systems prevailing in Mizoram and details some ongoing agroforestry research trials.

AGRO-ECOLOGICAL ZONES OF MIZORAM

Mizoram is one of the frontier state of Indian Union flanked by Myanmar in East and South, Assam and Manipur in the North and Bangladesh in the West. It lies between 21° 58′ and 24° 35′ North latitude and 92° 15′ East longitude. Geographically the state occupies an area of 21081 km^2. Nearly 6000 km^2 of the land area is under 'jhum' (slash and burn or shifting cultivation) cultivation reckoning to 28.46 per cent of the total geographical area. The state is featured with young mountainous hilly, sharp crest, ridges, and deep gorges with narrow valleys and is characterized by monsoonal climate with an average annual rainfall of 2500 mm and 85 per cent of relative humidity. Most of the precipitation (over 80 %) occurs during May to September. The slope/gradient limits the cropping pattern in the state.

The state can broadly be divided into four agro-climatic zones (Fig. 22.1) such as:

(a) Humid temperate sub-alpine zone having altitudinal range from1600-2360 m having crop growing period of 200-250 days, rainfall ranges from 1800-2000 mm.

(b) Humid subtropical Hill zone having altitudinal range from 1000-1800 mm, having crop growing period of 250-300 days, rainfall ranges from 2000-2700 mm.

(c) Humid mild-tropical Hill zone having altitudinal range from 400-1000 m having crop growing period throughout the year, rainfall from 2000-2500 mm.

(d) Humid mild tropical zone having altitudinal range from 40-400 m having crop growing period throughout the year, rainfall from 2000-2200 mm.

The hill ranges which are parallel to each other are generally oriented north to south having an increase elevation towards the east. In a given agro-climatic zone, the altitudes are higher in the central part and taper in the north and south. The elevation difference between the hilltop and valley floor is about 100-200m and sometimes 200-600m and normally the western aspects of hills are steeper than the eastern. The natural vegetation of the state differs markedly between zones, for example, humid temperate sub-alpine zone represent subtropical evergreen forest composing *Dipterocarpus turbinatus, Cinnamomum* spp., *Betula* spp., *Quercus* spp. *Chikrasia tubularis, Bischofia javanica, Castanopsis* spp., *Schima wallichii, Terminalia myriocarpa, Michelia champaca* etc. In the eastern region, north of Champhai, close to Mynamar border *Pinus insularis* is found in association with *Schima wallichii, Kayea assamica, Castonopsis* spp., *Myrica negi* and *Rhododendron* spp (Singh, 1996) while the agricultural food crops grown are paddy, maize, brinjal, chillies, cucumber, bitter gourd, pumpkin of which maize and paddy are the principal rainfed crops. The humid subtropical hill zone exhibit mostly pine forest comprising *Quercus dealbeta, Q. incana, Castonopsis* spp., *Schima wallichii, Prunus cerasoides,* while the important agricultural crops include paddy, maize, tapioca, sweet potato, potato, beans, mustard, chillies, turmeric etc. The humid mild-tropical hill zone characterized by tropical evergreen forest comprising *Duabanga sonneratiodes, Artocarpus chaplasha, Amoora* spp., *Sterculia* spp., *Mangifera indica, Tetrameles nudiflora, Podocarpus nerrfolia, Trewia nudiflora, Skeria assamica, Pterospermus* spp., *Mesua ferrea, Dipterocarpus tubinatus, D. macrocarpus* etc. and the agricultural crops widely prevalent are sugar cane, maize, ginger, turmeric, paddy, brinjal. The humid mild tropical zone comprise mostly the tropical deciduous forest type composing *Gmelina arborea, Caricarpa* spp., *Linnea grandis, Lagerstomia flos-regina, Vitex penducularis, Bombax ceiba, Morus alba, Albizzia lebbek, Emblica officinalis, Erythrina indica, Alstonia scholaris, Macaranga, Bauhnia* spp., *Bischofia javanica, Terminalia chebula* etc and the major agricultural crops include paddy, maize, mustard, cucumber, pumpkin, bitter gourd, chillies etc. Besides, the bamboo brakes differing widely in character are found scattered along with the evergreen trees or deciduous secondary stands, throughout the state but their distribution is restricted up to 1600m altitude. The muli bamboo (*Melocanna baccifera*) is the most dominant species found in these forests. The other type of vegetation in the state include the wide spread jhum fallows (current, old and abandoned) predominantly found in the eastern part of the State. The common species inhabiting the area are *Arundinaria* sp., *Eupatorium* sp., *Mikania micrantha* etc. intermingled with few other bamboo species.

AGROFORESTRY SYSTEMS

The common agroforestry systems prevalent in Mizoram can be grouped under two tier system. The first tier consists of two groups as traditional and induced practices, while the second tier refers to their location in agro-climatic zones whether on hills or plains and on altitudinal and rainfall gradients.

TRADITIONAL AGROFORESTRY PRACTICES

Shifting cultivation

It is the commonest form of cultivation among the different tribes of Mizoram and locally known as 'Lo' or 'jhum', practiced since time immemorial. The system involves slashing the vegetation, burning the dried slash, raising a mixture of crops on a temporarily nutrient enriched soil for a year or two, fallowing the plot for re-growth of natural vegetation and eventually returning to the same plot for cropping after few years. In the past, the 'jhum' cycle was as long as 15-20 years, but due to population pressure and reduced acreage of available land, the jhum cycle has now been reduced to merely 5-6 years and even in some places to 2-3 years resulting in serious soil erosion, declining soil fertility and crop productivity. This practice although is predominant in the eastern part of the state but is practiced almost everywhere irrespective of the agroclimatic zones. A wide variety of crops are grown in the system sequentially mostly by dibbling method. The system involves no or low technical inputs and is advantageous in the sense that it provides production of certain crops even if some crops get damaged and therefore ensure some kind of economic security. However, there is a growing awareness among public to shift from this system but as yet no suitable agroforestry system is found as alternative to be totally acceptable to the people.

Home gardens

Home gardens (locally known as 'Chuktuah huan') are very common and age old practice in every village irrespective of the agroclimatic zone. These gardens are intended primarily for household consumption and there is a close association of woody perennial crops and invariably livestock within the compound of individual house, with the whole crop-tree-animal unit being managed by family labour. These are multi strata systems usually 3-4 tiers having enormous species diversity with different life forms (Table 22.1). The first layer is about 2m height from the ground, composed of vegetables, tuber crops, pine apple and grasses. The second layer is mostly occupied by papaya, banana, tapioca, yam, while the third layer is occupied by *Parkia roxburghii*, jack fruits, different bamboo species. In the garden, vegetables are generally grown in areas adjacent to the kitchens while the multipurpose tree species are planted on the boundaries or in other places depending on the farmers intuitive knowledge. There is no specific time period of the year for tree planting in the garden. However, in recent years, due to frequent public campaign by non-governmental agencies, State forest departments, the farmers have developed a tendency to plant valuable trees on the available garden space on Green Mizoram Day. The family also raises few animals like cows, goats, pigs and birds like chicken and ducks in an integrated manner for whom the garden support the feeding base. These animals are invariably raised for meat and egg production to be domestically consumed partly and in part to be sold in the market. The objective of animal integration into the home gardens also find in the utility of animal by-products like dung and urine in raising crop production.

Banana based agroforestry

This is a very common system prevailed in most villages. The most important and known villages for this practice are Khumtung and Baktawng, the people here believe that planting banana can withstand all sorts of natural calamities, besides being providing high economic returns. Although banana predominate the system, some other horticulture crops like lemon, orange and leafy vegetables like 'lai patta' (wild radish) are also grown. The system provides good cash from banana besides providing some measures towards splash erosion.

Silvihortiagriculture system

In this system, teak is main tree component which is planted at 10 m or 20 m apart with 2 m between plants. Field crops (paddy) are grown in between two teak plants. Fodder is planted on either side of teak row as soil conservative measure and to obtain green fodder. The combination may vary but the most successful model prevalent consists of Subabul or Pigeon pea + Teak + Lemon + Paddy. This system was very popular in the recent past because of assured high return from teak but now due to lesser demand of teak coupled with its high transportation cost, the farmers are not very much motivated to grow teak and adopt this system. Besides, the umbrella type large leaves are reported to cause splash erosion, further de-motivating the teak growers to adopt the system.

Silviagriculture system

This system has features of cultivation of tung, an oil yielding tree (*Alurites fordii* and *A. montane*) planted ideally at 5m x 5m spacing, sometime at 10 m apart where maize and paddy (upland variety) are grown as agricultural components. *A. fordii* is cultivated in the tropical hill zones while *A. montane* is practiced in the subtropical and sub-alpine zones along with other crops like french beans, chillies, ginger, turmeric, tapioca are grown during the first 3-4 years. This practice is well adopted in Aizawl and Lunglei districts of the state. Tung oil is used as an ingredient for waterproofing mansions for protective coating, enamels and electrical insulators, besides in leather dressing and in manufacturing of local soaps, therefore has more demand as a suitable agroforestry species in degraded lands.

Hortisilivicultural system

In this system, pine apple, orange, guava, mango are grown in the steep hills along with teak, *Bauhinia spp.*, very often subabul is incorporated with horticrops in different spacial pattern in a view to enrich nutrient content in the soil through nitrogen fixation. The system yields fruits from 5th year whereas, fodder and fuel wood from subabul are available from 3rd year onwards.

Alley or Hedgerow intercropping

This system is practiced by farmers of all hill zones where mostly nitrogen fixing hedges viz., subabul and sometimes pigeon pea are grown along the contour lines especially in the paddy field.

INDUCED AGROFORESTRY PRACTICES

Organic farming

This is a relatively new system of cultivation designed by the State Directorate of Agriculture for demonstration purpose in which the herbaceous and semi woody vegetation including bamboos are often felled and left un-burnt for nearly a year, however, the large trees (timber and fruit species) are left as such. Contour trenches are dug so as to conserve soil from run off by rain water. In the following year during the month of March - April, the debris and stack are removed so as to make room for cultivation. Simultaneously paddy, maize and other cash crops like sunflower, soya bean and ground nut are cultivated. No chemical fertilizer and farmyard manure are applied to enrich the crop. Occasionally, however, green leaves of subabul, pigeon pea and paddy straw are mulched when ginger and turmeric are grown in the system. This system is observed to be an improvement over the shifting cultivation in the sense that it reduces soil erosion and improve soil moisture and land productivity. Further, the system is some what like 'jhum' (except burning process) involving existing tradition and culture and thus has the entities of social acceptance. The fast growing and biomass yielding tree species (Sahoo and Jha, 2000) and profitable legumes should be tried in order to make the system economically viable for wider acceptance (Table 22.2).

Contour trench farming

This is one of the permanent farming systems introduced recently and is found to be quite promising on the hill slope. In this system, trenches (1½ feet wide and 1 foot deep) are dug along the contour across the slope. Grasses (preferably the fodder grasses) and sometimes, cowpea or stylo grasses are planted to bind the soil on lower edge of the trench line and also to support this, a row of leguminous species (preferably subabul or pigeon pea) are planted as hedges. In some places, a row of banana or sugarcane or pineapple is also planted. The trenches are utilized to dump the weed and the decayed biomass can be used as compost manure to the crops. But the system is yet to gain its popularity owing to its relatively high cost involved compared to other systems (Table 22.2).

Terrace farming

This is the permanent farming system adopted in all climatic zones but with a lesser degree by all tribes. Gamar and banana are mostly grown on the edge of terrace (bench terrace) and sometimes, pine apple is practiced. In the terraced land, numerous crops like rice, maize, mustard, cabbage, potato, tomato, chilies etc. are grown. Since terrace cutting is a costly affair, the system is not well adopted.

SALT and SWEET models

The SALT (Sloping Agricultural Land Technology) and SWEET (Sloping Watershed Engineering Technology) are two modified models of Agroforestry systems introduced in the state in recent past, still on trial by the state Directorate of Agriculture, the former is demonstrated in the subtropical and sub-alpine zone while the latter in the tropical hill zone. Both the systems have components of contour hedgerow intercropping. In the SALT, nitrogen fixing plants such as subabul (*Leucaena leucocephala* and *L. diversifolia*), Pigeon pea (*Cajanus cajan*) are used as hedge rows as barriers against soil loss from the hill slope and to retain soil moisture. The hedges are periodically pruned and the pruned leaves, twigs and branches are used as mulches to suffice the process of soil and water conservation and to build soil fertility in due course of time. In SWEET models, however, a number of other species such as tapioca, Assam lemon, bamboo in addition to subabul and pigeon pea are tried. The strips (spaces between two hedgerows) are used for cultivation of rice, maize, ground nut, millet and other desirable horticultural crops like orange, lemon and pine apple and of some wood values like gamar and neem. Although the principle of both the models is same, the SWEET models are introduced more in the areas where the elevation difference between valley and hills exceeds 200-300 m and above (Table 22.2).

Bamboo/Ratan based agroforestry

The state is known for its rich bamboo diversity and cover. The muli bamboo (*Melocanna bacifera*) is the predominant bamboo species which cover more than 90 per cent bamboo cover. Its rhizome is very popular for its delicacy. The bamboo species are tried for their fast growth and easy propagation, no management and versatile use. The models that have been successfully practiced are Bamboo-Ginger, Bamboo-Turmeric. Bamboo-Soya bean, Bamboo-Mustard etc. Bamboo conserves soil against erosion, creates a large shade under its foliage reducing evapo-transpiration loss. As more soil moisture exists under and around bamboo groves, a congenial atmosphere is created for microbial activities resulting better nutrient turn over and crop production. The ratan species tried in some systems include *Calamus spp.*

Coffee based agroforestry

Coffee based agroforestry system are found in Bualpui, Kolasib District of Mizoram on mild to steep hill slopes and established by the State soil conservation department. In this system the existing tree growths were utilized as shade and coffee were grown successfully. The shade trees in use include *Gmelina arborea, Schima wallichi, Teminalia* sp. *and Michelia champaca, Toona ciliata, Erythrina indica, Bauhinia variegata, Melia azedarach* etc. Recently, the farmers have started cultivating coffee in patches and the number of enterprising farmers is gradually increasing. The system has proved to be crucial in preventing soil erosion and increasing land productivity.

IMPACT OF AGROFORESTRY SYSTEMS ON SOIL FERTILITY AND MOISTURE CONSERVATION

The introduction of various multipurpose tree species either at different spacing or as alley cropping has shown positive impact on soil fertility over different years (Table 22.3). Under the tree of *Michelia champaca* canopies, soil organic carbon was enhanced by 40.0-50.0% and available nitrogen by 6.0-

14.0% over control in 0-20 cm soil during a 6-yr period. The respective values under *Toona ciliata* were 20.0-30.0% and 7.5-11.6% respectively. However, under *Leucaena leucocephala*, there was significantly higher fertility build up when maize was intercropped (Table 22.3), besides it resulted in better growth and productivity of paddy (Mohan *et al.*, 2005) and maize (Sahoo *et al.*, 2005). The fertility improvement was low in sole cropping land use and in neem-turmeric and neem-ginger systems and higher in subabul based systems. This could well be attributed to the leguminous nature of the latter while ginger and turmeric are highly nutrient depleting crops, therefore, the systems in which they were the crop components fail to build up nutrients. During a 15-20 year tung cultivation, the system added significant amount of organic carbon and in available N, P, K. The effect of different tree-crop combinations such as Alnus-ginger, Alnus-turmeric, neem-ginger, neem-turmeric etc. on soil fertility changes could not be judged due to a very short duration studies. In all the systems, introduction of tree components resulted in a marginal decrease in soil pH. Different mulches were also tried for conserving soil moisture. Among the various mulches, subabul mulch retained more soil moisture than paddy straw or bamboo stakes (Mohan *et al.*, 2004). In another study with different tree species, we observed the order of moisture retention capacity as subabul>paddy straw>weeds>polythene. Similarly, 8 tonnes of mulches/ha in general provided better yield of ginger and turmeric than 6 and 10 tonnes of mulches/ha irrespective of mulch type. Among the various tree species tried, plots having Alnus showed a tendency for more soil moisture retention and crop yield although no significant variation in either moisture or yield was noticed during this short period of 2-years research.

PROJECTED ECONOMICS OF SOME AGROFORESTRY SYSTEMS OF MIZORAM

The cost of agroforestry model is an important aspect for its acceptability. Generally, Agroforestry gives more income to the farmers than pure agriculture or forestry. Our projection highlights that the small farmers can maximize return from multistoried cropping where multipurpose tree species like subabul, bamboo etc. are intercropped with maize, soyabean and paddy. Similarly, bamboo (*Dendrocalamus strictus*) and *Bambusa tulda* are estimated to provide a gross return of Rs. 25000 per ha in ten years time. The cost benefit ratio in these system is 4.03 each (Table 22.4). The productivity and survival rate of plant species are worked out under optimal conditions (Rocky and Sahoo, 2002, Rocky and Sahoo, 2004, Rocky *et al.*, 2004). There are many tree species viz *Schima wallichii, Erythrina indica, Tectona grandis, Michelia champaca, Toona ciliata* and medicinal plants like citronella, pacholi etc. which may be tried in agroforestry systems for better return. The spices like large cardamom, black pepper which have shown very promising results elsewhere and which too have high demand in the local market could too be tried in the system. The bamboo (both edible and non-edible) can bring substantial income to the farmers. The tribal decision making, however, is mostly influenced by the short term gains and thus, the farmers prefer traditional practices over the induced agroforestry systems. Therefore, introduction of fast growing, remunerative tree and crop species offer viable proposition in providing return at short intervals. Since herbal medicinal and aromatic plants have a large market potential and the people here are highly NTFP-dependent, these plants could well be introduced into the agroforestry systems to make it more income oriented than the present subsistence farming.

FUTURE RESEARCH THRUSTS

The fragile hill ecosystem coupled with traditional jhum cultivation has been the dominating factor responsible for the low crop productivity and consequent low economic return in this hilly state. Therefore sustainable agroforestry research in Mizoram should focus on:

(a) Identification of traditional agroforestry systems which are more productivity oriented resulting in higher adaptativity by the farmers. The benefit cost ratio of the traditional agroforestry system should be worked out and compared with the scientific or induced agroforestry systems.

(b) An in-depth study on tree-crop compatibility (competition, allelopathy etc) from soil fertility, soil moisture and crop and land productivity point of view should be made to prescribe suitable models for different agro-climatic zones. Quantification of various competitive effects for natural resources be made for proper designing of Agroforestry systems and prescribing better management practices.

(c) Rehabilitation of degraded hills could be carried out using economic species like bamboo, cane, broom grass, various spices, condiments, medicinal and aromatic plants and orchids.

(d) Research on relative efficacy of different quality and quantity mulches to improve soil-water-plant relationship and build up soil fertility in the steep hills.

(e) There is a need to study growth performance of various multipurpose tree species and fuel wood characteristic of energy plantation to prescribe species to be grown in the agroforestry systems.

Table 22.1. Important trees, crops and other species grown in the homestead gardens of Mizoram

Scientific name	Common/Local Name	Family	Remark
Tree species			
Albizzia lebbek	Thingri	Mimosaceae	Rare
Aleurites fordii	Tung	Euphorbiaceae	Occasional
Artocarpus integrifolia	Lamkhuang	Moraceae	Common
Bambusa tulda	Mau	Bambusaceae	Common
Bauhnia purpurea	Vau-favang	Ceasalpinaceae	Common
Cinnamomum zeylanicum	Khiangzo	Lauraceae	Common
Dendrocalamus hamiltonii	Tursing	Bambusaceae	Common
Ficus cunea	Hmang	Moraceae	Occasional
Gmelina arborea	Thlanvawng	Verbenaceae	Common
Leucaena leucocephala	Japan zawntah	Mimosaceae	Common
Litchi chinensis	Theifeimung	Sapindaceae	Common
Mangifera indica	Theihai	Anacardiaceae	Common
Melia azadiratcha	Neem	Meliaceae	Common
Melocanna baccifera	Muli bamboo	Bambusaceae	Common
Mesua ferrae	Herhse	Guttiferae	Occasional
Michelia champaca	Ngiau	Mangoliaceae	Occasional
Morus alba	Thingtheihmu	Moraceae	Common
Parkia roxburghii	Zawngta	Mimosaceae	Common
Prunus cerasoides	Tlaizawng	Rosaceae	Common
Schima wallichii	Khiang	Theaceae	Common
Crops & vegetables			
Abelmoschus esculentus (L)	Lady's finger	Malvaceae	Common
Allium cepa L.	Onion/Purun sen	Amaryllidacea	Occassional
Allium sativum L.	Garlic/Purunvar	Amaryllidaceae	Occassional
Ananas comosus (L)	Pine apple	Bromrliaceae	Common.
Arachis hypogaea L.	Ground nut/ Badam	Papilionacea	Occassional
Blencine corocana (L)	Finger millet	Poaceae	Rare

Contd...

Table 22.1. Contd...

Brassica juncea (L) var. *cuneifolis*	Leaf mustard/ Antam	Brassicaceae	Common
Brassica oleracea var. *capitata*	Brocole	Brassicaceae	Occassional
Brassica oleracea L. var. *acephala*	Knol Knol/ Bulbawk	Brassicaceae	Occassional
Brassica oleraceae L. var. *acephala*	Cabbage/ Zikhlum	Brassicaceae	Occassional
Brassica rapa L.	Turnip/Bulbawk	Brassicacea	Occassional
Capsicum annuum L	Chilly/Hmarcha	Solanaceae	Common
Cucumis sativus L.	Cucumber	Cucurbitacea.	Common.
Cucurbita maxima Duch	Pumkin/Mai	Cucurbitaceae	Common
Curcuma longa L.	Turmeric/Aieng	Zingiberaceae	Common.
Dancus carota L.	Carrot	Umbelliferae	Occassional
Dioscoriea a lata L.	White yam/Bahra	Dioscoreacea	Common
Dolichos lablab L.	Broab bean/ Bepui	Papilionacea	Common
Glycine max (L)	Soyabean/Bekang	Papilionacea	Common
Hibiscus sabdarira L.	Anthur	Malvaceae	Common.
Ipomoea batata (L)	Sweet potato/ Kawlbahra	Convolvulaceae	Common
Luffa acutangula (L)	Ribguard	Cucurbitaceae	Common.
Luffa cylindrica (L)	Veg.Sponge	Cucurbitacea.	Common.
Lycopersicon esculentum	Tomato	Solanceae	Common
Manihot esculents	Tapioca/Pangbal	Euphorbiacea	Common
Momordica charantia L.	Bitter guard	Cucurbitaceae	Common.
Momordica cochichinensis	Maitamtawk	Cucirbitaceae	Common.
Nicotiana tabacum L.	Tobacco/Vaihlo	Solanaceae	Occssional
Oryza sativa L.	Rice/Buh	Poacea	Common
Zea mays L.	Maize/Vaimim	Poacea	Common
Phaseolus vulgaris L.	French bean	Papilionacea	Common
Raphanus sativus L.	Radish/Buluih	Brassicace	Common
Sorghum vulgare (L)	Sorghum/Faisa	Poaceae	Occassional
Vigna sinensis L.	Lobia/Bete	Papilionacea	Common
Zingiber officinale Rose.	Ginger/ Sawhthing	Zingiberaceae	Common.

Table 22.2. Few common agroforestry (induced) practices prevailing in Mizoram

Agroclimatic zone	Agroforestry practices	Status
Humid temperate sub-alpine	Tung with vegetables	Successful
	Teak with vegetables	Successful
	Subabul with rice	Successful
	Pigeon pea with rice	Successful
	Tapioca-Assam orange-zinger	on-trial
	Bamboo-soyabean	on-trial
	Contour trench farming	Successful
	Organic farming	Successful
Humid subtropical Hill	SALT	on-trial
	Subabul with rice	Successful
	Alnus -large Cardamom	Successful
	Erythrina-coffee	Successful
	Organic farming	Successful
	Mango-squash-french bean	Successful
Humid mild tropical Hill	Alnus -large Cardamom	Successful
	Tung-ginger, soyabean/lady's finger	Successful
	SALT	on-trial
	SWEET	on-trial
	Alnus-ginger/turmeric	on-trial
	Gamar-ginger/turmeric	on-trial
	Neem-ginger/turmeric	on-trial
	Erythrina-ginger/turmeric	on-trial
	Organic farming	Successful
	Contour trench farming	Successful
	Squash-french bean with different MPTs	Successful
Humid mild tropical	Erythrina with coffee and black pepper	Successful
	Khasi Mandarine with pine apple	Successful
	Broom grass-french bean/ other vegetables with tapioca	Successful
	Organic farming	Successful

Table 22.3. Change in soil properties (0-20 cm) under different land use/ agroforestry practices in Mizoram.

Agroforestry/ Landuse	Year of establishment	At the time of establishment*					In 2004				
		pH	OC (%)	N	P	K	pH	OC (%)	N	P	K
				(Kg/ha)					(Kg/ha)		
Sole paddy cropping	1997[1]	5.2	1.4	228.0	5.8	155.4	5.2	1.3	200.0	7.2	150.4
Sole squash cropping	1997[1]	4.8	1.46	256.4	12.5	108.8	4.8	1.4	197.4	11.5	120.0
Sole banana plantation	1997[1]	5.2	2.7	432.4	2.4	232.7	5.0	3.2	450.0	3.0	250.0
Paddy-pigeonpea cropping	1997[1]	5.1	1.6	311.7	2.2	142.3	4.9	2.0	325.0	3.5	165.0
Toona-subabul-maize	1999[2]	5.3	2.3	200.2	7.5	153.2	5.0	3.0	254.0	9.2	168.0
Toona-pigeonpea-maize	1999[2]	5.3	2.5	198.5	10.2	186.0	5.0	3.5	214.0	13.8	201.0
Michelia-subabul-maize	1999[2]	5.4	2.4	186.6	7.3	176.0	5.0	3.2	210.0	10.5	195.0
Michelia-pigeonpea-maize	1999[2]	5.4	2.4	198.0	7.3	182.0	5.0	2.5	220.0	13.5	200.0
Subabul-maize	1999[3]	5.5	1.2	186.6	17.05	102.7	5.0	2.5	248.0	15.0	117.5
Subabul-ginger	2000[4]	5.5	1.2	176.5	17.0	102.5	5.2	1.5	170.0	15.0	100.0
Tung-soyabean	2000[5]	5.62	1.3	188.0	18.06	127.24	5.0	2.0	230.0	14.0	136.0
Tung-soyabean	1985[6]	5.2	1.4	200.0	18.0	187.0	4.8	2.6	278.0	36.0	211.0
Tung-soyabean	1990[7]	5.3	1.3	185.0	19.55	188.32	4.8	2.2	265.0	22.0	205.0
Subabul-rice	2003[8]	5.2	1.3	190.0	21.40	182.2	5.1	1.4	199.0	20.0	187.0
Alnus-ginger	2003[9]	5.1	1.3	176.0	36.32	149.0	5.0	1.5	186.0	38.0	152.0
Alnus-turmeric	2003[9]	5.2	1.3	167.0	39.41	156.2	5.1	1.4	176.0	40.0	157.0
Erythrina-maize	2003[9]	5.3	1.2	186.2	39.50	176.0	5.0	1.6	201.0	46.0	192.0
Gamar-maize	2003[9]	5.3	1.2	180.0	39.50	159.12	5.0	1.6	196.0	48.5	174.0
Neem-maize	2003[9]	5.3	1.3	174.0	39.40	159.12	4.9	1.3	178.0	40.2	156.0
Neem-turmeric	2003[9]	5.2	1.2	165.0	39.40	155.19	5.1	1.3	170.0	35.0	157.0
Neem-ginger	2003[9]	5.1	1.2	165.0	39.42	151.78	5.0	1.2	160.0	35.0	157.0
Gamar-frenchbean	2003[9]	5.2	1.3	171.0	39.20	152.0	4.8	1.6	184.0	37.0	199.0

* [1] Annual Research Report, NWDPRA, Govt. of Mizoram, [2] Lalchhandami (1999), [3] Laldikpuia (1999), [4] Lalrinsangi (2000), [5] Lalremruata (2001), [6,7] Sailo (2003), [8] Mohan etal. (2005), [9] ongoing icar project

Table 22.4. Projected Economics of different Agroforestry models of Mizoram

Tree	Crop	Years	Expenditure (Rs/ha)	DF	IRR (%)	Cost:Benefit ratio
Parkia roxburghii	maize	10	33280	10	69.36896	3.09
Morus alba	maize	10	236175	10	3.514121	1.03
Bambusa tulda	soyabean	10	6200	10	51.86139	4.03
Azadiratcha indica	paddy	10	66826	10	43.88	2.196
Leucaena leucocephala	paddy	10	31000	10	38.055	2.5968
Dendrocalamus strictus	soyabean	10	17000	10	51.86	4.0323

Source: Rocky and Sahoo (2004)

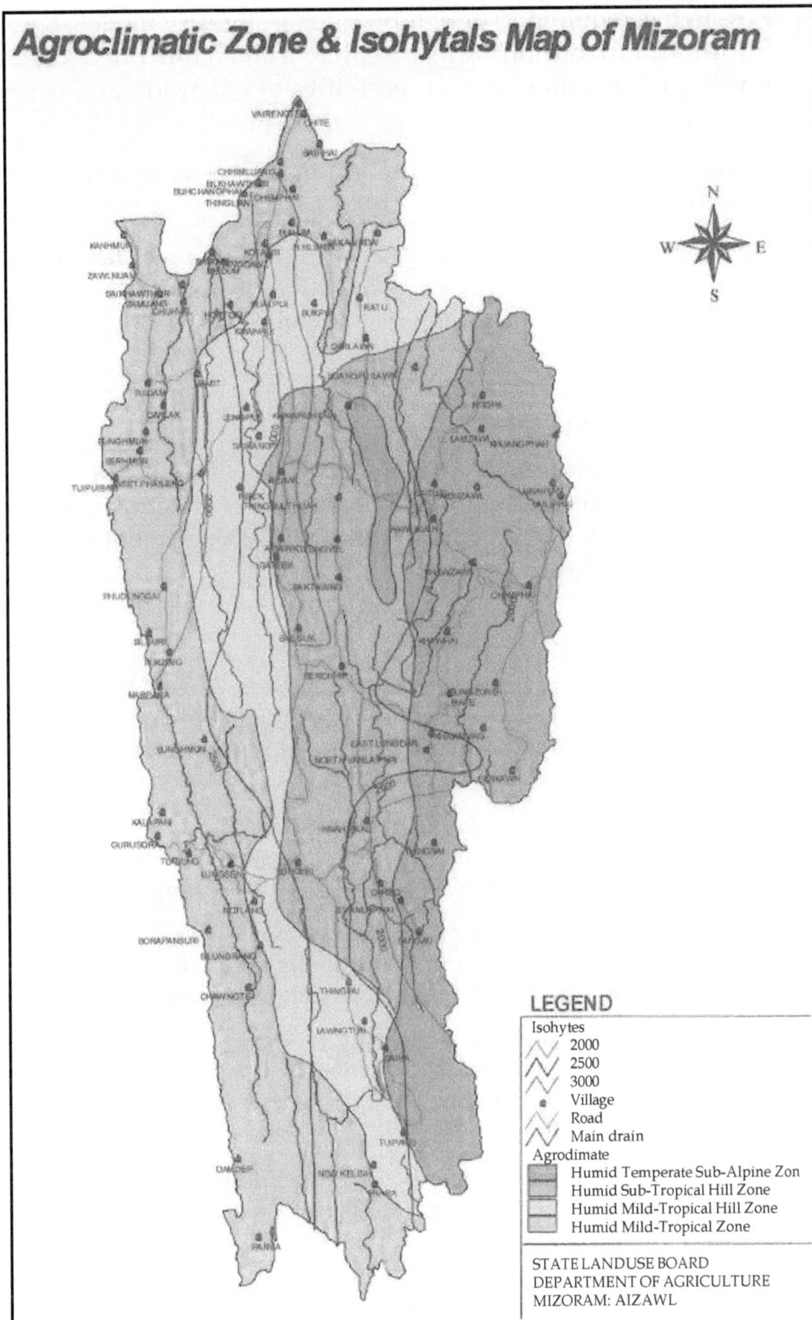

Fig. 22.1. Map showing Agroclimatic zones of Mizoram.

(f) Research is required to design proper agroforestry models based on the information from on-farm research trials, controlled on-stationed investigations and eventual feasibility of extension and acceptability approaches.

(g) Bio-economic modeling of different agroforestry systems to facilitate decision making.

(h) Research on superior provenances (high economic, fast growing and beneficial to crop components) of different multipurpose tree species should be carried out to enhance income, based on profitability and marketability.

ACKNOWLEDGEMENTS

Financial assistance received from ICAR, New Delhi (Grant F.No. 5(29)/97-SW&DF) to establish few agroforestry models is gratefully acknowledged. Thanks are due to Mr. P.C. Vanlalhluna, SRF, Department of Forestry for analysis of soil samples.

REFERENCES

Lalchhandami, C. (1999). Growth behaviour of *Michelia champaca* L. and *Toona ciliata* M.Roem along with hedges of *Cajanus cajan* L. and *Leucaena leucocephala* (Lam) de Wit in an agroforestry system of Mizoram. *M.Sc dissertation*. North Eastern Hill University, Mizoram Campus, Aizawl, pp.

Laldikpuia, B. (1999). Growth behaviour of *Leucaena leucocephala* (Lam) de Wit with Zea mays L. as affected by N,P, K fertilizers in an agroforestry system of Mizoram. *M.Sc dissertation*. North Eastern Hill University, Mizoram Campus, Aizawl, 69pg.

Lalremruata, I. (2001). Effect of various spacing of tung (*Aleurites fordii* Hemsl) and the yield of soyabean (*Glycine max* L.Merr.) under agroforestry system of Mizoram. *M.Sc dissertation*. North Eastern Hill University, Mizoram Campus, Aizawl.

Lalrinsangi (2000). Effect of different spacing on the growth behaviour of *Leucaena leucocephala* (Lam) de Wit and its effect on *Zingiber officinale* Rosc. productivity in agroforestry system of Mizoram. *M.Sc dissertation*. North Eastern Hill University, Mizoram Campus, Aizawl, 73pg.

Mohan, R. Lalsanglura, C. and Sahoo, U.K. (2004). Effect of different mulches on soil moisture and productivity of maize in an agroforestry system of Mizoram. *Science Vision*, IV (4), 14-17.

Mohan, R, Sailo, R. H. and Sahoo, U.K. (2005). Effect of subabul (*Leucaena leucocephala* (Lam.) de Wit) spacing on the growth and productivity of paddy (*Oryza sativa* L.) in Mizoram, India. *Journal of Nature Conservation*, 17(1). In press.

Rocky, P. and Sahoo, U.K. (2002). Tree bean (Parkia roxburghii G. Don): A promising NTFP of Manipur. *MFP News*, XII (2): 4-5.

Rocky, P. and Sahoo, U.K. (2004). Social forestry - A viable option for restoration of degraded land in Manipur. *In*: Bejurbarua, K.M. and Bhatt, B.P. (eds). *Agroforestry for NEH region*, ICAR Publication, Shillong, in press.

Rocky, P., Sahoo, U.K. and Thapa, H.S. (2004). Livelihood generation through tree bean (Parkia roxburghii G. Don) in Imphal West district of Manipur. *Journal of Non-timber Forest Products*, 11(2): 135-139.

Sailo, R.H. (2003). Effect of different spacing of *Leucaena leucocephala* (Lam) de Wit on the growth and yield performance of paddy (*Oryza sativa* L.) under agroforestry system of Mizoram. *M.Sc dissertation*. Mizoram University, Aizawl.

Sahoo, U.K, Mohan, R., Vanlalhluna, P.C. and Sangma, M.M. (2005). Effect of *Leucaena leucocephala* (Lam.) de Wit) on the growth and yield of *Zea mays* L. in Mizoram, India. *Journal of Nature Conservation*, 17(1), in press.

Sahoo, U.K. and Jha, L.K. (2000). Growth of biomass production of seedlings of some multipurpose tree species under nursery conditions. *In*: Jha *et al.*.(eds). *Agroforestry and forest products*, proceedings of International workshop, Department of forestry, North Eastern Hill University, Mizoram Campus, Aizawl. pp. 187-190.

Singh, D. (1996). *The last frontier people and forests in Mizoram*. Tata Energy Research Institute, New Delhi, pp. 301.

23

Agroforestry Systems and Practices Prevailing in Arunachal Pradesh, their Constraints and Potentials

KAMAL KISHOR SOOD, P. HANDIQUE AND B. SINGH
Faculty of Forestry, North Eastern Regional Institute of Science and Technology, Nirjuli, Arunachal Pradesh - 791 109

ABSTRACT

Arunachal Pradesh, the largest state of North-East India is unique state of the country in the sense that its altitude range between 70 – 7000 m, have 26 major tribes besides 110 sub-tribes, have 82 per cent of its land under forest and most of the area is under the control of local tribes which practice shifting cultivation. The paper presents the potentialities of agroforestry in Arunachal Pradesh. The different agroforestry systems prevailing in the state are also given in detail. Agroforestry systems which are scientifically tested and the direction in which research are to be done in future to extend agroforestry has also been presented.

INTRODUCTION

Indian Eastern Himalaya, covers the states of Assam, Sikkim, Arunachal Pradesh, Nagaland, Mizoram, Manipur, Tripura and Meghalaya (Kukreti, 2000). The erstwhile North East Frontier Agency, now Arunachal Pradesh, the largest state among the North-eastern states of India lies at latitude 26°28′- 29°30 N and longitude 91°30′- 97°30 E with a geographical area of 83,743 km^2. The human population of the state is 10,91,113 (GOI, 2002). The recorded forest area in the state is 68602 km^2 which is 82.0 per cent of the geographical area of Arunachal Pradesh (Kukreti, 2000). The forest cover in the state forms the highest amongst all the states of the country and state is least densely populated. The state is bounded by Bhutan to the west, Myanmar to the east, China to the north and north east, and the plains of Assam to the south.

There are 26 major tribes and 110 sub tribes which inhabit different parts of the state, with a life style totally dependent on the forest resources. Moreover, almost all the major tribes and sub tribes has been practicing shifting cultivation from the ancient time in this region which is integral part of socio-cultural life of the tribal people.

For the administrative purpose the state is divided into 15 districts: Tawang, West Kameng, East Kameng, Papum Pare, Lower Subansiri, Upper Subansiri, West Siang, East Siang, Upper Siang, Dibang Valley, Lohit, Changlang and Tirap, Kurung Kumey (Anonymous, 05).

The land of Arunachal Pradesh consists of plain and mountainous terrain. Five main rivers Baroli, Rana, Lohit, Dibang and Tirap dissect the state transversely. The extent of plain area is large in Dibang, Lohit and Tirap basins. The elevation ranges from 70 m from above mean sea level to the elevation of 7000 m. The land cover pattern of Arunachal Pradesh can be divided into four broad categories: (i) the land under barren, rocky, including snow covered areas of conical hills and steep slopes of mountain ridges of Greater Himalaya, (ii) the land under forest cover in most parts of the Lesser Himalaya including denudational Siwalik areas (iii) the moderate slopes of main river valley tracts including the land under degraded forest and shifting cultivation, and (iv) the alluvial plains of Siang, Sissri, Dibang and Lohit rivers under settled cultivation. The major land use of the state has been shown in Table 23.1

Arunachal Pradesh comprises a wide range of climates from subtropical to temperate, varied soil types, abundant rainfalls and mild to cool temperature. This trait is attributed to frequent variation in altitude. The temperature falls below freezing point to - 37 ^{0}C depending on the altitude. The total rainfall varies from 1000 to 5000 mm per annum. Distribution of different climatic zones in state has been presented in Table 23.2. The tropical zone covers altitude up to 900 m, subtropical from 800 m to 1900 m. Temperate from 1900 m to 3500 m and alpine above 3500 m (Kaul and Haridasan, 1987).

The state is endowed with good agro-climatic conditions and widespread vegetation due to heavy rainfall spread over several months of the year. The forest vegetation includes tropical forest, subtropical forest, pine forest, temperate forest, broad leaved forest and alpine forests (Hegde, 2002). The floristic features of Arunachal Pradesh are such that it has maximum diversity in comparison to the other states of North Eastern region of India. This state is considered as the cradle of speciation and centre of origin of some useful plants which are source for crop improvement (Hegde, 2002). Arunachal Pradesh with in the north-eastern region is one of the eighteen biodiversity hotspots of the world (Hegde, 2002). A lot of different valuable forest trees and forest products are available in the whole region of the state. Some other endangered plant species are also available in Arunachal Pradesh. Out of these *Magnolia pterocarpa, Amentotaxus assamica, Gymnocladus* spp, *Rhododendron dalhousiae, Tetracentron sinense, Lithocarpus kamengesis, Rhynchoglossum lazulinum, Dendrocalmus sahnii, Hypericum griffithii, Coptis teeta, Sapria himalayana* are found in different parts of Arunachal Pradesh. The state is a store house of different medicinal plant available in different hilly parts. *Abroma augusta, Abrus precatorius, Adhatoda zeylanica, Acorus calamus, Aloe barbadense, Alpinia galanga, Caesalpinia* spp, *Centella asiatica, Costus speciosus, Datura stramonium, Embilia ribes, Gmelina arborea, Gracinia pedunculata* are important tropical and subtropical medicinal plants. *Artemesia nilagirica, Berberis asiatica Bergenia ligulata, Hoya globulosa, Gaultheria fragrantissima, Illicium griffithii, Mahonia nepalensis* are temperate medicinal plants abundantly grown in Arunachal Pradesh. The sub alpine medicinal plants are *Aconitum ferox, A. heterophyllum, Coptis teeta, Swertia chiraita, S. hookeri, Podophyllum hexandrum, Helinia elliptica, Taxus wallichiana,* etc.

POTENTIALS OF AGROFORESTRY

Agroforestry as a sustainable production system for food and trees products on farmland and rural development has gained importance through out the world (Sood *et al.*, 2003). Mountainous forests provide productive and protective benefits not only to the mountain inhabitants but also to a large number of lowland people. The brunt of the losses of deforestation in mountains is not only limited to the region but also has consequences for other regions. For instance, the annual cost of deforestation in the Himalayas was estimated to be $1 billion in the Ganges valley in 1980 (Myers, 1999). The greatest value of mountains is as source of world's major rivers (Price, 1998). Half of the world's population is dependent on mountains for water for various uses. The Himalayas, one of the most important mountainous ranges of the world, are ecologically fragile and subsistence agriculture constitutes the main source of livelihood (Tucker, 1983). The Himalayan forests with a rich biodiversity are presently decreasing in area due to large-scale exploitation of fuelwood, fodder and timber (Khosla and Khurana, 1996). Indian Eastern Himalayas, with Arunachal Himalayas as its integral component, are one of the biodiversity hot spots of the world. Deforestation and unsustainable use of natural resources in Arunachal Himalayas, causes floods in the adjacent plains of Assam and other adjoining countries and have regional as well as global consequences in terms of floods, droughts, loss of biodiversity and other environmental hazards like sea level rise through melting of permanent glaciers. Moreover tree and tree products play an important role in livelihoods of the local people. Thus it is important to conserve forests and increase the tree cover in this region for the benefit of the people of the state, country and international community as a whole. Agroforestry is one the approaches which has potential for sustainable use of the land in the state. The potential uses of scientific agroforestry for Arunachal where 66.7 per cent of the total population is tribal, includes: (i) maximisation of over all production of the crops and degraded lands, (ii) reduction of loss of soil as well as nutrients through reduction of run off, (iii) addition of organic manure to soils, (iv) nitrogen enrichment by fixation of nitrogen by nitrogen fixing trees and shrubs, (v) utilisation of the waste lands, rehabilitation of abandoned and current shifting cultivation area, improvement of physical conditions of soils, water holding capacity (vi) release and recycling of elements by affecting biochemical nutrients cycling, (viii) moderating effect on extreme conditions particularly the floods in Arunachal and surrounding areas of Assam (ix) lowering effect on the water table in areas where water table is high, (x) provision of employment opportunities to rural tribal people, (xi) increase of farm income of the tribals of Arunachal Pradesh, and (xii) reduction of the pressure of existing forests, increase in effective tree cover and conservation of plant biodiversity (Singh, 1993; Sood *et al.*, 2000). At present the efforts are to conserve the forests, provide ecological security and meet the subsistence requirement of the tribals for food, fuelwood, fodder and necessary plant products and agroforestry is now emerging as an appropriate alternative for the state like Arunachal Pradesh. Table 23.1 shows potential areas available for agroforestry development in the state. In all the areas marked for agroforestry (Table 23.1), various agroforestry systems can be designed and adopted depending on needs of the farmers. However, marshy and swampy areas holds potential for agrisilvifishery system (Sood *et al.*, 2000).

AGROFORESTRY SYSTEMS AND PRACTICES

The systematic work on documentation of traditional agroforestry systems and development of scientifically designed agroforestry systems in Arunachal Pradesh is almost non-existent. Trend to understand and develop scientifically tested agroforestry system is now gaining importance and studies in this direction have already been started (Singh, 1998; Singh *et al.*, 1994; Singh *et al.*, 1997; Singh *et al.*, 2000). Based on review of literature and author's observation the agroforestry systems existing in the state has been delineated below:

Agrisilviculture

Agriculture is the major economic activity in Arunachal Pradesh. It plays an important role in the overall socio- economic development in the state.

Shifting cultivation (jhum) a primitive sequential form of agroforestry system is commonly practised to more or less extent by the people in all agroclimatic zones of the state. However its extent in tropical zone, where wet rice cultivation is very common, is very less compared to other agroclimatic zones of the state (Table 23.2). Different tribes in the state like Adi, Galo, Monpa, Tangsa, Apatani, Memba, Khamba, Miniyoung, Pasi, Bokar, Bori, Panngi, Aka, Ramo, Mismi, Mizi, Nocte, Wancho, etc. are involved in shifting cultivation over an area of 7812 km^2. The farmers maintain high species diversity in jhum agro-ecosystem (Majumdar and Arunachalam, 2004; Ramakrishnan, 1992). The whole socio economic structure, organization and institutions of various population group of the state are woven around shifting cultivation (Sood, 1999a; Sood, 1999b; Sood *et al.*, 2003; Sood *et al.*, 2000). It is argued that in shifting cultivation there is no deliberate choice of planting woody perennials (Lundgren and Nair, 1983), hence this type of land use has become ecologically unsound due to decrease in shifting cultivation cycle as a consequence of increase in population pressure.

Sood *et al.* (2002b) based on their systematic field observations and literature review has delineated the following causes of shifting cultivation:

- High land: man ratio
- Physical, socio-cultural isolation of the state from the rest of the world
- Community ownership of the land acting as disincentive for permanent cultivation
- Integration of jhuming with social and cultural aspects of the tribes
- Lack of knowledge of consequences of jhuming
- Higher returns to labour in shifting cultivation
- Lack of know-how, equipments, credit and other facilities for settled cultivation
- Higher agricultural production of agricultural crops in initial 2-3 years
- Lack of infrastructural facilities like input sale centres, lack of markets, etc.

Das and Pradhan (1990) reported that jhuming is not only means of livelihoods of tribes but way of life. Jhuming is so closely interwoven with social structure that any dislocation in the setting may send shock-waves to the entire society. Sood (1999b) in his study noted similar reasons for the existence of shifting cultivation. In this scenario and with increased demand of food and other material in the state, there is a need to take corrective measures carefully to prevent extension of jhum and control it. Now with spread of education, more job opportunities with increased development, increased opportunities for marketing tree products, out-migration for education and jobs and consequent decreasing hands in the family to work on various labour intensive jhum operations like felling of trees, a sense of sedentary cultivation is getting inculcated among the shifting cultivators. Now many existing plants are retained around the jhum fields, banana and tapioca are being planted on the boundary of jhum fields and inside many agricultural crops are grown (Singh and Bag, 2002). This type of practice has now become common in tropical and subtropical zones of the state where population density is high and demand for value tree products is high. Further planting of *Terminalia myriocarpa* and *Morus laevigata* along the periphery of jhum areas is also becoming popular in tropical and subtropical zones.

Besides this simultaneous form of agroforestry systems are developing in Arunachal Pradesh and this type of systems are still in their infancy. This type of systems is mainly limited to tropical zone and some of subtropical zone where few scattered trees, bamboos and banana, mainly around the periphery of crop land are grown. Raising of *Piper nigrum* with *Erythrina indica* forms one the common agrisilviculture system in the state.

i. Agrisilvipastoral system

In this type of system the tribal people of the state cultivate agricultural crops along with grasses/ rearing of livestock and the forest trees to meet their food, fodder and other tree-based needs. For example *Thysanolaena maxima* grass is raised along the margin of terrace of cropland mainly in Subansiri district of the state. Agrisilvipastoral systems have been developed by the tribals of the state based on long trial and error approach over centuries. This type of the systems is common in tropical and subtropical zones of the Arunachal Pradesh but less common than sequential agroforestry (shifting cultivation). It protects soil from erosion as well as provides fodder. *Saccharum spontaneum* is naturally occurring grass in Arunachal Pradesh. Homesteads with rearing of livestock is one of agrisilvipastoral type of agroforestry systems which exists in the state (Sood *et al.*, 2000). These homesteads are used for growing vegetables, spices, ornamental plants, medicinal plants, fruit trees and other forest trees. Homesteads are prevalent in almost all agroclimatic zones of the state. The crop diversity in such homesteads is very high. Bamboos are also one of the important non timber forest product cultivated in and around the homestead gardens.

ii. Hortisilviculture system

This is specialised form of agrisilviculture system which occurs in the state. The most common form of this system is cultivation of tea (*Camellia sinesis*) interspersed with mainly with trees of *Dipterocarpus* sp, *Shorea robusta, Tectona grandis, Albizzia lebbeck, Macarnage denticulate, Schima wallichii* and *Acacia* spp. Besides grasses of various kinds are grown around the periphery of tea plantations. This type of system is of frequent occurrence up to an elevation of about 1500 m above mean sea level. These fast growing trees not only provide fuel, timber, shade to tea but also some of them fixes nitrogen and enrich the soil.

Orange and banana and some other forestry species trees are planted for shade along with pineapple and its occurrence is common upto lower limits of temperate zones (Singh *et al.*, 1994). Such type of hortisilviculure system is successfully practiced in Arunachal Pradesh mainly in the tropical, subtropical zones. Coffee is also one of the important crop raised under the shade tree *of Dipterocarpus macrocarpus.* Forestry tree species are grown on periphery of fruit trees to a limited extent almost in all agroclimatic zones of the state.

iii. Silvipastoral practices

It is also one of the common agroforestry systems in Arunachal Pradesh. Bamboo groves, cane, pine groves and tree groves surrounded by the pasture land with their natural vegetation is found in almost all the agroclimatic zones of the state. This system is mainly dominated in Apatani Plateau of Upper Subansiri district and adjoining temperate and subtropical parts of Arunachal Pradesh. The most common bamboo species grown in this system are: *Dendrocalamus hamiltonii, Dendrocalmus strictus, Melocanna bambusoides, Arundinaria grifithii, and Arundinaria callosa*. The common cane species under are *Calamus erectus, Calamus flugellum and Calamus floribundus*. The common grasses in silvipastoral system include *Cynodon dactylon, Imperata cylindrica, Thysanolaena maxima, Chrysopogon* species, *Paspalum* spp and *Setaria* spp. The Apatani people of the area freely graze the Mithun (*Bos frontalis*) and other livestock in the pasture (Sood *et al.*, 2000).

iv. Agrihorticultural practices

In this type of the system agricultural crops along with horticultural woody perennials have concurrently grown in various parts of the state. Aracanut with agricultural crops and on cropland also forms agrihorticultural system which is common in lower altitudes of the state. In Apatani plateau in Upper Subansiri district of Arunachal Pradesh, vegetables are cultivated in home garden along with the fruit trees. The common vegetables and cereals cultivated are cucumber, millet, pumpkin, maize, tomato, laisag, ginger, soybean, onion, and French bean. They also grow maize alongwith tobacco, millets and fruit trees mainly plum, apple, and pear (Rethy *et al.*, 2000a).

v. Other Agroforestry practices

Besides above other practices are also prevalent in some parts of Arunachal Pradesh. Agrisilvifishery system is also most prevalent in Apatani Valley of Upper Subansiri district of Arunachal Pradesh (Rethy *et al.*, 2000b). The farmers reared fish in paddy fields surrounded by scattered trees. The cocoon of muga or eri is most delicious food among the tribal peoples of Arunachal Pradesh. To meet this type food requirement tribals rear silkworm under agrisilvisericulture system (Rethy *et al.*, 2000a).

SCIENTIFIC EXPLORATION OF AGROFORESTRY

A few scientifically designed agroforestry systems or scientifically tested traditional agroforestry systems exist in various parts of North Eastern India. Notwithstanding this scientifically designed site specific agroforestry systems are very few in Arunachal Pradesh. Further scientific studies on existing traditional agroforestry systems in Arunachal Pradesh are now gaining importance.

A number of new agroforestry systems have been developed by ICAR research complex for Northeast India by North Eastern Hill Region, Shillong and other research agencies (Singh, 1999). These include sericulture based agroforestry, silvipasture on waste land, agrihorticulture systems and bamboo based agroforestry, however, their potential for Arunchal Pradesh is yet to be explored. Besides, a few scientific investigations on existing traditional agroforestry systems in and around Arunchal Pradesh has been done by researchers at North East Regional Institute of Science and Technology (NERIST), Arunachal Pradesh (Bora *et al.*, 2000; Rethy *et al.*, 2000; Sood, 1999b; Sood *et al.*, 2000a). These studies included categorisation of existing agroforestry systems, identification of their systems units, biological and economic productivity of different existing agroforestry systems, species preferences in the tropical zone of Arunachal Pradesh. Further constraints and potentials of different agroforestry systems were identified and strategies were suggested to improve the adoption and productivity of the systems. However, tree-crop interactions were not the component of these studies. Singh *et al.* (1997), and Singh (1998) at NERIST designed and evaluated some scientific agroforestry interventions for sloppy degraded lands in tropical agroecological zones of the state. They tested agrisilivicultural, silvipatoral, agrihorticultural, agrisilvihorticultural, and pure silvicultural models with different forestry species, horticultural species and crop combinations to fulfil the objectives of soil conservation and economic profitability. The growth and survival of all tree species was higher in all agroforestry models compared to pure silviculture. This enhanced biomass and economic yield under these agroforestry models. The main crops included were maize, millets, upland paddy, cabbage, cauliflower, pea, radish, ginger and turmeric. Trees included *Anthocephalus cadamba, Melia azedarach, Leucaena leucocephala, Dalbergia sissoo, Terminalia bellirica, Gmelina arborea, Morus* spp. Horticultural component included *Emblica* spp, Kinnow, Peach, Pear, Guava, Pine apple, Jamun, Papaya. The pastoral species grown were *Scyto santhes, Medicago sativa,* Guinea grass (Singh *et al.*, 1997). This study did not include social acceptability of agroforestry systems by the tribal people of the region. Sood *et al.* (2002a) designed *Anthocephalus cadamba* based silvopastoral system with local grasses for lower altitudes in the tropical region and studied growth response of *Anthocephalus cadamba* in the system with respect to nitrogen application. However this study did not investigate the tree-grass interaction.

Keeping in view the above, it can be concluded that comprehensive studies to delineate existing agroforestry systems, their constraints and potentials, species preference are yet lacking in the state.

IMPORTANT SPECIES SUITABLE FOR AGROFORESTRY SYSTEMS

There are no comprehensive studies to document existing and potential agroforestry species (herbs, shrubs and trees) for different climatic zones of Arunachal Pradesh (Sood *et al.*, 2000). However a broad list is given below (Table 23.3).

In the lower limits of alpine zone a few woody perennials like *Taxus baccata, Populus ciliata, Betula* spp, *Rhododendron* spp, *Berberis asiatica, Berberis wallichiana, Eurya acuminata, Vaccinium venosum* are grown in silvipastoral form. The alpine zone above 4000 m remains covered with snow for major part of the year and is considered as tree line (Hegde, 2002).

FUTURE RESEARCH NEEDS OF AGROFORESTRY

Agroforestry as a sustainable land use strategy of rural development has received peripheral attention in Arunachal Pradesh until recently. Now this strategy of farming is attaining attention of the policy makers, planners, academicians and researchers in Arunachal Pradesh (Singh and Bag, 2002). It has been now realised that there is a vast scope to introduce new agroforestry system and to improve the old agroforestry system in Arunachal Pradesh which will help in sustainable land use, sustain and improve the livelihoods of the tribal farmers of the state and control and improve the shifting cultivation practices (Singh *et al.*, 2000). Sood (1999a) and Singh *et al.* (1999) proposed some new agroforestry models based on different improved varieties of crops and horticultural species for Arunachal Pradesh. However, agroforestry systems are very site specific and there is need to study their constraints and potentials case-to-case basis. Notwithstanding, Sood *et al.*, (2000) has listed the common constraints of agroforestry land use in Arunachal Pradesh:

- Lack of proper documentation of agroforestry systems, systems units, species preference in existing agroforestry systems in different agroclimatic zones of the state.
- Lack of knowledge of constraints and potentials of existing agroforestry systems with respect to their biological and economic productivity.
- Farming practices are based on the experience of the tribals and lacks scientific inputs.
- Lack of sufficient technical know-how of agroforestry amongst the planners and executors of agroforestry programmes.
- Community ownership of the land acting as disincentives for settled agriculture and agroforestry.
- Lack of systematic wide-scale research endeavours in agroforestry in the state.
- Lack of extension facilities for agroforestry.
- Lack of systematic endeavour to design scientifically viable agroforestry systems for different zones of the state.

To improve the existing agroforestry systems with respect to their biological and economic productivity and sustainability the following are broad areas which merit immediate attention (built on Sood *et al.*, 2000):

i. Survey, Diagnosis and Design

- Survey and evaluation of existing land use problems and diagnosing agroforestry alternatives to solve the land use problem.
- Diagnosis of land area suitable for agroforestry.
- Design of site specific agroforestry system having high productivity, suitability and adaptability to the cultural aspects of tribal.

ii. Identification and evaluation of plant species for agroforestry

- Preparation of inventory of existing agroforestry plants species (agriculture, horticulture, forestry and grass species) for different climatic zone of the state.
- Evaluation of potential trees species for their suitability for different agro climatic condition.
- Breeding and genetic improvement of the plant species.

iii. Agroforestry management

- Studies of trees crop interactions to find suitable tree-crop combinations for different climatic zones of the state.
- Identification of nitrogen fixing trees for different agroforestry systems in different agroclimatic zones.
- Evaluation of effect of nitrogen fixing tree species on yields of crops.
- Nutrient dynamics, water regimes and soil fertility status in various tree-crop combination under various agroforestry systems.
- Effect of adoption of different agroforestry systems on plant biodiversity
- Nodulation ecology and microbial population studies on different potential agroforestry systems.

iv. Socio-economic, cultural aspects and extension

- Knowledge of socio-economic status of the tribals, their attitude and perceptions towards trees and tree cultivation.
- Socio economic and cultural constraints for practising agroforestry.
- Traditional belief, ethos, and taboos of different tribal communities associated with cultivation of trees and their conservation should be evaluated.
- Availability of various resources with the tribals for the practicing agroforestry.
- Cost-benefit estimation of various agroforestry systems and their environmental impacts.
- Education and training of forestry planners and managers in biophysical, socio-economic and extension aspects of agroforestry.

In Arunachal Pradesh, three areas has become important for considering the domain of agroforestry (Handique *et al.*, 2004; Singh, 1999)

- Development of alternative to shifting cultivation
- Improvement in traditional agroforestry systems
- Rehabilitation of degraded hill slopes.

According to Pathak and Singh (2000), the major components of the priority agroforestry systems are agrisilviculture (jhum land), and agrihorticulture. Sundriyal (2001) suggested contour-hedgerow agroforestry with multipurpose fast growing trees for hilly regions of Arunachal Pradesh. Sood (2003) has argued that agroforestry uptake should not be looked upon in isolation of livelihood strategies of the farmers. Trees play an important role in the livelihoods of the farmers: secure provision of food and essential subsistence goods, provide cash for purchase of outside goods and services, savings (resource accumulated to meet future planned needs and emergencies), provide social security (secure access to subsistence goods and productive resources), and reduce the risks to livelihood. Tree management is an integral part of rural livelihood strategies. The issues facing growing of trees on farmland cannot be seen in isolation of farmers overall livelihood strategies (Malla, 2000). The role of the trees in the livelihoods of the farmers is dependent on the biophysical status of the farms owned; tree resource use, socio-economic, psychological and attitudinal aspects of the farmers (Sood, 2003; Sood and Mitchell, 2004). Thus to design biophysical suitable and socially acceptable agroforestry systems all the biophysical, socio-economic, perceptional and attitudinal aspects of the farmers should be taken into account. In Arunachal Pradesh these aspects needs careful attention due to frequent variation in biophysical as well as socio-economic aspects in different areas and between different tribes. Further agroforestry has long gestation period and thus requires long period of experimentation and study. Traditional agroforestry systems are products of centuries of trial and error by the farmers that have adapted them to long ecological, socio-cultural and economic factors (Christanty and Iskandar, 1985). Thus knowledge of existing agroforestry can act as useful resource to reduce the long period of experimentation and study to design new agroforestry systems, particularly in Arunachal Pradesh where traditional agroforestry exists in various forms.

CONCLUSION

Agroforestry as sustainable land use and alternative to jhum cultivation is now gaining importance in Arunachal Pradesh. Different types of traditional agroforestry systems and practices are prevalent in Arunachal Pradesh.

Table 23.1. Major land use of Arunachal Pradesh and potential areas for agroforestry

S. No	Land use	Area (km²)	Area (%)
1	Built up	99.12	0.12
2	Net sown area	2063.80	2.56*
3	Evergreen forest	58133.2	69.25
4	Degraded area	10937.19	13.06*
5	Forest blank	35.83	0.04*
6	Marshy/Swampy	166.73	0.19*
7	Barren rocky	980.59	1.27
8	River/ stream	803.30	0.96
9	Shifting cultivation	3811.54	4.55*
10	Grassland	20.56	0.02*
11	Snow cover	6446.49	7.96

Adapted from : Upreti, 2002, * potential areas for agroforestry development

Table 23.2 Agro-climatic zones of Arunachal Pradesh

Name of Agroclimatic zone	Climatic conditions	Area under zone	Major land use
Alpine zone	Annual precipitation 2000 mm, mean temp. of coldest month < 0°C, Winter months 4-5, Dry months 5-6	Gourichen, Upper Tawang, Tulungla, Shela pass areas of West Kameng district, Jidu and adjoining areas of Northern Siang	Pasture land, jhum cultivation, Lower parts of alpine zone form the upper limit of potato cultivation
Temperate zone	Precipitation 2000 mm, Temp. of coldest month < 10°C, Winter months 1-2, Dry months 4, High humidity	Tawang, Dirang, Bomdila, Shergaon areas of West Kameng district of Dibang valley, Northern parts of East Siang, Upper Subansiri, parts of West Siang, around Anini and North Eastern part of Lohit district.	Jhum prominent Upper limit of successful dry maize, rice/millet
Sub-Tropical zone	Precipitation 2000-4000 mm, Temperature of coldest month 15°C, High humidity, Dry months 3-4	Changlang, Naga & Khonsa areas of Tirap district, Basar areas of Siang district.	Wet rice, low intensity of jhum
Mid-Tropical hill zone	Precipitation 2500-5500 mm, Temperature of coldest month 20°C, High humidity, Dry months 2-3	Southern part of lower Subansiri district	Wet rice cultivation
Mid-Tropical plain zone		Pasighat area, Singphow areas of Tirap district and Lower part of Lohit district	

Based on: Borthakur, D.N., 1992; Kaul and Haridasan, 1987; Sundriyal, 2001

Table 23.3. Important Species Suitable for Agroforestry in Arunachal Pradesh

Tropical Zone up to 900m Elevation		
Herbs	**Shrubs**	**Trees**
Acorus calamus	*Phyllostachys bambusoides*	*Erythrina stricta*
Agave Species	*Phylogacanthus* spp.	*Litsea monopetala*
Amaranthus spp	*Calamus erectus*	*Bambusa polymorpha*
Ananus cosmosus	*Calamus tenuis*	*Bambusa tulda*
Costus speciosus	*Callicarpa arborea*	*Bambusa vulgaris*
Cymbopogon citratus	*Musa spp.*	*Emblica officinalis*
Dioscorea floribundus	*Polymorphum* spp	*Macranga denticulate*
Diplazium esculentum	*Maesa chisia*	*Catanopsis indica*
Musa spp.	*Dendrocalamus hamiltonii*	*Cephalostachyum pergracile*
Phrynium pubinerve	*Gloriosa superba*	*Dendrocalamus giganteus*
Piper longum	*Murraya koenighii*	*Elaeocarpus floribundus*
Vetiveria zizanoides	*Manihot esculenta*	*Dillenia indica*
	Pseudostachyum spp	*Moringa oleifera*
	Sarcochlamys pulcherima	*Parkia roxburghii*
	Solanum torrum	*Sapindus mukorossi*
	Zanthoxylum alatum	*Spondias pinnata*
		Spondia axillaries
		Bamusa arundinacea
		Bambusa balocoa
Subtropical and temperate Zone 900-2000m		
Herbs	**Trees**	**Trees**
Aconitum ferox	*Albizzia mollis*	*Populus gamblei*
Coptis teeta	*Alnus nepalensis*	*Pyrus* spp
Dichranopteris linearis	*Erythrina indica*	*Salix daptoniana*
Fagopyrum cymosum	*Exbucklandia populnea*	*Dendrocalamus hamiltonii*
Panax pseudoginseng	*Juglans regia*	*Dendrocalamus membranceous*
Podophyllum hexardrum	*Lindera* spp	*Dendrocalamus strictus*
Pterdium acquilinum	*Malus* spp	*Melocanna bambusoides*
Shrubs	*Myrica esculenta*	*Oxytenanthera albociliata*
Artemisia spp	*Pinus* spp	*Oxytenanthera albociliata*
Zanthoxylum spp	*Populus ciliata*	*Arundinaria maling*

Based on: Choudhary and Haridasan, 1993; Rethy *et al.*, 2000a; Singh, 1993

Documentation of different existing agroforestry systems, system units and their biophysical and socio-economic evaluation would be an important strategy to improve the existing and design new agroforestry systems. This would also reduce the long experimentation period required to develop the new systems. There is also a need to study the role of the trees in the livelihoods of different tribes of Arunachal Pradesh. Understanding of biophysical and socio-economic aspects of the farms and tribes would help in designing biophysically suitable and socially acceptable agroforestry systems for Arunachal Pradesh. Training of forestry planners and executors coupled with right extension approach would boost the wider acceptability of agroforestry in Arunachal Pradesh.

REFERENCES

Anonymous. (2005). List of Indian districts (http://en.wikipedia.org/wiki/List_of_Indian_districts, accessed on 3/07/05).

Bora, D., Kumar, H.K., Sood, K.K., Singh, B. and Rethy, P. (2000). Diagnosis of existing agroforestry systems in Lakhimpur, Assam (India). *In: International Conference on Forest Ecosystem: Ecology, Conservation and Sustainable Management*. August 15-21, 2000. Chengdu, Sichuan, China.

Borthakur, D.N. (1992). *Agriculture of Northeastern region with special reference to hill agriculture*. Beecee Prakashan, Guwahati.

Choudhury, S.S. and Haridasan, K. (1993). Water conservation via-a-vis forestry in Arunachal Pradesh. *Arunachal Pradesh Forest News*. 11(1): 5-17.

Christanty, L. and Iskandar, J. (1985). Development of decision making and management in traditional agroforestry: Examples in West Java. In: Rao, Y.S., Vergara, N.T. and Lovelace, G.W. (Eds) *Community forestry: socio-economic aspects*. Regional Office for Asia and Pacific (RAPA), FAO, Bangkok. pp. 199-214.

Das, A.P. and Pradhan, S.K. (1990). Jhuming in Arunachal Pradesh. *In: Shifting cultivation in North East India*. pp. 198. (cited in Sood *et al.*, 2002).

GOI. (2002). *Basic statistics of North Eastern Region*. North Eastern Council Secretariat, Shillong, Maghalaya.

Handique, P. Sood, K.K., Singh, B. and Rethy, P. (2004). Improvement of jhum cultivation for sustainable development in Arunachal Pradesh. *In*: Pant, R.M. and Pandey, A. (Eds) *Proceedings of seminar on Environment, Agriculture and sustainable life support systems: Perspectives of NER*, Organised by North Eastern Regional Institute of Science and Technology, Nirjuli on November 6, 2004. pp. 42-44.

Hegde, S.N. (2002). *Arunachal Pradesh State Biodiversity strategy and action plan*. State Forest Research Institute, Itanagar, Arunachal Pradesh.

Kaul, R.N. and Haridasan, A. (1987). Forest types of Arunachal Pradesh - A preliminary study. *J. Eco. Tax. Botany*. 9 (2): 379-389.

Kukreti, M.C. (2000). Vegetation and its status in Himayan region. *Indian Forester* 126: 175-179.

Khosla, P.K. and Khurana, D.K. (1996). Agroforestry for sustainable development in the Himalayas: Potential and scope. In: GOI (Ed) *Farming systems in the Himalayas*. Department of Agriculture and Cooperation, Ministry of Agriculture, Government of India, New Delhi. pp. 85-110.

Lundgren, B.O. and Nair, P.K.R. (1983). Agroforestry for soil conservation. *In*: Proc, 2nd *International Conference on soil erosion and soil conservation*, Hawaii.

Majumder, M. and Arunchalam, A. (2004). Agricultural crop diversity and management of jhum cultivation systems in Northeast India. *In: Pant, R.M. and Pandey, A. (Eds)*

Proceedings of seminar on Environment, Agriculture and sustainable life support systems: Perspectives of NER, Organised by North Eastern Regional Institute of Science and Technology, Nirjuli on November 6, 2004. pp. 45-62.

Malla, Y.B. (2000). Farmer's tree management strategies in a changing rural economy and factors influencing decisions on tree growing in Nepal. *International Tree Crop Journal* 10: 247-266.

Myers, N. 1999. The next green revolution: its environmental underpinnings. *Current Science.* (http://www.iisc.ernet.in/~currsci/feb25/articles16.htm)

Pathak, P.S. and Singh, K.A. (2000). Sustainable farming systems in the hilly regions of India with special reference to North Eastern Region. *In*: Kohli, R.K., Singh, H.P., Dhir, K.K., Batish, D.R. and Khurana, D.K (Eds). *Man and Forests.* ISTS, Solan and Punjab University, Chandigarh. pp. 313-346.

Price, M.F. (1998). Mountains globally important ecosystems. *Unasylva* 49(195). (http://www.fao.org/docrep/w9300e/w9300e03.htm).

Ramakrishnan, P.S. (1992). *Shifting agriculture and sustainable development.* Man and Biosphere Series: V.10. The Parthenon Publishing Group, Paris. 424 pp.

Rethy, P., Sood, K.K. and Singh, B. (2000a). Agroforestry systems as alternative to shifting cultivation for flood prone areas of Arunachal Pradesh and adjoining regions. In : Patwary, B.C, Kumar, S.R., Mani, P. and Panigrahy, N (Eds) *Proceedings of Brain Storming Session on Water Resources Problems of North Eastern Region Organised by North Eastern Regional Centre,* NIH, Roorkee. pp. 59-63.

Rethy, P., Sood, K.K., Singh, B. Chakrabarty, A. and Kago, D. (2000b). Unique Agri piscicultural practice of Apatani tribe in Zero valley of Arunachal Pradesh. *Journal of Ecobiology.* 12(3):185-188.

Singh, B. (1998). *Final technical report of project financed under Himalayan Eco-development research programmes (IERP),* by GB Pant Institute of Himalayan Environment and Development, Almora.

Singh, B., Rethy, P. and Arunachalam A.S. (1994). Prospects of agroforestry at Arunachal Pradesh. *Indian Journal of Hill Farming* 7(2):119-123.

Singh, B., Rethy, P. and Singh, A.B. (1997). Agroforestry intervention for sustainable productivity in Arunachal Pradesh. *Journal of Nature Conservation* 9(1): 97-106.

Singh, B., Rethy, P. and Sood, K.K. (2000). Role of agroforestry systems in the reclamation of wasteland in Arunachal Pradesh. *In*: Arunachalam, A. and Khan, M.L. (Eds) *Sustainable Management of Forests.* International Book Distributors, Dehra Dun, Uttranchal, India. pp. 387-393.

Singh, B., Solanki, G.S., Sood, K.K. and Rethy, P. (1999). Agro-horticultural practices and suitable management of degraded lands in North East India. Paper Presented in National Seminar on Natural Disaster Management, NERIST, Nijruli, October 13, 1999.

Singh, K.A. (1999). Resource management and productivity enhancement through agroforestry in the Eastern hilly agroecosystem in India. *Indian Journal of Agroforestry* 1(1):63-72

Singh, K.A. And Bag, T.K. (2002). Option for sustained Agricultural Production in Arunachal Pradesh. *In* Arunachal Pradesh-Environmental planning and sustainable development – opportunities and challenges. In: Sundriyal, R.C., Singh, T. and Sinha, G.N. (Eds) *Arunachal Pradesh: Environmental Planning and Sustainable Development*. HIMAVIKAS Occasional Publication, G.B. Pant Institute of Himalayan Environment and Development, Almora. pp. 311-330

Singh, N.B. (1993). Present and future of agroforestry in Arunachal Pradesh. *J. Eco. Tax. Bot.* 17: 519-533.

Sood, K.K. (1999a). Agroforestry for minimisation of ecological hazards in degraded watersheds of Arunachal Pradesh. Paper presented in National Seminar on Natural Disaster Management, NERIST, Nijruli, October 13, 1999.

Sood, K.K. (1999b). Forest resource use in relation to tribal development in Arunachal Pradesh. *Thesis for Post-Graduate Diploma in Rural Development* (Unpublished). Indira Gandhi National University, New Delhi.

Sood, K.K. (2003). Factors affecting tree growing in traditional agroforestry systems of Western Himalaya, India. *Ph.D. Thesis* (Unpublished). University of Aberdeen, Aberdeen, U.K.

Sood, K.K., Meiti, T.C., Pangging, G., Upadhayaya, K. and Singh, B. (2000a). Dominant Indigenous agroforestry systems of biodiversity rich North Lakhimpur District of Assam, India. *Journal of Ecobiology* 12(3): 189-198.

Sood, K.K., Singh, B. and Rethy, P. (2000). Agroforestry approaches for sustainable development in Arunachal Pradesh in next millennium. In: Kohli, R.K., Singh, H.P., Dhir, K.K., Batish, D.R. and Khurana, D.K (Eds). *Man and Forests*. ISTS, Solan and Punjab University, Chandigarh. pp. 461-470.

Sood, K.K., Singh, B. and Rethy, P. (2002a). Growth performance of *Anthocephalus cadamba* plantation under nitrogen application under silvopastoral system. *Journal of Nature Conservation* 14 (1): 67-74.

Sood, K.K., Singh, B. and Rethy, P. (2002b). Shifting cultivation in Arunachal Pradesh – Causes and control. *In*: Sundriyal, R.C., Singh, T. and Sinha, G.N. (Eds.). *Arunachal Pradesh: Environmental Planning and Sustainable Development*. HIMAVIKAS Occasional Publication, G.B. Pant Institute of Himalayan Environment and Development, Almora. pp. 161-165.

Sood, K.K., Singh, B. and Rethy, P. (2003). Traditional forest resource use and conservation practices of the tribals in Arunachal Pradesh. *In*: Rethy, P., Dabral, P.P., Singh, B. and Sood, K.K. (Eds) *Forest Conservation and Management*. International Book Distributors, Uttranchal, India. pp. 527-539.

Sood, K.K, Soni, J.C. and Singh, B. (2000). Cultural aspects of tribals and their awareness in relation to forest conservation in Papumpare district of Arunachal Pradesh. *Journal of Ecobiology* 12(3): 179-182.

Sood, K.K. and Mitchell, C.P. (2004). Do socio-psychological factors matter in agroforestry planning: Lessons from smallholder traditional agroforestry systems? Small-scale *Forest Economics, Management and Policy* 3(2): 239-255.

Sundriyal, S.C. (2001). Hill agro-ecosystem and sustainable development in Arunachal Pradesh. *Arunachal Forest News* 19: 186-206.

Tucker, R.P. (1983). The British colonial system and the forests of Western Himalayas 1815-1914. In: Tucker, R.P. and Richards, J.F. (Eds) *Global deforestation and the nineteenth century world economy*. Duke Press Policy Studies, Durham, N.C. pp. 146-166.

Upreti, T.C. (2002). Arunachal Pradesh- A physiographic profile. *In*: Arunachal Pradesh- Environmental planning and sustainable development - opportunities and challenges. In: Sundriyal, R.C., Singh, T. and Sinha, G.N. (Eds.). *Arunachal Pradesh: Environmental Planning and Sustainable Development*. HIMAVIKAS Occasional Publication, G.B. Pant Institute of Himalayan Environment and Development, Almora. pp. 51-64.

24

Homegarden Systems in the Humid Tropics of North-Eastern India

A. ARUNACHALAM, K. ARUNACHALAM, S. TANGJANG AND S. DEB
Department of Forestry, North Eastern Regional Institute of Science and Technology, Nirjuli-791 109, Arunachal Pradesh

ABSTRACT

Homegardens are traditional agroforestry systems present throughout the northeast India especially in foothill region. The high diversity and the intensive care that is given to the homegarden in and around Arunachal Pradesh results in a unique combination of high levels of productivity, stability and sustainability of the system. The homegardens studied here were vertically distributed in four different layers viz, floor, middle, sub-canopy and canopy. Such stratification has given rise to the distribution of a variety of species based on light requirement in the system. Density and diversity index of tree species in the homegardens were highest in Harmutty in the foothills (468.5 plants ha^{-1} and 1.17 respectively), whereas lower values were recorded in Ziro in the upper elevation (16 plants ha^{-1}, 0.651 respectively). Soil nutrients were greater in Harmutty, followed by Doimukh, Nirjuli and then Ziro. Maximum plant detrital (litter and fine roots) biomass and their subsequent decomposition in the homegardens of Harmutty (10185.67 kg ha^{-1}) might be the probable reason for the efficient nutrient cycling and generation of more available nutrients.

INTRODUCTION

Homegardens, the dominant form of traditional Agroforestry, probably originated more than 10,000 years ago, when hunters and gatherers discarded domestic refuse, containing seeds and other propagules, in the vicinity of their dwelling and then tended and protected that plants that appeared (Hutterer, 1984). The term 'homegarden' here refers to the intimate association of multipurpose trees and shrubs with annual and perennial crops and invariably livestock within the compound of individual houses, with the whole crop-tree-animal unit being managed by the family labours (Fernandes and Nair, 1986). The homegardens form an important land use system in the northeast India and the ethnic groups are utilizing the homegarden products both for their subsistence and to some extent for commercial needs. The people of the region practise multistrata-homegarden systems, which is the result of a continuous selection process in which the farmers usually prefer fruit trees of high production. The survival of the system is based on its adoption to a demand for different products or a change in the mixture of plants and animals. Understanding the controlling biological factors and interaction within these systems will help formulating strategies towards the potential increase in the yield *per se*.

The homegardens present in and around Arunachal Pradesh are usually carefully structured systems with every component having a specific place and function. The layered structure and species selection is never static, the pool of replacement species results in a productive structure that is always dynamic while the overall structure and function of the system are maintained. However, there is a lack of information on the structure, and nutrient viability of such dynamic systems adopted traditionally by the people of the far-eastern India. The present study consummates the structure, vegetation and soil fertility status of the traditional homegarden systems practiced by three ethnic groups in and around Arunachal Pradesh, north-eastern India.

METHODS AND METHODOLOGY

Study sites

The study was conducted in traditional homegarden systems of three villages in Arunachal Pradesh namely, Nirjuli (118m asl) and Doimukh (126m asl) of Papum Pare district, and Ziro (1700m asl) of Lower Subansiri district and in Harmutty village (120m asl) of North Lakhimpur district of Assam, bordering Arunachal Pradesh in the northeastern India. The Kalita group of people dominate Harmutty (site I), Nyishi in Doimukh and Nirjuli (site II and III respectively) and Apatani in Ziro (site IV). Three traditional homegarden systems practiced by the three different ethnic groups were selected for this study. The average area of the homegarden varied between 200-400 m^2 in Nirjuli/Doimukh, 320-490 m^2 in Harmutty and over 400 m^2 in Ziro.

The mean annual temperature varies between a minimum of 5^0C to a maximum of 37^0C in all the study sites experiencing a humid tropical climate, except in Ziro where the mean temperature varies from -5^0C to 15^oC (sub-tropical climate). Mean annual precipitation for the past five years varied between 1100 mm and 1600 mm with the maximum rainfall occurring during

June-September. The regional climate has two seasons, a cool and dry (RH 54%) winter (November-February) and a warm and wet (RH=80-98%) summer (March-October). The rocks of Ziro valley belong to Hapoli formation and are considered to represent the lacustrine deposits in the inland basins. The peat has given ^{14}C dating of 40,000 years. Doimukh and Nirjuli area composed of newer alluvium (newer terrace deposits) represented by valley field deposits comprising of sediments, while Harmutty area is composed of alluvium belonging to Pleistocene and recent times (Kumar, 1997).

Methods

Plant species in the traditional homegarden systems were measured using ten randomly placed quadrats (10m x 10m) for trees, twenty quadrats for (5m x 5m) shrubs and herbaceous (1m x 1m) species. Nomenclature of the plant species followed Hooker (1872-1897). Important value index (IVI) of plant species was calculated as the sum of relative frequency, relative dominance and relative density of an individual species (Curtis, 1959). Species diversity (Margalef, 1968) was calculated as: $H' = -\Sigma \{(n_i/N) \log_e (n_i/N)\}$, where H′=Shannon index of general diversity, n_i =IVI of a species, N= Total IVI of the community (i.e. 300). The dominance index (Simpson, 1949) of the community was calculated as: $C=\Sigma \{(n_i/N)^2\}$, where C=dominance index, n_i and N are same as for Shannon's index. Similarity index (community coefficient; Sorenson, 1948) among different sites was calculated as: (2C x 100)/ (A+B), where C= Number of species common to both sites, A= total number of species in site A, and B= total number of species in site B. Pielou's, 1966 evenness index (e) was calculated as: e= H′/logS, where H′= Shannon's index of diversity, and S= Total number of species.

In each site, five randomly located soil samples were collected from top 0-10 cm soil layer using a soil corer (5.5 cm inner diameter). The soil samples were thoroughly mixed and sieved to obtain a composite sample. The soil pH and moisture content were determined in fresh soil samples following potentiometric and gravimetric methods, respectively. The rest of the soil samples were air-dried for the analysis of water holding capacity using the Keen's box method, soil texture by hydrometer method, available phosphorous (P) by molybdenum blue method, total nitrogen (N) by digestion and distillation method, nitrate-N by disulphonic acid method, ammonium N by indophenol blue method and organic carbon (C) by Walkley and Black's rapid titration method given in Anderson and Ingram (1993). Litter biomass was studied based on experimental designs given in Maithani *et al.* (1997). Fineroot (<2 mm) biomass in the top soil layer (0-15, 15-30cm depth) was studied by sampling ten soil cores (5.5 cm diameter) and the roots were separated by wet-sieving method as described in Arunachalam *et al.* (1996).

To understand the structure of the traditional homegarden systems, twenty farming households belonging to each of the three ethnic groups were also selected, and relevant information on the vertical distribution of species and its indigenous categorization on the basis of light requirement were collected through personal observation and/or personal interview by using standardized questionnaire.

RESULTS AND DISCUSSION

Types of homegardens

The homegardens found in and around Arunachal Pradesh can be classified in to five different types on the basis of dominant plants found in the system. These are arecanut-based homegarden (site I), pineapple-based homegarden (site I & II), bamboo-based homegarden (Site I, II, III & IV), apple-based homegarden (site IV) and vegetable-based homegarden (Site I, II, III & IV) systems. All these homegardens were found in our study sites and are populated by a wide variety of plants, varying from small herbs to tall trees.

Species composition and vegetal stratification in traditional homegardens

The homegardens found in study sites are multi-storeyed having four layers (Table 24.1). In most of the homegardens, the top storey was occupied by *Areca catechu* Linn. except in Ziro where *Altingia* sp. was dominant in the upper storey. *Artocarpus, Mangifera*, etc. were dominant in the sub-canopy layer of the first three sites and *Pyrus* sp. in Ziro site. Papaya, guava and citrus are some of the commonly observed middle-storey species.

The pre-dominant floor species in Doimukh & Nirjuli was *Ananas comosus*.Vegetable crops were the mostly preferred floor species in Harmutty and Ziro. Okafor and Fernandes, 1987 also reported a four-layered canopy in the homegardens in West Africa. In south India, particularly in Kerala, cereals like rice and finger millets, pulses like green gram (mung), black gram and pigeon pea (*Cajanus cajan*), root crops like cassava, sweet potato, yam, etc. produced reasonable yields as intercrops with coconut (CPCRI, 1976). The upper storey plants in the homegarden gives shade on the crop plants and therefore only the shade loving plants with low light compensation point and light saturation can grow better in such environment (Bohning and Burnside, 1956). The farmers of the traditional homegardens in and around Arunachal Pradesh have poor scientific knowledge on the selection of shade loving plants. However, they categorize the plants indigenously on the basis of light requirement (Table 24.2). Over all, the number of medium light demanding and shade loving plants were more than the strong and moderate light adaptable plants in the traditional homegarden system.

Vegetation in traditional homegardens

Homegarden are typically populated by a wide variety of plants varying from small herbs to tall trees. In this study 42 tree species, 45 shrubs and 53 herbs were recorded in the homegardens practised by the three local communities in and around Arunachal Pradesh. Many species are represented by several strains; some are partly domesticated, but farmers clearly recognize its long-term importance for genetic diversity. They use many plants for medicinal and other ethnobotanical purposes. The homegardens present in this region also act a good habitat for some domesticated animals such as mithun (*Bos frontalis)*, pig, cattle and poultry.

The diversity of trees and herbs were higher in Harmutty than in Doimukh and Nirjuli (Table 24.3). The high elevation site (i.e. Ziro) registered the lowest diversity. At lower altitudes arecanut, coconut and fruit trees predominate, whereas in higher altitudes the local people prefer to grow vegetables. Karyono (1981) reported that plant associations in homegardens reflect a variety of complementary functions, and its design and composition are influenced by climatic, edaphic and economic factors as well as cultural and traditional uses. In this study also there were significant differences in the plant categories. The Nyishi and Apatani people prefer medicinal plants in their garden, whereas the Kalitas in Harmutty grow more vegetable crops.

The high diversity of plants in the tropical homegarden has been fostered by year round climatic conditions that generally are favourable for plant growth. However, the diversity in homegardens arises primarily as a conscious response to socioeconomic forces. By daily use of family labour close to home and by recycling household waste, homegarden owners can satisfy a wide range of household requirements and local markets. This however requires use of the usually limited amount of land around dwellings. The tree diversity was greater in Harmutty as compared to those in other sites. This could be due to the cultural advancement of the Kalitas who have been practicing a kind of sub-commercial homegardening. For instance, areca-nut tree has been returning good profits to the Kalitas in the region (Arunachalam *et al.* 2003). The species richness and evenness index of trees and shrubs were highest in site II, but the herbaceous species richness was more in Ziro. Tree diversity was more in site I, while that of shrubs in site II and herbaceous species in site III. Such variations might be due to socio-cultural variations between the ethnic groups studied and/or variation in climatic, topographic and altitudinal factors. Similarity of trees and herbs highest between the two Nyshi villages (sites II & III) 83.21 & 88.79 respectively. Due to altitudinal variations, there was low similarity between the homegardens of Ziro and the rest.

Plant detritus in traditional homegardens

One of the most important intrinsic features of the homegarden in this hilly region is its ability to withstand the natural forces of soil erosion. Several interacting factors are important. The almost closed canopy provides some protection, atleast from intensive rainfall and the different stratified layer of vegetation helps to reduce the velocity of raindrops, but by far the most important factor in preventing erosion is the protective layer of litter (Young, 1986). Most homegardens found in this region have a heavy litter layer and total litter biomass varied from 590 to 1498 kg ha^{-1} (Table 24.4). Variation of litter mass in this traditional homegarden system was due to a number of fruit trees, which produced high amount of plant detritus. The fewer amounts of litter in site II and III than in site I was due to the cleaning and/or occasional burning of plant litter (leaves, flowers, fruits and branches) mainly to reduce insect population in the garden and to maintain the cleanliness of the homegarden. These activities possibly contribute to energy being lost from the system through carbon liberation while disrupting the microbial decomposition process, thus affecting the soil nutrient cycling. Roots also play an important role in enhancing and maintaining the soil fertility in the homegarden systems and information regarding fine and coarse root biomass has become essential in designing and management of homegardens for optimum resource use. In the top soil (0-15cm), fineroot biomass was highest in site I (5986 kg ha^{-1}) due mainly to the practice of more agricultural crops, whereas coarse roots biomass was more in site III (2806 kg ha^{-1}), which had more timber-yielding tree species.

Soil nutrients in traditional homegardens

Soils in the different traditional homegarden system were mostly loamy (Table 24.5). The concentration of plant available nutrients in soil (ammonium N, nitrate N and available P) was maximum in homegardens at Harmutty. This could be attributed to greater litter availability and accumulation, as confounded by higher plants species richness and diversity in this system. Earlier it was reported that the higher fertility status might be due to efficient cycling by the presence of more nitrogen fixing plants in the system (Arunachalam *et al.* 1997). Moreover, the Harmutty site is also having N_2-fixing plants like *Moringa oleifera, Cajanus cajan, Zea mays, Mimosa pudica* and some pulses. Soils of Site II and III recorded low available nutrients, particularly ammonium and nitrate-N as compared to site I (Table 24.5). This may be due to higher rates of soil erosion as these two sites are on the Dikrong riverbank. However, the confounding effects of cleaning and/or burning of litter in this regard cannot be completely ruled out. On the other hand, abundance of

Table 24.1 Vertical distribution of species in the traditional homegardens.

Stratification	Plant Species
Storey (floor species) (<1000 lux)	*Zea mays* Linn., *Ananas comosus* (L) Merr., *Oscimum sanctum* Linn., *Solanum tuberosum* Linn., *Spinacea oleracea* Linn., *Brassica oleracea* Linn., *Daucus carrota* Linn., *Raphanus sativus* Linn., *Solanum melongena* Linn., *Phaseolus vulgaris* Linn., *Dolichos lablab* Linn., *Capsicum annum* Linn., *Abelmoschus esculentus* Linn, *Lycopersicum esculentum* Linn., *Vigna sesquipedalis*(L), *Lageneria siceraria* (Molina) Standl., *Momordica charantia* Linn., *Cucurbita moschata* Poir., *Luffa cylindrica* Roxb., *Cucumis sativus* Linn., *Colocasia esculenta* (L) Schott., *Allium cepa* Linn., *Zingiber officinale* Rosc., *Coriandrum sativum* Linn., *Pisum sativum* Linn., *Curcuma longa* Linn.
Storey (middle layer) (35990 lux)	*Carica papaya* Linn., *Moringa oleifera* Lamk., *Musa* spp., *Psidium guajava* Linn., *Citrus limon* (L.) Burm., *Cinnamomum tamala* Buch-Ham., *Nyctanthes arbor-tristis* Linn., *Cassia fistula* Linn., *Caesalpinia pulcherrima* Swartz., *Cajanas cajan* Mill. *Tabernamontanae coronaria* R. Br., *Nerium indicum* Mill., *Averrhoa carambola* Linn., *Hibiscus rosa sinensis* Linn., *Manihot esculenta* Crantz.
Storey (sub-canopy) (39325.50 lux)	*Artocarpus heterophyllus* Lam., *Azadirachta indica* A. Juss., *Mangifera indica* Linn., *Zizyphus jujuba* Lamk., *Callistemon linearis* Linn.
Storey (Canopy) (46465 lux)	*Areca catechu* Linn., *Bambusa* spp., *Cocos nucifera* Linn., *Eucalyptus globosa* Labil.

Table 24.2. List of plants categorized by farmers on the basis of light requirements in the traditional homegardens.

Light requirement	Plant species
Strong light- demanding species	*Artocarpus heterophyllus* Lam., *Mangifera indica* Linn., *Cocos nucifera* Linn., *Bambusa* spp. Linn.
Medium light- demanding species	*Averrhoa carambola* Linn., *Azadirachta indica* A. Juss., *Zizyphus jujuba* Lamk., *Areca catechu* Linn., *Lycopersicum esculentum* Linn., *Brassica oleracea* Linn., *Daucas carrota* Linn., *Raphanus sativus* Linn., *Allium cepa* Linn., *Coriandrum sativum* Linn., *Pisum sativum* Linn., *Solanum melongena* Linn.
Moderate light- demanding species	*Carica papya* Linn., *Moringa oleifera* Lamk., *Musa* spp., *Psidium guajava* Linn., *Citrus limon* (L.) Burm., *Cinnamomum tamala* Buch-Ham., *Nyctanthes arbor-tristis* Linn., *Cassia fistula* Linn., *Caesalpinia pulcherrima* Swartz., *Nerium indicum* Mill., *Averrhoa carambola* Linn., *Hibiscus rosa-sinensis* Linn., *Manihot esculenta* Crantz., *Zea mays* Linn., *Capsicum annum* Linn., *Cajanus cajan* Mill.
Shade-loving species	*Hibiscus rosa-sinensis* Linn., *Nerium indicum* Mill., *Tabernamontanae coronaria* R. Br., *Nyctanthes arbor-tristis* Linn., *Ananas comosus* (L)Merr., *Oscimum sanctum* Linn., *Spinacea oleracea* Linn., *Brassica oleracea* Linn., *Phaseolus vulgaris* Linn., *Dolichos lablab* Linn., *Abelmoschus esculentus* Linn, *Vigna sesquipedalis* (L)Fruw., *Lageneria siceraria* Standl., *Momordica charantia* Linn., *Cucurbita moschata* Poir., *Luffa cylindrica* Roxb., *Cucumis sativus* Linn. *Colocasia esculenta* (L) Schott., *Zingiber officinale* Rosc., *Pisum sativum* Linn., *Curcuma longa* Linn.

Table 24.3 Vegetational characteristics of traditional homegardens.

Parameters	Site	Tree	Shrubs	Herbs
Density (Plants ha^{-1})	I	468.5	360.84	40140
	II	142.5	330.42	16870
	III	170.0	616.09	23520
	IV	16.0	152.20	60500
Basal Area (m^2 ha^{-1})	I	27.74	0.62	528300*
	II	8.42	0.31	522300*
	III	5.28	0.43	63010*
	IV	4.55	0.22	175400*
Number of species	I	19	12	17
	II	10	11	11
	III	8	12	14
	IV	5	10	11
Species Richness Index	I	1.45	0.84	0.47
	II	1.56	0.96	0.41
	III	1.40	0.79	0.44
	IV	1.25	0.79	0.45
Species Diversity Index	I	1.171	1.987	1.127
	II	1.165	1.106	1.297
	III	1.104	1.149	1.188
	IV	0.651	0.99	0.98
Dominance Index	I	1.146	0.101	0.221
	II	0.142	0.102	0.279
	III	1.103	0.130	0.162
	IV	0.240	0.11	0.12
Evenness Index	I	0.91	0.91	1.03
	II	1.29	1.063	1.25
	III	1.10	1.06	1.03
	IV	0.93	0.99	0.94
Similarity Index	I vs II	63.01	88.02	82.31
	I vs III	65.27	84.77	80.52
	I vs IV	-	16.67	8.33
	II vs III	83.21	71.19	82.79
	II vs IV	12.5	14.28	13.04
	III vsIV	-	17.39	12.50

Site I = Harmutty; Site II = Doimukh; Site III = Nirjuli; Site IV = Ziro

* For herbaceous plants, abundance (ha^{-1}) value was used for calculating IVI

Table 24.4. Litter and root biomass (kg ha^{-1}) in traditional homegarden

Biomass	Component		Site-I	Site-II	Site-III	Site-IV
Litter	Leaf		1031.33	524.33	767.50	432.55
	Stem		199.17	199.17	85.84	61.24
	Misc.		267.84	80.83	129.67	96.46
	Total		1498.34	804.33	983.01	590.25
Roots	Fine-root	0-15	4267.00	1486.17	1461.17	1040.32
	(<2mm)	15-30	1718.50	1188.33	640.50	662.80
	Total	0-30	5985.5	2674.5	2101.67	1703.12
	Coarse	0-15	827.17	956.33	921.84	510.08
	(>2mm)	15-30	1874.66	1778.83	1884.33	487.48
	Total	0-30	2701.83	2840.66	2806.17	997.56

Table 24.5 Soil physico-chemical properties in traditional homegardens.

Parameters	Site-I	Site-II	Site-III	Site-IV
Texture				
Clay (%)	11.21 ±0.30	6.09 ±0.29	3.71 ±0..08	9.27 ±0.35
Silt (%)	11.33 ±0.31	11.16 ±0.39	7.14 ±0.28	7.09 ±0.14
Sand (%)	76.69 ±0.59	82.75 ±0.07	89.14 ±0.53	85.76 ±1.58
Texture class	Sandy loam	Loamy sand	Loamy sand	Loamy sand
WHC (%)	36.36 ±0.76	42.26 ±0.76	42.70 ±0.63	57.99 ±1.55
Soil moisture (%)	22.56 ±1.01	33.59 ±0.93	26.49 ±1.83	39.82 ±1.81
Soil pH (1:2.5 w/v H_2O)	5.68 ±0.02	6.09 ±0.03	5.82 ±1.24	6.10 ±0.44
Organic C (%)	3.88 ±0.15	3.30 ±0.16	2.36 ±0.29	2.77 ±0.13
Organic matter (%)	6.69 ±0.28	5.69 ±0.29	4.07 ±0.51	4.78 ±0..23
TKN (%)	0.40 ±0.01	0.35 ±0.03	0.38 ±0.04	0.38 ±0.01
C/N	9.70	9.43	6.21	7.29
Ammonium N (µg g^{-1})	6.93 ±0.03	5.45 ±0.48	6.19 ±0.13	4.76 ±0.14
Nitrate N (µg g^{-1})	1.02 ±0.04	0.83 ±0..01	0.94 ±0.01	1.01 ±0.03
Available P (µg g^{-1})	32.40 ±1.41	14.31 ±0.51	25.33 ±0.44	3.20 ±0.36

WHC-Water Holding Capacity; TKN-Total Kjeldahl Nitrogen
±S.E. (n=5)

slow-decomposing litter of species such as *Prunus, Pyrus* and *Citrus* sp. and gentle sloping nature of the area affected the soil nutrient mineralization in site IV.

CONCLUSION

The present study revealed that the traditional homegarden of north-east India maintained a multi-storeyed system like in other parts of India. Identification and classification of plant species by the farmers grown in traditional homegarden system on the basis of dominant plants, relative height and light requirements indicated that the farmers have considerable ecological knowledge of aboveground interaction of the species. Species diversity seems to be managed in the homegardens in accordance with the family needs primarily linked to tradition. Significant difference in species selection for homegarden has been observed between lower and upper elevation sites, perhaps due to the change in the climatic regime. The vegetational study, detrital biomass and soil physico-chemical properties of Harmutty village showed that it is relatively more sustainable than the other sites. However, these systems may not be productive on a long-term basis in general, as it is suffering from lack of proper designing and management, and therefore much work has to be conducted in this direction for improvement of these low to no input traditional systems.

ACKNOWLEDGEMENT

The authors are thankful to the Indian Council of Agricultural Research for financial support.

REFERENCES

Anderson, J.M. and Ingram, J.S.I. (1993). Tropical Soil Biology and Fertility: A Handbook of Methods, CAB International, Wallingford, U.K.

Arunachalam, A., John, B., Maithani, K., Pandey, H.N. and Tripathi, R.S. (1997). Vegetational and edaphic during secondary succession in a subtropical pine and broadleaved forests in northeast India. *Ekologia* 16: 253-264.

Arunachalam, A., Pandey, H.N., Tripathi, R.S. and Maithani, K. (1996). Biomass and production of fine and coarse roots during regrowth of a disturbed subtropical humid forest in northeast India. *Vegetation* 123: 73-80.

Arunachalam, A., Melkania, U., Arunachalam, K., Srivastava, K. (2003). An ecosystem level study in litter and belowground dynamics in agroforestry and homegarden systems in and around Arunachal Pradesh, *Annual Progress Report, Indian Council of Agricultural Research Institute,* New Delhi.

Bohning, R.H. & Burnside, C.A. (1956). The effect of light intensity on rate of apparent photosynthesis in leaves of sun and shade plants. *American Journal of Botany*.3: 557-561.

Central Plantation Crops Research Institute, CPCRI (1976). Annual Report for 1975. CPCRI, Kasargod, Kerala, India. pp.24-27.

Curtis, J. K. (1959). The Vegetation of Wisconsin: An Ordination of Plant communities. University of Wisconsin Press, Madison, Wisconsin.

Fernandes, E.C.M. & Nair, P.K.R. (1985). An evaluation of the structure and function of tropical homegardens. *Agricultural systems* 21: 279-310.

Hooker, J. D. (1872-1897) The Flora of British India, 7 volumes, L. Reeve & Co., London.

Hutterer, K.L. (1984). Ecology and evolution of agriculture in Southeast Asia. pp.75-97 *In* T.A. Rambo and P.E. Sajise, eds., *An Introduction to Human Ecology Research on Agricultural Systems in Southeast Asia.* University of Philippines, Los Banos.

Karyono, (1981). Homegarden structure in areas of the Citarum watershad, West Java. Ph.D. Dissertation, Padjadjaran University, Bandung. Indonesia.

Kumar, G. (1997). Geology of Arunachal Pradesh. Geological Society of India, Bangalore.

Maithani, K., Arunachalam, A., Pandey, H.N. and Tripathi, R.S. (1997). Dry matter and nutrient dynamics of litter during forest regrowth in humid subtropics. *Ekologia* 16: 49-57.

Margalef, R. (1968). Perspective in Ecological Theory. University of Chicago Press, Chicago.

Okafor, J.C. & Fernandez, E.C.M. (1987). The compound farms at southeaster Nigeria: A predominant agroforestry homegarden system with crop and small livestock. *Agroforestry Systems* 5: 153-168.

Pielou, E. C. (1966). The measurement of diversity in different types of biological collections. *Journal of Theoretical Biology* 13:131-144.

Simpson E.H. (1949). Measurement of diversity. Nature (London) 163:688.

Sorenson, T. (1948). A Method of establishing groups of equal amplitude in plant sociology based on similarity of species content. Det. Kong. Danske Vidensk, Selsk Biology Skr (Copenhagen) 5: 1-34.

Young, A. (1986). The potential of agroforestry for soil conservation. Part I. Erosion Control. *ICRAF working paper* No. 42. International Council for Research in Agroforestry, Nairobi, Kenya.

25

Structure and Function of Homegardens of Andaman and Nicobar

C. B. PANDEY AND R. B. RAI

Central Agricultural Research Institute, Post Box 181
Port Blair- 744 101 Andaman and Nicobar Islands

ABSTRACT

Homegarden is a least studied agroecosystem of the world. Present study, conducted in 430 homegardens before Tsunami in five island groups namely North Andaman, Middle Andaman, South Andaman, Car Nicobar and Noncowry (former three, Andaman group and later two, Nicobar group of islands), India, reports composition, species structure and its distribution and functioning of the system. The homegarden is comprised of 6 components i.e., garden, livestock, poultry, fishpond, homesteads, and rice fields. Coconut and arecanut trees predominate in the garden. Fruits, spices and agroforestry trees are other ligneous species, which together with the former trees form multistorey canopy structure. But, composition and dominance of the tree species vary across the islands. Arecanut to coconut ratio is greatest invariably in all the homegardens in the Andaman. Family members do maximum 70 per cent labour in the homegarden and consume 35-65 per cent fruits, poultry, livestock and fishpond products, however, surplus is sold. Free range grazing in nearby protected forests is an age old practice of animal feeding whereas poultry birds are fed among the garden trees and household refuge. Livestock includes cattle, buffaloes and goats in Andaman and pig and goat in Nicobar. Cattle are used for milk, he buffaloes as a draft animal and pig and goat for meat. Cattle, mostly local breed, yield a little amount of milk, an average, 1 liter cow^{-1}. Plantation crops production is low: coconut 18-30 nuts palm^{-1} and arecanut 0.75 kg palm^{-1}. The homegardens serve the subsistence as well as earn income, but low (US$ 2357) to the house. In Nicobar, Nicobarese huts interspersed in a big natural coconut plantation form the homegarden. Tribe Nicobarese do not know crop husbandry, rather they pursue round the year sea fishing and pig rearing. They relish most pork, unripe boiled wild banana, Nicobari tuber, fruits and seeds of Pendanus and fried fish.

INTRODUCTION

Homegardens are the basic production units and the centers of social and cultural well being of farm families (Rugalema *et al.* 1994). It forms a dominant and promising land use system in many parts of tropics (Michon *et al.* 1986; Padoch and De Jong, 1991; Singh 1987) and maintains high levels of diversity, productivity and sustainability (Michon *et al.*, 1986; Thaman,1988; Soemarwoto and Conway, 1991).

Andaman and Nicobar archipelago is a chain of 572 islands, isle and reefs, but so far only 36 are inhabited (Figure 25.1). Homegardens are most prevalent agroforestry system in these islands, which owe their origin from forests. The islands were thickly covered with forests before the British colonial rule. During the late mid 19th century the British rule established penal colonies with the mutineers, rebellions and convicts of murder, robberies and frauds from occupied India and Burma that could not last ever long and was abolished an year before the India was set free and those willing were repatriated. Subsequently, the Govt. of independent India rehabilitated these islands, in phases, from mid 20th century with the ex-convicts who resorted to stay back, refugees of East Pakistan, ex-servicemen and poor landless people of India from different parts of the country (Kloss, 1971; Chak, 1967; Majumdar, 1975). The settlers were provided with 2 ha of forested hilly land, 2ha of valley (low lying rice fields) and 0.4ha land for house construction. However, size of the allotted land varied across the islands being greater in the beginning. Settlers cleared the forests, constructed houses and raised coconut, arecanut, fruit and other trees in their premises to serve their maximum basic needs as the transportation and organized market were completely lacking. This led to the development of homegardens, a permanent system of rural economy. Moreover, the settlement was not homogeneous resulting in the predominance of a community in a particular island or in a region of islands. For example, Nicobarese, one of the six aboriginals, dominates in Nicobar whereas Bengalies, descended from east Pakistan and Bengal province of India, in Diglipur and Hut bay, who have their different inherent food habits and socio- cultural structures that caused variation in the composition and structure of the homegardens across the islands. Many homegarden systems around the world have been already analyzed (Michon *et al.*, 1986; De Clerck and Negreros-Castillo, 2000; Millate-E-Mustafa *et al.*, 1996; Wezel and Bender, 2003), but knowledge about the homegardens of Andaman and Nicobar is very small. Present study reports composition, structure, productivity, sustainability and socio-economic characters of the homegardens in Andman and Nicobar Islands, India before Tsunami.

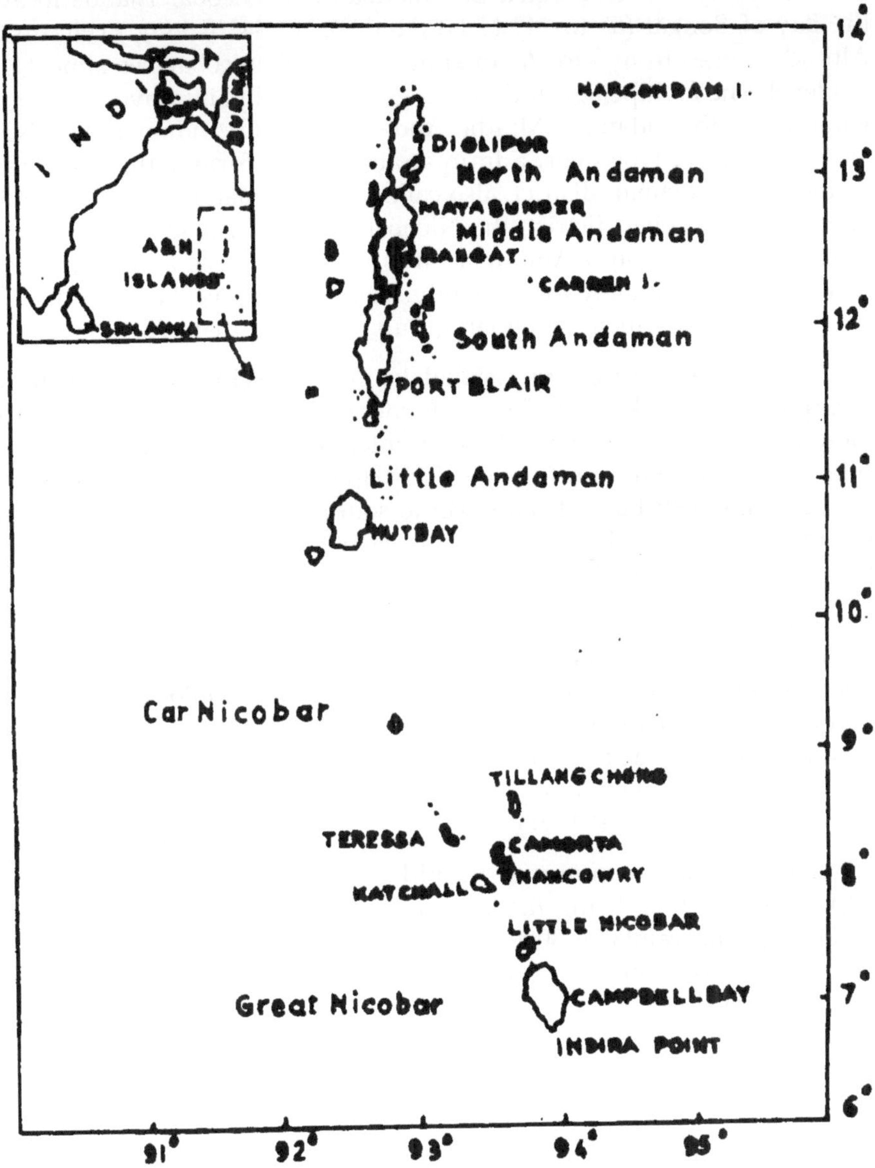

Fig. 25.1. Location of Island groups of Andaman and Nicobar

GENERAL DESCRIPTION OF AREA

The study was conducted in Andaman and Nicobar islands located in the Bay of Bengal (6° to 14° N lat., and 92° to 94° E long.) (Figure 25.1). Altitude ranges from 0 to 365 m above sea level. Nicobar lies almost on the sea level. The archipelago is divided, geographically, into five island groups namely, North Andaman, Middle Andaman, South Andaman, Car Nicobar and Nancowry. However, administratively the former three groups are included in Andaman district, also called as Andaman group of islands, and later two in Nicobar district, the Nicobar group of islands. Little Andaman though falls into south Andaman, it is well known as a vegetable island, hence treated separately for the study. Total geographical area of the islands is 8249 sq km of which Andaman group of islands share 78%.

The islands are characterized by hills, having moderate to steep escarpments; hillocks, with low to moderate slopes; and valleys, narrow to large (200 m to 2 km); flat bottomed, interspersed in between the hills and hillocks. The terrain is undulating rising up to 732 m (Saddle peak). The hills are generally distributed from north to south with ridges and spurs striking out in different directions. Some of the islands like Little Andaman, Neil and Car Nicobar are entirely flat.

Predominant soils in the islands are alfisols, entisols, inceptisols and mollisols. Inceptisols and entisols are most important agricultural soils found abundantly on the hillocks and valleys. Soils on the hillocks are shallow to moderately deep owing to erosion, well drained, generally gravelly loamy in texture, acidic in reaction and low to moderate in nutrients. However, soils of the valleys are alluvial, relatively deeper, imperfectly drained and clayey to clay-loamy (NBSS LUP 1991).

The climate is equatorial humid tropical. Temperature varies from 23.1 ^{0}C to 30.10 ^{0}C, being maximum in May and minimum in December. Average annual rainfall is 3000 mm, distributed over 8-9 months. High rainfall (73 %) prevails from May to Oct., whereas four months from December to March are relatively dry. The rainfall is most frequent from 2 to 6 am and 2 to 6 pm and least frequent shortly before the mid night. Humidity ranging from 71 per cent to 85 per cent is maximum in September and minimum in February.

Potential natural vegetation in the islands is tropical rainforest. Natural forest trees are occasionally found in the villages. Moist deciduous forest represented by *Pterocarpus dalbergiodes* DC., *Lagerstoemia hypoleuca* Kurz, *Terminalia bialata* Steuiel etc., occupies mostly the foothill whereas semi-evergreen forest dominates on the hill slope. Few evergreen species *Dipterocapus grandiflora* (Blanco) Blanco, *Artocarpus chaplasha* Roxb., *Hopea odorata* Roxb. etc. is interspersed among the deciduous species, particularly in the clay rich depressions, and increases with the altitude on the hill slope and replacing others, *D. alatus* Roxb. ex G. Don, *Amoora wallichii* King, *D. gracilis* Blume, *Calophyllum* L. *spp.* etc. occupy the hilltop. Hills of north Andaman, having springs, streams and rivulets are covered from foot hill to hill top with *Dipterocarpus alatus* and its associates. Whereas, hillocks of South Andaman inhabit mostly moist deciduous forests. On the interface of the sea and land mass mangrove vegetation is found which protect the islands from sea invasion. Mangrove includes *Rhizophora mucronata* Lamk. *R. apiculata* Bl., *Bruguiera cylindrical* (L.) Bl., *B. parviflora* (Roxb.) W. and A., *Avicenia officinalis* Linn., *Heritiera littoralis* Dryander, *Ceriops tagal* (Per.) Robins., *Kandelia candal* (L.) Druce, *Xylocarpus mollucensis* (Lamk.) Roem., *Sonneratia caseolaris* (L.) Engl., *Excoecaris* L. spp. Littoral forests, found on the sea beeches, are dominated by salt tolerant *Manilkara litoralis* (Kurz) Dub., *Erythrina variegata* L., *Pandanus tectorius* Parkinson, *P. leram* Jones ex Fontana, and *P. odoratissimus* L. F. (Forest Working Plan, 1980).

MATERIAL AND METHODS

A total of 43 villages, 10 from each of North Andaman, Middle Andaman, South Andaman, and 10 villages, 5 from each Car Nicobar and Nancowry, and 3 villages in Little Andaman were selected randomly. Ten random homegardens in the each village were sampled. Thus, total 430 homegardens were studied. Five 10 x 10 m quadrats were laid randomly in the each garden and number of individuals was counted and height was measured, and density and frequency of individual plant were calculated following Curtis and McIntosh (1950) and presented on per hectare basis. Vegetable's habit included climber as well as erect herbs; therefore, their relative cover was measured to evaluate their relative dominance. Ten long transects of 100 m was laid at random across the vegetable fields, presence was noted of individual crop and cover was assessed by point intercept method (Goodall 1953) at 5 m interval using 1 x 1 m^2 random quadrats for pin contacts with 100 pins placed in a grid pattern over each quadrat. Each grid consisted of 10 transects placed 10 cm apart. Total point: crop contact was noted for individual crop and percent cover was computed. The formulae used are as follows:

$$\text{Frequency (\%)} = \frac{\text{Total number of quadrats in which the species occurred}}{\text{Total number of quadrats studied}} \times 100$$

$$\text{Density} = \frac{\text{Total number of individuals of a species}}{\text{Total number of quadrats studied}}$$

$$\text{Relative frequency\%(RF)} = \frac{\text{Frequency of the species}}{\text{Total frequency of all species}} \times 100$$

$$\text{Relative density \% (RD)} = \frac{\text{Density of the species}}{\text{Total density of all species}} \times 100$$

Importance value index (IVI) was calculated for all the species in the garden by summing up the values of their RF and RD. To know the relative dominance of plant species in homegardens and vegetable crops among the islands their IVI and relative cover, respectively have been grouped into 6 dominance ranks (I-VI) where rank I is assigned maximum value and rank VI the lowest. Two family members (one male and one female) and agriculture officer of the area were interviewed for the socio-economic aspects of livestock, poultry, fishpond, paddy and vegetable crops. However, secondary data were obtained from Basic Statistics (2001).

RESULTS AND DISCUSSION

Forest (87%), homegarden (4.6%) and Rice fields (1.3%), covering together around 93 per cent of the total geographical area, are three major land uses in the islands (Basic Statistics, 2001). Homegardens are located on the hill slopes and hillocks whereas rice fields in the valleys (Figure 25.2). Size of the homegarden ranges from 0.05 to 2 ha in Andaman and 0.5 to 5 ha in Nicobar. It is fragmented with time in both the island groups mainly due to population growth and partly because the land is saleable commodity similar to that found in North- Western Tanzania (Rugalama *et al.*, 1994). Most of the hills are still virgin covered with thick tropical rain forests. Forests are divided in two groups, reserved forest (41%) and protected forest (59%) from the biodiversity conservation viewpoint and are managed by the Forest Dept. of Andaman and Nicobar administration. Until year 2002, the bona fide inhabitants of the villages involved in agriculture on the allotted land were provided with license on the payment of nominal amount of royalty, to extract 15 cubic meters of timber for construction of a new house in life, 6 cubic meters of timber for repair a house and 1 cubic meter of timber for the construction of agricultural implements, once in five years, to be used solely for domestic purposes. Unauthorized timber extraction, saw milling, and

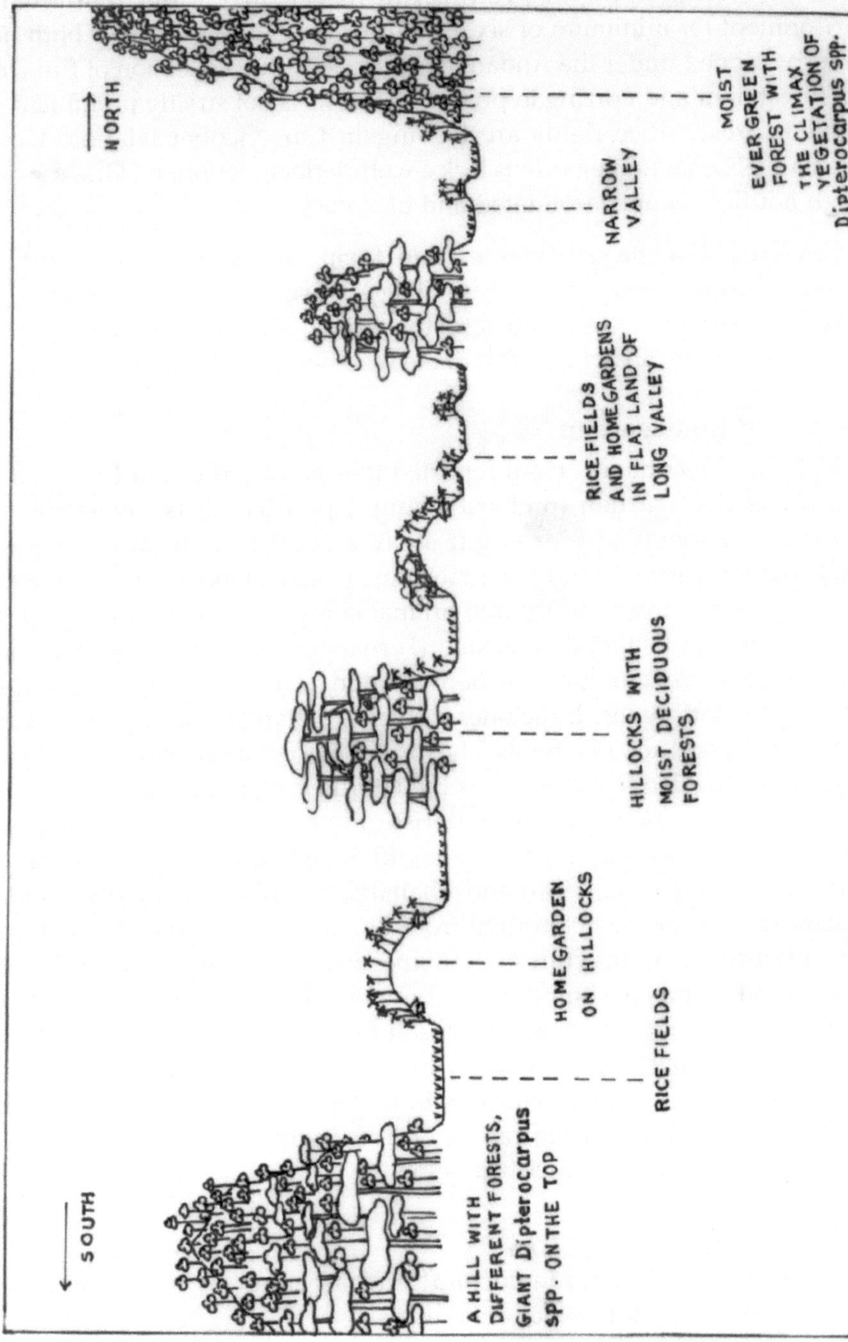

Fig. 25.2 Location of Uomegardens in AXN Islands

collection of minor forest produce, however, are offences for which imprisonment for minimum of six months or fine up to Rs. 500 or both have been provisioned under the Andaman Forest Act. But, collection of fire wood from forest floor and grazing in protected forests is not strictly prohibited like reserved forests. Rice fields are lacking in Car Nicobar islands. Unlike Andaman, Nicobar homegardens lack a well-defined boundary. Grazing land, though notified, is unconsolidated and of no use.

Sea infringes the landmasses and form creeks in the islands. It is abundantly found near the villages. Villagers use the creeks for fishing and mud crab harvesting. Mud crab is costly and exported to the central market. Villagers in Car Nicobar, however, make sea fishing as the creeks are lacking.

Structure of homegarden

Nair and Sreedharan (1986) reported that homegardens of Kerala, India are composed of garden (orchard), house, poultry birds, livestock, and fishpond. Rice fields also exist but away from the homegarden. Chagga homegarden in Tanzania and Javanese homegarden in Indonesia include food crops, cash crops, trees, shrubs and animal components like cattle, goat, pig and chickens, but fishpond is lacking (Fernandes and Nair, 1986; Michon *et al.*, 1986). However, the present homegarden is, infact, unique envisaging characters of both types. It includes garden (orchard), house, poultry birds, livestock, fishpond and rice-fields. The former five components are well fenced in with *Gliricidia* live fence, whereas the later lies adjacent to the garden, but always down to garden. However, floristically our homegardens are similar to that found in Kerala, India (Nair and Sreedharan, 1986), Bangladesh (Millate-E-Mustafa *et al.*, 1996) and Thailand (Boonkird *et al.*, 1984), but all the plant species are exotic brought from Burma and different states of India during the settlement. Structure of the homegarden is influenced due to holding size, food habit and consumption vs. commercial needs. The homegarden in the Nancowry group of islands is structurally similar to that in Andaman group of islands, but it differs entirely in Car Nicobar. Dominance (Important Value Index IVI) of plant species within a garden and in the gardens across the island groups is depicted in Table 25.1. Trees, which are commonly planted, cared and harvested, are described below.

Plantation crops

Plantation trees include coconut and arecanut, which serve house consumption and earn income to the house as well.

Coconut (Cocos nucifera L.)

Coconut is second dominant tree crop after arecanut in the homegardens (Table 25.1). It is reported to dominate also in the homegarden of Kerala, India (Nair and Sreedharan 1986), Bangladesh (Millate-E-Mustafa *et al.*, 1996), Mexico (De Clerck and Negreros- Castillo, 2000), Saparua islands, Indonesia (Kaya *et al.*, 2002), Thailand (Boonkird *et al.*, 1984). It is an indispensable part of food in all the societies in the islands, in one form or other, therefore, found in all the homegardens, an average, 128 tree ha^{-1}. The old trees are found wide (around 8.8 m) apart from one another, however, the spacing is not uniform due to new growth at random places from fallen seeds. Michon *et al.* (1986) also reported that fallen seeds always supplement propagation in the homegarden of Indonesia. Tall cultivars (Andaman tall and Katchal tall) are most common in the old plantation whereas dwarf cultivars in new orchards. However, in Nicobar maximum 98 per cent coconut trees are found in the homegardens (Table 25.1). Coconut plantation in Nicobar is natural from seeds flown in from Burma. Nicobarese neither make new plantation nor supplement regeneration by planting in the gaps formed due to death of old trees.

Arecanut *(Areca catechu* L.)

Density of arecanut is highest (an average, 1100 palm $tree^{-1}$) in all the homegardens. Like coconut, spacing in arecanut is also not uniform; rather it is too close (60 cm) partly due to natural regeneration and partly due to intention of the farmer to maximize the yields. Two high yielding cultivars, Mangala and Samurudhi are found in the homegardens, but former invariably greater in all the islands. Like coconut, new plantation is raised with a year old seedlings procured from State Agricultural Dept. In Nicobar islands, arecanut is very few scattered in the coconut plantation and does not merit elaboration.

Fruit trees

Fruit trees are an important component of homegardens in Asia (Millate-E-Mustafa *et al.* 1996). Fruit trees contributed maximum 50 per cent to the total species encountered in the homegardens. They are banana (*Musa paradisiaca* Linn.), Mango (*Mangifera indica* Linn.), papaya (*Carica papaya* Linn.), sapota (*Achras sapota* Linn.), guava (*Psydium guajava* Linn.), and lemon (*Citrus limon* Burm f.). These fruit trees are reported to found also in Cuban homegardens (Wezel and Bender, 2003) and African homegardens. Number of fruit trees in the present homegardens is very low, mango 2-3, lemon 1-2, sapota 1-2, jack fruit 0-1, orange 0-1 and banana 5-8 culm per home garden, indicating that the peasants are more intended to house consumption. Number of banana and mango is relatively greater as these are preferred most. All the fruit trees are planted with the seedlings obtained from State Agriculture Dept. Twenty cultivated and 6 wild types of banana are found in the islands, but champa (*Musa balbisiana* var. *andamanica* (Colla), is most common in the homegarden (Damodaran *et al.*, 2001). All the varieties are grown from suckers from old clump. Family members consume maximum (60 %) of the fruits. Banana grows wild in Car Nicobar. Fruit trees are found in meager in Nancowry group of islands.

Tree spices

Tree spices include clove (*Syzygium aromaticum* Thunb.), nutmeg (*Myristica fragrans* Houtt.), bay leaf (*Cinnamomum tamala* (Buch. -Ham.) Nees & Eberm.), cinnamon (*Cinnamomum zeylanicum* Nees.) and black pepper (*Piper nigrum* L.). These spices are reported to found in the homegardens of Indonesian archipelago (Kaya *et al.*, 2002). Maximum (95 %) peasants grow these tree spices for house consumption, hence, low in density, an average 1-2 trees in each homegarden. Cinnamon is a relatively greater 3-4 trees per homegarden. But, a few peasants (8 %) grow them for commercial purpose. Subsistence farming of the tree spices seems due mainly to high income from the arecanut at minimum management. Partly it may also be attributed to high labour input. Clove and nutmeg are raised from one year old seedling obtained from the State Agriculture Dept. Clove after 4 years and nutmeg after 5 years start flowering and bear fruits for around 50 years. Flowering in clove occur in April - May and harvesting in October. But, in nutmeg flowering and fruiting occurs round the year with a peak in July similar to that in Indonesia (Michon *et al.*, 1986). Nutmeg produces 1000 nuts per tree and clove 800g buds (flower) per tree. Macis of the nutmeg are also dried and used as spice. Like clove and nutmeg cinnamon is also grown with seedling. Two years old cinnamon stems (girth (cbh) = 17 cm, length = 3m) are harvested, debarked and dried. No diseases are observed in cinnamon may be due to short rotation. Michon *et al.* (1986) observed in Indonesia that more that 6 years old stems are infested with diseases. An average, 4-5 arecanuts in a garden are grown with the black pepper. Four rooted cuttings of black pepper are planted close to the palm and vines are trailed on the trunk. The vine grows for two years and bear fruits from third year for consecutive 20 years. Flowering occurs in June and harvesting in October with average production of 500 g vine^{-1}. Tedious and labour intensive harvesting of pepper berries and severe damage to the vines during arecanut harvesting that causes high yield loss, which distracts the peasants from its cultivation. Tree spices are almost absent in Car Nicobar and Nan Cowry group of islands.

Other trees

Agroforestry tree, *Ceiba pentendra* (L.) Gaertn. is abundantly found, an average 1 tree per homegarden, in Andaman. Flosses from the tree are used for making beds and pillows. Giant rain tree (*Samanea saman* (Jacq.) Kunth ex Walp.) is sparsely distributed mostly along the roads. The villagers use bark of the tree for wine preparation. Some natural forest trees are encountered occasionally in the homegarden (Table 25.1).

Annual crops

Rice is the most cherished food crop in the Andaman and Nicobar islands. Rice is imported maximum in Nicobar and partly in Andaman from the mainland territory. In Andaman, peasants grow mostly long duration rice (6 months) cultivar C14-8, in July-Aug and harvest in December. Farmers avoid short duration hybrid rice to overcome unfavorable long rainy months for post harvest operations. Average production of the rice is 2.69 t ha^{-1}. Rice is mostly transplanted with one-month-old seedlings raised in nursery, but only few (2 %) peasants pursue the weeding. However, fertilization in any case is lacking. Between two rice crops vegetables are grown in the limited rice fields. *Vigna unguiculata* L., *Luffa acutangula* (Roxb.) L., *Momordica charantia* L., *Cucumis sativus* L., *Trichosanthes cucumerina* L., *Lagenaria siceraria* (Mol) Standl., *Abelmoschus esculentus* L. and leafy vegetables like *Ameranthus tricolor* L. and *Basella alba* L. are grown round the year whereas cole crops from Nov. to Jan. In South Andaman landless new immigrants from the main territory in want of job perform maximum (95 %) vegetable and 20 per cent rice cultivation on the leased lands. Govt. service is a major source of income here hence most of the landowners lease their lands for half of the rice produced. But, for vegetables, land is leased only for money. However, in Middle, Little and North Andaman maximum peasants cultivate vegetables as well as rice of their own. But, in few islands like Neil, Diglipur and Hut bay, vegetable cultivation is a major source of income to homesteads and is grown in 20-25 per cent rice fields round the year. Vegetables are transported by ship to other islands particularly more to Andaman. Nicobarese in Nicobar do not know vegetable and rice crop husbandry. Paddy cultivation in Nancowry group is similar to that in Andaman group.

Livestock

Cattle, buffaloes, goats and pigs constitute the livestock. Size of livestock is 3 animals $homegarden^{-1}$. Cattle contribute maximum 83 per cent and buffaloes lowest 4 per cent to the livestock. In West Java livestock includes mostly wild fauna like monkey, civets and squirrels and wild pigs (Michon *et al.*, 1986). Milking cows are tethered in cattle shed, close to house, during night, whereas buffaloes under the trees in homegardens. Local breeds of cattle, buffaloes and goats are common in the islands which produce small amount of milk, an average, 1.0 L cow^{-1}, however, crossbred produce 6 L cow^{-1}. But, crossbreds are susceptible to mastitis, hump sore and reproductive problems; hence very few peasants (10 %) keep them. Cattle are used for milk, he buffaloes as a draft animal and goats for meat. Maximum (35 - 50 %) milk is used for house consumption but surplus is sold. Goat is one of the income sources to the homesteads. An adult goat costs around Rs.800 (Rs. 45.40= 1 $ US). In North- Western Tanzania cattle is primarily kept for manure because local breeds produce small amount of milk and meat (Rugalema *et al.*, 1994).

Free range grazing is a most prevalent animal feeding system in the islands. But, unlike North- Western Tanzania grazing herd in the present study include crossbred also (Rugalema *et al.* 1994). Mostly, grazing is done in near by protected forests. But, in Indonesia some buffaloes are kept for grazing in the agroforest (Michon *et al.*, 1986). After milking cattle, buffaloes and goats are let loose. They come back during evening as they are trained with grains and oil cake. But, during rice growing season community grazing occurs. One member from each cattle-keeping family joins the grazers and leads the herd to the forest. Nevertheless, in urban and close to urban areas cattle are made free to graze around round the year. Dry cows are left uncared to roam around until she is conceived. In middle Andaman, cattle and buffaloes are tethered during night in the rice fields particularly in dry months. Urine and dung's of the animals enriches the fields with nutrients.

In Car Nicobar, pig is a major constituent (98 %) of livestock. Pigs are fed with coconut kernel. Pigs are trained to rest on the ground floor of the hut whereas family members dwell on the first floor. The hut is constructed on the pillers of *Dipterocarpus* spp. First floor is walled with the panels of bamboo mat whereas ground floor have no wall.

Poultry

Chicken is an only component, which is common in almost all homegardens world over (Dury *et al.*, 1996; Nair and Sreedharan, 1986; Millate - E- Mustafa *et al.*, 1996). A poultry shed behind the house is a common feature of Andaman's homegardens. Poultry birds include chicken and duck. Poultry size ranging from 5 to 20, averages 15 homegarden^{-1}. Poultry birds are fed among the trees in orchard and household refuge. Maximum (65 %) poultry products like eggs and meat are used for house consumption. Eggs, cheaper in cost, mitigate the vegetable requirement, which is always in deficit due to high rainfall. Duck is always a constituent of poultry, but low (2-3 homegarden^{-1}) in number. Duck was introduced in the island as a bio-control to giant African snail (*Achatina filuca* Bouudich), which is a menace to agricultural crops. It is also one of the reasons why farmers do not grow annual crops in the homegarden. Ducks are fed in the ponds, rice fields and among the trees in homegardens. Egg and meat of ducks are mostly relish by family members. Surplus eggs and meat of both the birds are sold. Duck is lacking in the homegardens of Nicobar.

Fishpond

A typical homegarden of Andaman envisages a small pond called 'diggi' (r = 0.60 m, depth = 2.5 m) mostly nearer to the house. Bengali families relish fish most and own a diggi for fish cultivation for house consumption. But, 8 per cent homegardens have a big pond measuring 10 m x 10 m in size generally at the interface of garden and rice fields. Dury *et al.* (1996) also reported fishpond in Pekarangan homegardens of Indonesia. *Catla catla* (Hamilton), *Cirrhinus mrigala (Hamilton)*, *Labeo rohita* (Hamilton), *Pagasius pangasius* (Hamilton), *Clarias betrachus* (Linn.), *Heteropneustes fossilis* (Bloch), *Anabas testudineus* (Bloch), *Channa striatus* (Bloch) and *Maerobraehium rosenbergii* (de Man) are common fishes found in the fishpond of studied homegardens. Govt. is providing subsidies for the big size pond particularly for the conservation of rainwater, which serves also the irrigation need to vegetables during dry months.

Arrangement of components

Most distinguishing feature of the homegarden is a mixed distribution of species in three distinct tiers like other tropical homegardens (Nair and Sreedharan, 1986; Salam and Sreekumar, 1991; Millate-E-Mustafa *et al.*, 1996; Michon *et al.*, 1986; Kaya *et al.*, 2002; Boonkird *et al.* 1984) (Figure 25.3). Top storey (12 to 16 m) is always occupied by arecanut and coconut. Arecanut to coconut ratio is highest in all the homegarden indicating that arecanut is most favoured crop perhaps owing to its high price. The vertical stratification provides a gradients in light and relative humidity, which creates different niche for enabling various species groups to exploit them (Kumar and Nair, 2004). Until the last 6-7 years, when trade liberalization with Asian countries was not effected, the price of arecanut was Rs. 120 kg^{-1}. In bigger size homegardens a separate block of arecanut are grown mainly for the maximization of house income. Coconuts are planted generally at wider spacing in the beginning and arecanut in its interspaces. Trees like mango, jack fruits, neem, tamarind and *Ceiba pentendra* form the second storey (4.50

to 9.50 m). Fruit trees are mostly closer to the house. Spices like nutmeg, cinnamon, papaya and lemon occupying first storey are grown mostly under the coconut and occasionally under arecanut. Clove is found in both the first and second storey as well. Cinnamon is communally grown under the arecanut. Like homegardens of Indonesia banana requiring greater water and sunlight are grown relatively in open where water of house drains (Michon *et al.*, 1986). Distribution of the house has no specific pattern but is preferred on the up lands to avoid stagnation of rainwater. Generally ground cover is devoid of annual crops mainly due to giant African snails which being nocturnal and voracious causes severe loss to annuals and tender seedlings of trees and shrubs. Nonetheless, it is never a threat to the homegardens. It, infact, controls the stocking rate in the homegardens as the peasants never remove natural growth from the fallen seeds. Major threat to the homegardens is from soil erosion, which is estimated 12 t ha^{-1} under the plantation trees. However, tilth preparation on the slope (> 45^0) provokes the soil erosion to 122 t ha^{-1} yr^{-1} in open (Pandey and Venkatesh, 2003). Thus, zero cultivation on the ground floor of the homegardens seems advantageous. Few peasants (3 %) grow *Curcuma longa* (L.) Roxb, *Zingiber officinalis* Adams. (Boehm.), *Manihot esculenta* Crantz, *Ammorphophallus campanulatus* Blume and *Ananas comosus* (L.) Merr. in the garden in South Andaman and 20 per cent in North and Little Andaman mainly for house consumption. In old traditional Javanese homegardens also ground cover is devoid of crops. Vegetables are grown with papaya in open fields (Michon and Mary, 1994). Seasonal ephemeral grasses and legumes are also found in the garden particularly more during rainy season. Forest trees, an average one per home garden, are found occasionally in the homegardens. Flowers like crossandra (*Crossandra infundibuliformis* (L.) Nees), tuberose (*Polyanthes tuberosa* L.), marigold (*Tagetes erecta* L.) and jasmin (*Jasminum grandiflorum* L.) are grown a little in front of house and *Ixora parviflora* L. on the garden's boundary, but their contribution to the structure and economy of the homegarden is extremely low, hence not studied in detail. Unlike Andaman, single storey coconut garden predominates in Nicobar where wild tuber (*Tacca lentipetiloides* (Linn.) Kuntze.) and grasses constitute the ground flora.

Management

Peasants adopt agricultural practices particularly weeding and hoeing around the newly planted trees and undertake integration of natural processes of organic matter and nutrient cycling and tree regeneration similar to that in Javanese homegardens (Michon *et al.*, 1986). Soil working, weeding and an occasional little fertilization (150 g $tree^{-1}$) are performed in all the newly planted plantation and fruit trees in the garden, until flowering starts (4-5 years).

Once the bearing starts only weeding with local made swords is done to avoid soil losses. During weeding fallen leaves of respective trees are cut and put around the trunk for decomposition. Major factors determining the functioning of homegardens are the interactions among the soil, human, plant and animals. Peasants have years of accumulated knowledge of the ecological requirements of the trees. Cinnamon, clove and nutmeg requiring shade is planted as an understorey of coconut and arecanut palms. Banana and guava requiring greater light are planted comparatively in open. Regular planting is not exclusive way of reproduction in the homegardens as in West Sumatra, Indonesia (Michon *et al.,* 1986). Planting is always supplemented by the natural regeneration particularly of arecanut, coconut and cinnamon. Cinnamon is harvested and debarked and to avoid invasion in the gaps formed due to harvesting, weeding is done particularly during wet rainy season. It promotes good regeneration of the cinnamon. Generally diseases are not reported in the garden trees except occasional appearance of stem borer of coconut caused by rhinoceros beetle. Though not very common, peasants resort to bio-control measure by putting pheromone traps. Unlike North - Western Tanzania tending operations in the trees are not performed (Rugalema *et al.,* 1994) which helps change in the microclimate suitable to spices. Peasants generally do not cut the trees; rather let them die naturally like homegardens in West-Sumatra, Indonesia (Michon *et al.,* 1986). Gap formed due to death of the trees is filled with the natural growth. Peasants seem to have perception of nutrient cycling; therefore, they let the litter decompose on the garden floor. But, inorganic fertilizers are avoided, as they are scared of soil fertility to be spoiled. However, weeding, soil working, insecticide and fertilizer application are done frequently for vegetable cultivation.

Functioning of home garden

An average family size in Andaman is five; similar to that in Northwestern Tanzania, of which 1-2 are young men of age between 20-35 yrs (Rugalema *et al.,* 1994). In homegardens mostly family members do the works. But, in a community descended from Ranchi province of India, peasants mutually exchange the labour for rice cultivation. Peak season of farm operations is during paddy cultivation. Most of the tasks in the homegarden have no fixed period and can be staggered if needed. Transplanting, harvesting, winnowing are done maximum (80%) by women. Men prepare land, collect seedlings from the nursery and transport it to rice fields. Men and women both make fire wood collection from the nearby-protected forests. In vegetable cultivation, men do cumbersome works like plowing, fertilizer and pesticides application and marketing. Women perform weeding and picking. Nevertheless, men and women do irrigation jointly during dry spell. The men do cleaning of ground cover and harvesting of nuts, fruits, and spices. Mostly women carry out collection of nuts, dehusking of arecanut and post harvest operations of clove, black pepper and nutmeg. But, exclusively men do coconut dehusking.

Capital and other input

Mechanization is lacking in the homegardens. Tools used in the homegardens are spate, pickaxe, axe, local made sword and *dav*. Tilth preparation in the rice fields is done maximum 80 per cent by draft animals (he buffalos) and 20 per cent by hired tractor. Either no or a very little amount of fertilizer is used for paddy cultivation. However, fertilizers as well as insecticides are applied for vegetable cultivation. Humid climate of the island is very congenial for breeding of insects, which causes severe losses to vegetables.

COMPONENT INTERACTIONS

Rugalema *et al.* (1994) classified components interactions in North - Western Tanzania's homegardens into three groups i.e. cyclic interaction between crop residues, cattle manure, and tree litter; direct interaction between crops, livestock and trees / shrubs; competitive interaction between various livestock and trees / shrubs. In the present study components of homegardens are inter- dependant and inter- related (Figure 25.4). Dead roots, leaf, inflorescence and nut husks provide substratum to bio- chemical cycle similar to that of natural ecosystem which is further supplemented with the organic matter input from poultry and livestock and humus carried from the adjoining forests. It is observed that pool size of nutrients in the homegarden is either at par or greater than that of nearby forests depending upon the altitudinal location (C, N, P, K, Na = 0.91%, 0.32%, 25.46 μg^{-1}, 0.034 % and 0.027 %, respectively in homegardens and 0.96%, 0.41%, 21.18 μg^{-1}, 0.01 %, 0.013 %, respectively in evergreen forest) (C B. Pandey; pers. Comm.). Complementary interaction is evident from changed microclimatic condition under coconut and arecanut trees which helps better growth and yields in tree spices like clove, nutmeg, cinnamon and black pepper. Temperature is 2 ^{0}C lower (30 ^{o}C vs 28 ^{o}C) compared to open (C B. Pandey; pers. Comm.). (Intra- specific competition is common in arecanut patch owing to closer spacing (60 cm), which causes reduction in nut size and yields. Inter-specific competition occurs between coconut and fruit trees for soil- water, nutrients and light but the later greater, which change the geometry of fruit trees (lower canopy and greater height). Light is reduced 4.8 times under the trees in the homegarden compared to that in open (2439 Lux in homegardens vs 11636

Lux in open, C B. Pandey; pers. Comm.). However, components of homegarden i.e. trees, crops, livestock and fishpond interact together in space and time and result in steady flow of energy which leads to the cycling of nutrients that distinguish it from other farming system.

ECONOMIC CHARACTERISTICS

Homegardening is characterized of subsistence as well as commercial farming like other tropical homegardens (Dury *et al.* 1996; Alvarez-Buylla Roces *et al.*, 1989; Michon *et al.*, 1986). Arecanut and coconut are major commercial crops in the homegardens, production figures for which have been obtained from peasants whereas for consumed items it has been estimated (Table 25.2). Maximum 70 per cent coconut and 98 per cent arecanut of total production are sold to local vendors. The family members consume remaining amount. Arecanut is used for chewing whereas coconut as oil seed and several food items like sweets and sauce. An average production of coconut is 30 nuts palm^{-1} (10-50 nuts palm^{-1}) in Andaman and 18 nuts palm^{-1} (6-30 nuts palm^{-1}) in Nicobars. Arecanut production is, an average, 0.75 kg palm^{-1} (0.5 - 2.8 kg palm^{-1}). Total income from the all components, including rice and vegetables is estimated 2357 $ US which is within the range reported in Mexican (0 to 6525 $ US) (Alvarez- Buylla Roces 1989) and Javenese homegarden, Indonesia (365 to 5000 $ US) (Dury *et al.*, 1996). Coconut and fish are major commercial commodities, which earn livelihood to the Nicobarese in Nicobar. They prepare and sale coconut kernel. Nicobarese either do not eat or eat little coconut carnel but feed (25 %) always to their pigs. Pigs never contribute to the house income. Nicobarese celebrates pig festival once in a year where maximum pigs produced during the year in the village is, in fact, gifted to guests from neighbouring villages. It is a long tradition of mass propagation of pigs to cope up with the adversaries. Round the year they are engaged in sea fishing. Generally they use powder of *Heritiera littoralis* as a sedative to catch the fish. Boiled unripe banana, Nicobari tuber (*Tacca lentipetiloides*), *Pandanus* spp. and fried fish are their favourate food. They sell fish occasionally for the purchase of rice. Banana and Nicobari tuber are wild and the later grow relatively more abundantly everywhere in the islands.

Homegardens provide multiple products like fruits, food and spices for several years once established to house consumption without outside inputs like forest. Though price of arecanut has dropped drastically (4 times) in recent past, nonetheless, the peasants are not inclined to remove them, rather they are serious to compensate the loss by growing highly remunerative crops like black pepper and vanilla on *Gliricidia* standards in the interspaces of coconuts. Introduction of these crops undoubtedly will increase the diversity of the

Table 25.1. Dominanace (IVI) of plant species in a homegarden, averaged a cross the island groups, and among the islands groups of Andaman and Nicobar archipelago, India.

Common name	Scientific name	Dominance within a homegarden (IVI,%)	# Dominance rank among the islands groups					
			South Andaman	North Andaman	Middle Andaman	Little Andaman	Car Nicobar	Nancowry
Fruits								
Coconut	*Cocos nucifera* L.	25.30	III	V	IV	VI	I	II
Arecanut	*Areca catechu* L.	93.99	II	III	III	I	IV	IV
Cashewnut	*Anacardium occidentale* L.	2.15	I	II	II	II	-	-
Mango	*Mangifera indica* Linn.	12.53	I	I	II	III	IV	IV
Lemon	*Citrus lemon* Burm f.	2.09	II	I	III	III	-	-
Banana	*Musa paradisiaca* Linn.	13.60	I	II	IV	III	V	V
Papaya	*Carica papaya* Linn.	2.15	I	III	II	III	III	III
Guava	*Psydium guajava* Linn.	0.63	I	III	II	III	III	III
Sapota	*Achrus sapota* Linn.	1.03	I	III	II	III	-	-
Bael	*Aegle marmelos (L.)*	0.53	I	I	I	-	-	-
Custard apple	*Annona squamosa* L.	0.10	II	III	III	I	-	-
Sweet orange	*Citrus aurantiifolia* (Christm.) Swingle	0.45	-	I	II	-	-	-
Aowla	Phyllanthus niruri L.	0.32	I	III	II	III	III	III
Passion fruit	*Passiflora edulis* Sims f.	0.11	I	-	-	-	-	-
Pummelo	*Citrus grandis* Osbeck.	0.13	II	I	I	III	-	-
Tamarind	*Tamarindus indica* Linn.	1.28	IV	II	II	III	-	-
Jack fruit	*Artocarpus hetrophyllus* Lamk.	2.24	III	II	II	II	II	II
Pineapple	*Ananas comosus* (L.) Merr.	15.50	I	III	II	III	-	-
Spices								
Clove	*Eugenia cariophyllata* Thunb	4.31	I	II	II	III	-	-

Contd...

Table 25.1. Contd...

Cinnamon	*Cinnamomum zylanicum* Nees.	2.11	I	III	II	III	-	III
Nutmeg	*Myristica fragrans* Houtt.Nees.	2.07	I	III	II	III	-	-
Black pepper	*Piper nig*Lr.	0.85	I	III	II	III	IV	VI
Bay leaf	*Cinnamomum tamala* (Buch-Ham.) Nees & Eberm.	0.61	I	II	II	II	-	-
Agroforest and forest trees								
Silk cotton	*Ceiba pentandra* (L.) Gaertn.	4.22	I	I	I	I	II	II
Neem	*Azadirachta indica* A. Juss.	1.96	I	II	II	II	III	III
Kari leaf	*Murraya koenigii* (L.) Spreng.	2.09	I	I	I	I	II	II
Rain tree	*Samanea saman* (Jacq.) Merr.	0.06	I	I	I	I	-	-
Bamboo	*Bamboosa vulgaris* Schrad.exJ.C.Wendl.	0.03	II	I	II	III	-	-
Jindaballi	*Gliricidia sepium* (Jacq.)Kunth ex Walp.	13.84	I	II	II	II	III	III
Pyima	*Lagerstroemia hypoleuca* Kurz	2.15	I	I	I	I	-	I
Badam	*Terminalia procera* Roxb.	2.10	I	I	I	I	-	I
Padauk	*Pterocarpus dalbergioides* DC.	1.01	I	I	I	I	-	I
Garjan	*Dipterocarpus grandiflora* (Blanco) Blanco	0.31	I	I	I	I		I
Toungpinne	*Artocarpus chaplasha* Roxb.	0.24	I	I	I	I		I

Contd...

Table 25.1. Contd...

@Vegetables								
Drumstick	*Moringa oleifera* Lam.	0.13	I	II	II	I	III	IV
Bottle gourd	*Lajenaria siceraria* (Mol) Standl.	4.91	III	I	II	-	-	-
Bitter gourd	*Momordica charantia* L.	2.34	I	I	I	I	-	II
Little gourd	*Coccinia grandis* L.	1.12	I	I	II	I	-	III
Ridge gourd	*Luffa acutangula*(Roxb.)L.	8.55	I	I	I	I	II	II
Snake gourd	*Trichosanthes cucumerina* L.	0.97	I	I	I	I	-	II
Cucumber	*Cucumis stivas* L.	5.41	I	I	I	I	-	-
Pumpkin	*Cucurbita moschata* Duch ex Poir.	5.59	I	I	I	I	-	-
Pointed gourd	*Trichosanthes dioica* L.	1.61	III	II	IV	I	-	-
Cowpea	*Vigna unguiculata* L.	9.59	I	I	I	I	-	-
French bean	*Phaseolus vulgaris* L.	0.40	I	III	III	II	-	-
Okra	*Abelmoschus esculentus* L.	9.89	I	II	II	I	III	IV
Bringal	*Solanum melongina* L.	1.25	I	I	I	I	II	II
Tomato	*Lycopersicon esculentum* Mill.,nom.cons.	1.76	I	II	IV	III		
Chilli	*Capsicum anuum* vr. *Capitata* L.	4.83	II	I	III	IV	V	V
Cabbage	*Brassica oleracea* vr. *Capitata* L.	0.48	I	I	I	-	-	-
Cauliflower	*Brassica oleracea var. botrytis* L.	0.13	I	I	II	-	-	-
Radish	*Raphanus sativus* Linn.	0.35	I	I	I	II	-	-
Amananthus	*Amaranthus tricilor* L.	10.37	I	II	I	I	III	III
Poi Bhangi	*Basella alba* L.	0.13	I	II	II	I	-	-
Ginger	*Zingiber officinalis* Adams.(Boehm.)	4.82	I	II	II	II	-	-
Haldi	*Curcuma longa* (L.)Roxb	4.31	I	I	I	-	-	-

Contd...

Table 25.1. Contd...

Tapioca	*Manihot esculenta* crantz	0.22	I	II	II	III	-	-
Elephant foot yam	*Amorphophallus campanulatus* Blueme	1.50	I	I	II	II	III	III
Sweet potato	*Ipomoea batatus* (L.)Poir	0.24	II	III	III	I	-	II
Taro	*Colocasia esculenta* Schott	0.21	II	III	III	I	-	-
Greater yam	*Discorea alata* L.	0.12	II	III	III	III	I	I
Nicobari tuber	*Tecca lentipetiloides* (Linn.) Kuntze.	8.60	-	-	-	-	I	-

@ Relative cover is used to know relative dominance of vegetable crops

Dominance rank is arranged in decreasing chronosequence (rank I = height, rank VI = lowest)

system indicating no threat to the system's sustainability. Species and structural diversity are critical to the sustainability of homegardens, allowing efficient use and transfer of Carbon, nutrients and water (Benzamin *et al.,* 2001).

Table 25.2. Economic characteristics of homegardens of Andaman and Nicobar archipelago, India.

Components of @Home garden	Andaman			Nicobar		
	Gross income †(Rs)	#Input (Rs)	Net income (Rs)	Gross income (Rs)	Input (Rs)	Net income (Rs)
Arecanut	61495	10000	51495	0	0	0
Coconut	14000	5116	8884	12000	4000	8000
Fruit & spices	10492	1559	8933	0	0	0
Poultry	2684	0	2684	800	0	0
Dairy	10542	4032	6510	0	0	0
Pigery	0	0	0	7000	1000	0
Vegetables	48370	30721	17649	0	0	0
Fish - Pond	1728	198	1530	0	0	0
Sea fishing	8430	6590	1840	6000	5000	1000
Rice	13350	5875	7475	12500	6750	5750
Total	1,71,091	64,091	1,07,000	38,300	16,750	14,750

@Land management unit is ha^{-1}

Family labour is included in "Input." Input in rice include transplanting , harvesting, threshing, tilth preparation.

†1$ (US) = Rs.45.40, August 04

High diversity of species and their functions and the low input nature in the homegarden have a high degree of biological as well as socioeconomic sustainability (Michon *et al.,* 1986; Bezamin *et al.,* 2001). The trees in the homegarden input organic matter through leaf and twig litter and the death and decay of roots which is further supplemented with the organic matter input from livestock and poultry which maintain the productive potential of the soil similar to that of natural ecosystems. In rice field's organic rich fine soil particles carried from forests and homegardens along the run off is a major source of nutrients. Interviews of the peasants revealed that coconut produces 25-30 nuts $palm^{-1}$ and arecanut 0.75 kg nut $palm^{-1}$ and rice produces constant but low yields (2.69 t ha^{-1}) since past several years at minimum management. Greater ligneous life form, stable yields, high species richness and deep peasant's knowledge of tree positioning according to their ecological needs are the factors which seem to have made the homegardens of Andaman and Nicobar islands sustainable.

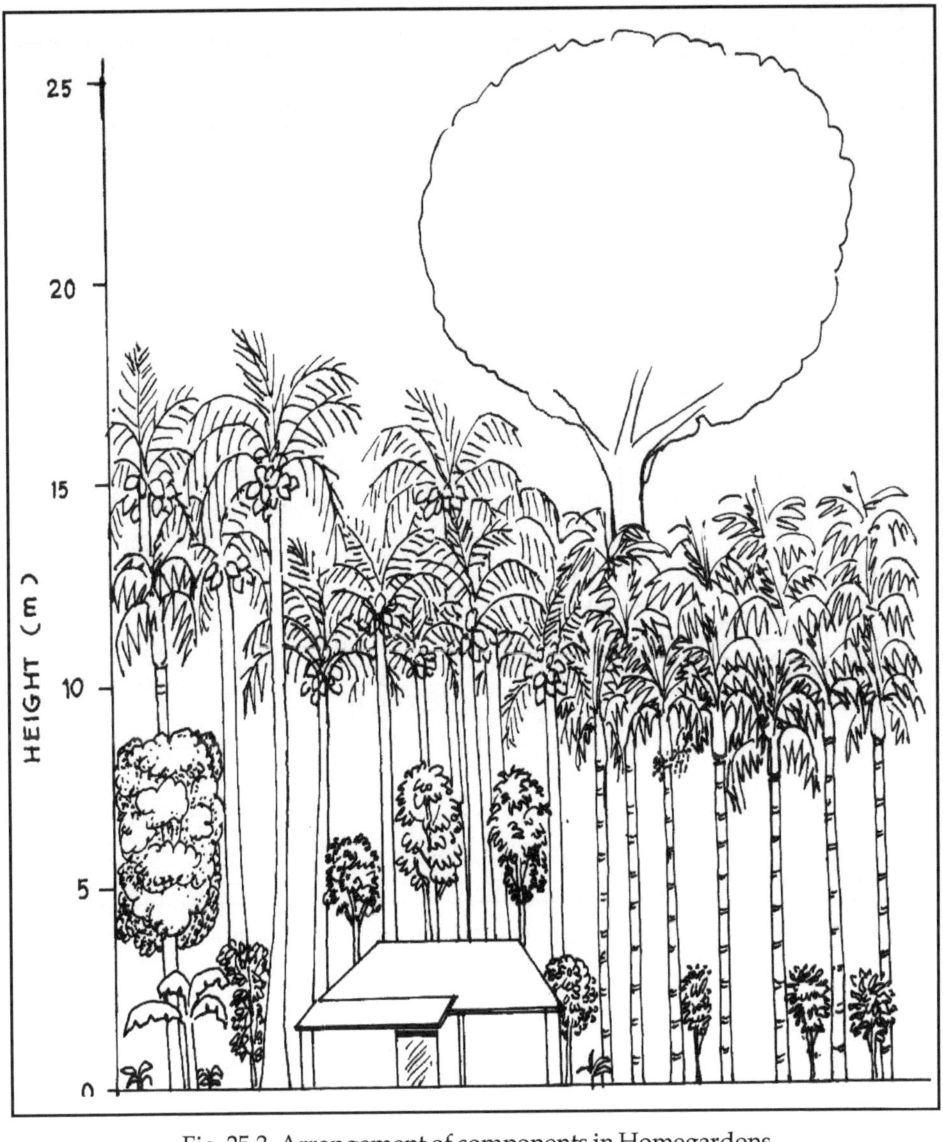

Fig. 25.3. Arrangement of components in Homegardens

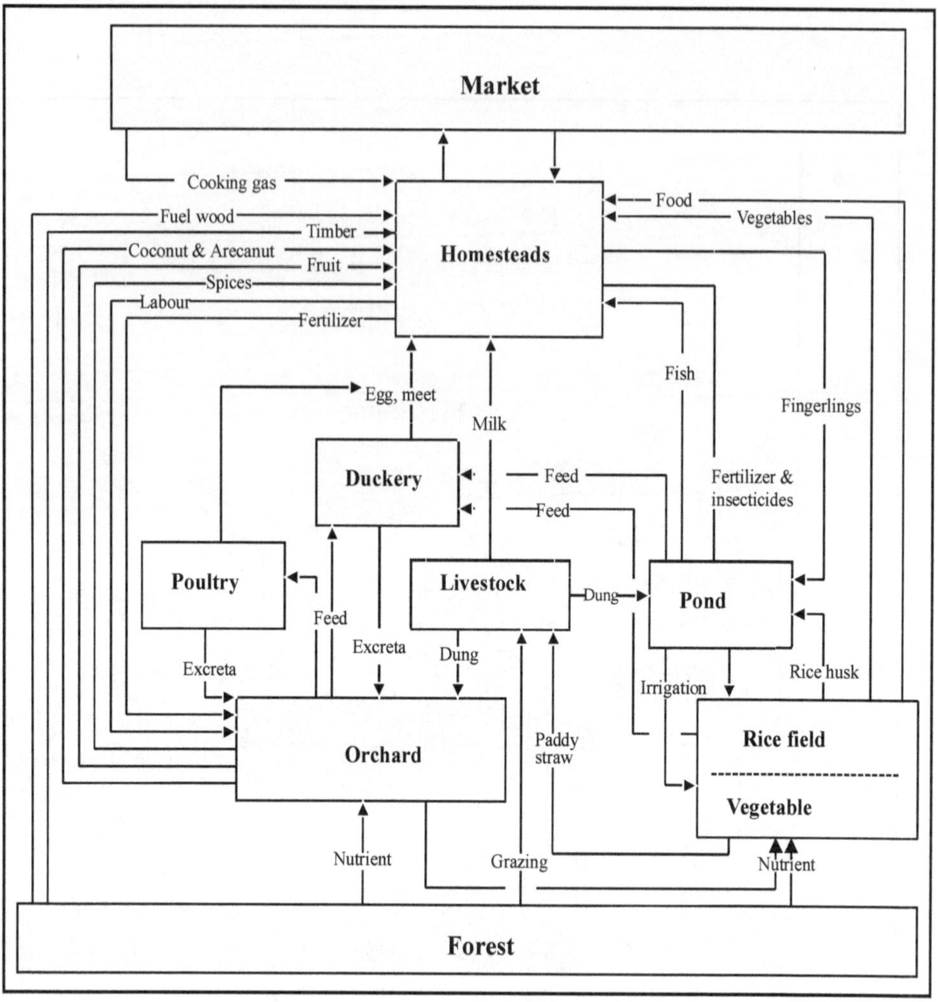

Fig. 25.4 Component interaction in Homegardens

ACKNOWLEDGEMENTS

Financial support from AP Cess Fund, ICAR, is thankfully acknowledged.

REFERENCES

Alvarez-Buylla Roces M. E., Chavero E. L. and Garcia Barrios J. R. (1989). Home gardens of a humid tropical region in southeast Mexico: an example of an agroforestry cropping system in a recently established community. *Agroforestry Systems* 8: 133-156.

Basic Statistics (2001). Andaman and Nicobar Islands: Basic Statistics. Directorate of Economics and Statistics, Andaman and Nicobar Administration, Port Blair, India, 461 pp.

Benjamin T. J., Montanez P. I., Jimenez J. J. M. and Gillespie A. R. (2001). Carbon, water and nutrient flux in Maya homegardens in the Yucatan peninsula of Mexico. *Agroforestry Systems* 53: 103-111.

Boonkird S., Fernandes E. and Nair, P. (1984). Forest villages: an agroforestry approach to rehabilitating forest land degraded by shifting cultivation in Thailand. *Agroforestry Systems* 2: 87-102.

Chak, B. L. (1967). Green Islands in the Sea. Ministry of Information and Broadcasting, Govt. of India Press, Nasik, India, 83 pp.

Curtis, J. T. and McIntosh, R. P. (1950). The interrelations of certain analytic and synthetic phytosociological characters. *Ecology* 31 : 434 – 455.

Damodaran, T., Medhi, R. P., Attri, B. L. and Suryanarayan M. A. (2001). Conservation of banana diversity in Bay Islands: An approach to prevent genetic erosion. *Indian Journal of Plant Genetic Resources* 14: 292 – 293.

De Clerck, F. A. J. and Negreros-Castillo, P. (2000). Plant species of traditional Mayan homegardens of Mexico as analogs for multistrata agroforests. *Agroforestry Systems* 48: 303-317.

Dury S.,Vilcosqui, L. and Mary, F. (1996). Durian trees (*Durio zibethinus* Murr.) in Javanese home gardens: Their importance in informal financial system. *Agroforestry Systems* 33: 215-230.

Fernandes, E.C.M. and Nair, P.K.R. (1986). An evaluation of the structure and function of tropical home gardens. *Agroforestry Systems* 21:279-310.

Forest Working Plan (1980). Working Plan for the Middle Andaman Forest Division. Port Blair, India, 238 pp.

Goodall, D.W. (1953). Some considerations in the use of point quadrats for the analysis of vegetation. *Australian Journal of Science Research Series* B 5:1-41.

Kaya, M., Kammesheidt, L. and Weidelt, H.J. (2002). The forest garden system of Saparua island, Central Maluka, Indonesia, and its role in maintaining tree species diversity. *Agroforestry Systems* 54: 225-234.

Kloss, C. B. (1971). In the Andamans and Nicobars: The Narrative of a Cruise in the Schooner " Terrapin" with Notices of the Islands, Their Fauna, Ethnology, etc. Vivek Publishing House, Delhi, India, 373 pp.

Kumar, B. M. and Nair, P. K. R. (2004). The enigma of tropical homegardens. *Agroforestry Systems* 61: 135-152.

Majumdar, R. C. (1975). Penal Settlement in Andamans. Ministry of Education and Social Welfare, Govt. of India Press, New Delhi, India, 339 pp.

Michon, G. and Mary, F. (1994). Conversion of traditional gardens and new economic strategies of rural household in the area of Bogor, Indonesia. *Agroforestry Systems* 25: 31-58.

Michon, G., Mary, F. and Bompard, J. (1986). Multistoried agroforestry garden system in West Sumatra, Indonesia. *Agroforestry Systems* 4: 315-338.

Millate -E- Mustafa, M.D., Hall, J.B. and Teklehaimanot, Z. (1996). Structure and floristics of Bangladesh homegardens *Agroforestry Systems* 33: 263-280.

Nair, M. A. and Sreedharan, C. (1986). Agroforestry farming systems in the homegardens of Kerala, Southern India. *Agroforestry Systems* 4: 339-363.

NBSS LUP (1991). Soil Resource Atlas: Andaman and Nicobar Islands. National Bureau of Soil Survey and Land Use Planning (ICAR) and Directorate of Agriculture, Andaman and Nicobar Administration, Port Blair, India, 64 pp.

Padoch, C. and De Jong, W. (1991). The house gardens of Santa Rosa: diversity and variability in Amazonian agricultural systems. *Economic Botany* 45: 166-175.

Pandey, C. B. and Venkatesh, A. (2003). Tree-soil-crop interactions in Andaman and Nicobar islands. In: Rai R. B., Sharma T. V. R. S., Suadararajan R., Sarangi N., Medhi R. P., Chaudhuri S. G., and Pramanic S. C. (eds.) Annual Report , Central Agricultural Research Institute, Port Blair, India.pp.15-16.

Rugalema, G. H., Johnsen, F. H., Oktingati, A. and Minjas, A. (1994). The homegarden agroforestry system of Bukoba district, North - Western Tanzania. 3. An economic appraisal of possible solutions to falling productivity. *Agroforestry Systems* 28: 227-236.

Salam, M.A. and Sreekumar, D. (1991). Kerala homegardens: A traditional agroforestry system from India. *Agroforestry Today* 3:10.

Singh, G. B. (1987). Agroforestry in Indian sub-continent: past, present and future. In: Steppler H. A. and Nair P. K. R. (eds) Agroforestry: A Decade of Development. ICRAF, Nairobi, Kenya, pp. 117-140.

Soemarwoto, O. and Conway, G. R. (1991). The Javanese home garden. *Journal of Farming System Research and Extention* 2 : 95 – 118.

Thaman, R. R. (1988). By the people and for the people : homegardens and national development in Pacific Islands. In: Hrist J., Ovington J., Allen B. and Byron Y. (eds) Small Scale Agriculture. Australian National University Canberra, Australia, pp. 167 – 181.

Wezel, A. and Bender, S. (2003). Plant species diversity of homegardens of Cuba and its significance for household food supply. *Agroforestry Systems* 57: 39-49.

26

Agroforestry Systems and Practices Adopted in Andhra Pradesh

B. SREEMANNARAYANA, B. JOSEPH AND L.G. GIRI RAO

AICRP on Agroforestry, Acharya N. G. Ranga Agricultural University, Rajendernagar, Hyderabad - 500 030

ABSTRACT

Andhra Pradesh is predominently an agrarian state, where 70 per cent of the people live in rural areas. Andhra Pradesh has 7 Agroecological zones spread over in coastal Andhra, Telangana and Rayalaseema regions with varied climatic situations. The existing agroforestry systems in costal Andhra are Mango-based Agrihorticulture system, Betelvine-sesbania system, Borrasus flabellefer, Glyrecidia boundary plantations, Eucalyptus, Causarina and Subabul block plantations. Besides Neem, Tamarind and jamuns are also found in isolated manner. In Telangana region, Custard apple based Agrihorticulture system, Mango based Agrihorticulture system, Boundary plantations of Neem, teak, Jamun etc., are present. Eucalyptus based Agrisilviculture system is propagated by ITC, Bhadrachalam in Khammam and elsewhere in the state. In Rayalaseema region in all the districts tamarind, mango and sweet orange based agroforestry systems are prominently seen. The betelvine - sesbania system alone is fetching an income of 100 crores/year to the states' exchequer. The common trees and tree systems available in the state are : *Azadirachta indica, Tamarindus indica, Acacia nilotica, Pithecellobium dulce, Parkinsonia aculeata, Pterocarpus santalinus, Dalbergia latifolia* and many others.

INTRODUCTION

The practice of agroforestry in Andhra Pradesh is as old as agriculture. Farmers in paddy growing areas, inclusive of those under rainfed and other irrigated areas, have been using glyricidia as green manure crop since time immemorial. In places where fruit growing is in vouge, trees such as Tamarind, Bamboo, Neem, and Jamun etc, are being grown as wind breaks to prevent damage due to fruit drop at the time of gales, especially during summer months. These include mango, cashew and coconut based agroforestry systems. A brief account of economic analyses of Betelvine -sesbania system shows that for an area of 3750 acres an amount of 100 crores/year is realized. Casuarina is grown as a shelterbelt in the coastal region of Andhra Pradesh. These shelterbelts prevent the drift of sand, protect the valuable agriculture land, roads and buildings from sand casting. The economic value of agriculture land with shelterbelts abutting has gone up and has helped to increase the crop yield by more than 30 per cent. The yield of casuarina varies from 10 (1x1m) to 20 (2x2m) tonnes/ha after 5 to 6 years, which in terms of revenue amount to Rs. 7000 to Rs.15,000/ha/year. In high altitude and tribal zone, the total yield of black pepper was 48 Mt/year under silver oak based Agroforestry system and is becoming as an important practice in this zone.

CLIMATE

Andhra Pradesh represents a transition climatic zone from tropical to sub tropical India. The climate is predominantly semi-arid to arid except in the coastal belt, which has humid to sub humid climate. Occurrence of rainfall in Andhra Pradesh is influenced by the south west and north-east monsoon. The annual average normal rainfall of the state is 941mm, varying from about 500 mm in Anantapur district of Rayalaseema region to 1200 mm in North Coastal and Northern Telangana districts of the state. The south-west monsoon over Andhra Pradesh comes during the second week of June (normally by June 10). Withdrawal of south-west monsoon begins from last week of September and it retreats south of Andhra Pradesh by the end of October. At about the same time, the north-east monsoon sets in and the activity of the north-east monsoon is generally confined to Coastal and Rayalaseema regions of Andhra Pradesh. The distribution of annual rainfall in the state as a whole is about 68 per cent during south-west monsoon, 22 per cent during north-east monsoon and 9 per cent during winter and hot weather months. Region-wise rainfall analysis indicates that Telangana, Coastal and Rayalaseema regions of the state receive 82, 59 and 55 per cent of the annual rainfall, respectively, during south-west monsoon. The north-east monsoon contributes only 11 per cent in Telangana region while the Coastal and Rayalaseema region receive over 30 per cent of the annual rainfall during this period.

SOILS

Andhra Pradesh is endowed with a wide variety of soils ranging from fertile coastal sands to highly fertile and highly productive deltaic alluviums of Godavari, Krishna and Pennar rivers. Besides there are red and black soils spreadover in Telangana and Rayalaseema regions. These soils are developed from different parent materials. The six major soil groups present are : i) Red soils, (65 %) ii) Black soils (25 %) iii) Alluvial soils (5 %), iv) Coastal sands (3 %) v) Latrite and Lateritic soils (1%) and vi) Problem soils (1 %) including saline, saline alkali and non saline alkali soils.

LAND USE PATTERN

Land use pattern of state shows that net area sown is 40.2 per cent and the area under forests is 22.7 per cent. The area under miscellaneous tree and grooves which is not included in net area sown is very marginal (1.01 %). There is an area of 2.8 per cent under culturable waste, that can be brought under cultivation of tree based systems, besides there is an area of 7.5 per cent under barren and unculturable land, that can also be brought under tree cover through which forest area can be substantially increased from the present 22.7 per cent (Table 26.1).

AGROFORESTRY IN DIFFERENT ZONES

1. North coastal zone

This zone comprises parts of Srikakulam, Vizianagarm and Visakhapatnam districts. The irrigation facilities available are confined to Vamsadhara, Sarada and Bahuda river systems where paddy and sugarcane are predominantly grown. In Srikakulam district, Mesta is the commercial crop from which fiber is extracted. Agroforestry systems prevailing in these districts is Cashewnut based agrihorticulture system, where horsegram, cowpea, and some vegetables are grown as intercrops. Cashew industry is the main stay of people in some parts of these districts. Besides, coconut gardens also exist wherein several intercrops are grown, which forms another type of agrihorticulture system.

The D and D survey revealed that *Prosophis juliflora* is used as fuelwood. Recently teak plantations have came up in a big way. These can be seen as future source of timber. In some of the areas, particularly in Vizianagarm, Betelvine gardens with support tree as sesbania is prevalent. This has been the main source of income to the farming community.

2. High altitude and tribal zone

This zone comprises of hilly terrain carved out from the districts of Srikakulam, Viziagagarm, Visakhapatnam and East Godavari. The area is dominated by tribal people. Taungya cultivation is very much in practice as in other states of the country. Tribal farmers depend on the cultivation of paddy using wells and streams as major irrigation source. They also grow pulse crops like Rajmash and vegetables. *Madhuca longifolia* is the tree most tribal farmers like. Tamarind is another tree which gives them income through sale of processed tamarind. They also depend on different tree species for their firewood requirements. As in North Coastal Andhra Pradesh, Cashewnut based agroforestry is also prevalent in this zone. Tribals grow hill banana, which is used as food. The Andhra Pradesh Forest Development Corporation has introduced cultivation of coffee and black pepper in these areas Black pepper cultivation is becoming popular, in which silver oak has been introduced as support tree. This practice is catching up with farmers.

3. Krishna - Godvari zone

Krishna - Godavari zones is the richest zone in Andhra Pradesh endowed with fertile soils, and with well-laid irrigation system. Paddy is the main crop, followed by sugarcane and other commercial crops. Banana is grown extensively. Mango and coconut based agroforestry practices are prevalent. As in north coastal zone, in this zone also, Betelvine cultivation is practiced in almost all the districts, with support tree as sesbania. Trees such as neem and Tamarind are also present as scattered trees in different land use systems. Neem tree is utilized for many purposes such as for furniture, fuelwood and for house building. *Prosopis juliflora* is also very much popular, the wood of which is used as fuel. In Guntur district and other coastal regions of the zone, *Borrasus flabellifer* is a popular tree seen in and around fields and on canal bunds.This tree has multiple benefits. The leaves of these trees are used for roofing houses. Toddy drawn from the tree is used as drink. Its trunk is used for house building and as a boat by fisherman. *Casuarina equisetifolia* is a traditional tree grown in these areas. In the olden days when cooking gas and kerosene were not in use, casuarina was very much in use as fuelwood. With the increasing demand from paper industry, this tree also became a source of raw material for that industry. Presently Eucalyptus plantations are encouraged in a big way by the paper industry, particularly by ITC, Bhadrachalam, who are supplying clones to the farmers to raise plantations with a buy back agreement.

Cashewnut cultivation is also a livelihood venture for many. Cashewnut processing industry is located at Vetapalam. Mango based agrihorticulture system is also found in this zone and much of the area is confined to Nuziveedu in Krishna district (Photo 26.1 and 26.2). Contiguous to this zone, Prakasam and Nellore districts are on southern most part of the state, which come under southern zone agro-ecologically. In these zones, large concentration of subabul and eucalyptus plantations are present. In the upland or rainfed areas of these districts, prior to introduction of subabul and Eucalyptus, Chillies, Tobacco, and cotton were the main crops. The areas under these crops have come down drastically as the corporate sector, mainly from paper and Rayon industry, entered the field exploiting the situation to their advantage and introduced tree cultivation under buy back agreement. The farmers were jubiliant to accept the offer, and at present the area under tree cultivation has increased to anywhere around 1,00,000 to 1,50,000 acres or more. The profits obtained by farmers through cultivation of Casuarina, Subabul and Eucalyptus are given below through some successful stories.

Subabul Plantations - Success story

District	:	Prakasam
Village	:	Tangutur
Area Planted	:	8.4 ha
Spacing	:	8′ x2′ (7, 500 plants/ha)

Two years after planting, trees with a girth of 8 cm and below were sold to tobacco barns which fetched them Rs. 1000/ha. Subsequently, after four years of plantation, 1000 trees were sold to M/s A.P. Ryons, which fetched them Rs.12, 625/ha. Income from remaining trees after five years was Rs. 12,500/ha. The net returns after five years was Rs. 37,500/ha. Subabul leaves are used as fodder, and the income from fodder is not included.

Eucalyptus plantations

The ample evidence provided by the research results of AICRP on Agroforestry on the growth patterns of *Eucalyptus camaldulensis* and its sustainability to semi-arid situations provided much needed encouragement to paper industry. Encouraged by the response from the farmers, some companies have developed their own tissue culture clones and achieved success. The area under Eucalyptus plantations stood at approximately 50 hectares in Andhra Pradesh, and an average yield of 100 t ha^{-1} could be obtained from these plantations in first and second rotations.

Income generated by a farmer in West Godavari district of Andhra Pradesh is quoted here under.

Name of the Village	:	Chintalapudi
Area planted	:	4 ha.
Spacing	:	2m x 2m
Plantations sold	:	Rs. 30,000/ha
Expenditure /ha	:	Rs.5,000
Net income	:	Rs. 5,000/ha/ annum

Eucalyptus field bund plantations

District	:	Chittoor
No. of trees planted on field bunds	:	1,000
No. Survived	:	800
After 13th year Price fetched/tree,	:	800x100 = 80,000

After cutting, the coppices were allowed to grow which would have fetched some income after 5 to 6 years.

The results obtained in ANGRAU, on the benefits of bund plantations as wind breaks encouraged farmers. The wind break benefits shown by ANGRAU are:

1. It reduced wind velocity by 30 - 50 per cent.
2. It increased humidity by 2 - 50 per cent
3. It increased yield of Bajra, Jowar and Groundnut by 10 - 25 per cent and Redgram by 46 per cent.

4. Southern zone

The southern most part of Andhra Pradesh is characterized by dry climate. Rainfall is mostly through north-east monsoon. In this zone, the districts of Praksam (Part), Nellore, Chittoor and part of Caddapah are included. Agroforestry as Acid lime gardens are present in this zone with intercrops as korra and Ragi. Mango based agroforestry is also present. Inland fish farming is intensively practiced with coconut trees on the bunds, which is also prevalent in other coastal districts of the state. The trees which are growing since many years started drying up due to pollution created by fishponds. A survey undertaken in these districts has clearly indicted that fish farming has affected ecosystem. In Chittoor district of southern zone, groundnut is the major crop. In groundnut fields, sparse populations of tamarind trees exist. *Prosopis* is the source of fuel wood for many of the farmers in these districts, as in other parts of the state.

In southern zone, in Chaudepalli village processed Tamarind business is a highly fetching enterprise with the farmers. Farmers are so much socially attached to this tree that at the time their daughters' marriage they present to grooms tamarind tree as dowry. A well-grown tree of tamarind can fetch any where around Rs.750 –1000 / year. For some of the farmers, this is the only income generation in these districts.

5. Scarce rain fall zone

The three Rayalasema districts of Cuddapah, Kurnool and Anantapur come under the purview of scarce rainfall zone. This is a drought prone area. The average annual rainfall of the zone is 300-500 mm. The main crop in this zone is groundnut. Among vegetables, tomato is grown in extensive areas. The area under tomato is so extensive that in peak season, due to lack of remunerative price, produce is left on the fields to be ploughed down at a later stage. Sweet orange gardens and mango gardens also exist. Fruit tree based agroforestry systems and sweet orange gardens are main source of income for some of the farmers. Trees like *Ficus bengalensis* and *Acacia nilotica* serve as fodder for sheep and goats. *Prosopis juliflora* and *Pithocellobium dulce* are also seen extensively. The Betelvine sebania system is very much popular in the zone, which is fetching quite a good amount to the farmer. The scarce rainfall zone is covered with Nallamala forest areas. In these areas, *Pterocarpes santalinum* and *Dalbergia latifolia* are extensively seen.

6. Southern Telangana zone

The Southern Telangane zone comprises the districts of Rangareddy, Mahaboobnagar, part of Nalgonda, part of Medak and Hyderabad districts. In Ranga reddy district, most of the areas are nearer to Hyderabad. The advantage of nearness to Hyderabad city is a boon to the farmers Fruit trees like custard apple and Amla, are systematically grown, which has got lot of local market and export potential. In these gardens, vegetables are grown and good profit is obtained by farmers. Custard apple is available in all parts of the zone. The tribes like Sugalis earn quiet a bit of income from these fruits during peak season by selling them in Hyderabad City. Custard apple based Agrihorticulture system is an age-old practice in this zone. Generally rainfed castor is grown as intercrop. In Nalgonda district, and adjoining Warangal district of the zone, *Borrasus* is a money earning tree which is seen in cultivated fields either as boundary or as isolated. This tree is a bread earner to a particular sector of people who belong to gowda community. There is a toddy association in the zone, which looks after the interests of these farmers. The excise department of government of Andhra Pradesh looks after the affairs of palmyrah tree growers association. Plantations of Eucalyptus, Subabul and teak came up in a big way around Hyderabad and other adjoining districts. These plantations are serving as raw material to the mini power plants that were installed around Hyderabad city. Eucalyptus based agro forestry systems is gaining popularity with pastures and medicinal plants as intercrops. *Prosopis* spp. is found in extensive areas in the zone and is also now being seen as potential source of raw material for mini power plants. In Nalgonda

Photo 26.1 Mango + Cowpea based agrihorticulture agroforestry system

Photo 26.2 Mango + *Sorghum* based agrihorticulture agroforestry system

and Mahboobnagar districts, Mango gardens are grown with intercrops as vegetables and flowers that have good market in Hyderabad city.

7. Northern Telangana zone

Northern Telangana zone being border to Maharastra and Chhattisgarh States, are covered by forests. Many trees of commercial and economic importance are available. The Eucalyptus and Subabul plantations are taken up on large scale by State Forest Development Corporation in degraded forest areas. Mango based Agrihorticulture system is a common practice in this zone. *Borrasus flabelleifer* is found in many areas. *Acacia nilotica, Prosopis juliflora* and Jamun trees are seen on field boundaries. Karimnagar district, as canal system from Sriram Sagar Project built on Godavari river. Sugarcane and maize are important crops besides paddy. Turmeric and ginger are grown extensively some times as intercrops in mango and other fruit gardens.

Table 26.1. Land use pattern of Andhra Pradesh

S. No.	Item	Area (ha)	Total (%)
1	Total area	274.4	100
2.	Forests	62.81	22.7
3.	Barren & uncultureable land	20.73	7.5
4.	Land put to non- agril. Use	23.52	8.6
5.	Culturable waste	7.67	2.8
6.	Pastures & grazing land	8.31	3.0
7.	Misc. tree crops & grooves not included in the net area sown	2.62	1.0
8.	Current fallows	13.66	5.0
9.	Fallow lands other than current fallows	24.68	9.0
10.	Net area sown	110.40	40.2

27

Agroforestry Systems of Karnataka — An Overview

B.N. SATHIS AND C.G KUSHALAPPA

College of Forestry, University of Agricultural Sciences
Bangalore, Ponnampet - 571 216, Kodagu, Karnataka, India

ABSTRACT

Agroforestry a sustainable land use system being practiced since long back through out the country. A diverse climatic, edaphic, and environmental conditions prevailing in the country has made it possible to practice different kinds of agroforestry systems. The local research organizations and individuals have undertook various studies for the betterment of the local agroforestry systems for getting higher yields to meet the need of food, fodder, fuel wood etc. It was found that Jowar+Teak+Papaya followed by Chilli+Teak+Papaya combination gave more economic returns compared to pure agricultural crop. *Casuarina equisetifolia, Grevillea robusta, Dalbergia sissoo and Melia dubia* have been found to be suitable for agroforestry systems in dryland. Coffee based agroforestry system and other tree-crop combination tried have been elaborated in this paper.

INTRODUCTION

Agroforestry is a new word for the old practice of growing woody plants with agricultural crops and/or livestock together on the same land. Various experts and scientists working on agroforestry or involved in land use related sciences have defined agroforestry according to their individual understanding of what they call agroforestry.

"Agroforestry is sustainable management system for land that increases total production, combines agricultural crops, tree crops and forest plants and or animals simultaneously or sequentially and applies management practices that are compatible with the cultural patterns of the local population" (Chundawat and Gautham, 1993).

Karnataka, is situated in south west India with an area of 191791 sq. km (19.179 million ha). Most part of the state lies on a plateau bordered in the west by costal strip along the Arabian sea, (about 30 km wide; average annual rainfall 3000 to 3700 mm) and the western Ghats which are a chain of low hills with altitude of 1500 m and average annual rainfall of 2500 mm (maximum of 7000 mm in a place called Agumbe). There are however considerable variations in both quantity and distribution of monsoon rains. The south eastern part of the state also receives rainfall from north-eastern monsoons. Nearly two third of the state in the north-eastern part form the dry zone and is classified as drought prone (Chalawadi and Jayakumar, 1987).

LAND USE AND MAJOR CROPPING SYSTEMS

Out of the total geographical area of 19.179 million ha, 11.92 million ha equivalent to 62.50 per cent is under cultivation (net cropped area 10.61 million ha and fallow 1.38 million ha) of which about 1.59 million ha is under irrigation (irrigated by canals 0.66 million ha, by tank 0.32 million ha, by wells 0.43 million ha, and through tube wells and other sources 0.18 million ha). The major crops grown are Jowar (*Sorghum bicolor*), Paddy (*Oryza sativa*), Ragi (*Eleusine coracana*), Cotton (*Gossypium species*), Ground nut (*Arachis hypogea*), Bajra (*Pennisetum typhideum*), Wheat (*Triticum species*), Sugarcane (*Saccharum species*), pulses and minor millets. Of these Jowar and pulses form the major part of agricultural crops. Nearly 3.45 million ha lie uncultivated (including cultivated and uncultivated fallow and waste lands) and about 1.9 million ha is not available for cultivation (for reason being contiguous to roads, canal banks, railway tracks, bridges public buildings etc.). Of the uncultivable and fallow lands (3.45 million ha) 1.38 million ha is cultivable waste land and about 2.07 million ha is barren uncultivable land (Chalawadi and Jayakumar, 1987).

AGRO ECOLOGICAL REGIONS

Karnataka has four agro ecological regions, which fall under the following ecosystems:

1. *Arid - ecosystem* : Parts of the state mainly comprising portion of Raichur, Bellary, and Chitradurga districts falls in this category and can be described as *Deccan Plateau hot arid eco-region* with mixed red and black soils. Bund planting is the most common agroforestry system with *Azadirachta indica* a dominant tree species

2. *Semi arid - ecosystem* : Is a region which is described as *Deccan Plateau hot semi arid ecoregion,* which has shallow medium black soil. It has most

parts of districts of Bidar, Gulbarga, Bijapur, Belgaum, Darwar and parts of Raichur and Uttara Kannada districts. Strip and bund planting are most common systems in addition to alley cropping system and very few agrihorticulture systems.

3. Another ecoregion which is again a semi - arid ecosystem is *Eastern Ghats and Daccan Plateau hot semi arid ecoregion* with red loamy soils. Entire districts of Bangalore urban, Bangalore rural, Mandya, Hasan and major parts of Mysore, Tumkur, Chitradurga, Shimoga, Chikmagalore and small portion of Kodagu, Bellary form this eco region. This zone is dominated by most of the agroforestry system. The common agroforestry systems are Agrisilvipasture, Agrisilvihorticulture, and Home gardens.

4. *Coastal Eco -system :* This category is described as *Western Ghats coastal plains hot humid Pre humid eco-region* with red soil and Alluvial derived soils. Home gardens, Agrisilvipasture systems are the common systems in this zone.

RESEARCH ACTIVITIES UNDER TAKEN IN AGROFORESTRY SYSTEMS

Dendrocalamus strictus, D. brandisii (Burma bamboo) and *Bambusa arundanaceae* have been found to be suitable for dry land farm forestry. Under dry land and poor site condition *Eucalyptus* hybrid exhibits adverse affects such as suppressing under growth or adjoining crops by reducing soil moisture regime.

An agroforestry research project implemented in three research stations of UAS, has shown that *Eucalyptus hybrid,* causes maximum deleterious effects on field crops under rain fed conditions followed by *Dalbergia sissoo, Casuarina* and *Acacia nilotica*. However, this agroforestry model is found to be economically more viable than pure agriculture crop.

Studies conducted in University of Agricultural sciences, Darwar, has shown that growing agricultural crops along with teak and papaya yielded higher net returns as compared to growing only agricultural crops. Similarly NPV (Net present Value) and IRR (Internal Rate of Returns) were higher with this treatment (Table 27.1). However, in addition to Teak and Papaya if grass or subabul for pasture are included the net return per ha per year would be more. Hence adopting agroforestry system is economically viable also.

In studies on collection and evaluation of promising tree species, *Leucaena leucocephala* a multi purpose tree species, recorded highest survival percentage

(95.8) followed by *Casuarina equisetifolia* (93.4) *Anogeissus latifolia* (91.04), *Albizzia lebbek* (89.9), *Eucalyptus tereticornis* (88.3) and *Dalbergia sissoo* (87.7). The lowest survival was of *Grevillea robusta*.

Further, studies on agrisilvihortipastoral system in red gravel soils revealed that grain and stock yield of jowar (sorghum) was highest in sole crop while the dry fruit and stalk yield of chilli was not so. In general, crop yield were low in different agroforestry combinations as compared to sole crop. In another study, it has been reported that, among two pasture crops, grass(green) yields were higher than Subabul green fodder. Similarly it has been reported that, performance of Teak trees (height and dbh) was better without pasture crops. In view of this, the estimated value of Teak trees (pole purpose) at the end of seven years was higher to an extent of 40-60 percent in treatment without pasture crops. In a study to analyze the economics of agroforestry systems, it has been reported that the total income during the year from all components was maximum for Jowar + Teak + Papaya followed by Chilli + Teak + Papaya, compared to pure agricultural crop returns. Similarly in another study, it has been reported that, the arable crop (chilli and jowar) yields were gradually less in up to five individual rows with more impact on western side as compared to eastern side. It is interesting to note that resource rich farmers have started Agrisilvihorticulture systems in cultivated lands.

In an experiment conducted at Gandi Krishi Vignana Kendra, Bangalore, it was reported that, *Casuarina equisetifolia, Grevillea robusta, Dalbergia sissoo* and *Melia dubia* have been found to be suitable for agroforestry systems in dryland. Whereas, a number of other species including Eucalyptus have been found suitable for irrigated lands. In Eucalyptus initial 1 x 2 m spacing with one thinning in the third year and soil working every year with application of complex fertilizer at the rate of 10 g/plant has been found to increase growth rate and reduce rotation period from 8 years to 6 years.

Chaturvedi (1990) worked out the economic factors with reference to Eucalyptus, Poplars, Subabul, shade trees in Tea gardens, Bamboo in agroforestry and concluded that incorporation of trees with agriculture will increase the over all productivity and the marginal fall in agricultural production in some cases will be compensated by the general beneficial environmental impacts over large areas. For individual farmer the financial returns from sale of wood will compensate the reduction in agricultural yields.

Based on the previous experience at the Dry land Research Station at G.K.V.K., four species *viz., Acacia auriculiformis, Casuarina equisitifolia,* Silver oak (*Grevellia robusta*), and *Calliandra* were selected for planting on the bunds in the dry land at Chokkanahalli. The inter row distance maintained was 2 m between plants accommodating 2500 plants per ha. and planting was done in 1987. Though, initially all the species have behaved almost uniformly, subsequently *A. auriculiformis,* expressed an adverse effect on crop growth due to its faster growth. At its sixth year of growth, this species affected the crop rows on its either side up to 20 rows incase of Ragi and 15 rows in case of Groundnut. Reduction in crop yields has been partially compensated by the value of fuel wood to the tune of 5.6 tons at the final harvest of trees at 8-9 years of growth which is estimated to fetch Rs. 5000 per ha (Table 27.2). The other two species viz., Silver oak and *Casuarina* affected the crop up to 7 and 5 number of rows, but the fuel wood yield due to side loping was not high. However, the four years average of grain yield in case of *Casuarina* plot was only marginally reduced compared to sole crop. *Calliandra* did not make satisfactory growth and the fodder yield obtained by the periodical trimming was only three tons per hactare.

Another observation made in these studies was that the tree species on bunds should be preferably planted in East - West direction to minimize shade effects. In another study in sivipastoral system an area of 15 ha wasteland/ gomal (grazing land) in Chokkanahalli area of watershed was developed into silvipastoral system by planting mixed species of *Acacia auriculiformis, Pongamia pinnata, Dalbergia sissoo* and *Tamarindus indica* in contour trenches at 5 m spacing during 1987. Subsequently a mixture of *Stylosanthes hamata* and *Stylosanthes scabra* was broadcasted in the inter space (Table 27.3). At the end of 6 years the area has been fully covered with *Acacia auriculiformis* but the growth of other species was completely stunted. This indicates the mixing of slow and fast growing species will be only at the cost of slow growing species. The pasture component has been yielding valuable green fodder. The estimated fodder yield at present is 18 tons per ha. and lopping of trees is 3.3 tons per ha. and thus the total income will be higher by adoption of alternate crop in these lands rather than growing Ragi.

In Eastern Ghats and Daccan Plateau hot semi arid ecoregion, especially in Coorg nearly 30 percent of its geographical area is under agrisilviculture system (crop combination with plantation crops). A recent study with respect to diversity of trees and economic value of these systems shows that they act as home for very rich diversity of tree species and trees can fetch Rs. 5-6 lakhs per ha (Sathish, 2005).

Coffee based agroforestry systems of Kodagu

Kodagu is one of the second smallest district in Karnataka which falls under Eastern *Ghats and Daccan Plateau hot semi arid ecoregion.* The district is highly forested and nearly 30 percent of its geographical area is under coffee

Table 27.1 Income from an agroforestry system experiment

Treatment		Average net return per ha per year after seven years	NPV at 20 per cent discount rate (Rs/ha)	IRR per cent
1	Only agricultural crops	Rs.4,409	–	–
2	Crops + teak + Papaya	Rs.7,894	18,967	117.59
3	Crops + teak + Papaya + grass	6,754	15,862	98.79
4	Crops + teak + Papaya + grass	7,035	16,362	98.91

Table 27.2 Yield of Ragi and fuel wood in agroforestry system with different forest tree species.

S No	Year of sudy	*A. auriculiformis*		*C. equsitifolia*		*Grevellia robusta*		Sole crop
		Grain	Fuel wood	Grain	Fuel wood	Grain	Fuel wood	Grai n
1	1988 (2 year)	11.3	15.0	12.6	3.0	12.0	4.2	13.0
2	1989 (3 year)	16.8	28.0	18.5	5.0	17.8	6.5	19.2
3	1990 (4 year)	9.8	44.0	10.4	7.0	10.2	8.1	11.2
4	1991 (5 year)	16.8	56.0	18.1	11.0	17.9	12.1	19.0
Average of four years		13.7	35.7	14.9	6.5	14.5	7.7	15.6

Table 27.3 Yield of crops in alternate land use systems in poor oils.

S. No	System	Yield per ha.				
		Green forage	Lopping for fuel wood	Grain (Ragi)	Straw	Total value (Rs/ha)
1.	Ragi	-	-	9.4	12.8	3080
2.	Silvipasture *A.Auriculiformis* + *S. hamata*	180	33.0	-	-	3930
3.	Hortipasture system (Mango + *S. scabra*)	200	-	-	-	4000
4.	Pasture system (*S. scabra*)	230	-	-	-	4600

Note: Cost of commodities at the rate of Rs. 300 per quintal, straw Rs. 600 per ton, green fodder Rs. 200 per ton, fuel wood (side loppings) Rs. 100 per ton based agroforestry. In these coffee plantations majority of the tree used for the shade are natural and are native and very recently due to decrease in coffee price, planters have planted few exotic tree species viz., *Grevelia robusta, Acrocarpus fraxinifolius*. Recent studies conducted by Muthappa (2000) and Sathish (2005) has shown that they are supporting high levels of biodiversity, as diverse and well stocked tree cover is retained. In addition to this, on an average rupees seven lakh timber is present in these coffee plantations per hactare. These coffee based systems therefore can be considered as managed agroforests.

REFERENCES

Chaturvedi, A.N. (1990). Agroforestry in India with particular reference to economic factors.

Chalawadi, S.M., and Jayakumar, M.N. (1987). Land use system and agroforestry research in Karnataka (India). Paper presented at ICRAF, Nairobi, Kenya.

Chundawat, B.S., and Gautham, S.K. (1993). Text book of Agroforestry. Oxford and IBH publication, New Delhi.

Muthappa, P.P. (2000) A Resource economic study on tree diversity in coffee based plantations in the Western Ghats region of Karnataka. *M.Sc. Thesis*, University of Agricultural Sciences, Bangalore.

Sathish, B.N. (2005) Assessment of tree diversity in coffee plantations under different land tenure systems in Virajpet taluk, Kodagu. *M. Sc. Thesis* submitted to University of Agricultural Sciences, Bangalore.

28

Agroforestry Systems and Practices of Kerala

B. MOHAN KUMAR

Department of Silviculture and Agroforestry, College of Forestry
Kerala Agricultural University, KAU PO, Thrissur - 680 656, Kerala

ABSTRACT

Agroecologically the state is divided into 13 zones based on altitudinal variations, precipitation pattern, soil type and topographic features. Land use systems that integrate woody perennials with other life forms abound in all these zones, along with a great diversity of trees and field crops. Although these practices are apparently specific to the topographic and edaphic factors, a complex pattern of overlapping which transgresses over a wide range of agroecological zones is obvious. An attempt is made to describe the important agroforestry systems of Kerala, both traditional as well as those potentially suitable.

INTRODUCTION

The state of Kerala with many woody perennial-based land use practices is often regarded as the "Mecca" of agroforestry. Most of these are, however, traditional systems evolved over centuries of cultural and biological transformations and may represent the accrued wisdom and insights of farmers who have interacted with environment, without access to exogenous inputs, capital or scientific skills. Over long periods in the history of land use in these highly populated humid tropical lands, such diverse land use systems have apparently remained as engines of food security, besides sourcing multiple products and services including certain intangible benefits (e.g., microclimatic modification, soil and water conservation); thus providing the rationale for the co-existence of an array of pluralistic land use orders often in close juxtaposition with one another. Yet another plausible explanation for the occurrence of many diverse woody perennial-based systems and practices over a relatively small geographical range is perhaps the brute heterogeneity of the Kerala landscape.

The state has a total geographical area of 39 lakh ha (8°18' N to 12°48' N latitude and 74°52' E to 77°22' E longitude), much of this is well endowed with abundant precipitation (>3000 mm per annum) and/or bio-resources. Physiographically, the land falls under the coastal plains consisting of beaches, level lands with valleys, submerged lands, swamps and marshes, reclaimed lands, coastal laterites, midland laterites and the valleys, hills and uplands of Sahyadri and the Nigiris; with profound altitudinal variations. Soils are predominantly well-drained and infertile: acidic, kaolintic and gravelly soils with low CEC, low water holding capacity and high phosphate fixing capacity are frequent. On the basis of the morphological features and physico-chemical properties, the soils of Kerala have been classified into red loam, laterite coastal alluvium, riverine alluvium, *Onattukara* alluvium, brown hydromorphic, saline hydromorphic, *Kuttanad* alluvium, black soil and forest loam; mostly falling under the orders of oxisols, alfisols, inceptisols and mollisols. Defined in USDA soil taxonomy terms, these are areas with an udic soil moisture regime and an isohyperthermic soil temperature regime. A bimodal distribution pattern with most rains during the south-west and north-east monsoon seasons is a distinctive feature of the state's bioclimatic regime. Rainfall distribution is also variable; it is relatively well distributed in the southern half (with a peak in June) and rather ill-distributed in the northern region (July peak). Based on altitudinal variations, precipitation pattern, soil type and topographic features, the state has been delineated into 13 agroecological zones (Table 28.1; Fig 28.1). Crop management recommendations are often made considering these agroecological zones as the basic unit (KAU, 2002).

AGROFORESTRY SYSTEMS AND PRACTICES FOR DIFFERENT AGROECOLOGICAL ZONES

Biophysical heterogeneity—which abounds even within a given ecological zone—underscores the need for location-specific systems, components and/or their management. Coincidentally, site-specific traditional agroforestry practices and the modern agroforestry technologies (Note: the term 'agroforestry technology' refers to agroforestry practices that have been developed recently with research backing–involving either improvement of traditional practices or introduction of new ones; Nair, 1993), are typical of Kerala (Table 28.2), as it is elsewhere. Yet, a constant radiation surplus, the general thermic uniformity and high rainfall availability, intrinsic to a range of ecological zones, make the eco-climatic variations of Kerala virtually continuous. This, in turn, results in a fair amount of overlapping among the agroforestry systems of different eco-zones. That is, distinctive inter-zonal variations in land use practices of Kerala are seldom manifested, except for those fundamentally different eco-regions such as 'Onattukara' and the 'high ranges'; partly because the socio-economic stratification transcends the eco-climatic boundaries too (Kumar *et al.*, 1992). However, profound 'within zone' variations are customary in all the eco-zones of Kerala.

Overall, agrisilvicultural, silvopastoral, agrosilvopastoral systems constitute the dominant agroforestry types of Kerala. They also involve many disparate and intricate species-mixes, often specific to the topographic and edaphic niches. And include irregular intercropping and/or mixed cropping, sequential cropping and many other forms of polycultures. Examples are numerous: tropical homegardens, plantation crop-food crop combinations, jackfruit tree and palm-based food production systems, integrated agriculture-aquaculture systems—agrosilvofishery systems (e.g. integrated rice-fish production systems, trenches/ponds in orchards, mangroves and brackish water fish/shrimp production), small holder livestock production systems and so on. Furthermore, the system components are variable and often involve cereals, pulses, tubers, fruits, vegetables, latex and oil yielding tree and annual crops, beverages, condiments and spices, sugar crops, forages, medicinal plants, green manure crops and timber yielding species (Tables 28.2 and 28.3). An attempt is made here to briefly describe the major agroforestry systems and practices of Kerala.

Tropical homegardens

Presumably, homegardening is the oldest land use activity next only to shifting cultivation. It is widely practiced in all agroecological zones of the state and is probably the most important and unique agroforestry system of Kerala (Kumar, 2003). The concept refers to "intimate, multi-story combinations of various trees and crops, sometimes in association with domestic animals, around homesteads." It evolved through generations of gradual intensification of cropping in response to increasing human population pressure and the corresponding shortage of arable lands (Kumar and Nair, 2004). Homegardening has been a way of life for centuries and is still critical to the local subsistence economy and food security in Kerala that has about 5.4 million small gardens (mostly less than 0.5 ha in area; KSLUB 1995).

A fundamental attribute of homegardening is the great diversity of species with several life forms varying from those creeping on the ground, such as the sweet potato, to tall trees (e.g. coconut palm), and vines climbing on poles and trees (e.g. pepper vine). Physiognomically, the homegardens exhibit a multi-tiered canopy structure somewhat similar to that of the tropical evergreen forest formations. Dominant trees include coconut (*Cocos nucifera*), arecanut (*Areca catechu*) and para-rubber (*Hevea brasiliensis*), besides a large number of multipurpose trees (MPTs) such as *Ailanthus triphysa*, mango (*Mangifera indica*), jack (*Artocarpus heterophyllus*), teak (*Tectona grandis*), cashew (*Anacardium occidentale*), wild jack (*Artocarpus hirsutus*), tamarind (*Tamarindus indicus*), *Erythrina* spp., *Macaranga peltata, Gliricidia sepium* and so on. In a study of the Kerala homegardens, Kumar *et al.* (1994) documented a total of 127 woody species with a mean Simpson's diversity index ranging from 0.251 to 0.739.

Food production being a dominant objective, food crops including herbaceous plants and vegetable and fruit yielding trees and shrubs flourish in most homegardens. Homegardens, however, seldom meet the entire basic-staple-food needs of the family in any given area, and are thus, a component of the larger farming system of the household. In general, dietary supplies from homegardens do not exceed 3 per cent to 44 per cent of the total calorie and 4 per cent to 32 per cent of the protein intake (Torquebiau, 1992). Homegardens are also sources of minerals and nutrients for improving household nutritional security especially for at-risk populations. The diverse products available year-round contribute to food security especially during the "lean" seasons. As little or no chemical inputs are used, the products can be expected to be of superior quality. Yet another key dimension of homegardening is the equitable distribution of the produce within the community. While a large proportion of the production is consumed domestically, many products such as fruits, vegetables and medicinal/ornamental plants are also generously shared within the local communities. This egalitarian distribution of the agricultural produce is significant in a social milieu and endows the homegardens a unique social disposition (Kumar and Nair, 2004).

In addition to providing food products, homegardens are recognized repositories of non-timber products such as medicinal and aromatic plants, ornamentals, bamboos, gums, resins, chemical extractives and green leaf manure. They also constitute a principal source of biofuels for the rural households (Shanavas and Kumar, 2003), and provide farmers with materials for fence posts, poles and farm implements. The average standing stock of commercial timber in the Kerala homegardens ranged from 6.6 to 50.8 m^3 ha^{-1} and that of fuelwood from 23 to 86 $m^3 ha^{-1}$ (Kumar *et al.*, 1994), implying that a substantial portion of the society's timber and firewood demands can be met from these homesteads (74.4 to 83.5% according to Krishnankutty, 1990). Products such as rubber, fruit, nuts, resins, medicines, spices and oils, as well as the materials to make household, hunting, fishing and agricultural implements are also important for promoting food security either directly (used for consumption) or indirectly (enhancing the purchasing power). Homegardens also support a variety of animals - cows, buffaloes, bullocks, goats, sheep, and birds such as chicken and ducks. Milk and egg production, both for domestic consumption and for sale, constitute the dominant objectives of integrating animals into these production systems. In addition, the large ruminants traditionally provided draught power for land preparation, soil conservation practices and haulage. Animals are also important for the maintenance of soil fertility and sustainability of the production system.

Temporal complementarity in labour allocation is yet another significant advantage of homegardening. For example, labour demands of agroforests in general and homegardens in particular seldom show sharp peaks and troughs and that they are flexible and distributed throughout the year. In many places, women play a vital role in the design and management of homegardens. Activities such as vegetable growing, harvesting the products (fruits, nuts, vegetables and medicinal plants), fuel collection and animal rearing, especially in the smaller gardens are exclusively the domain of females, with or without assistance from the male members of the family. Also, the possibility of gender equality for participating in homegarden management and sharing of benefits is perhaps one of the major stimuli for continued household security enjoyed by homegardeners for generations (Kumar and Nair, 2004).

The ultimate challenge in the 21st century is to find ways to sustain the provision of goods and services that society derives from forests in ways that "meet the needs of the present without compromising the ability of future generations to meet their own needs" (Bruntland Commission, 1987). Significantly, homegardens are often regarded as an 'epitome of sustainability' — both in a biophysical and socioeconomic perspective (Kumar and Nair, 2004). In addition, they provide many goods and services, for which the mankind depends on the forests heavily, and *inter alia* eases the anthropogenic pressure on tropical forests, which are our "doomed warehouses of global biodiversity".

PLANTATION CROP COMBINATIONS

Woody perennial plantation agriculture is yet another chief land use activity in the state of Kerala. Coconut, arecanut, rubber, tea, coffee, cardamom, and oil palm are important in this respect. For many plantation crops, the combination of crown architecture and the wide spacing often facilitate growing a broad spectrum of inter crops in association with them (Nair, 1984). Shade tolerance, however, is an important criterion for choosing the field crops to be included under such mixed species systems. Shade relationships of several field/plantation crops were evaluated at the Kerala Agricultural University, with a view to optimise understorey productivity and have been summarised in Kumar (1994) as follows:

- shade sensitive: sweet potato (*Ipomoea batatas*), cowpea (*Vigna unguiculata*), groundnut (*Arachis hypogaea*), black gram (*Vigna mungo*), Soybean (*Glycine max*) and red gram (*Cajanus cajan*)
- shade intolerant: coleus (*Coleus parviflorus*)
- shade tolerant: colocasia (*Colocasia esculenta*)
- shade loving: ginger (*Zingiber officinale*) and turmeric (*Curcuma longa*)

Coconut-based cropping systems

In India, the state of Kerala represents an important coconut growing region, with an area of about 8,99,200 ha (Farm Guide, 2005). Seventies and eighties saw a remarkable boom in the area under this crop. Polycultural systems involving several species are a unique feature of coconut farming in Kerala. This can be rationalized based on the incomplete utilization of the incoming solar radiation by the palms (Abraham, 1993). Furthermore, the limited lateral spread of the coconut roots leads to sub-optimal utilisation of below ground resources (Anilkumar and Wahid, 1988). As a result, annual, seasonal and perennial crops abound in coconut plantations. In particular, perennial spices like pepper, nutmeg, clove; beverage crops such as cacao, coffee; fruit/nut yielding crops like arecanut, sapota, mango, citrus, guava pineapple, jack, arecanut, papaya; annuals like banana, turmeric, ginger; tuber crops, fodder crops, pulses and oilseeds are frequent in these systems.

Traditionally, dicot MPTs were included in the coconut gardens for green manure and fodder purposes and/or as support trees for trailing pepper vines. *Gliricidia sepium, Erythrina indica, Pajanelia rheedii* and *Leucaena leucocephala* are prominent in this respect (Nair and Sreedharan, 1986; Ghosh *et al.*, 1989; Liyanage *et al.*, 1990). Lately, however, with a view to developing low input sustainable production systems numerous MPTs are systematically inter-planted in the coconut plantations (Nair, 1983; 1993; Kumar and Kumar, 2002). Such mixed planting systems are particularly relevant for the disease affected and/or senile plantations. This is of special significance in a state like Kerala where an estimated 80 lakh palm trees are affected by root (wilt), a serious debilitating disease. Several MPTs have been recommended for border planting or interplanting in the coconut gardens. Prominent ones include: *Ailanthus triphysa, Azadirachta indica, Casuarina equisetifolia, Ceiba pentandra,*

Calliandra calothyrsus, Gmelina arborea, Grevillea robusta, Sesbania grandiflora, Tamarindus indica, Tectona grandis, Vateria indica and so on (Jambulingam and Fernandes, 1986; Nair and Sreedharan, 1986, Liyanage *et al.*, 1988; Kumar *et al.*, 1994; 1999).

Although intensive, integrated land use systems involving MPTs, plantation crops are ideally suited for the small and medium growers, information on the compatibility of dicot MPTs in a coconut plantation is scarce. Many aspects of the functional dynamics such as system productivity, competitive interactions and resource sharing are also less known. Farmers frequently apprehend competition for light, water and nutrients between coconut and the inter-planted MPTs. Concentration of feeder roots of many trees in the surface horizons of the soil profile (Jamaludheen *et al.*, 1997; Thomas *et al.*, 1998; Gowda, 2002) further complicates the situation. In certain cases, inter planting MPTs in the coconut plantations led to altered ^{32}P recovery patterns (Kumar *et al.*, 1999). For instance, *Grevillea* exerted a favourable influence on ^{32}P uptake by coconut, while both *Ailanthus* and *Vateria* had a depressing effect on it, implying variations in nutrient uptake owing to differences in root architecture and the need to select the associated tree components based on below-ground competitiveness (Gowda, 2002).

Rubber-based cropping systems

Rubber (*Hevea brasiliensis*) is a principal plantation crop in Kerala. Area under rubber cultivation increased during the last 3-4 decades and the current extent of rubber cultivation in Kerala is 4,76,050 ha (Farm Guide, 2005). Bulk of Indian rubber is produced in tiny gardens of one acre or so, by small holders, who produce rubber with household labour to meet part of their household income requirements. Faced with uncertain markets and discriminatory treatment from government, however, the most obvious survival strategy for these small holders has been to diversify so that they are not dependent on a single crop alone. Intercropping holds promise especially during the early phase of rubber growth. Potential intercrops include banana, vegetables, tapioca, yams etc., besides repeated cropping of upland rice for subsistence to continuous cultivation of pineapples on a commercial basis. Leguminous cover crops, maize, peanuts, pineapple, hedgerows and banana are also frequent in the rubber plantations of Kerala. Sri Lankan experiments (Yogaratnam and Iqbal, 1995) have shown that rubber and tea are suitable for intercropping, which, however, would require some adjustment of the spacing of the rubber plants to allow for the growth of perennial intercrops.

Arecanut and oil palm-based cropping systems

Like coconut, due to its height, crown architecture and wide spacing adopted for planting, the arecanut palms (*Areca catechu*) and oil palm trees (*Elaeis guineensis*) permit the growing of a wide variety of annual, biennial and perennial intercrops. Banana, cacao, black pepper, pineapple, betel vine, elephant foot yam, tapioca, paddy, sorghum, cowpea, vegetables, turmeric, ginger and guinea grass are prominent in this respect.

Commercial crops under the shade of planted trees: beverage crops and tree spices

Beverage crops such as tea (*Camellia sinensis*), coffee (*Coffea* spp), cacao (*Theobroma cacao*), spices like clove (*Syzigium aromaticum*), nutmeg (*Myristica fragrans*) etc., often require varying levels of shade for optimum growth and production. Although a number of shade trees are grown in coffee and tea plantations, there is very little documented information regarding the requirement of shade by these crops. Presumably the tea plant requires shade only during the establishment phase. Among the shade trees of tea, *Grevillea robusta, Albizia* spp., *Erythrina* spp. and *Acacia* spp. are important. Dominant shade trees in the coffee plantations of Wayanad in Kerala include: *Erythrina indica, Mangifera indica, Artocarpus heterophyllus, Grevillea robusta, Melia dubia, Areca catechu* and *Tectona grandis*. As regards to cacao, shade stimulates vegetative growth and it suppresses reproductive growth (KAU, 1993). A dense overstorey, which intercepted incoming solar radiation to the extent of 25 to 75 per cent of that in the open, drastically reduced pod yield. Nevertheless, clove, nutmeg and cacao require shade for optimum growth and production. Coincidentally, they are all grown as components of homesteads as well as coconut, arecanut and coffee plantations (Kumar, 1994; Tejwani, 1994).

Commercial crops grown under the shade of trees in the natural forests: small cardamom

Small cardamom (*Elettaria cardamomum*) is indigenous to the medium elevation wet evergreen and semi-evergreen forests of the Western Ghats of peninsular India. In ancient times cardamom was collected form forest, but later on man started cultivating it as an understorey crop in the tropical forests. It is perhaps the most important cash crop of the Indian wet tropics, probably after rubber. The traditional system of cardamom production in the Western Ghats represents a unique, but little studied agroforestry system. It involves growing a sciophytic commercial crop under the shade of trees in the natural forests, with little or no reliance on external inputs such as fertilisers and/or chemicals.

The process of cardamom cultivation involves selective removal of trees to facilitate light infiltration to the understorey and planting of the crop. Clearing the forests for cardamom cultivation, however, may result in disturbance to the forest ecosystem and cause drastic changes in vegetation structure and physiognomy (Kumar *et al.*, 1995). Transforming the multi-tiered stand structure of the wet evergreen forests into a monolayered structure is perhaps the most dramatic effect in this respect. Indiscriminate exploitation of natural forests and illicit felling are also common in the cardamom growing tracts of Western Ghats. Eroded and degraded soils and poor fertility status, besides pest and disease incidence, are consequential problems.

Loss of forest canopy may result in sub-optimal levels of shade for the cardamom crop leading to its lower productivity. In the past, the Government of Kerala treated such degraded forestlands (on lease to private growers) as 'cardamom hill reserves' (CHRs). Cardamom productivity is particularly low in these leased areas, which account for about 20 per cent of the cardamom plantations in Kerala, owing to poor soil management and shade regulation. The ownership pattern of CHRs (land belonging to State Revenue Department, trees on the land belonging to the State Forest Department and the cardamom crop being owned by the growers) also acts as a disincentive towards good practice.

Although in the traditional system of cardamom production little or no chemical fertilisers were applied, with the advent of modern technology, farmers have started using chemical fertilisers to supplement the inherent site fertility. Furthermore, to ensure reasonable levels of cardamom productivity, a regular schedule of cultural practices consisting of weeding, mulching, trashing, shade regulation, fertiliser application and irrigation is considered essential (Kumar *et al.*, 1995). Maintenance of optimal levels of shade in the cardamom areas is critical for ensuring sustained productivity of cardamom. Unless pruning, lopping, pollarding and/or planting quick growing tree species in the larger canopy gaps regulate overstorey shade, cardamom productivity cannot be sustained. However, shade regulation of cardamom plantation is a grossly neglected aspect.

Although shade trees abound in cardamom plantations, there is very little documented information on the relative performance of these species on cardamom productivity. Shade trees often vary from place to place depending on local preferences and the silvicultural requirements of trees. Important shade trees in CHRs include: *Actinodeaphne malabarica, Artocarpus heterophyllus, Cedrela toona, Canarium strictum, Cullenia exarillata, Erythrina* spp, *Macaranga peltata, Mangifera indica, Mesua nagassarium, Palaquium ellipticum, Persea macrantha, Vernonia arborea,* etc.

Commercial crops trailed on support trees: black pepper

Black pepper (*Piper nigrum* L.) is a major foreign exchange earning commercial crop of India. The state of Kerala accounts for about 97 per cent of total Indian area of 1,95,050 ha and production of 53,110 tonnes. Crop productivity is, however, low in comparison to other pepper growing countries such as Indonesia, Malaysia and Brazil, owing to poor management (Cheeran *et al.*, 1992). The cultural system involving pepper is unique in that it represents a commercial crop being trailed on a variety of support trees. *Erythrina indica* forms the dominant support trees in pepper plantations. Many other species such as jack (*Artocarpus heterophyllus*), coconut (*Cocos nucifera*), arecanut (*Areca catechu*), mango (*Mangifera indica*) and timber species such as *Tectona grandis, Artocarpus hirsutus* and the like are also popular.

Although numerous MPT species are used as supports for trailing pepper vines, only anecdotal information is exists concerning their relative merits. Mathew *et al.* (1996) evaluated seven fast growing trees and found that *Erythrina stricta, E. indica* and *Garuga pinnata* are promising in this respect. An ideal support tree should have the following traits: fast growth, ease of propagation, light crowns, ability to produce multiple outputs and tolerance to pests and diseases. Typically such trees should possess rough, non-exfoliating bark and deep root systems. Nitrogen fixing potential and ability to retain foliage during summer (when shade is essential for the vines) and shed the same during the rainy season are other desirable traits in this respect.

Pepper vines are also trailed on various dead supports too including teak poles. Kumar and Cheeran (1981) compared teak poles (non-living supports) with *Erythrina indica* and found that in juvenile stands of a relatively shade intolerant pepper variety (Panniyoor-1), *Erythrina* supports gave lower yields. Consistent with this, Sankar *et al.* (1988) reported interspecific competition for applied ^{32}P between pepper vines and the surface feeding *Erythrina* roots. Likewise, Cheeran *et al.* (1992) observed that competition-related yield reduction was higher in vines trailed on *Erythrina indica* and *Garuga pinnata* supports, compared to teak poles. Despite such advantages, teak poles are seldom used in commercial pepper production, owing to the high cost and durability problems.

MULTIPURPOSE TREES ON CROPLANDS

Trees, often intentionally planted and/or allowed to persist from natural regeneration, are common features of the Kerala landscape. Agroforestry systems inventory prepared under the Diagnosis and Design Project of the All India Co-ordinated Research Project on Agroforestry catalogued several

MPTs (Table 28.3); including coconut, arecanut and rubber—three most abundant trees of the state (Kumar *et al.*, 1992). The MPTs generally follow a random or irregular geometry, except in the case of plantations and orchards, and serve multiple objectives such as providing green manure, fodder, fuel, fruits, nuts, small timber and so on. Root competition, however, is a major constraint in this respect and roots of many species grown on the same land management unit frequently intermingle and often this overlap of the roots can be extensive, especially in older stands (Jamaludheen *et al.*, 1997; Divakara *et al.*, 2001).

Trees on Farm Boundaries

Borders around farmlands are a preferred niche for tree planting, especially for the small-scale farmers. Borders include those with neighbouring lands and internal boundaries that subdivide the farm. When planted on field risers (*bunds*) and/or borders, quite apart from the specific uses for which trees are planted on farmlands, trees and shrubs serve to demarcate the field boundaries and also serve as windbreaks/shelterbelts/live fences. Farmers in different parts of the state use diverse kinds of trees for this purpose (Kumar, 1994). Examples include: *Ailanthus triphysa, Artocarpus* spp., *Erythrina* spp., *Gliricidia sepium, Macaranga peltata* and the like. In the coastal regions, however, mangrove species such as *Rhizophora* sp., *Avicennia officinalis* are important. Likewise, on paddy field risers, usually trees such as *Borassus flabellifer, Casuarina equisetifolia, Eucalyptus tereticornis, Gliricidia sepium* etc. are frequent. Many such trees (e.g. *Casuarina, Ailanthus, Borassus*) do not intercept much of the incoming solar radiation and may yield substantial quantities of green leaf manure on lopping (e.g. *Gliricidia sepium*). Commercial timber species (e.g., *Tectona grandis, Acacia mangium*) are also being increasingly planted on farm boundaries.

Hedges and living fence posts

Hedges protect crops from roaming wildlife, domestic animals and human interference. In addition, hedges are planted and managed for fuel, fodder and green manure production. Trees on hedges are also planted close together (0.5 m apart in single rows or paired rows at 1 m apart) and require frequent pruning to maintain the desired form (height, width and shape) and/or to encourage secondary branching to create an impenetrable barrier. Nitrogen fixing tree genera of particular importance for hedges include *Acacia, Casuarina,* and *Pithecellobium.*

Thorny species such as bamboos are remarkable for keeping out large animals and are, therefore, popular as hedgerows in the traditional land use systems of Kerala (Kumar, 1997). Yet, there is a tacit assumption that the profusely growing surface (fibrous) root systems of bamboo may out-compete the field/tree crops grown in association. In this respect, Divakara *et al.* (2001) tested the following three hypotheses: (1) Effective root foraging space is a function of crown spread, (2) Proximity of trees depress lateral spread of roots in mixed species systems and (3) Closer the trees are located greater will be the subsoil root activity which in turn facilitates active absorption of nutrients from deeper layers of the soil profile. As expected larger clumps had a wider root spread. Highest root counts were found in the 10-20 cm layer with nearly 30% of total roots. Proximity of bamboo clumps depressed root activity of associated teak and Malabar white pine in the surface layers of the soil profile, while root activity in the deeper layers was stimulated. Root competitiveness in polycultural systems is, therefore, a function of the proximity of the associated trees/crops, which in turn, decides the rooting intensity in individual soil layers (Kumar and Divakara, 2001). Coincidentally, trenching to spatially isolate bamboo roots has been recommended as a management strategy to reduce root competition.

Living fence posts are similar to hedges in function, however, living fence post trees are planted further apart and managed less intensively (e.g. cut less often) than hedges, and the tree is used together with other materials (barbed wire, wooden slats etc.) to form a barrier. Often living fence-post trees are allowed to grow to larger sizes than the hedges. Nitrogen fixing trees particular importance as living fence posts include *Erythrina* and *Gliricidia*, which can serve as fences, besides serving as sources of fuelwood and fodder.

Windbreaks/shelterbelts

Windbreaks are single or multiple rows of trees planted along the windward field boundaries. By slowing wind speeds, windbreaks help conserve soil moisture and prevent wind erosion and increase crop yields. Crops immediately next to the windbreak, however, may be adversely affected by competition, especially for light. Regarding windbreak design, it can be planted in a variety of spacing and arrangements. A single row of trees can form a reasonable windbreak for small farms with limited space, particularly when planted with a hedge. For windy areas, a three-tiered windbreak is recommended (a hedge, a row of medium sized trees and a row of larger trees). Windbreak trees can also be pollarded to encourage lateral growth and provide the needed tree products. Trees can be harvested from windbreaks but this should be planned to avoid gaps in the windbreak, especially during crop season. Important genera for windbreak include *Acacia, Calliandra, Erythrina, Leucaena, Pithecellobium, Pongamia, Albizia, Casuarina, Dalbergia, Paraserianthes, Pterocarpus, Tectona* and the like.

Energy plantations and woodlots

Development of tropical tree plantations has been on the rise in recent decades, presumably to meet the ever-increasing demands for timber and firewood. Cultivation of oil-bearing crops is also gaining importance and a wide spectrum of MPTs is involved in such programmes (e.g. *Simaruba glauca, Jatropha curcas* etc.). Important attributes of MPTs include rapid juvenile growth, efficient dry matter production in terms of water and nutrient inputs, crown characteristics to maximise interception of solar radiation and ease of regeneration by coppicing. Objectives of tree planting also vary correspondingly, from multiple use of perpetually 'natural looking forests', development of high yielding and sustainable industrial plantations for wood production, control of land degradation and development of agroforestry systems. High biomass production obviously is an important consideration in all tropical tree planting programmes. Biomass productivity of MPTs, however, differs enormously with species, site characteristics and stand management practices. Kumar *et al.* (1998a) evaluated the biomass production potential of nine fast growing multipurpose tropical taxa. Rate of biomass accumulation and nutrient accumulation was highest for *Acacia auriculiformis* followed by *Paraserianthes falcataria*. Differences in age, stocking, species, stage of stand development and site generally account for such differences. Tree management practices such as planting, inter-cultivation, application of chemical fertilisers, resultant soil fertility changes and/or lopping are also key determinants in this respect.

NUTRIENT EXPORT THROUGH HARVESTED BIOMASS

Short rotation tropical plantations that couple intensive management and rapid growth rates are also characterised by high rates of nutrient removal in the harvested biomass, which in turn, raises concerns about long-term site quality and sustainable production. The potential nutrient export, especially with whole tree harvesting may deplete the site nutrient capital. Nutrient accumulation and export from the site have, therefore, become important considerations in short-rotation, high-yield plantation systems, where nutrients removed through frequent harvests may exceed the natural rates of nutrient inputs such as mineral weathering, atmospheric inputs and biological fixation. Selection of tree species and parts of the tree to be removed from the site will determine the nutrient 'cost' of biomass extraction. However, only isolated attempts were made until now to characterise the magnitude of nutrient losses associated with selection of either tree species or their parts in the Indian context. Altering the rate of nutrient removal in products is probably an important design criterion in intensive short rotation silvicultural systems. A slight reduction in the tree parts removed from the site would definitely alter the rate of nutrient export (Kumar *et al.*, 1998a). Bole fraction, generally accounts for only about 50 per cent of the total nutrient export. Therefore, returning leaves and small twigs to the site at the time of harvest may be a worthwhile option to restrain nutrient export from the site. In planning sustainable short-rotation, intensive cultural systems, biomass productivity and/or site nutrient removal must be considered.

Mixed species woodlots involving N_2-fixing trees

On many sites in Kerala, production is limited by low nutrient availability, particularly N_2. Chemical sources of nitrogen, however, are dependent on supplies and prices of non-renewable fossil fuels. Nitrogen fixing plants form an attractive option, as they are non-polluting sources of cheap N. They bring in substantial quantities of atmospheric nitrogen into a combined form. And a significant portion of this nitrogen is also released into the rhizosphere through leaf litter and/or fine root turnover, besides the sloughing off mechanism of root nodules. Associated crops may, thus, utilise the nitrogen so released (Mathew *et al.* 1992; Kumar *et al.*, 2001a). Nonetheless, nitrogen-fixing trees are only rarely used to supplement the nutrient requirement of tree plantations. In this respect, an experiment was conducted to examine the growth of teak and *Leucaena* in various combinations and to evaluate the soil fertility aspects of such mixed species stands (Kumar *et al.*, 1998b). Intercropping treatments exerted a profound influence on teak growth and improved soil fertility. In particular, increasing relative proportions of *Leucaena* in the mixture had a consistently favourable effect on both height and radial growth of teak sapling up to 3 to 4 years of age. Using N_2 fixing trees could, therefore, be a viable silvicultural option for stimulating growth of tree on unfertilised and/or degraded sites.

Facilitation

Planted forests also accelerate natural forest regeneration on degraded sites where persistent ecological barriers to succession would otherwise preclude recolonization by native forest species (e.g., Kumar *et al.*, 2005b). This facilitative role of planted forests is due to their influence on understorey microclimatic conditions, vegetation structure, and development of litter and humus layers during the early years of plantation growth (Jamaludheen and Kumar, 1999; Kumar *et al.*, 2001a, b). Instances of such "catalytic effect" of forest plantings in stimulating regeneration of native species on degraded landscapes have been demonstrated in many tropical and subtropical countries (e.g., George *et al.* 1993). Furthermore, species mixtures may be considered as one-step closer in this transformation of barren landscape to "perpetually natural looking forests".

GROWING HERBACEOUS INTERCROPS IN WOODLOTS

Woody perennial plantation agricultural systems and MPT woodlots seldom utilise the site resources completely, especially during the early stages of stand development. Therefore, shade tolerant field crops such as ginger (*Zingiber officinale*), turmeric (*Curcuma longa*), medicinal herbs, colocasia, yams and forage grasses can be successfully interplanted in such plantations. Literature relating to the performance of such crops and the magnitude of interspecific competition, is however, scarce. Available reports, nevertheless, indicate that presence or absence of *Ailanthus triphysa* trees did not affect the growth and yield (Kumar *et al.*, 2001b) or ^{32}P uptake pattern (Thomas *et al.*, 1998) of intercropped ginger, until the trees were four years-old and when the intercrop was adequately fertilised.

Medicinal and aromatic plants under the shade of trees

Shade tolerant medicinal and aromatic plants such as *koduveli* (*Plumbago rosea*). *Nilappana* (*Curculigo orchioides*), arrowroot (*Maranta arundinacea*), *chittaratha* (*Alpinia calcarata*), *adapathian* (*Holostemma annulare*), *thippali* (*Piper longum*), *Dioscorea* sp, *patchouli* (*Pogostemon cablin*) etc. are suitable intercrops in coconut and other plantations. Galangal (*Kaempferia galanga*) is a promising shade tolerant medicinal and aromatic oil-yielding herb that can be profitably grown in the subcanopy of multistrata production systems in the tropics (Nair *et al.*, 1991). Kumar *et al.* (1999; 2005a) evaluated the effects of differing light transmission levels on its growth, yield, quality and nutrient dynamics under a solitary canopy (coconut palms), six multistrata canopies (i.e., coconut+dicot trees: *Vateria indica, Ailanthus triphysa* or *Grevillea robusta* grown in two planting geometries—single row and double row) and 'no-over-canopy'. Presence or absence of over-canopy, however, seemed to have little effect on galangal rhizome yield; as yield responses under 'no-over-canopy', single-strata and multistrata systems were essentially similar (mean daily subcanopy photon flux density 95 to 646 m moles $s^{-1} m^{-2}$, as against 968 m moles $s^{-1} m^{-2}$ in the open). Likewise, rhizome quality also did not exhibit any remarkable trends with respect to canopy structure.

Silvopastoral system

Silvopastoralism denotes land use systems in which pasture and livestock production are integrated with woody perennials. They broadly cover grazing systems, 'cut and carry' fodder production practices and various types of tree and grassland management systems. In general, silvopastoral systems maintain high productivity in a sustained manner at relatively low input levels even on poor substrates. High productivity levels are primarily on account of the vertical stratification of the shoot and root systems of different components (Mathew *et al.*, 1992). Consistent with this, George et al. (1996) found that most of the absorbing roots were concentrated near the trunk in the case of *Casuarina equisetifolia*, indicating the superiority of mixtures involving this species.

Adapted tree+grass combinations, therefore, can make a valuable contribution to forage production of Kerala, but knowledge of the interactive effects between trees and grasses on their production is limited. To evaluate this, Kumar *et al.* (2001a) conducted a field experiment involving combinations of four trees and grasses, besides monospecific grass controls, for seven years. Grass productivity in association with leguminous and non-leguminous multipurpose trees (MPT) having disparate canopy architecture were compared. Understorey herbage production increased until three years in all tree+grass combinations, but declined subsequently, as tree crowns expanded. Overall, casuarina among MPTs, and hybrid napier and guinea grass among forage crops, were more productive than others. Careful selection of the tree and grass components is, therefore, crucial for optimising herbage productivity in silvopastoral systems. Tree management practices such as pruning, pollarding, thinning etc. also can be employed to regulate potential interspecific competition.

Fodder banks

Fodder banks are intensive plantings of fodder trees to maximise leaf production and provide a source of "cut and carry" fodder. Trees with nutritious fodder can be either planted alone or intercropped with other fodder plants, such as grasses. Trees in fodder banks are planted closely. Systems can be designed to provide fodder only during the dry season, or to be harvested on a rotational basis providing year-round fodder. Important genera for fodder banks are *Leucaena, Gliricidia, Calliandra calothyrsus* and *Sesbania.*

OTHER TREE-BASED PRODUCTION SYSTEMS

Apiculture and sericulture and many forest-based production systems for cultivation/collection of spices, fruits, nuts and other non-timber forest products are important in the Kerala context. There are several examples for such land use activities that integrate trees either at the landscape or plot level with other life forms, which are central to food security and biodiversity conservation (Muraleedharan *et al.*, 2005). With industrialisation, however, many such systems are considered unimportant. Integrated paddy and fishery

Table 28.1. Agro-ecological zones of Kerala: main features (source: KAU, 2002)

Zones[1] No.	Name	Soil type	Topographic class[2]
I	*Onattukara*	Sandy loam	I
II	Coastal sandy	Sandy loam	I
III	Southern midlands	Laterite without B horizon	III
IV	Central midlands	Laterite	IIa
V	Northern midlands	Laterite	IIb
VI	Malappuram type	Laterite	IIc
VII	*Malayoram*	Laterite without B horizon	III
VIII	Palakkad plains	Red loam	II
IX	Red loam	Red loam	III
X	Chittoor black soil	Black soil	IIa
XI	*Kuttanad*	Peat (*kari*)	I
XII	Riverbank alluvium	Alluvium	I
XIII	High ranges	Red loam	III

[1] Based on altitude, precipitation pattern, soil type and topographic features: All except high ranges (zone XIII) are below 500 m above mean sea level; rainfall is distributed poorly in zones V, VI, VIII, X and partly in IV and XIII, i.e., northeast monsoon relatively weak; all other zones have both the southwest and northeast monsoons and moderately distributed (mean annual precipitation for the state is >3000 mm).

[2] Topographic class I: extensive valleys with level but raised garden lands; IIa: less extensive valleys and hills/slopes with moderate gradients; IIb: less extensive valleys, hills with moderate gradients, top oval and steep slopes; IIc, less extensive valleys, hills with table tops and steep slopes and III: narrow valleys and hills/slopes with steep gradients

Table 28.2. Prominent agroforestry systems for different agro-ecological zones of Kerala

Zones		Agroforestry systems
I	*Onattukara*	• animal-based agroforestry • coconut based multispecies production systems • commercial crops on support trees-black pepper • energy plantations and woodlots • hedgerows/biofences • herbaceous intercrops in woodlots • homegardens • multipurpose trees on croplands/farm boundaries

Contd...

Table 28.2. Contd...

		• plantation crop combinations • scattered fruit/nut and timber yielding trees
II	Coastal sandy	• coastal agroforestry • integrated rice-fish production systems • systems and practices mentioned for zone I
III	Southern midlands	• silvopastoralism • systems and practices mentioned for zone I
IV	Central midlands	• silvopastoralism • systems and practices mentioned for zone I
V	Northern midlands	• silvopastoralism • systems and practices mentioned for zone I
VI	Malappuram type	• silvopastoralism • systems and practices mentioned for zone I
VII	*Malayoram*	• silvopastoralism • stream bank stabilization • systems and practices mentioned for zone I
VIII	Palakkad plains	• systems and practices mentioned for zone I • mango-based agri-horticulture systems
IX	Red loam	• plantation crop combinations • systems and practices mentioned for zone I
X	Chittoor black soil	• systems and practices mentioned for zone I
XI	*Kuttanad*	• integrated rice-fish production systems • systems and practices mentioned for zone I
XII	Riverbank alluvium	• stream bank stabilization • systems and practices mentioned for zone I
XIII	High ranges	• animal-based agroforestry • commercial crops grown under the shade of trees: natural forests/planted shade trees – small cardamom, coffee and tea • commercial crops on support trees-black pepper • energy plantations and woodlots • hedgerows/biofences • homegardens • multipurpose trees on croplands/on farm boundaries • plantation crop combinations • scattered fruit/nut and timber yielding trees • silvopastoralism • stream bank stabilization

Table 28.3. Major agroforestry systems in the humid tropical zones of Kerala showing dominant components

Zone	Agroforestry systems	Predominant multipurpose trees	Fruit/nut/latex yielding/othert rees	Crops
All zones	Homegardens[1]	*Acacia mangium* *Ailanthus triphysa* *Artocarpus hirsutus* *Bambusa bambos* *Cassia fistula* *Casuarina equisetifolia* *Ceiba pentandra* *Dalbergia latifolia* *Erythrina* spp. *Gliricidia sepium* *Grevillea robusta* *Macaranga peltata* *Pterocarpus marsupium* *Tectona grandis* *Thespesia populnea*	arecanut bilimbi camboge coconut guava Indian hog plum jackfruit mango moringa Indian gooseberry para-rubber[2] papaya sapota tamarind	banana black pepper cassava cocoyam colocasia (taro) cowpea ginger guinea grass medicinal plants pine apple ornamental plants turmeric vegetables vanilla yams
	Farm boundary	*Acacia auriculiformis* *Acaia mangium* *Albizia lebbeck* *Albizia procera* *Artocarpus heterophyllus* *Bambusa* spp. *Bombax ceiba* *Ceiba pentandra* *Erythrina indica* *Gliricidia sepium* *Gmelina arborea* *Grevillea robusta* *Leucaena leucocephala* *Macaranga peltata* *Morus alba* *Pterocarpus marsupium* *Sesbania sesban* *Swietenia macrophylla* *Tectona grandis* *Thespesia populnea* *Vatica chinensis*	mango Indian gooseberry tamarind	
	Agrisilviculture Silvipasture Silvihorticulture	*Acacia auriculiformis* *Acacia mangium* *Albizia lebbeck* *Artocarpus hirsutus* *Azadirachta indica* *Bombax ceiba* *Casuarina equisetifolia* *Dalbergia latifolia* *Dendrocalamus strictus* *Erythrina indica* *Eucalyptus tereticornis*	arecanut bilimbi cacao cashew citrus clove coffee coconut coffee custard-apple guava	banana black pepper Congo signal cardamom cocoyam colocasia (taro) cowpea elephant foot yam fodder grasses groundnut

Contd...

Table 28.3. Contd...

	Garuga pinnata *Gliricidia sepium* *Grevillea robusta* *Leucaena leucocephalaa* *Macaranga peltata* *Paraserianthes falcataria* *Syzygium cumini* *Tamarindus indica* *Tamrarindus indica* *Tectona grandis* *Vateria indica*	jackfruit lime mandarin orange mango Indian gooseberry nutmeg para-rubber papaya pomegranate sapota tamarind tea	horsegram hybrid napier mung paddy paragrass sorghum stylo sweet potato til vanilla vegetables
Energy plantations/commercial woodlots	*Acacia auriculiformis* *A. mangium* *Ailanthus triphysa* *Albizia lebbeck* *Albizia procera* *Artocarpus heterophyllus* *Casuarina equisetifolia* *Eucalyptus tereticornis* *Gliricidia sepium* *Gmelina arborea* *Leucaena leucocephalaa* *Paraserianthes falcataria* *Tectona grandis*		

[1] For a comprehensive list of trees in the homegardens, please see Kumar *et al.* (1994) where more than 127 woody perennial species have been listed.

[2] usually as monospecific stands except in early stages

production system in the brackish water environments of the coastal regions of Kerala is a case in point. Although the paddy stubbles left in the field aid the microfauna which are beneficial to shrimps, and the organic wastes deposited by prawns provide nourishment to the paddy crop, such integrated systems have not received adequate scientific attention in the past.

FUTURE DIRECTIONS

This review has outlined the trends in agroforestry research in the humid tropics of Kerala during the past 15-20 years. Evidently all systems and technologies have not received ample attention in this treatment. Many aspects of functional dynamics such as system productivity, competitive interactions and resource sharing are probably less known. Although considerable research has been done on system management of agroforestry, resource partitioning in such systems and aspects related to tree management to manoeuvre the resource allocation strategies are not adequate. The choice of tree species is

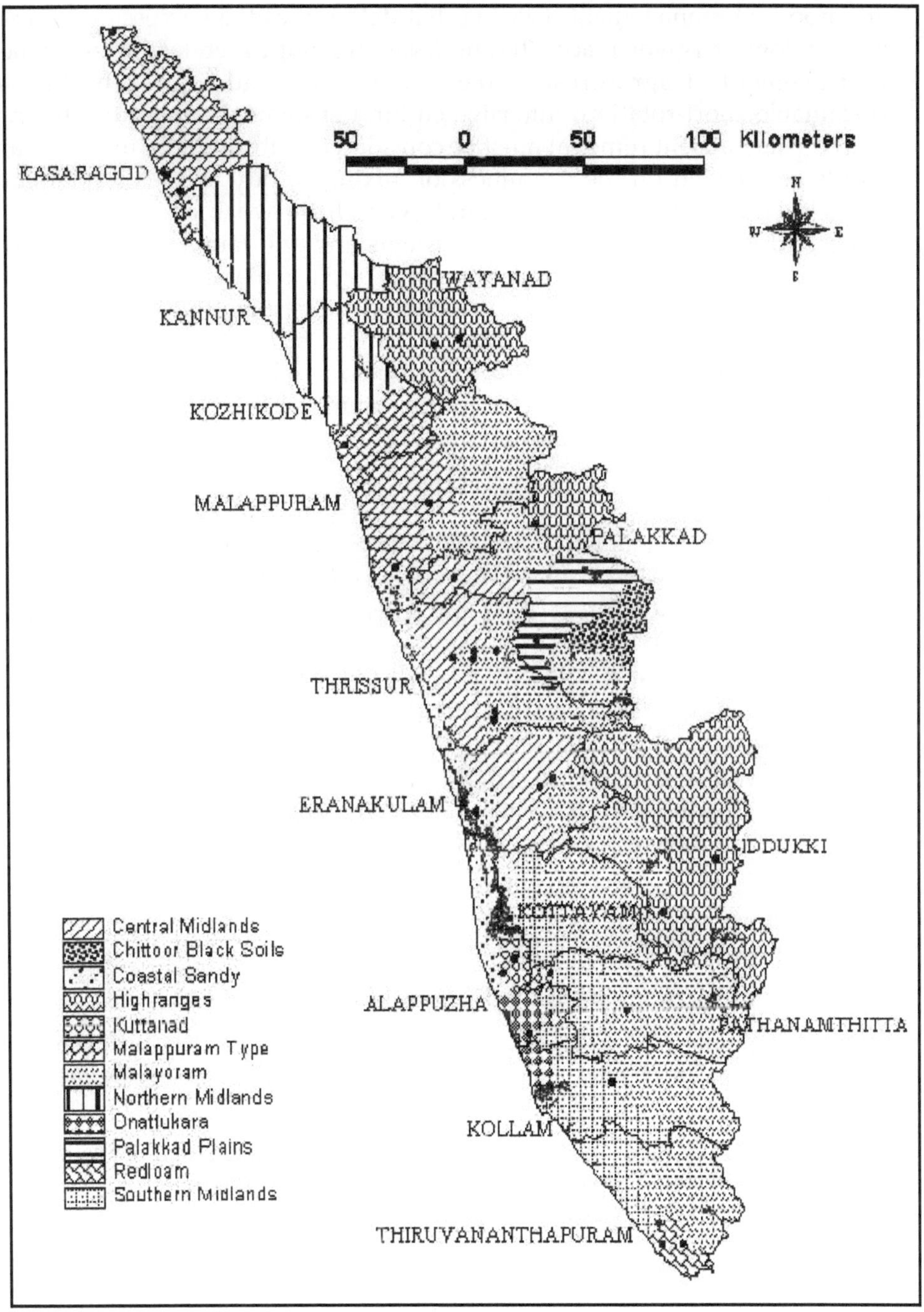

Fig. 28.1. Agroecological zones of Kerala (source: KAU, 2002)

an important management decision that determines the success or failure of a technology at a given place. The emphasis on canopy architecture and genetic improvement of agroforestry trees has not been adequate. In planning sustainable short-rotation, intensive cultural systems, biomass productivity and/or site nutrient removal must be considered. Substantially more research efforts are needed on the usefulness of mixing woody species in plantation crops such as coconut and arecanut. Very little work has been done on integrated fishery-crop production systems and soil fauna. Further effort is required to develop management practices in participation with farmers that will maximise complementary interactions and resolve the location-specific constraints to wide spread adoption of the technologies. Considering competitive relationships, the distribution and degree of overlap of neighbouring root systems are of particular interest. Hence future research should focus on below ground interactions in tropical agroforestry.

REFERENCES

Abraham, K. (1993). Light infiltration studies in coconut-based homestead situations. *MSc thesis*. Kerala Agricultural University, Thrissur, pp.60-61.

Anilkumar, K.S. and Wahid, P.A. (1988). Root activity pattern of coconut palm. *Oleagineux*, 43: 337-342.

Bruntland Commission (1987). Our Common Future. The World Commission on Environment and Development. Oxford University Press. Oxford, United Kingdom. 43p.

Cheeran, A. Wahid, P.A., Kamalam, N.V., Kurien, S. and Mathew, L. (1992). Effect of variety, spacing and support material on nutrition and yield of black pepper (*Piper nigrum* L.). *Agric. Res. J. Kerala*. 30:11-16.

Divakara, B.N., Kumar, B.M., Balachandran, P.V. and Kamalam, N.V. (2001). Bamboo hedgerow systems in Kerala, India: Root distribution and competition with trees for phosphorus. *Agroforest. Syst.*, 51: 189-200.

Farm Guide (2005). Farm Information Bureau, Thiruvananthapuram, Kerala.

George, S.J., Kumar, B.M. and Rajiv, G.R. (1993). Nature of secondary succession in the abandoned Eucalyptus plantations of Neyyar (Kerala) in the peninsular *India. J. Trop. For. Sci.*, 5: 372-386.

George, S.J., Kumar, B.M., Wahid, P.A. and Kamalam, N.V. (1996). Root competition between the tree and herbaceous components of silvopastoral systems of Kerala. India. *Plant Soil*, 179: 189-196.

Ghosh, S.P., Kumar, B.M., Kabeerathumma, S. and Nair, G.M. (1989). Productivity, soil fertility and soil erosion under cassava based agroforestry systems. *Agroforest. Syst.*, 8: 67-82.

Gowda, H.B.S. (2002). Root competition between coconut palms and multipurpose trees under varying nutrient management regimes. *MSc (Forestry) thesis* submitted to the Kerala Agricultural University, Vellanikkara, India, 82p.

Jamaludheen, V. and Kumar, B.M. (1999). Litter of nine multipurpose trees in Kerala, India- variations in the amount, quality, decay rates and release of nutrients. *For. Ecol. Manage.*, 115: 1-11.

Jamaludheen, V., Kumar, B.M., Wahid, P.A. and Kamalam, N.V. (1997). Root distribution pattern of the Jack tree (*Artocarpus hirsutus* Lamk.) as studied by ^{32}P soil injection method. *Agroforest. Syst.*, 35: 329-336.

Jambulingam, R. and Fernandes, E.C.M. (1986). Multipurpose trees and shrubs on farmlands in Tamil Nadu State (India). *Agroforest. System*, 4: 17-32.

Kerala Agricultural University (KAU) (1993). Annual Research Report 1992-93. KAU-Cadbury Cocoa Research Project. Kerala Agricultural University, Vellanikkara.

Kerala Agricultural University (KAU) (2002). Package of Practices Recommendations: Crops. 12th ed., Kerala Agricultural University, Trichur 278p.

Kerala State Land Use Board (KSLUB) (1995). Land Resources of Kerala State, Vikas Bhavan, Thiruvananthapuram, Kerala, pp 199-203.

Krishnankutty, C.N. (1990). Demand and supply of wood in Kerala and their future trends. KFRI Research Report 67. Peechi 66p.

Kumar, B.M. (1994). Agroforestry principles and practices. *In* Trees and Tree Farming. P.K. Thampan (ed.). Peekay Tree Research Foundation. Cochin. pp 25-64.

Kumar, B.M. (1997). Bamboos in the homegardens of Kerala- a shrinking resource base. *J. Non-Timber Forest Products*, 4(3/4): 156-59.

Kumar, B.M. (2003). Homegardens as a livelihood security system in the humid tropics with special reference to Kerala. *In: Agroforestry – Potentials and Opportunities.* P.S. Pathak and Ram Newaj (eds.). Agrobios, Jodhpur, pp 121-142.

Kumar, B.M. and Cheeran, A. (1981). Nutrient requirements of pepper vines trained on live and dead standards. *Agric. Res. J. Kerala*, 19: 21-26.

Kumar, B.M. and Divakara, B.N. (2001). Proximity, clump size and root distribution pattern in bamboo: A case study of *Bambusa arundinacea* (Retz.) Willd. Poaceae in the Ultisols of Kerala, *India. J. Bamboo and Rattan*, 1: 43-58.

Kumar, B.M. and Kumar, S.S. (2002). Coconut+multipurpose tree production systems in Kerala, India: Influence of species and planting geometry on early growth of trees and coconut productivity. *Indian J. Agroforest.*, 4: 9-16.

Kumar, B.M. and Nair, P.K.R. (2004). The enigma of tropical homegardens. *Agroforest. Syst.*, 61: 135-152.

Kumar, B.M., George, S.J. and Chinnamani, S. (1994). Diversity, Structure and standing stock of trees in the homegardens of Kerala in peninsular India, *Agroforest. Syst.*, 25: 243-262.

Kumar, B.M., Sajikumar, V. and Mathew, T. (1995). Floristic attributes of small cardamom (*Elettaria cardamomum* (L) Maton) areas in the Western Ghats of peninsular India. *Agroforest. Syst.*, 31: 275-289.

Kumar, B.M., George S.J., Jamaludheen V. and Suresh T.K. (1998a). Comparison of biomass production, tree allometry and nutrient use efficiency of multipurpose trees grown in wood lot and silvopastoral experiments in Kerala, India. *For. Ecol. Manage.*, 112: 145-163.

Kumar, B.M., Kumar, S.S. and Fisher, R.F. 1998b. Intercropping teak with *Leucaena* increases tree growth and modifies soil characteristics. *Agroforest. Syst.*, 42: 81-89.

Kumar, B.M. George, S.J. and Suresh, T.K. (2001a). Fodder grass productivity and soil fertility changes under four grass+tree associations in Kerala, India. *Agroforest. Syst.*, 52: 91-106.

Kumar, B.M., Thomas, J. and. Fisher, R.F. (2001b). *Ailanthus triphysa* at different density and fertiliser levels in Kerala, India: tree growth, light transmittance and understorey ginger yield. *Agroforest. Syst.*, 52: 133-144.

Kumar, B.M., Kumar, S.S. and Fisher, R.F. (2005a). Galangal growth and productivity related to light transmission in single-strata, multistrata and 'no-over-canopy' systems. *J. New Seeds*, 7: 111-126.

Kumar, B.M., Peter, K.V. and Babu, L.C. (2005b). Carbon Sequestration Potential of Mixed-species Plantations in the Tropics. Lead paper/special lecture. National Workshop on *Status and Future of Mixed Forest Plantations in India*. Kerala Forest Research Institute, Peechi. 30 May to 1 June 2005.

Kumar, B.M., Babu, K.V.S., Sasidharan, N.K. and Mathew, T. (1992). Agroforestry practices of central Kerala in a socio-economic milieu. *In*: Proceedings of the Seminar on *Socio-economic Research in Forestry* held at the Kerala Forest Research Institute, Peechi, May 1992. Kerala Forest Research Institute, Peechi, India. pp 209-220.

Kumar, S.S., Kumar B.M., Wahid P.A., Kamalam N.V. and Fisher R.F. (1999). Root competition for phosphorus between coconut, multipurpose trees and kacholam (*Kaempferia galanga*) in Kerala, India. *Agroforest. Syst.*, 46: 131-146.

Liyanage M. de S., Jayasundara, H.P.S. and Perera, K.C.P. (1990). Studies on the biomass productivity of selected multipurpose trees. *Proc. 1st Regional Workshop on Multipurpose Trees*, Gunasena H.P.M. (ed). Kandy, Sri Lanka, pp. 27-36.

Liyanage, L.V.K., Jayasundara, H.P.S. and Gunasekara, T.G.L.G. (1988). Potential uses of N-fixing trees on small coconut plantations in Sri Lanka. *Multipurpose Tree Species for Small Farm Use. Proc.* Int. Workshop, Nov 2-5, 1987, Thailand, pp. 251-253.

Mathew, T., Kumar, B.M., Babu, K.V.S. and Umamaheswaran, K. (1996). Evaluation of some live standards for black pepper. *J. Plant. Crops*, 24: 86-91

Mathew, T., Kumar, B.M., Babu, K.V.S. and Umamaheswaran, K. (1992). Comparative performance of four multipurpose trees associated with four grass species in the humid regions of southern India. *Agroforest. Syst.*, 17: 205-218

Muraleedharan, P.K., Sasidharan, N., Kumar, B.M., Sreenivasan, M.A. and Seethalakshmi, K.K. (2005). Non-Wood forest products in the Western Ghats of Kerala, India: floristic attributes, extraction, regeneration and prospects for sustainable use. *J. Trop. For. Sci.*, 18: 121-135

Nair, G.S., Sudhadevi, P.K. and Kurian, A. (1991). Introduction of medicinal and aromatic plants as intercrops in coconut plantations. *In: Recent Advances in Medicinal, Aromatic and Spice Crops* (Vol. 1). S.P. Raychaudhuri (ed.). Today and Tomorrow's Printers & Publishers, New Delhi, India. pp. 163-165.

Nair, M.A. and Sreedharan, C. (1986). Agroforestry farming systems in the homesteads of Kerala, southern India. *Agroforest. Syst.*, 4: 339-363

Nair, P.K.R. (1983). Agroforestry with coconuts and other tropical plantation crops. *In: Plant Research and Agroforestry*, P.A. Huxley (ed.), ICRAF, Nairobi, Kenya, pp. 79-102.

Nair, P.K.R. (1984). Soil productivity aspects of Agroforestry. International Council for Research in Agroforestry, Nairobi, Kenya, 85p.

Nair, P.K.R. (1993). An Introduction to Agroforestry. Kluwer Academic Publishers, Dordrecht, Netherlands, 499p.

Sankar, S.J., Wahid, P.A. and Kamalam, N.V. (1988). Absorption of soil applied radiophosphorus by black pepper vine and support tree in relation to their root activities. *J. Plant. Crops*, 16: 73-87.

Shanavas A. and Kumar B.M. (2003). Fuelwood characteristics of tree species in the homegardens of Kerala, India. *Agroforest. Syst.*, 58: 11-24.

Tejwani, K.G. (1994). Agroforestry in India. Oxford & IBH Publishing Co, Pvt. Ltd. New Delhi etc. 233p.

Thomas, J., Kumar, B.M., Wahid, P.A., Kamalam, N.V. and Fisher, R.F. (1998). Root competition for phosphorus between ginger and *Ailanthus triphysa* in Kerala, India. *Agroforest. Syst.*, 41: 293-305.

Torquebiau, E. (1992). Are tropical agroforestry homegardens sustainable? *Agric. Ecosyst. Envt.*, 41: 189-207.

Yogaratnam, N. and Iqbal, S.M. (1995). Multicropping of rubber lands with tea. *J. Nat. Inst. Plant. Manage.*, 11: 1-2.

29

Agroforestry Systems and Practices of Tamil Nadu

M. PARAMATHMA, R. JUDE SUDHAGAR AND P. RATHAKRISHNAN

Centre of Excellence in Biofuels, Agricultural Engineering College and Research Institute, Tamil Nadu Agricultural University, Coimbatore

ABSTRACT

Tamil Nadu, the southern most state of India has 8 per cent of the country's population and its net sown area is only 4-5 per cent of the country's area. The state is endowed with varied physiography from hills to coastal lands and hence varied agroclimatic and edaphic conditions. This paper details the characteristics of different agroclimate and provides varied perennial and annual components being practiced in different combinations in these agroforestry systems.

INTRODUCTION

Land is a major non renewable natural resource. With the increase in population, the demand of land for all activities increases. The requirement of food, fibre, fodder and fuel of the ever increasing population has to be met with the finite land resources. In India, the pattern of land use in agriculture and forestry took a drastic turn during the period from fifties to eighties. Conservation, development and management of the available land is basic for ensuring food, ecological, environmental and economic security of our country. Tamil Nadu's population is around 8 per cent of the total country and the net sown area is only a meager 4-5 per cent of that of the country. With the ever increasing population and expanding industrialization, steps need to be taken to utilize the land to the optimum extent possible and to identify and utilize lands effectively and efficiently in order to make full use of their maximum capability.

AGROCLIMATIC ZONES OF TAMIL NADU

Tamil Nadu is classified into seven distinct agro climatic zones based on climate, rainfall and other physical and ecological characteristics. The various zones are as follows:

1. North Eastern zone
2. North Western zone
3. Western zone
4. Cauvery delta zone
5. Southern zone
6. High rainfall zone
7. Hilly zone

AGROFORESTRY PRACTICES IN TAMIL NADU

The various agroforestry systems and practices followed predominantly in Tamil Nadu are described agroclimatic zonewise in the paragraphs below.

I. North Eastern zone

This zone comprises of Chengalpattu, North Arcot, (excluding hilly areas), Tiruvannamalai (excluding hilly areas), South Arcot (excluding Chidambaram and Kattumannarkoil taluks) and Ariyalur and Perambalur situated between 8^0 5′ and 13^0 2′ of North latitude and 76^0 15′ and 80^0 22′ East longitude. North East monsoon season provides the maximum amount of rainfall while the major soil types are red loam and clay loam. Black soils are also present in limited extent and coastal alluvial soils occur along the sea coast. Saline and alkaline soils are found over 12,000 ha in patches. The climate is basically semi arid tropical. Agrisilviculture, Silvipasture and Agrisilvipasture are the three major systems practiced in this zone.

1. *Agrisilviculture*

a) Fuelwood plantations

Vast stretches of coastal alluvium and red lateritic soils in this region accommodate monocropping of *Casuarina* and *Eucalyptus*. The yield of *Casuarina* in sandy loam in rainfed areas is about 25-30 tonnes per ha over ten year rotation whereas in coastal alluvial soils, the yield increases as much as 100 tonnes per ha depending on rainfall and groundwater availability. Cropping of *Casuarina* increases the fertility status of the soil by means of nitrogen fixation.

In red soils, the yield of *Eucalyptus* ranges from 40 to 100 tonnes at 4 x 2 m spacing over eight year rotation. During the first three years of the plantation, agricultural crops like sorghum, fingermillet and ragi can be grown. *Acacia holosericea* which was introduced in Tamil Nadu is also planted as a monocrop for fuelwood and poles. It survives even in hard laterite soil and agricultural crops like sorghum, groundnut can be grown in the interspaces of the trees for the first two to three years.Yet another species of Acacia viz. *A. auriculiformis* is also grown as a fuelwood plantation in this zone. For use as fuelwood, it can be felled on seven year rotation at a spacing of 4 x 2 m. The yield ranges from 20 -25 tonnes of fuelwood per ha. The tree sheds most of its leaves and presence of thick mat of root make cultivation of intercrops during the initial years difficult. Cashew (*Anacardium occidentale*) is found widely in this zone and this is integrated with short duration *Acacia holosericea*. The normal spacing adopted for cashew is 7 x 7 m which permits intercropping with gingelley, cowpea and fingermillet during the initial period.

In the saline and alkaline patches of this zone, trees like *Prosopis juliflora, Acacia leucophloea, Albizia amara, Cassia spectabilis, Thespesia populnea* are grown as monocrops for fuelwood. *Casuarina equisetifolia* can be successfully grown in these areas by deep furrow plouging and planting.

b) Tree gardens

In the coastal tracts of this zone, horticultural fruit trees like mango and citrus species can be successfully grown. In mango, Bangalora and Neelam varieties are preferred by farmers. Under these fruit trees sorghum and groundnut are grown as intercrops for the first three years. In addition to these, *Thespesia populnea* and *Calophyllum inophyllum* are grown singly and in blocks in the coastal areas of this zone.

c) Multipurpose trees and shrubs

As an age old practice, cashew has been cultivated at 7 x 7 m spacing in the lateritic soils. The fertility status of these soils is very low and hence intercropping with legumes such as greengram (*Phaseolus mungo*), lab-lab (*Doichos lablab*), groundnut (*Arachis hypogea*) and red gram (*Cicer aurietinum*) is followed till the canopy closes.

d) Shelterbelts

Casuarina is the major species which is used for the above mentioned system. Several pockets of *Casuarina* growing areas all along the coastline and inland form a discontinuous stretch of shelterbelt. Agricultural crops grown on the leeward side give a higher yield than those in the exposed areas and allow the farmer to grow cash crops like sugarcane instead of groundnut. Besides this, these shelterbelts protect vast cultivated areas during times of inclement weather and cyclones.

2. Silvipasture

This region does not have well established silvipasture system and the fodder demands are met from agricultural wastes and grain legumes. However, there is a great potential for the development of silvipastural systems in this zone. Trees like *Acacia leucophloea, Thespesia populnea, Cassia* sp. Can be used in these system.

3. Agrisilvipasture

Woody hedges of *Delonix regia, Erythrina indica* are common features along the inland and coastal areas in agricultural fields. Presence of *Thespesia populnea* is a common occurrence in paddy fields. All the above mentioned species provide Greenleaf manure, fuelwood and fodder for the cattle.

II. North Western zone

This zone includes Salem district (excluding hill areas and Tiruchengode taluk), Dharmapuri district (excluding hill areas) and Trichy district. It is situated between 11^0 and 12^0 55′ north latitudes and 77^0 28′ and 78^0 50′ east longitudes. The major soil types are red loam (non calcareous red or brown) and black soils. Saline and alkaline soils cover about 4000 ha. Climate ranges from semi arid to sub humid with frequent occurrence of drought. About 43 per cent of the rainfall is received during the Southwest monsoon and the remaining during the Northeast monsoon. Due to even distribution of rainfall, two crops are raised in the rainfed regions. Agrisilviculture, Silvipasture and Agrisilvipasture are predominant agroforestry system in this zone.

1. Agrisliviculture

a. Tree gardens

A distinct feature of this zone is the dominance of mango crop wherein growing of intercrop of agricultural crops like sorghum, fingermillet and groundnut during the first three years is a common practice.

b. Multipurpose trees and shrubs on farmlands

Trees like Karanja (*Derris indica*), Mahua (*Bassia latifolia*) and Imli (*Tamarindus indica*) are grown either as block plantations or as scattered trees in the farmlands. Agricultural crops like sorghum, groundnut are grown during the initial period.

Silver oak (*Grevillea robusta*) and Jack (*Artocarpus heterophyllus*) are found in the western part of the zone adjoining Karnataka at an espacement of 5 x 5 m in which *Dolichos lablab* and *Elusine coracana* (Ragi) are integrated. Similarly, *Derris indica* (Karanja) is also found dominantly in the dry lands of this zone.

Rocky shallow soils are found in the eastern part of this zone and *Agave sisalana* is grown at an espacement of 2 x 2 m. Agave is an excellent soil binder and is used for soil conservation as well as for fibre. Crops like horsegram, castor etc. are grown in the interspaces during the initial period.

Intercropping in *Eucalyptus* in the northern parts of the zone is a common feature. *Eucalyptus* is planted at a spacing of 4 x 2 m which facilitates intercropping with agricultural crops like sorghum, fingermillet.

2. *Silvipasture*

a) Trees and shrubs on pasture lands

The eastern and northern parts of the zone receive low rainfall during the north east monsoon. These areas are not suitable for horticultural crops and hence support growth of pastures of spear grass. Integration with occasional tamarind is also seen in some locations. A silvipasture system comprising of *Cenchrus* sp. grown in association with *Acacia ferruginea* (Parambai), *Derris indica* (Karanja) and *Hardwickia binata* (Aacha) for fodder, small timber and fuelwood is also found in many locations.

3. *Agrisilvipasture*

a) Multipurpose woody hedge rows

Cassia siamea is the most preferred species as multipurpose woody hedgerow for green leaf manure besides *Albizia lebbek* (Siris) which also gives fodder. In some areas of the zone, *Delonix regia* is exclusively grown for green leaf manure either in the village common lands or along the field fences.

III. Western zone

This zone includes Coimbatore and Erode districts (excluding hilly areas), Nilakottai and Palani taluks of Dindigul district, Thiruchengode taluk of Salem district, Usilampatti, Uthamapalayam and Periyakulam taluks of Theni district and Karur and Manaparai taluks of Trichy district. It is situated between 9^0 30′ and 12^0 north latitudes and 70^0 30′ and 78^0 east longitudes. This zone has undulating topography sloping from west to east with small hillocks. The soil consists mainly of shallow red soils and deep black soils. Red soils occur side by side with black soils. Deep loamy red soils occur in some areas of southern parts of Coimbatore district. The climate in the zone ranges from semi arid to sub humid with frequent occurrence of drought. Majority of the rain is received during the Northeast monsoon and hence October - December is the major cropping season in this zone. Agrisilviculture and silvipasture are the major agroforestry systems in vogue in this zone.

1. *Agrisilviculture*

a. Multipurpose trees and shrubs on farmlands

Trees like Imli (*Tamarindus indica*), Neem (*Azadirachta indica*), Perumaram (*Ailanthus excelsa*) and Karanja (*Derris indica*) are grown commonly either scattered or in blocks in fields or along field bunds under dryland cropping. In the plains, *Acacia ferruginea* and *A. nilotica* are occasional trees. The above mentioned trees are also found as block plantations wherein agricultural crops like sorghum, fingermillet, ragi are grown for the first three years.

In dry lands and in areas where soil is shallow, trees like *Acacia ferruginea*, *Acacia planifrons* and *Zizyphus jujuba* are grown. Fruit tree species like *Anona squamosa* and *Carisa carandas* are also found in some areas.

Rich red loam soil is found in the lands bordering the Western Ghats which support the growth of silkcotton (*Ceiba pentandra*) at 6 x 6 m spacing in block plantations. Bamboo species like *Dendrocalamus strictus* at 5 x 5 m spacing are also found in this tract where agricultural crops like groundnut, sorghum, ragi are grown initially.

b) Fuelwood plantations

In black clay loam soils, *Acacia nilotica* is the dominant tree species found in a scattered manner as well as in block at 4 x 4 m spacing. These soils support the growth of *Prosopis juliflora* which is used for manufacture of charcoal. In the low rainfall areas, *Acacia leucophloea* and *Albizia amara* are the two fuelwood species grown at 5 x 5 m spacing.

2. *Silvipasture*

a. Trees and shrubs on pastures

Silvipasture systems include *Acacia leucophloea* and fodder grass *Cenchrus ciliaris* and *C. setigerus*. Aridity of the area along with the presence of cankar layer of Calcium carbonate interferes with the growth of other tree species. The tree provides fodder, timber and fuelwood while the yield of grass ranges from 30 to 40 tonnes per ha per year.

IV. Cauvery Delta zone

This zone covers the Cauvery delta areas in Thanjavur, Nagapatinnam districts, Musiri, Lalgudi, Thuraiyur and Kulithalai taluks of Trichy district, Aranthangi taluk of Pudukkottai district and Chidambaram and Kattumannarkoil taluks of South Arcot district. The soil of this zone mostly consists of river alluvium which is characterized by layers of fine and coarse sand which lie alternately. Saline coastal alluvium varying in texture from sandy to pure sand occurs along the coastal regions of this zone. Nearly 50 per cent of the annual rainfall is recorded during the north east monsoon. This zone is mostly irrigated by the river Cauvery and its canal system.

This region which is known for its rich soil and adequate water supply is the rice bowl of Tamil Nadu. Because of this very reason, the integration of trees with farm crops is very low.

1. Agrisilviculture

a) Multipurpose trees and shrubs on farmland

Acacia nilotica is grown along paddy field bunds for providing farm timber. Similarly, *Thespesia populnea* and *Casuarina* are planted to supply house timber in this zone.

Trees like Teak (*Tectona grandis*), *Syzygium cuminii*, Sissoo (*Dalbergia sissoo*), Neer Marudhu (*Terminalia arjuna*) and Hollow bamboo (*Bambusa arundinacea*) are grown along the canals and as block plantations in 5 m x 5 m spacing and more. Agricultural crops like ragi, groundnut are grown for first two years as intercrops in these block plantations.

In areas where the soil is clay loam and slightly alkaline, *Acacia leucophloea* is encouraged whereas in sodic pockets of the zone, *Derris indica* and *Cassia siamea* are found. In the dry regions of this zone, Acha (*Hardwickia binata*) and Red Sanders (*Pterocarpus santalinus*) are grown in block plantations or in a scattered manner and in between the trees, crops like sorghum, ragi and fingermillet are grown as intercrops.

b) Hedgerows

Lannea coramandelica can be grown for green leaf manure as well as a live fence in the alkaline tracts of this zone. Similarly, *Casuarina equisetifolia* can be raised alternating with paddy for improving fertility and fuelwood for small industries of this zone.

2. Silvipasture

a) Protein banks

This zone being a region of cereal crops needs supplementation with proteinaceous legume fodder species like *Leucaena leucocephala, Desmanthis virgatus, Erythrina indica, Acacia nilotica* and *Glyricidia sepium*. In addition to these species, bajra, napier hybrids can also be recommended for this zone.

V. Southern zone

This is the largest zone comprising of Pudukottai district (excluding Aranthani taluk), Melur, Tirumangalam Madurai north and Madurai south taluks of Madurai district, Dindigul and Natham taluks of Dindigul district, Ramanathapuram, Virudhunagar, Tuticorin and Tirunelveli districts. It is situated between 8^0 and 10^0 55′ north latitudes and 77^0 and 79^0 50′ east longitudes. It comprises of flat terrains and intermittent hills at varying latitudes. The topography is undulating with the gradient sloping towards the east. The zone consists of saline coastal alluvium and river alluvium along the east coast, black soil in the middle of the zone, red sandy soil in the northeastern and southwestern regions with deep red soil on the western side. The climate is generally semi arid and majority of the rainfall (50-60 %) is attributed to the North East monsoon.

1. Agrisilviculture

a) Multipurpose trees and shrubs in farmlands

Red soil tract of this zone support essentially *Eucalyptus* in a large area and *Casuarina* in some patches. *Eucalyptus* plantations are often found in the form of blocks at a spacing ranging from 2 x 2 m to 5 x 5 m. In fact, the *Eucalyptus* belt of Tamil Nadu in the plains is found in this zone. In the southern part of Pudukottai district, *Acacia leucophloea* and *Tamarindus indica* are found to be dominating, whereas, in Natham taluk, *Mangifera indica* is found to be very common. Tamarind in association with annual crops is a common occurrence in the south and southwestern areas of this zone. In the black cotton soils of Virudhunagar district, neem and Babul (*Acacia nilotica*) are found scattered in farmlands besides *Morinda tinctoria*. In the Thiruchendur taluk, sandy teri soils are found which support *Dodonea viscosa, Tecomella undulata* and palmyrah.

In the western part of the zone along the Western ghats, *Tamarindus indica, Albizia lebbek* are extensively grown. In addition to these species, silk cotton (*Ceiba pentandra*) is also recommended. In all the above cases, agricultural crops like sorghum, fingermillet, ragi and groundnut are grown as intercrops for the first two to three years. Cultivation of Imli (*Tamarindus indica*) is widely found in the Gudalur and Cumbum area along with groundnut as intercrop.

In the windprone areas of this zone viz. Watrap and Rajapalayam, windbreaks using *Casuarina equisetifolia, Dalbergia sissoo* and *Tecomella undulata* have proved to be successful. These species provide timber as well as fuelwood. Deep loamy soils of the zone are suitable for cultivation of orchard crops. Fruit trees like mango, guava and annona can be encouraged for cultivation in these areas.

b) Fuelwood plantations

The sandy coastal tracts of Tiruchendur, Kayalpattinam and Veerapandiapattinam support best growth of *Casuarina equisetifolia* which is used for fuelwood and scaffolding. *Prosopis juliflora* (Vilayathi babul) is found extensively in the zone which is meant for fuelwood and for making charcoal out of the wood. Though this species is found in this zone, it needs to be cultivated by following suitable silvicultural practices with necessary inputs like ploughing, seeding, proper spacing and pruning to increase the rate of growth and stocking should be recommended to capture the full potential of the crop. In areas with high wind velocity of this zone like the Valiyoor region, *Acacia mellifera* is found suitable given its high survival percentage and its use as a live fence as well as fuelwood.

2. Silvipasture

a) Trees and shrubs on pasture lands

Majority of the regions under this zone are rainfed and depend upon the north east monsoon and hence silvipastoral options are also encouraged in this zone. The most prominently found tree species in silvipasture is *Acacia planifrons* (Umbrella thorn tree). It is often associated with grass species like *Dicanthium annulatum* and *Cenchrus.* In addition to this, in areas with assured irrigation, legume fodders like *Stylosanthus* can also be grown.

VI. High rainfall zone

It includes the entire Kanyakumari district lying between 8^0 03′ and 8^0 35′ north latitudes and 77^0 5′ and 77^0 36′ east longitudes. This region is characterized by three major soil types. A narrow belt of sandy alluvial soil is found along the coast while the plains have a deep red loam and the hilly regions have lateritic soil. Acidic soils are found in Thovalai, Thuckalay and Thiruvattar.

Due to the proximity of the Western Ghats and the sea, the climate is influenced by South West and North East monsoons. The average annual rainfall in this zone is above 1400 mm and the distribution of rainfall is equal during the two monsoon seasons.

1. *Agrisilvipasture*

a. Homegardens

The most prominent agroforestry system in practice in this zone is homesteads. Many species of trees, shrubs, vegetables and herbaceous plants are grown in dense and apparently random arrangement. Most homesteads are supported with a variety of animals like cows, bullock, sheep and pigs. Generally, the homesteads vary in size from 0.75 to 1.5 acres. This is the best suited system to meet the needs of the farming community as well as caters to the commercial means. These are characterized by high species diversity and usually three to four canopy strata. In general, all the homegardens consist of a herabaceous layer near the ground, a tree layer at the top.

In this zone, the woody species commonly found in homegardens are *Anacardium occidentale, Artocarpus heterophyllus, Haevea brasiliensis, Azadirachta indica, Syzygium cuminii*, Citrus sp. Spices like pepper (*Piper nigrum*), clove, cardamom (*Elateria cardamomum*), nutmeg are grown in the lower layer. Horticultural crops like banana (*Musa* sp.), arecanut (*Areca catechu*), pineapple are cultivated. In addition to these, vegetables like beans, bhendi, onion are cultivated on the lowermost stratum. Another conspicuous feature of this zone is the extensive spread of rubber plantations.

VII. Hilly zone

This zone consists of the hilly areas of Tamil Nadu viz., Kollimalai and Shervroys of Salem district, the Nilgiris district, Elagiri and Javadi of North Arcot district, Pachamalai of Thiruchirappalli district, Anamalais of Coimbatore district, Palni of Dindigul district and Podhigaimalai of Tirunelveli district.

The soil is mainly laterite in nature with varying depths dependent upon the topography. Organic matter content is more in these soils and the mean annual rainfall is above 1300 mm.

1. *Silvihorticulture*

Depending on the situation and the local climatic conditions, horticultural crops like cardamom, tea, coffee, cocoa, grapes, peaches, apricot are grown along with tree crops like silver oak (*Grevillea robusta*), *Erythrina indica*, teak, rosewood (*Dalbergia latifolia*), *Mesua ferrea*.

2. *Agrisilviculture*

a) Alley cropping

Leucaena leucocephala, Glyricidia sepium are used in alley cropping which are maintained as hedges at around 1 -1.5 m height above the ground level. Their main purpose is to provide fodder for the cattle in addition to their use as green leaf manure. Vegetables like cabbage, carrot, potato, beetroot are cultivated in terraces in between the hedges. This system helps to reduce soil erosion due to the formation of terraces as well as the presence of hedges.

b) Multispecies tree gardens

In this system, different kinds of tree species are grown mixed. In this zone, tree species like *Artocarpus heterophyllus, Mesua ferrea, Areca catechu, Syzygium jambolana, S. cuminii, Cocos nucifera, Mangifera indica* are grown mixed together. The main function of this system is production of food, fodder and wood products for home consumption and sale for cash.

One more conspicuous agroforestry system of this zone is the woodlots of exotic species like *Eucalyptus*, Pines (*Pinus patula, P. carribiensis*) and Wattle (*Acacia mearnsii, A. dealbata, A. decurrens*). Such woodlots / plantations are found in a large scale in the Nilgiris and in the Palnis.

Section 4

Allied Topics Related to Agroforestry

30

Agroforestry Modelling

AJIT, A. K. HANDA AND UMA
National Research Centre for Agroforestry
Jhansi - 284 003 (UP), India

ABSTRACT

Modelling has been used in natural sciences since centuries in one-way or the other. However, system modelling as a discipline in itself is established in the recent years. Agroforestry modelling is a complex process that essentially involves at least two elements viz. tree and crop, where tree is the long term and crop is the short-term component. Agroforestry models, usually, predict tree growth and yield and the reduction/enhancement in crop yield as influenced by the tree component in temporal and spatial sequence. In this paper the need of modeling, stages of model development, its validation and classification of models are given. The models, which are relevant to agroforestry, have also been given.

INTRODUCTION

Agroforestry combines agriculture and forestry technologies to create more integrated, diverse, productive, profitable, healthy and sustainable land use systems. According to the Association for Temperate Agroforestry - "Agroforestry practices are *intentional* combinations of trees with crops and/or livestock that involve *intensive* management of the *interactions* between the components as an *integrated* agroecosystem". These key characteristics are the essence of agroforestry and are what distinguish it from other farming or forestry practices. To be called agroforestry, a land use practice must satisfy all of these criteria:

Intentional : Combination of trees, crops and/or animals are intentionally designed and managed as a whole unit, rather than as individual elements that may occur in close proximity but are controlled separately.

Intensive : Agroforestry practices are intensively managed to maintain their productive and protective functions; these practices often involve annual operations such as cultivation and fertilization.

Interactive : Agroforestry management seeks to actively manipulate the biological and physical interactions between tree, crop and animal components. The goal is to enhance the production of more than one harvestable component at a time, while also providing conservation benefits.

Integrated : The tree, crop and/or livestock components are structurally and functionally combined into a single, integrated management unit. Integration may be horizontal or vertical, and above or below ground. Such interaction utilizes more of the productive capacity of the land and helps balance economic production with resource conservation.

MODELLING

Modelling has been used in natural sciences since centuries in one way or the other. However, system modelling as a discipline in itself is established in the current century only. Models have been built in the physical, biological and social sciences. The building blocks for these models have been taken from calculus, algebra, statistics, geometry, probability and nearly every other field within mathematics. Various researchers have defined modelling in their own way, some prominent and worth mentioning are :

Mclone and Andersons (1976) described modelling as the representation of our so called 'real-world' in mathematical terms so that we may gain a more precise understanding of its significant properties, and which may hopefully allow some form of prediction of future events.

Hall (1963) has said "The goal of modelling is to understand reality mathematically".

Meyer (1985) defined model to be an object or concept that is used to represent something else. It is reality scaled down and converted to a form we can comprehend. He further adds that a mathematical model is a model whose parts are mathematical concepts, such as constants, variables, functions, equations, inequalities, etc.

A model is a way to represent, in mathematical terms, the relationship that exists in the real world. You can use models to describe what might happen in future based on certain conditions and assumptions. You intuitively use models every day although you probably do not think about it consciously. For example you know how much time you must allow to get home from your work place. If you choose a mini bus it usually requires less time than a regular bus. If you wanted you could have described that relationship with a mathematical model. Models are handy tools because they parallel the way we human beings think.

Once you describe relationships that exists in the real world you can then extend the model you developed to predict what will happen if those relationships does not remain the same. Thus, modeling is done to aid the conceptualization and measurement of complex systems and some times to predict the consequences of an action that would be expensive, difficult or destructive to do with the real system. Thus, you can easily do "what-if" modeling. Just change your model assumptions and conditions, repeat the calculations, and observe the predicted outcomes.

In structuring a model's complexity, one must take trade-offs between the models generality, its realism, and its precision. The types of trade-offs one may consider are :

Simple model

Advantage	(General, Easy to use, Few parameters to estimate)
Disadvantage	(Unrealistic or over simplified assumptions, Imprecise predictions)

Complex model

Advantage	(Realistic, Precise)
Disadvantage	(Difficult to use for general problems, Many parameters to measure, Expensive to use)

Usually one or more of these qualities is sacrificed in favour of the others. All models are imperfect: that is, they are built to suit some purposes but not for all purposes. If two models work equally well for a stated purpose, one should use whichever one is simpler.

Purpose of modelling

Situations are modelled for various purposes.

A. Foremost among these is the need to predict new results or new features, which may be of the form of extension of existing results, or of a more radical nature. The predictions are often of conditions likely to exist at some future date. There may on the other hand, be predictions of events for which direct experimental evidence is unobtainable.

B. Not all situations are modelled for this purpose; In some it is sufficient to be able to describe the working of the system by mathematical means in order to obtain a greater understanding - many of great physical theories do this, although they make predictions as well. What is not usually involved in such descriptions is an element of control.

C. Indeed, many models in forestry/agroforestry studies are designed to assist management in the making of decisions.

Artistry of modelling

Development of a model, essentially involves two stages

- Model ESTIMATION
- Model VALIDATION

Construction of a precise model, not only includes sound mathematical /statistical formulation based on theoretical considerations of the process being modelled but equally important are practical observations and from this view point, it is desirable that one should have two independent data sets ; one to be used for model estimation and the other for model validation. However, if the practical situations does not permit to have two independent data sets, then the single data set should be judiciously utilized for model estimation and validation. Geisser (1975) suggested that out of a single data set, a random sample (with out replacement) of about 80% data points should be selected and it should be utilized for model estimation and the remaining 20% data points may be kept as the second independent data set to be used for model validation.

Model Estimation

At this stage the model is formally developed on the basis of theoretical considerations and experimental observations. Model estimation involves three steps;

Simplification or idealization

It includes translation of a real life situation into a problem amenable to mathematical treatment. This is a crucial stage since the real situation is usually complex, involving many processes. Some features will appear significant, many others irrelevant. This stage essentially involves variable identification. A major decision often taken at the very onset of modelling concerns the nature of mathematical variables involved. Basically variables are of two types. One class represent known features, that is quantities capable, at least in theory, of precise measurement and control; they are called deterministic variables. The other representing the unknown features, that is quantities the measurement of which can never be known precisely, but which illustrate a random feature; these are called stochastic variables. A model which contains stochastic variables must by definition be concerned with mathematical techniques appropriate to probability theory; deterministic variables often, but by no means exclusively, calls for the use of calculus, differential equations and stochastic techniques. Some situations do not reveal their nature at the onset; others have variables of both types in them. It is fundamental to the construction of the model that the nature of variables be properly diagnosed.

Construction or formulation

Once the significant features have been identified, the next stage is to translate these into mathematical entities, and to postulate relations between these identities. This is generally the most difficult stage and one in which it is virtually impossible to define a formal course of action. At this stage the equations and other mathematical relations are set up in the model.

The mathematical model thus constructed represents a simplification of the actual situation. A recognized stage of simplification is obtained when, by ignoring insignificant features, an originally complex problem is idealized to one which is mathematically tractable. Such abstractions, as a part of modeler's idealization, which nevertheless can often usefully be regarded as a close approximation to real life situation.

Parameter estimation

After the biological problem has been translated into mathematical one i.e once the model has been constructed/formulated in terms of equations or functions, the next step is to estimate the parameters of the model.

Model Validation

Once a model is constructed it needs to be validated. Indeed, some form of validation is usually carried out through out the formulation; the equation or other mathematical relations set up in the model are continually checked with the initial situation. Validation takes a number of forms. In the first place, the mathematics itself which constitutes the model must be self consistent and obey all usual laws of mathematical logic. In the second place, a model validity rests in its ability to represent the situation initially described; however, whether a model passes this test is to a large extent subjective. A model may have to represent reality, but it is not 'itself' reality.

Validity also appears in other forms. For example, a solution of the same problem at a different time, depends upon the criteria assessed by the modeler as much as on identification of features of initial situation. Much time can be invested in refining the solution to the model to an extent not justified by the formulation of the problem itself. An aspect of this concerns the empirical data. It may also be the case that an approximate answer that can be obtained quickly may be more effective than a more accurate answer that takes a longer time - which often means a straight forward numerical approximation without spending time looking for the most elegant analytical solution. Model validation also involves two steps:

Optimization/refinement

Once the model has been constructed, its parameters are to be estimated from the experimental observations. In agroforestry/forestry modelling studies, the growth equations usually involves three or four parameters. The estimates of the parameter along with confidence interval and asymptotic standard errors (ASE) are computed. Smaller the confidence interval, more accurate the parameter estimates. For assessing the certainty of the parameter estimates, Wald confidence region and C-W confidence curves (Cook and Weisberg, 1990) should be plotted. The agreement between the two methods indicates the certainty of estimates. However, if the parameters in the model are highly inter correlated, this might indicate that the model has too many parameters or the model is over parametized. The effect of over parametization in the full model may be solved by fixing one parameter (usually the one for which Wald and C-W curves coincides) and estimating the others for which C-W curve shows asymmetry in distribution. Thus in the partial model (with one parameter fixed) the estimates of the parameters gets refined leading to optimized values and minimum possible confidence intervals.

Model diagnostics or Statistical validation techniques

Once the model has been optimized, the last phase of model development is the statistical testing procedures that can ensure, to a certain extent, the safe use of model for predictive (interpolative/ extrapolative) purposes. Standard statistical texts (Gomez and Gomez, 1984 ; Mead *et al.*, 1993), in context of statistical inference and hypothesis testing, recommends BLUE (Best Linear Unbiased Estimate) i.e an estimator that is unbiased and has the minimum variance in the class of unbiased estimators. On the similar lines we seek for unbiased model with minimum value of MAR (Mean Absolute Residual) and RSS (Residual Sums of Square). Leaving aside usual residual diagnostics, numerous other statistical procedures have been devised for validating a model. For example, Residual diagnostics, to be performed as a compulsory measure with any proposed model, in addition to the usual coefficient of determination R^2(obs. vs pred) and MSE (Mean Square Error) it should include the following statistics and graphical plot.

1. To verify that the residuals are normally distributed, The residuals should be plotted against their expected values for normal distribution.
2. To check that the residuals are independent, the autocorrelation plot of the residual should be prepared.
3. To validate that the residuals have constant variance, the residual should be plotted against the estimate
4. To confirm that the residuals are not continuously being over/under estimated, the residuals should be plotted against the independent or explanatory variate.

In continuance to the above mentioned classical procedures, some of the most recent and worth mentioning statistical model validation procedures are

1. Neter *et al.*, (1996) proposed that to test the unbiasedness of a model 't' test should be used.
2. Chen *et al.*, (1998) ; Nigh and Sit (1996) proposed polynomial regression procedure on the deviation of the predicted value from the test value. The procedure applies a fifth order polynomial regression analysis, followed by the back word elimination procedure to examine whether the mean of the residuals derived from fitting is associated with a linear, or second, third, fourth or fifth order polynomial pattern. This procedure tests whether residuals differ significantly from zero.
3. Amaro(1998) proposed that a linear regression between observed value (y-obs) and estimated or predicted (y-pred) value should be fitted i.e

   (observed) = a+b*(predicted), and for a perfect model, we expect that, 'a' should be zero and 'b' should be one. $R^2_{adjusted}$ should also be one.
4. Kayahara *et al.*, (1998) used the statistic root-MSPR to measure the precision of the model. The term root-MSPR is defined as the square root of the mean squared prediction error and lesser the value of root-MSPR, better is the precision.

Goelz and Burk (1998) suggested that for proper validation of a model the long term residuals must be examined in details. The examination may suggest an inappropriate form for the equation was used, or may suggest that important variable are missing from the model. An additional benefit of detailed examination of trends in residuals is the ability to uncover misidentification of models. Although the methodology of Goelz and Burk (1992) was intended to remove any trend in residuals related to the predicted or predictor variable, this was only approximately achieved. Goelz and Burk(1998) proposed to plot the subset of residuals, so that any pattern, if present, could be uncovered. They further suggested that the residuals, before plotting, should be adjusted for autocorrelation and heteroscedasticity (i.e non-constant variance) so that they are unit less. Non-constant variance was corrected by weighting, auto correlation was addressed by expanding the error term.

Classification of models

Agroforestry models can be classified in many ways. There is frequently some degree of overlap and fuzziness between the different classifications, and between the classes within one classification. Attempts to classify models frequently confuse as much as they clarify. It is, perhaps, more useful to think of model classification in relation to pure examples of each class, with many models sharing the attributes of more than one class, rather than to expect that the classification will result in a clear-cut partitioning of all models into discrete groups. One useful classification, proposed by Davey *et al.* (1991) is into descriptive, prescriptive and predictive type of model.

Descriptive Models

Descriptive models are primarily intended for structuring and communicating knowledge about the system. Clearly, the main function of this type of models is to describe the system. However, one could conceivably use even this type of information for making inferences: for example, on the changes that one might expect over some short period of time, or on the relative susceptibility of two such systems to disturbance.

Prescriptive Models

Prescriptive models are primarily intended for inferring what action a manager of the system should take under given conditions, in order to achieve some goal. In other words, the role of prescriptive model is to generate a set of management recommendations given a goal and a set of values characterizing a given system. The management recommendations relate to the decisions that a manager can take. They may be quantitative (e.g. spacing between crop rows, amount of mulch to apply or time to prune the lower branches of trees), or they may be qualitative (e.g. species of hedge to plant in an alley cropping system or type of erosion control method to apply) various forms of prescriptive models are as follows.

(a) Prescriptive equations

These consists of an equation (or a set of equations) whose inputs are site conditions and whose output is a management variable. Prescriptive equation models generally have no explanatory component. However, they may be derived mathematically from a mechanistically-based or theoretical model, as in optimum harvesting models. The goal is usually implicit (for example, to maximize crop yield) rather than an explicit input to model. Prescriptive equation models have potential for use in delivering the result of the analysis of complex simulation models to managers.

b) Mathematical programming methods

The most familiar type of programming method is linear programming (LP), but there are a number of related methods, including non-linear programming, dynamic programming, multiple-objective linear programming, and goal programming. These are characterized by having as input

- Parameters characterizing relationships within the system
- Constraints
- The objective(s)

The methods use an algorithm which is able to find the optimum combination of inputs to maximize (or minimize) the value for the objective function subject to constraints. Given goals and physical constraints, it attempts to minimize deviation from goals without needing to satisfy the constraints completely. It requires an explicit statement of the relative weighting between goals. While mathematical programming techniques have considerable sophistication, they constrain the formulation of model that can be used, and this limitation may prohibit their use.

(c) Prescriptive use of predictive models

As noted above, one type of input to a predictive model is management decision variable. A predictive model could, therefore, be embedded in a software system that:

- Searches over the range of possible values that the decision variable can take.
- Determines the values for the output variables for each set of values for the decision variables

- Determines the optimum set of values for the decision variables (i.e that set which maximizes or minimizes the values for the output variables.)

The approaches has been advocated and used extensively in the area of resource management, however not much examples of the use of predictive agroforestry models based on this approach could be traced. This technique is a logical extension of exploratory, 'what-if' use of management models to evaluate alternative management strategies. The approach, in common with other techniques which are based on exercising a model over a range of conditions, such as sensitivity analysis, treats the model as a black box, embedded in a software system that solves the model repeatedly. It requires no modification to the model itself, provided that it must be possible to call it from the higher level software system. There are numerous search techniques available that facilitate efficient searching for an optimum over parameter space.

The prescriptive use of predictive models differs fundamentally from mathematical programming optimization methods. In the later, the form of underlying model is heavily constrained, whereas in the present context there is in principle no restriction on the form of model, except that it must be able to execute fast enough to permit plenty of runs to determine the optimal set of management inputs.

d) Decision trees and expert systems

Decision trees are branching structures that require the user to answer a question at each node (e.g on soil type, resources available, etc.). The answer takes the user down one branch or another, until arriving a terminal node, which gives management recommendation. Some decision trees have been computerized, and glorified with the name 'expert system'. However, this term most appropriately restricted to software that has a rule base of expert knowledge (rather more sophisticated than a set of branching points), the ability to reason with this knowledge in ways that mimic the reasoning ability of human experts, and the ability to explain how a recommendation was arrived at. There have been many applications in resource management. Expert systems undoubtedly have considerable potential for delivering expert management advice in a way that is easy to understand, but they are typically based on heuristic (rule-of-thumb) knowledge. They also have considerable potential for acting as a user interface to a modelling system, to guide in the choice of parameter values, or in model construction.

Predictive Models

A predictive model enables inferences to be drawn about some aspects of a system given information about other aspects. The most familiar case is when a set of one or more formulae is used to calculate the numeric values for some variable given values for others. However, there are other predictive modelling methods (using diagrams or qualitative knowledge) and in some circumstances predictive models constitute the most appropriate modelling method. Various forms of predictive models are

(a) Predictive modelling using diagrams

Diagrams are a common and powerful modelling tools. They can be used as

1. Descriptive models, to communicate ideas about system structure to others
2. Precursors to construction of mathematical models ('conceptual modelling')
3. Predictive models in their own right, permitting the inferences of things about the system

b) Qualitative predictive modelling

As a predictive tool, qualitative modelling allows prediction of qualitative rather than quantitative outcomes. For example:

- X will change
- X will increase
- X will change from 'medium' to 'high'; and
- X will have a higher value under these conditions than under those.

Given the fact that most people knowledge about how system function is qualitative (and it is very common to hear experts reason qualitatively about system behaviour), it is surprising that there are very few computer based approaches for qualitative modelling, in contrast with the very large number of one-off and generic programs for quantitative modelling. However, there is considerable potential in qualitative modelling approaches, and this may frequently be a more appropriate way of addressing modelling goals than the construction of quantitative models requiring mathematical formulae and extensive parameterization.

(c) Quantitative Predictive Models

The great majority of agroforestry/forestry models involve the numerical calculation of the response of a system given certain inputs, using a set of mathematical functions. The term 'quantitative predictive models' is primarily used to include these, and also some:

- Mathematical models which are analyzed algebraically rather than numerically.
- Models using values generated from some pseudo random number generator.

Analytical equation model

Some models consist of a set of one or more algebraic formulae (or equation), that enables the model output to be obtained by substituting values on the right hand side, and performing the necessary calculations. Empirical equations are those derived by some statistical curve fitting method, e.g. typical yield-table models for tree stands, where top height, for example, is related to age and yield class by a polynomial equation. The parameters (coefficients) in empirical models have no biological or physical meaning as such. Mechanistic models, on the other hand, incorporate some theory of underlying mechanism. Parameters in mechanistic models generally carry biological and physical meaning.

Simulation models

There is a frequent requirement to know how some variable of interest (e.g tree height or tree biomass) changes over time. This may typically be handled by a repetitive process of working out the change that is expected over some (short) time interval, then adding this change onto the variable of interest. Such models have one or more variables whose values are determined by a process of addition or subtraction of changes from one point in time to the next : these variables are called state variables, and correspond to amounts of something e.g tree volume. The state variables change through the effect of one or more processes. Models of this type are frequently specified in terms of differential equations (continuous change), but are still solved on a computer by some process of numerical integration. Simulation models almost invariably contain within them analytical equations.

Thus the above detailed description of modelling in general and agroforestry in particular clearly revealed that a 'model' means to mimic/ simulate the real situation in terms of mathematical equations or formulas and 'agroforestry' means land use systems in which woody perennials (trees, shrubs) are grown in association with herbaceous plants (crops, pastures). Thus the term 'agroforestry modelling' essentially involves the mathematical treatment of the biological process. A 'model' in context of 'agroforestry' may be used in three ways (i) to describe a system (descriptive models) (ii) to generate management recommendations (prescriptive models) and (iii) to predict system behaviour and consequences of management action (predictive models).

Truly speaking it takes a minimum span of 7-8 years to generate some agroforestry technology but modelling, may evolve such a technological recommendations in probably half a time on one hand while on the other hand can provide a way for integrating research knowledge derived from different disciplines in a common frame work and can be used to enable alternative agroforestry management options to be rapidly explored and evaluated. Agroforestry modelling may also help to define clear goals for future directed research. The importance of modelling in the strategic aspects of research has been recognized by the working group reports of agroforestry meetings (Huxley, 1985 ; Reifsnyder and Darnhofer, 1989).

Metamodelling

Metamodelling is a very new statistics based field of modelling and simulation, whereby mathematical approximations, equations are built between a model's set of input factors and its output response(s). Iterative order-reduction techniques are performed on the chosen model (and its subsequent reduced-order approximations) to eventually arrive at a "statistically acceptable" abstraction of the original, detailed model. It is an extremely difficult problem, and therefore, has not been widely explored. However, it is believed that it could have enormous application in resolving the multi-level fidelity problem; currently the biggest stumbling block in hierarchical simulation. In our country, the research work on meta modelling is yet to be initiated.

Modelling Techniques

The basic techniques start with the sampling of trees that represents the tree population in a system, then the harvesting of trees and finally the recording of observations on the harvested trees. Some of the basics of tree data recording are discussed below.

Sampling of trees

The most accurate method of assessment of biomass can be the complete harvest, however cutting and weighing every tree is not possible. Sampling of trees is necessary to avoid such difficulties. In laying out an agroforestry experiment it should be ascertain that at the end of experimental period the number of trees per treatment would be 9-12 for single site. This is the minimum number; we may select more when trees are clearly variable due to soil variability etc. Tree can be selected with stratification by size to ensure that the whole size range is covered. Although random sampling is easier to do than using recommended stratified sampling. The random method is potentially dangerous, because the selected trees may not cover the whole range. When using random sampling, we may take minimum 16 trees as sample per treatment. For tree selection, we rank all the trees according to CD (collar diameter) in descending order and delete the smallest tree (s) from list until remaining number of trees are divisible by 3 (i.e. if there are 17 trees in the plot, we should take larger 15 and divide them into tree groups. Within each group, randomly select 1,2,3 trees as per the minimum number of tree required for each treatment. All the trees should be cut 20 cm above the ground.

Measurement of height

In forestry concept, surveyor can measure the approximate height of a standing tree with a transit, but in case of destructive sampling when a tree has already been felled on the ground, a hypsometer is of course unnecessary, as its height can be measured directly with a measuring tape or a meter scale. This height is from base to maximum top in the canopy of the tree.

Measurement of diameter

Word diameter implies that trees are circular in cross section, but this is only approximation. In many cases, the section is much wider in one direction than in another. In such case, the long and short axes, average gives the diameter. When the cross section is irregular in shape, the convenient procedure is to find and measure the maximum axis and axis just right angle to the previous one. A tree caliper is required for the purpose. Another way is to measure the circumferences of the tree with a flexible tape. The circumference divided by (3.142) gives the diameter.

Diameter at breast height (dbh)

In case of standing tree, a standard position above than base diameter is desirable and is arbitrarily defined as diameter at 1.37 m above the ground.

When there are more than one stem at 1.37 m point than dbh = SQRT $(d_1^2 + d_2^2 + d_3^2 \ldots)$ where d_1, d_2, d_3, ... are the diameters of different stems at 1.37 m height.

Measurement of volume

Volume of logs can be computed through their dimensions. Volume of such logs is calculated by V= (A + a) L/2, where V is the volume, 'A' is the area of large end and 'a' is the cross sectional area at the smaller end and L its length. Volume of top of the tree, above the last log some times resembles a paraboloid and can be calculated by V= ½ AL Here 'A' is the area at the base of top section, and L is the length.

Biomass components

To evaluate useful biomass yields in an experiment, separate harvested biomass components of interest may be classified into well-defined size classes and can be converted into dry weight. Sampling of components within each size class has to be done for using it to estimate biomass production of each component.

In deciding, to partition biomass component, one can partition the trees according to the end use and market requirements for different products and economic feasibility of the trees species being assessed. Relevant practices use MPTS group of F/FRED Project of Winrock International that branch size would be categorize as

< 1cm diameter (small branches, twigs, foliage etc.)

1 cm and < 5cm (small branches for fuel)

5 cm and < 10 cm (large branches for fuel and stems for timber)

Moisture content and oven dry weight

Moisture content is an important variable to define and determine carefully. Oven dry weight is strongly recommended. The main advantage of oven dry weight is that, it permits the reliable comparison of materials with greatly differing moisture contents and is a widely used means of reporting wood weight. To convert wet weight to total dry weight, we calculate,

Total dry weight= (Sample dry weight/ sample fresh weight) x total fresh weight

A drying temperature of 50-103 °C is recommended for most of the samples, depending upon the objectives. If the object is to determine the oven dry weight of woody biomass, 103 °C (or just below the boiling point of water) is recommended. Leafy biomass, litter and twigs with less than 1 cm diameter normally reach their constant weight after 24 hours and dried at a temperature above 80 °C. Woody material with a diameter more than 1 cm and less than 5 cm requires about 48 hours. Larger material will require even more time.

Sample should be approximate 1 per cent of the total weight. 100 g sample for moisture content in bush or small tree is minimum and may be more in case of larger trees. To obtain adequate representation of variation with in tree, wood samples from different parts of tree within each size category may be mixed to give over all moisture content value.

MODELLING STRATEGIES

The above model classification depicts available models, however an agroforestry model basically involve two components namely tree and crop. The tree component being long duration and requiring prediction of tree growth/ yield at an early stage, more emphasis has to be accorded to the tree component and hence tree growth modelling. Tree growth models can be classified on the basis of growth function being used (linear or non-linear), approach of modelling (classical or simulation) and finally model optimization (simple R^2 criterion or thorough statistical validation and refinement).

The existing tree growth modelling research in our country may be broadly classified into four broader categories viz empirical; linear; sigmoid; and quadratic/parabolic. Empirical models, the simplest of all kinds involves maximising the R^2 values of the fitted model by merely increasing the variants of independent variable(s). The linear models may suffer from the problem of negative estimation of size for lower range of explanatory variate (Ajit *et al.*, 2000) that can be settled by imposing lower bound on explanatory variate ; the sigmoid models leads to constant estimation of size particularly for the higher range of explanatory variate (Ajit *et al.*, 1999) that may be tackled by using either a very wide range of explanatory variate or simulation modelling approach and the quadratic/parabolic models forbears with the illogical estimation of size particularly in the middle range of explanatory variate that may be resolve by defining the acceptable range of the explanatory variate. By "Negative Estimation", it is meant that the predictive values of size/ dependent variable (viz biomass or volume of tree) using the developed model comes out to be negative for a specific range of explanatory/ independent variate. "Constant Estimation" refers to case where the value of the explanatory variable (usually dbh, height etc.) increase but the predictive values of the size remains almost constant. "Illogical Estimation" meant that the predictive values of size decreases with the increase in the values of explanatory variate.

MODELS RELEVANT TO AGROFORESTRY

Agroforestry is an interdisciplinary subject area within which many models and modelling paradigms are relevant and have been applied, creating problems in attempting to review relevant models. There are large number of modelling paradigms within which the models are constructed. Steppler and Raintree (1988) emphasize the importance of paradigms for the development of agroforestry, but also the dynamic nature of the rise, acceptance and fall of particular paradigms. In modelling, there is the additional problem of multiple paradigms, but the compartment flow-influence paradigms is dominant. However, there are many other such as those based on matrix algebra, probability theory, chaos, fractals or object-oriented programming

There are many relevant models within a particular paradigm. This is especially true in agroforestry, with its links to many areas of ecology. Dale *et al.*, (1985) refer to there being several hundreds single tree forest growth models, and there are probably greater number in other areas, such as crop growth and water dynamics.

It is clear from the preceding description that it would have been neither possible nor productive to calculate all models relevant to agroforestry and so the approach has been to summarize the models along with modelling approaches that have been adopted and characterizing the tasks involved in agroforestry modelling.

MAESTRO : Light interception by a tree within an array of trees (Wang and Jarvis, 1990) may be used to predict absorption of radiation by a crown surrounded by the crowns of other trees. The inputs are site location, slope bearing, soil temperature, reflectance, leaf transmittances and reflectance, density distribution, physiological parameters, tree coordinates, crown dimensions. The outputs are radiation flux density at any point in the stand , hourly and daily amounts absorbed, photosynthesis and transpiration. The time base is hourly for one growing system. The model has been implemented as FORTRAN program.

SCUAF : Soil changes under agroforestry (Young and Maurya ,1991) may be used to predict the effects of soils of specified agroforestry system in given environments. The inputs are climate zones, soil class, slope and the outputs are soil erosion (mass, nutrients, organic matter), soil fertility, cycling of N and P. The time base is one year steps, for twenty years and has been implemented as PASCAL program. Model type involves compartments, flows and subsidiary equations. The model is biologically /physically based, but with growth rates supplied rather than calculated and with simple use of multipliers to capture interactions.

PARCH : Integrated modelling of semi-arid agriculture -*Crop component* (Bradley and Crout, 1992) may be used to simulate crop growth in response to soil water status. The inputs are weather data (daily temperature, rain, sunshine hours, evaporation, relative humidity), cultivar parameters (phenological thermal time periods, grain parameters) and soil water. The output is crop yield. The time base is daily for growing season and has been implemented on Microsoft Quick BASIC for PC-compatibles.

HYBRID : Hybrid (Friends *et.al.*, 1993) is canopy forest photosynthesis model. Four input files are required : a daily climate driver, a tree species library and two files containing details on individual site. The output is water and carbon dynamics. The time base is daily or annual and has been implemented as FORTRAN program.

ERIN : Evaporation and Radiation Interception by Neighbouring species (Wallace,1995) accounts for light interception and water use for two species grown in a mixture. It is also used to examine the effect of fluxes of heat and water on tree-canopy transmission.

AME : Agroforestry Modelling Environment (Muetzelfeldt *et al.*, 1996) One of the outcomes of AMP Phase 1 [Agroforestry Modelling Project funded by DFID Forestry research programme to produce models which can be used to evaluate alternate agroforestry systems, and thus to explore the conditions under which agroforestry is likely to be a favourable option (Lawson *et al.*, 1995)] is the view that that modelling is not just about designing a model and writing a computer programme that implements this model : rather, it involves many aspects concerned with the ease of construction, maintenance and communication of models, and the ease of modification, re-use and linking of independently developed submodels. As a result AMP Phase 2 included a project with the aim of developing an Agroforestry Modelling Environment (AME) to be used to facilitate the process of constructing models through a graphical user interface, and to support a high level of modularity so that the models could be constructed by selecting and linking pre-existing submodels. In AME, the modeller designs a model graphically, that is by drawing a diagram on the computer screen that corresponds to the structure of the required model. This diagram is constructed by selecting from a toolkit of basic units, much as one might do using a computer based drawing package. The model is completed by adding mathematical information, such as equations and parameter values, to the elements on the diagram.

CONCLUSION

This chapter on agroforestry modelling discusses at length the perspectives (*which model* i.e the scope of agroforestry models), needs (*why modelling* i.e is it necessary to go for modelling) and issues (*how modelling* i.e the artistry and intricacies of modelling along with validations approaches). Modelling strategies (i.e simplification or idealization of problem, construction/ formulation of the biological problem into mathematical equations/functions, estimation of parameters, optimization/ refinement of the model and finally the validation of the model) has been elaborated in detail. The pros and cons of tree growth modelling have been elaborated and the limitations of models (linear models - negative estimation ; sigmoid models - constant estimation ; and parabolic models - illogical estimation) along with the possible solutions have been discussed (simple linear models - lower bound on explanatory variate ; sigmoid model - simulation modelling approach ; and parabolic model - permissible prediction range). There is no doubt in accepting that even with the investment of huge money by many institutes in both research and extension, it is still difficult to provide predictive insight into which type of tree/crop combination is appropriate for a specific location and a particular farmers objective. However, it is endorsed that agroforestry models can provide an deep sagacity in the tree growth pattern and the effect of tree component on crop yield (reduction/promotion) in a temporal framework.

REFERENCES

Ajit, Srivastava, P.N. , Gupta, V.K. and Solanki, K.R. (1999). Sigmoid functions in tree growth modelling: limitation of constant estimation and simulation approach as an alternative. *Indian Journal of Agroforestry*. 1(2) : 135-140.

Ajit, Srivastava, P.N., Gupta, V.K. and Solanki, K.R. (2000). Linear tree growth models: A limitation of negative estimation of size. *Indian Forester*. 126(12) : 1336-1341.

Amaro, A. (1998). Modelling dominant height growth for *Eucalyptus. Forest Science*. 4 (1) : 37-46.

Bradlay, R.G. and Crout, N.M.J. (1992). PARCH-a user guide. Sutton Bonnington, Leicester, U.K.. Tropical Crops Research Unit, University of Nottingham. U.K.

Chen, H.Y.N.; Klinka, K. and Kabzens, R.(1998). Height growth and site-index models for *Populus temuloides* in north British Columbia. *For. Ecol. And Mngmt*. 102:157-165.

Cook, R.D. and Weisberg, S. (1990). Confidence curves in nonlinear regression. *Journal of American Statistical Association*. 85 : 544-551.

Dale, V.H., Doyle, T.W. and Shugart, H.H. (1985). A comparison of tree growth models . *Ecological Modelling*. 29 : 145-169.

Davey, S.M.; Prinsley, R. and White, D.H. (1991). Editorial. *Agriculture Systems and Information Technology Newsletter*. 3. pp 3-4.

Friend, A.D., Schugart, H.H. and Running, S.W. (1993). HYBRID-A physiology-based gap model of forest dynamics. *Ecology*. 74 : 792-797.

Geisser, S. (1975). The predictive sample reuse method with application. *Journal of American Statistical Association*, 70 : 320-328.

Goelz, J.C.G. and Burk, T.E.(1998). Long-term trends in height growth of Jack pine in north central ontario. *Forest Science*. 44(1):158-164.

Goelz, J.C.G. and Burk, T.E.(1992). Development of a well behaved site index equation: Jack pine in north central Ontario. *Canadian Journal of Forestry Research*. 27:776-784.

Gomez, K.A. and Gomez, A.A. (1984). Statistical procedures for agricultural research, 2nd edition. Johan Wiley & Sons, New York.

Hall, G.G. (1963). The application of mathematical thinking. University of Nottingham Press.

Huxley, P.A. (1985). The tree/crop interface - or simplifying the biological/environmental study of mixed cropping systems. *Agroforestry Systems*. 3, 251-266.

Kayahara, G.J.; Klinka, K. and Marshall, P.L.(1998). Testing site -index site factor relationships for *Pinus contorta* and *Picea engelmannic* x *P.glauca* productivity in central British Columbia, Canada. *Forest Ecology and Management*. 110(1-3):141-150.

Lawson, G.J., Crout, N.M.J., Levy, P.E., Mobs, D.C., Wallace, J.S., Cannell, M.G.R. and Bradly, R.G. (1995). The tree crop interface: representation by coupling of forest and crop process models. *Agroforestry Systems*. 30 : 199-221.

Mclone, R.R. and Andersons, J.G. (1976). Mathematical modelling. Butter worth group, London (UK).

Mead, R.;Curnow, R.N. and Hasted, A.M.(1993). Statistical methods in agriculture and experimental botany, 2nd edition. Chapman & Hall. NewYork

Meyer, W. J. (1985). Concepts of mathematical modelling. McGraw-Hill Book Company, New York.

Muetzelfeldt, R.I., Taylor, J. and Sharp, L. (1996). Modular approaches to agroforestry modelling. In Agroforestry modelling and research co-ordination *Annual Report*. 1995-1996. ODA Forestry Research Programme, Project R5652. Institute of Terrestrial Ecology, Edinburgh.

Neter, J.; Kutner, M.H.; Nachtsheim, C.J. and Wasserman, W. (1996). Applied linear statistical models. R. D. Irwin Inc., Chicago.

Nigh, G.D and Sit, V.(1996). Validation of forest height - age models. *Can. J. of For. Res.* 26:810-818.

Reifsnyder, W.S. and Darnhofer, T.O. (1989). (Editors). Meteorology and agroforestry. Proceedings of an international workshop on the application of meteorology to agroforestry systems planning and management, Nairobi 9-13 Feb., 1987. Nairobi, Kenya; ICRAF 546 pp.

Steppler, H.A. and Raintree, J.B. (1983). The ICRAF research strategy in relation to plant science research in agroforestry. In Plant research and agroforestry (Edt. by Huxley, P.A.} Nairobi, Kenya; ICRAF.: 297-305.

Wallace, J.S. (1995). ERIN : A coupled light partitioning and transpiration model for use in mixed plant communities. *In* Agroforestry Modelling and Research Co-ordination, Second annual report to UK Overseas Development Administration. Institute of Terrestrial Ecology, Edinberg, UK.

Wang, Y.P. and Jarvis, P.G. (1990). Description and validation of an array model - MAESTRO. *Agricultural and Forest Meteorology*. 5: 257-280.

Young, A. (1989). Agroforestry for soil conservation. *Science and Practice of Agroforestry*. No.4., 276 pp.

Young, A. and Maurya, P. (1991). SCUAF - soil change under agroforestry. A predictive model. Ver. 2, Computer Programm with User's Handbook : International Council for Research on Agroforestry, Nairobi, 124 pp.

31

Potential of Medicinal and Aromatic Plants in Agroforestry Systems

L.M. GUPTA, M. SALEEM, S. K. GUPTA AND MEENAKSHI GUPTA
Division of Agroforestry, Faculty of Agriculture, Sher-e-Kashmir University of Agricutural Sciences and Technology - Jammu, Main Campus Chatha P.O. Bhour Camp, Jammu - 180 009 (J&K)

ABSTRACT

The popularity and belief in natural and herbal healthcare systems is inseparable from traditional ethos and practices in India. India has been known to be a rich repository of medicinal plants since time immemorial. Now many countries like China, Cuba, India, Sri Lanka, Thailand etc. have endorsed the official use of traditional system of Medicines in their healthcare programmes. The western system of medicine (allopathy) too has a large component of plant based drugs and most potent medicines use specific chemical constituents which in turn has generated bulk demand. A dramatic increase in exports of MAP's (medicinal and aromatic plants) attests to global interest in their products as well as in the traditional health system. Past efforts of 27 years in agroforestry research and extension is now yielding results; Agroforestry is coming up in private as well as community lands in a big way for both subsistence as well as commercial purposes. Incorporation of medicinal plants in agroforestry will not only diversify the cropping system but will also lead to conserve and reduce pressure on medicinal plants in forest besides improving economic status of the farmers.

INTRODUCTION

Medicinal plants play vital role in healthcare programmes world wide. 'Health is Wealth' is an old saying and for providing better health to itself, humanity has strived hard over the centuries to overcome ailments with reasonable success. However, most of the health providing elements especially medicinal plants have come from the mother nature itself. About 80 per cent of the population of most of developing countries still rely on medicinal plants for treating human diseases (de silva, 1997). It is therefore important to know about the term 'Medicinal Plant'.

1. A species known to be beneficially modulate the physiology of sick mammals and that has been used by mankind is termed as medicinal plant (Fellow, 1991).
2. Only when medicinal properties are proven by western research (Farnsworth and Soejarto, 1991).
3. Medicinal plants are those that are commonly used in treating and preventing specific ailments and diseases, and that are generally considered to play a beneficial role in healthcare (Srivastava *et al.*,1995).

The popularity and belief in natural and herbal healthcare systems is inseparable from traditional ethos and practices in India. India has been known to be a rich repository of medicinal plants since time immemorial. The Rigveda (5000 years B.C.), Yajurveda and Athervaveda (4500-2500 years B.C.) mentioned 67, 81, 290 species of medicinal plants, respectively. Later, Charak Sanhita (700 years B.C.) and Sushrut Sanhita (200 years B.C.) have described properties and uses of 100 and 2700 plants, respectively. Now many countries like China, Cuba, India, Sri Lanka, Thailand etc. have endorsed the official use of traditional system of Medicines in their healthcare programmes. For example, the Indian Systems of Medicines 'Ayurveda', 'Sidha' and 'Unani' entirely and 'Homeopathy' to some extent, depend on plant materials or their derivatives for treating human ailments (Prajapati *et al.*,2003). Approximately 90 per cent of medicinal plants are still collected from their natural habitats (Chadha and Gupta, 1999).

The western system of medicine (allopathy) too has a large component of plant based drugs and most potent medicines use specific chemical constituents which in turn has generated bulk demand. These chemical compounds can not be synthesized cost effectively. In china, more than a billion of people or 40 per cent of the urban patients and more than 90% of the rural patients depend on plant based medicines. In South Asia, over 800 million people rely on herbal medicines. Up to 60 per cent of the drug prescribed in Eastern Europe consist of unmodified or slightly altered higher plant products(Lancet, 1994). There are number of plant species having therapeutic properties (Chadha and Gupta, 1999 and Farooqi and Sreeramu, 2001), a few important have been listed in table 31.1. About 12.5 per cent of the 4,22,000 plant species documented world wide are reported to have medicinal value, the proportion of medicinal plants to the total documented species in different countries varies from 4.4 per cent to 20 per cent (Schippman *et. al*, 2002).

Majority of the plants are higher flowering plants, out of which 33 per cent are trees, 32 per cent are herbs, 20 per cent are shrubs, 12 per cent climbers and 3 per cent are others.

Natural essential oils which are wonderful products of plant metabolism used for their aroma since the beginning of civilization. These plants bearing aroma are now termed as 'Aromatic Plants' and this aroma is due to the presence of essential oils. Essential oils are complex mixtures of odorous and steam volatile compounds which are deposited by plants in the sub-cuticular space of glandular hairs, in cell organelles (oil bodies of hepaticae), in ideoblasts, in excretory cavities and canals or exceptionally in heartwoods. These are generally distilled by steam distillation. Recent advances in chemistry have lead to development of technology for separation of individual odoriferous principles from these aromatic plant materials. Eventually large production units got established which are engaged not only for isolation of these stubble aromas but also blend them for use for perfumes industry and in making great variety of products ranging from food, medicines, cosmetics, toiletries, paints, insecticides, and a lot of several other products. India has become the largest producer of mint (*Mentha arvensis)* and basil (*Ocimum basilicum*) out of below mentioned 42 species grown in India for essential oil (Table 31.2).

Uses of Aromatic plants have been broadly categorized under four classes

1. Dental Care

World's consumption of essential oils and its fractions such as menthol, ecalyptol, carvone, anethole and methyl salicylate in dental care sector is estimated to be 10,000 to 12,000 tons valued at US$150 millions.

2. Flavours and fragrances

India's biodiversity in plants is a boon for the production of many natural flavours/fragrances which are needed for blending in natural flavous production. Consumption of flavours/fragrances was estimated to be of $10,000 millions worldwide.

3. Essential oils for Aromatherapy

Several products based on aroma are now claimed to mitigate the effects of stress on body and mind. Some of the Indian Essential oils used in aromatherapy are: Basil, Palmarosa, Sandalwood, Spice oils, Vetiver, Cedar oil, Geranium, Jasmine, Juniper berry, Lavender, rosemary, patchouli, clarysage etc.

4. Essential oils/natural extracts for cosmetics

Natural or herbal cosmetics is another sector, which has a high potential for use of essential oils and natural extracts. Now there are number of herbal products with claims of having moisturizing and anti-ageing properties like 'Ayur' cosmetic industry is using *Aloe barbadensis* in most of its products.

A dramatic increase in exports of MAP's(medicinal and aromatic plants) attests to global interest in their products as well as in the traditional health system. Commercially, these plant derived medicines, essential oils and products are worth about US$72 billion (Rs.3,60,000 crores) worldwide, which include global business of medicinal herbal material of US$60 billion (Rs.3,00,000 crores). According to WHO estimates the projected demand for medicinal plants alone by the year 2050 would be US$ 05 trillion (Rs. 250,000,00 crores). The domestic market of Indian Systems of Medicines and Homeopathy is of the order of Rs.4,000 crores (US$800 million). India's total export earning from the crude drugs, herbal extracts and finished products stands at meager Rs.800 crores (US$160 million). India contribute only 2 per cent in medicinal plants global trade whereas, 0.4 per cent in world export of essential oils and perfumery materials. China and India are the world's leading producing nations while Korea and Japan are the main consumers of medicinal plants. It is estimated that the total number of MAP's in international trade is around 2500 species worldwide(Schippman *et. al*, 2002).

MEDICINAL AND AROMATIC PLANTS: SOURCES AND THEIR CONSERVATION

Forests are the primary sources of medicinal and aromatic plants. Basically forest products are classified in two categories: Major and Minor forest products (now Non-Timber forest products). Technically, all vegetable, animal and mineral products other than timber and fuel wood (comes under major forest products) are collectively termed as 'Non-Timber Forest Products'. Non-timber forest products in addition to MAP's include fibre & flosses; Gum, resin & oleo-resin; Grasses, bamboos; Oil seeds; Tans & dyes; Leaves; Edible products; Lac & shellac and Animal, mineral & miscellaneous products. India is bestowed with one of the 12 bio-diversity centers and two hot spot regions having diverse range of plants and their wild relatives. Forests provide a wealth of highly prized medicinal and aromatic plants in the alpine, sub-alpine, intermediate and subtropical region of north-western Himalaya (Gupta, 1986; Joshi & Rawat, 1997). Though some of the important medicinal and aromatic plants (MAP's) currently cultivated on large scale in India viz. Mints, Damask Rose, *Ocimum* spp. (Tulsi) in the Indo-gangetic belt; lemongrass, rosemary, geranium and patchouli in Southern parts of India; lavender, Pyrethrum,

Clarysage in the temperate region of Northern India; Citronella in North-Eastern regions and Ashwagandha, Senna, Safed Musli and Isabgol in western and central parts of India, but large volume of these are still collected from forests due to meet out the growing demand of MAP's and their products (Khanuja *et al.,* 2005). In India, medicinal plants are harvested regularly in appreciable quantities and exported with government approval or usufruct rights are extended to the so-called indigenous communities at the forest margins (Gupta, 1988). In Jammu and Kashmir state, contractual extraction of the medicinal and aromatic plants is done with the permission of State Forest Department i.e. an area inhabiting a particular species is auctioned to the private contractors/entrepreneurs. In Himachal Pradesh, tapping of *Pinus roxburghii* and extraction of Katha from *Acacia catechu* are executed by private contractors. Approximately 90 per cent of medicinal plants are still collected from their natural habitats (Chadha and Gupta, 1999). Due to un-scientific and indiscriminate harvesting of MAP's from their natural habitats without any concerted efforts to regrow them, many plant species have become extinct or are on the verge of extinction. It is therefore imperative to boost cultivation of MAP's so as to increase production and conserve bio-diversity. Threat to wild medicinal plants have been put into seven categories:

1. Human interference 7.8 %
2. Fragmentation 5.0 %
3. Loss of Habitats 18.7 %
4. Over-exploitation 17.0 %
5. Harvest 19.8 %
6. Trade 24.6%
7. Others 7.0 %

Government of India have put below mentioned 29 species of medicinal and aromatic plants under negative list of exports under Foreign Trade Development and Regulation Act-1992 for *in-situ* conservation of these species (Table 31.3). However, it has been clarified in the notification that if these 29 plants are cultivated and the cultivation source is certified by the competent authority i.e. forest officials of Centre/State concerned and prescribed procedure is followed as per notification, it is allowed for export.

Ex-situ and *in-situ* conservation, development of agro-techniques of medicinal and aromatic plants, even trade bans and people support for conservation of these resources, built through communication and cooperation are some of the steps which will not only ensures conservation of the bio-diversity but also the sustained supply of raw material of MAP's to the users.

In order to coordinate activities pertaining to medicinal and aromatic plants including drawing up policies and strategies for conservation, proper harvesting, cost effective cultivation, research and development, processing, marketing of raw material etc. a national level nodal body "National Medicinal Plants Board" was set up under a Government Resolution notified on 24th November, 2000 under the Chairmanship of the Union Minister of Health & Family Welfare, Govt. of India, with its headquarter at 36, Janpath, Chanderlok Building, New Delhi. In the initial stage, 25 State Medicinal Plants Boards were set up to develop the medicinal plant sector in different agro-climatic zones across the country.

Various Govt. organizations/institutions, NGO's and pharmaceutical industry are dealing research and development activities pertaining to medicinal plants in India. Indian Council of Agricultural Research, New Delhi is one such agency which works through its All India Coordinated Project on MAP's having various centers located in different agro-climatic zones of the country with National Research Centre on MAP's at Boriavi, Anand (GJ). Other such agencies are Council of Scientific and Industrial Research (CSIR), Delhi; Indian council for Forestry Research and Education (ICFRE), Dehra Dun; State Agriculture Universities etc.

MEDICINAL AND AROMATIC PLANTS IN AGROFORESTRY SYSTEMS

Growing of traditional food crops with multipurpose tree species is a common practice in India for centuries (Puri and Nair, 2004). But due to exponential growth of population, there is a considerable reduction in landholdings and this trend is expected to continue in the time to come. The existing land use systems appear inadequate to meet the requirement for basic needs. Diversification of present cropping pattern seems to be the need of the day to cope with the ever increasing demand for diversified products and assured income. Medicinal and aromatic plants are one such option which can be introduced in existing agroforestry systems for achieving diversification and assured income.

Many plants which already exist in our traditional agroforestry systems have medicinal and aromatic values. There are number of medicinal and aromatic plants species which are present in such systems in Jammu & Kashmir and Himachal Pradesh, a few have been listed in table 31.4. Some medicinal and aromatic plants viz. *Ocimum sanctum* (Tulsi), *Murraya koenigii* (Karipata), *Azadirachta indica* (Neem), *Moringa oleifera* (Drum stick tree), Mints (*Mentha* spp.), *Catharanthus roseus* (Periwinkle), *Aloe barbadensis* (Aloe), *Coriandrum sativum* (Coriander) etc. are common in Indian households. *Jatropha curcas*

(Ratanjot)- an important source of bio-diesel having medicinal value(purgative) is commonly used as live fence in farmlands in Jammu and Kashmir. *Pinus roxburghii* (chir pine) is also an important, remunerative aromatic tree species found commonly in grasslands of J&K, Himachal Pradesh and Uttranchal in addition to forests and plantations. *Acacia catechu* (Khair) having high medicinal value in addition to its value for fodder and timber, is another one such species found commonly in grasslands and farmlands in north-west Himalayas. There are number of such species which exists in different agroforestry systems but sometimes their medicinal/aromatic uses become secondary when they are used for timber, small wood, fodder etc. So the tree species like *Azadirachta indica* (Neem), *Moringa oleifera* (Drum stick tree), *Sapindus mukorossi* (Retha), *Terminalia chebula* (Harad), *Terminalia bellerica* (Bahera), *Emblica officinalis* (Amla), *Aegle marmelos* (Bael), *Syzygium cuminii* (Jamun) etc. can be intercropped with annual crops(food/medicinal/aromatic) in the early years until the tree canopy covers the ground. The tree species can be managed at low height or may be planted at larger spacing so to obtain the annuals through out.

In India, the supply of seventy per cent of medicinal and aromatic plants is from tropical areas-dry and deciduous vegetation, viz. Western and Eastern Ghats, Aravalis and Sub-tropical regions of the Himalayas, where as the 30 per cent of medicinal plants which include species having high medicinal values, are found in the temperate and evergreen forests. Medicinal and aromatic plants growing in forest areas require partial shade, moist soil with high humus content, high humidity and mild temperature. For example, *Taxus baccata, Valeriana jatamansi, Bergenia, Viola, Picrorhiza, Dioscorea, Sausurea etc.* grow as understorey in the forests of deodar (*Cedrus deodara*), fir (*Abies pindrow*) and spruce (*Picea smithiana*). Other species like Sarpgandha (*Rauvolfia serpentina*), Ashwagandha (*Withania somnifera*), Kalihari (*Gloriosa superba*), Cowhedge (*Mucuna puriens*), Aloe (*Aloe barbedensis*) etc. are common in mixed forests of sub-tropical regions. Cultivation of such plants in their natural zone of occurrence can be taken up in thinned forests and as intercrops in new forest plantations (Table 31.5). In India, a few medicinal plants that have been successfully intercropped with trees(*Acacia auriculiformis, Albizia lebbeck, Eucalyptus tereticornis, Gmelina arborea and Leucaena leucocephala*) are *Chlorophytum borivilianum, Rauvolfia serpentina, Curcuma longa, Curcuma aromatica, Curculigo orchioides* and *Zingiber officnalis* (Chadhar and Sharma, 1996; Mishra and Pandey, 1998 and Prajapati *et al.*, 2003). In general shade tolerant species can be grown successfully under the canopy of trees. Another option is intercropping of medicinal and aromatic plants in plantation crops like *Hevea brasilensis, Areca catechu, Theobroma cacao, Cocos nucifera, Coffea arabica, Camellia* spp.etc. The intercropping of aromatic crops such as *Cymbopogon* spp., *Mentha* spp., *Pelargonium* spp. gave two to four times greater economic returns than sole crops indicate that aromatic crops can be grown in agroforestry systems (Table 31.6). There were records of significantly higher net returns from inter-cropping of medicinal and aromatic plants with commercial timber species (Chauhan, 2000 and Dutt and Thakur, 2004). In China, cultivation of medicinal plants has been an age-old practice under the name of 'Silvo-Medicinal System'. In NE China, Ginseng (*Panax ginseng*) and other medicinal plants are grown in Pine and Spruce forests; in Central China, many medicinal plants are planted with *Paulownia tomentosa*(Zou and Sanford, 1990). Such system can be introduced in India also. As in the taungya system, newly established forest plantations can be intercropped with MAP's similar to food crops until the trees cover the ground. The participation of the local people with right to share benefits of the plantations, especially ownership to crops, has helped governments to establish and protect large-scale tree plantations without conflict with the local people in many Asian Countries (Nair, 1993). Home Gardens are another one such option where medicinal and aromatic plants can be incorporated in addition to the food and fruit crops.

With a view to prioritize and encourage medicinal plants for development, National Medicinal Plant Board have identified and prioritized 32 species of medicinal plants at national level which includes: *Emblica officinalis, Saraca asoca, Withania somnifera, Aconitum heterophyllum, Aegle marmelos, Phyllanthus amarus, Bacopa monnieri, Santalum album, Swertia chirata, Tinospora cordifolia, Gymnema sylvestre, Commiphora wightii, Plantago ovata, Nardostachys jatamansi, Gloriosa superba, Andrographis paniculata, Garcinia indica, Saussurea costus, Picrorhiza kurroa, Solanum nigrum, Glycerrhiza glabra, Chlorophytum borivillianum, Coleus barbatus, Piper longum, Berbaris aristata, Crocus sativus, Rauvolfia serpentina, Cassia angustifolia, Asparagus racemosus, Ocimum sanctum, Embelica ribes* and *Aconitum ferox.* Board also provide funding for contractual farming of the above referred species.

CONCLUSION

Agriculture continues to form the backbone of Indian economy contributing approximately one third to the GDP and providing employment to two thirds of the national population. However, with exponential growth in population a continuous fall in land holding size is well understandable. Smaller land holdings, combined large family size, shrinking natural resources, increasing input costs, diminished net returns from conventional crops and unreliable weather conditions are gradually proving monoculture

Table 31.1. Therapeutic properties of some important medicinal plants

S. No	Plant Species	Official Parts	Medicinal Uses	Active Constituents
1	*Aloe barbadensis** (Gritkumari)	leaves	Cosmetics, purgative	Barbaloin, aloesin, aloe-emodin,
2	*Andrographis paniculata** (Kalmegh)	Entire plant	Fever	Bitters
3	*Asparagus abscendens** (Shatavari)	Roots	Anabolic, Lactagogue, general vitality	Steroidal seponins (Shatavarin I-IV)
4	*Bacopa monnieri* (Neer Brahmi)	Entire herb	Memory enhancing	Bacoside-A, Bacoside-B
5	*Boswellia serrata* (Salai guggal)	Tree trunk exudes gummy oleo-resin	Osteoarthritis, Rheumatoid arthritis, Ulcerative colitis, Burstitis	Boswellic acids
6	*Cassia angustifolia* Vahl. (Senna)	Leaves and Pods	Laxative	Sennosides
7	*Catharanthus roseus** (Periwinkle)	Leaves & roots	Anti-cancerous properties	Vinblastine, Vincristine
8	*Cephaelis ipecacuanha*	Root	Emetic & dysentery cure	Ipecac
9	*Chlorophytum borivilianum** (Safed musli)	Root	Aphrodisiac & Sex tonic properties	Sapogenin
10	*Cinchona* spp.	Bark	(i)Anti-malarial (ii) Suppresser of out-of – sequence of heart beats	(i)Quinine (ii)Quinidine
11	*Datura metel* (Datura)	Whole plant	Antispasmodic	Hyoscyamine, scopolamine
12	*Digitalis lanata*	Leaves	Heart problems	Digoxin, Digitalin
13	*Dioscorea deltoidea* (Shingli-mingli) *Dioscorea floribunda*	Roots	Steroidal drugs, contraceptive pills	Doisgenin
14	*Emblica officinalis* * (Amla)	Fruit	Constituent Trifala useful in stomachic problems	Vitamin C
15	*Gloriosa superba* * (Agnishikha)	Seeds and tubers	Gout, Rheumatism	Colchicine
16	*Glycerrhiza glabra* * (Mulathi)	Roots	Expectorant, laxative, peptic ulcers	Glycerrizine
17	*Mucuna puriens** (Cow hedge)	Seeds	Parkinson's disease	L-Dopa
18	*Psoralea corylifolia* (Babchi)	Seeds	Psoriasis , AIDS, amoebiasis, wound & ulcer healing	Psoralen
19	*Rauvolfia serpentina* * (Sarpagandha)	Roots	Hypertension, mental disorders, epilepsy, gynaecological disorders	Reserpine, Serpentine
20	*Sapindus mukorossi** (Retha)	Fruit	Soaps, detergents, male contraceptive pills	Saponins
21	*Solanum viarum**	Berries	Steroidal drugs, contraceptive pills	Solasodine
22	*Spilanthus acmella* (Akarkara)	Flower heads	Pyorrhea	Spilanthol
23	*Stevia rebaudiana* (Natural Sweetener)	Leaves	Insulin balancing properties	Stevioside, Rabaudisides
24	*Valeriana jatamansi* (Mushkbala)	Rootstock	Sedatives, tranquilizers	Valepotriates
25	*Withania somnifera** (Ashwagandha)	Roots	Antibiotic, Anti-tumour, general vitality	Somniferine, withanine
26	*Silybum marianum** (Holy-thistle)	Seeds	Anti-hepatotoxic properties	Silymarin, Silybin

*Suitable crops for dryland farming

Table 31.2. Major aromatic plant species used by essential oil industry in India

Abelmoschus moschatus	*Lavandula officinalis**	*Apium graveolens*	*Artimissia annua**
*Eucalyptus citriodora**	*Lavandula intermedia**	*Bursera delpechiana*	*Humulus lupulus**
Cinnamomum camphora	*Chamomilla recutita**	*Cedrus deodara*	*Mentha spicata**
Cymbopogon flexuosus	*Cananga odorata**	*Cymbopogon martini*	*Mentha arvensis**
Cymbopogon winterianus	*Cyperus scariosus*	*Artimissia pallens*	*Tagetus minuta*
Cymbopogon khasianus	*Jasminum grandiflorum*	*Acorus calamus*	*Artimissia vestita*
Cymbopogon pendulus	*Mentha citrata**	*Mentha gracilis**	*Mentha piperita**
*Cymbopogon citratus**	*Ocimum basilicum**	*Ocimum canum*	*Ocimum sanctum*
*Ocimum gratissimum**	*Pelargonium graveolens**	*Pogostemon patchouli**	*Rosa damascena**
*Rosemarinus officinalis**	*Salvia sclarea**	*Santalum album*	*Skimmia laureola*
Eucalyptus globulus	*Vetiveria zizanoides*		

*Exotic Introductions

Table 31.3. List of MAP's under negative list of exports in India

Cycas beddomei	*Vanda coerulea*	*Saussurea costus*	*Paphiopedilium* spp.
Nepenthes khasiana	*Renanthera imschootiana*	*Euphorbia* spp	*Ceropegia bulbosa*
Dioscorea deltoidea	*Podophyllum hexandrum*	*Cyatheaceae* spp.	*Cycas ciricinalis*
Aquilaria malaccensis	*Rauvolfia serpentina*	Orchidaceae spp.	*Frerea indica*
Taxus wallichiana, *T. baccata*	*Pterocarpus santalinus*	*Aconitum* spp.	*Coptis teeta*
Coscinium fenestratum	*Dactylorhiza hatagierea*	*Gentiana kurroo*	
Kaempferia galanga	*Nardostachys grandiflora*	*Gnetum montanum*	
Swertia chirayata	*Penax pseudoginseng*	*Picrorhiza kurrooa*	

Table 31.4. Examples of Tree species having medicinal/aromatic value grown in traditional Agroforestry System in States of J&K and Himachal Pradesh

Tree species (common name)	Agroforestry system/zone	Medicinal value/commercial use
Terminalia chebula(Harad)	Field bunds, scattered trees in croplands, grazing lands, waste/degraded lands. (Subtropical, Intermediate)	Constituent of 'Trifala', rich source of vitamin C. constipation, heart tonic, purgative
Terminalia bellerica (Bahera)	Field bunds, scattered trees (Sub-tropical)	Constituent of 'Trifala'.
Emblica officinalis(Amla)	Field bunds, scattered trees on croplands, Homegardens, orchards. (Sub-tropical)	Constituent of 'Trifala' and important source of vitamin C.
Syzygium cuminii(Jamun)	Field bunds, scattered trees on croplands.(Sub-tropical)	Anti-diabetic(insulin balancing properties)

Contd...

Table 31.4. Contd...

Acacia catechu (Khair)	Field bunds, waste/eroded lands scattered trees on croplands and grazing lands. (Subtropical, Intermediate)	Sore throat, ulcers
Aegle marmelos(Bael)	Homegardens, around temples and religious places.(Sub-tropical)	Chronic dysentery, diarrhea etc.
Azadirachta indica (Neem)	Scattered trees in croplands.(Sub-tropical)	Antiseptic, antipyretic, astringent, pesticidal properties
Cassia fistula(Amaltas)	Scattered trees in grazing lands.(Sub-tropical)	Boils, leprosy, ringworm, colic etc.
Olea cuspidita (Olive)	Scattered trees in croplands, degraded forest lands (Intermediate & Sub-temperate)	Joint pain in childrens, fever
Vitex negundo	Field bunds, eroded lands (Subtropical, Intermediate)	Asthama, lung diseases, urinary troubles, joint pains
Hippophae rhemnoides	Cultivated as live fence in Kargil , Leh(J&K) and Lahaul Spiti(H.P.)	Rich source of Vitamin C
Sapindus mukorossi(Retha)	Scattered trees in croplands, waste land and forest areas. (Intermediate & Sub-temperate)	Shampoo, soaps, male contraceptive pills.
Pinus roxburghii (chir pine)	Commonly scattered trees in grasslands (Sub-tropical & Intermediate)	Turpentine in paint, varnish etc. and rosin in paper industry
Cedrus deodara (deodar)	Scattered trees on field bunds(Temperate)	Oil used in perfumery, cosmetics, soaps and for treating skin diseases and ulcers

Table 31.5. Examples of commercially valuable medicinal plants under cultivation or that can be produced as under-story component(s) in forest and tree plantations

Species	Plant type	Location
Aconitum heterophyllum (Atis)	Tall herb	Alpine and sub-alpine Himalayas
Aloe barbadensis (Gritkumari)	Herb	Sub-tropical to intermediate zone, India
Cephaelis ipecacuanha (Raicilla, Ipecac)	Shrub	Darjeeling, Assam, Sikkim, Nilgiris
Chlorophytum borivilianum (Safed musli)	Annual herb	Central and Peninsular India
Colchicum luteum (Sarnjan-e-talak)	Herb	Alpine and sub-alpine Himalayas
Costus speciosus (Crepe ginger)	Cane	North eastern sub-Himalayan tracts and Western Ghats
Dioscorea deltoidea (Himalayan yam)	Vine	Alpine and sub-alpine Himalayas
Gentiana kurroo (karroo)	Herb	Alpine Himalayas
Gloriosa superba (Kalhari)	Climber	Sub-tropical to intermediate zone, India
Picrorhiza kurrooa (karroo)	Herb	Alpine and sub-alpine Himalayas

Contd...

Table 31.5. Contd...

Podophyllum hexandrum	Herb	Alpine and sub-alpine Himalayas
Rauvolfia serpentina (Rauvolfia, Sarpgandha)	Shrub	Sub-montane zone
Saussurea lappa (Kuth)	Herb	Alpine and sub-alpine Himalayas
Swertia chirayata (Chirayata)	Herb	Alpine and sub-alpine Himalayas
Viola odorata (Banafasha)	Herb	Alpine and sub-alpine Himalayas
Tinospora cordifolia (Giloe)	Climber	Mixed deciduous Sub-tropical forests
Withania somnifera (Ashwagandha)	Shrub	Sub-montane zone

Table 31.6. Yields of aromatic plants in intercropping with trees relative to those in sole cropping at a few locations in India.

Location	Tree species and spacing	Aromatic Plants	Intercrop yield as % of sole crop Herbage	Oil	Comments
Pantnagar, Uttaranchal	*Populus deltoides* 5 x 4 m *Eucalyptus* hybrid 2.5 x 2.5 m	*Cymbopogon flexuosus C. martini* *C. flexuosus* *C. martinii*	82 96 96 100	82 86 90 86	Results averaged over five years
Pantnagar, Uttaranchal	*Populus deltoides* 5 x 4 m	*Mentha arvensis* *Mentha piperita* *Mentha citrata* *Mentha spicata* *C. flexuosus* *C. martinii* *C. winterianus*	90 85 90 87 99 99 98	90 88 90 87 97 98 99	Results averaged over three years
Hyderabad, Andhra Pradesh	*Eucalyptus citriodora* 3 x 2m	*Cymbopogon flexuosus* *C. martini* *C. winterianus* *Pelargonium* sp.	87 87 112 91	87 87 91 90	Results averaged over two years

Adapted from: Rao *et al.*, 2004

uneconomical proposition leading to grown inclination towards diversification. Introduction of MAP's in various agroforestry systems may provide sustainable means of natural source of high value industrial raw material for pharmaceuticals, agri-chemicals, food and cosmetic industry and opening of new possibilities for significant progress in rural economy. The incorporation of MAP's in agroforestry systems will not only uplift the socio-economic status of rural people but also contribute towards conservation of bio-resources.

REFERENCES

Chadha, K.L. and Gupta, Rajendra (1999). Advances in Horticulture. Vol.-11. Malhotra Publishing House, Delhi. 932 p.

Chadhar, S.K. and Sharma, M.C. (1996). Survival and yield of four medicinal plant species grown under tree plantations of Bhataland. *Vaniki Sandesh* 20(4): 3-5

Chauhan, H.S. (2000). Performance of Poplar (*Populus deltoides*) based agroforestry system using aromatic crops. *Indian Journal of Agroforestry*. 2:17-21

De Silva, T. (1997). Industrial utilization of medicinal plants in developing countries. *In*: Medicinal Plants for Forest Conservation and Healthcare. Non-Wood Forest Products No. 11(G. Bodeker; K.K.S., Bhat; J. Barley and P. Vantomme eds.). FAO, Rome, Italy.

Dutt, Vaishnu and Thakur, P.S. (2004). Bio-diesel-economics of cropping systems combining medicinal and aromatic herbs with commercial timber tree species. *Indian Journal of Agroforestry*. 6(1):1-7

FAO (2003). State of the World Forests. FAO, Rome, Italy. 153 p.

Farooqi, A.A. and Sreeramu, B.S. (2001). Cultivation of Medicinal and Aromatic crops. Univ. Press (Ind) Ltd. Hyderabad.518 p.

Fellow, L.E. (1991). Pharmaceuticals from Traditional Medicinal Plants and Others: Future Prospects. Paper presented at the symposium "New Drugs from Natural Resources" sponsored by I.B.C. Technical Services Ltd. London, June 13-14,1991. Royal Botanic Garden, Kew.

Fransworth, N.R. and Soejarto,D.D. (1991). Global importance of Medicinal Plants. *In*: Conservation of Medicinal Plants(C. Akerale, V. Heywood and H. Sygne eds). Cambridge University Press, Cambridge.

Gupta, M.P. (1988). Present status and future prospects for the cultivation and collection of medicinal plants of Himachal Pradesh. *Indian Forester*.114: 19-25.

Gupta, R. (1986). Integration of medicinal plants cultivation in forest and forest plantations of northwestern Himalaya. *In*: Agroforestry Systems: A New Challenge. Indian Society of Tree Scientists, Solan, India. pp.59-67

Joshi, D.N. and Rawat, G.S. (1997). Need for conservation and propagation of alpine and sub-alpine medicinal plants of northwest Himalayas. *Indian Forester*.123: 811-814

Kanuja, S.P.S.; Kalra, Alok and Singh, A.K. (2005). Gains from entrepreneurship. *In*: The Hindu: Survey of Indian agriculture(N. Ram ed.). National Press Kasturi Building, Chennai. 191-95.

Lancet, The(1994). Pharmaceuticals from Plants: Great Potential, Few Funds. *The Lancet* 343: 1513-1515

Maiti, Satyabrata (2000). Importance of Aromatic and Medicinal Plants: Global and National Prospective. *Indian Journal of Arecanut, Spices & Medicinal Plants.* 2 (4):153-60

Mishra, R.K. and Pandey, V.K. (1998). Inter-cropping of turmeric under different tree species and their planting pattern in agroforestry systems. *Range Management & Agroforestry* 19: 199-202.

Nair, P.K.R. (1993). An introduction to agroforestry. Kluwer Academic Publishers, Dordrecht, The Netherlands. 499p.

Prajapati, N. D.; Purohit, S.S.; Sharma, A.K. and Kumar, T. (2003). A Handbook of Medicinal Plants. Agribios Publishers Pvt. Ltd., Jodhpur (India). 553 p.

Puri, S. and Nair, P.K.R. (2004). Agroforestry research for development in India: 25 years of experience of a national programme. *Agroforestry Systems.* 61:437-452

Rao, M.R.; Palada, M.C. and Becker, B.N. (2004). Medicinal and aromatic plants in agroforestry systems. *Agroforestry Systems.* 61:107-122

Schippmann, U.; Leaman, D.J. and Cunningham, A.B. (2002). Impact of cultivation and gathering of medicinal plants on biodiversity: Global trends and issues. *In*: Biodiversity and the Eco-system Approach in Agriculture, Forestry and Fisheries. Ninth Regular Session of the Commission on Genetic Resources for Food and Agriculture. FAO, Rome, Italy. pp. 1-21

Srivastava, J.; Lambert,J. and Vietmeye, N. (1995). Medicinal Plants: A growing role in development. agriculture and natural Resources. Department Agriculture and Forestry Systems. The World Bank, Washington, D.C.

Zuo, X. and Sanford, R.L. (1990). Agroforestry systems in China: a survey and classification. *Agroforestry Systems.* 11:85-94.

32

Canopy Managament in Agroforestry — Options and Strategies

P. S. THAKUR AND ANJU THAKUR
Department of Silviculture and Agroforestry
University of Horticulture and Forestry, Nauni-Solan - 173 230 (HP)

ABSTRACT

The integration of woody perennials on the farm lands for obvious reasons has been the common practice with the farming community for ages. It is true that the integration of trees on the farmland, albeit paves the way for the improvement of existing system, but nevertheless creates complex biological interactions, which may not necessarily result in yield advantage. Management of multipurpose tree species, especially agroforestry tree species, integrated on the farmland with crops, is essential for two reasons, i) because canopy size decides the transmission of photosynthetically active radiation to the understorey crops and ii) that the impact of canopy management practices some times have drastic implications. Canopy management practices like coppicing; pollarding and partial crown removal not only affects foliage and branch wood biomass production but also modifies physiological parameters and resource utilization pattern. This paper deals with the need and techniques of managing canopy as well as the research activities carried out in this regards.

WHY AGROFORESTRY

The exponential growth in human and livestock population over the years coupled with declining production, soil erosion, declining cultivable land have increased gaps between supply and demand for timber, fuel and fodder. The growing appetite for trees for fuel, fodder and construction in both rural and urban areas has largely been responsible for much of the deforestation. The scenario on fuel, fodder and timber production is not satisfactory. There is a chronic deficiency of timber, fuel wood and fodder in India and other Southeast Asian countries. Continuous degradation of upland mountain watersheds in the Asian region is increasingly becoming a serious problem. It is severely undermining the economic, social and environmental security of large population living in this part of the world. The Himalayas is a vast and diverse mountain system spreading through eight developing countries of Southeast Asia including Afghanistan, Bangladesh, Bhutan, China, India, Nepal, Myanmar and Pakistan. Today, there is a huge shortfall of basic needs besides an enormous damage to the resources, renewable as well as non-renewable on account of their abuse and overuse. So to meet the gap between supply and demand, efficient management of natural resources is the need of the hour. The existing allocation to agriculture and forestry are inadequate to meet the demands for food, timber, fuel, fodder and other minor products. This is the right time to exercise option to convert low productive and less exploited land into a productive gold mine by adopting agroforestry for diversification and sustainable biomass production. However, in order to transform present day practices into high-tech scientific adventure, there is an urgent need to formulate hypothesis through conductive reasoning, test and validate the hypothesis through deductive experimentation and strive to attain predictable understanding that will have wide applicability. The inclusion of trees on the farmland paves the way for the improvement and diversification of existing systems but creates complex biological interactions (Lawson and Kang 1990; Ong *et.al.* 1991; Khybri *et. al.* 1992; Ong and Huxley 1996; Rao *et.al.* 1998; Gillespie *et. al.* 2000; Thakur and Singh 2003; Thakur and Dutt 2004, Thakur *et. al.* 2005). Integration of traditional knowledge with scientific research is required for the development of improved technologies. Selection of proper multipurpose tree species for agroforestry followed by suitable management practices are key factors for improving overall productivity per unit of land. Agroforestry is based on the principle of 'Self Maintenance' so it can fit well in a most potential manner in marginal & small land holdings where monoculture of agriculture, horticulture or forestry is neither viable nor feasible. For the last two decades agroforestry has gained prominence as a part of farming system in India. While pursuing to achieve such tall targets, the efforts must be made to educate farming community regarding tangible and intangible benefits of agroforestry. The Indian Council of Agricultural Research (ICAR), fully realizing the seriousness of the major constraints and technological gaps in making agroforestry as promising alternative to the existing landuse systems, has established 38 centers all over the country. These agroforestry centers have been established to achieve the twin objectives of promoting cash economy and environmental conservation. In addition, agroforestry is being promoted and popularized by the

Government and wood based industries in many parts of India in order to address the crisis of food, wood, fuel and fodder and as well to preserve the fragile ecosystem. After Eucalyptus, popular based agroforestry has become very popular in plains of Northern India.

The components of agroforestry system often differ greatly in size, texture, nature etc. It is true that the integration of trees on the farmland, albeit paves the way for the improvement of existing system, but nevertheless creates complex biological interactions, which may not necessarily result in yield advantage. The components of the system interact with one another or influence the microclimate and soil conditions. The interactions between the components are of dynamic nature and may prove competitive, complementary or supplementary at any phenological phase. The intensity and duration of interaction depend upon the availability of critical resources like light, water, nutrients, space etc. Light is one of the main climatic factors, which drives production. The quantity as well as quality of solar radiation transmitted by tree canopy decides growth and productivity potential of the field crops; whereas the amount of solar radiation intercepted by the tree canopy is responsible for production of foliage and branch wood biomass. Major agricultural crops grown under agroforestry trees are shade sensitive, so are bound to be adversely affected on account of shading. Shade intensity and duration both will have strong negative effects on the performance of understorey crops. Many studies have concluded that shading caused due to the presence of tree canopies is a major factor causing crop yield reduction (Lawson and Kang 1990; Corlett *et al* 1992; Thakur and Dutt 2004). Any agroforestry system consisting trees having leaf area index (LAI) between 4 and 6 may render it less productive and/or uneconomical due to developing canopies. On the expected lines, unmanaged tree canopy not only reduces the productivity of agricultural crops, but nevertheless deteriorates the quality of produce as well.

CANOPY MANAGEMENT

Management of multipurpose tree species, especially agroforestry tree species, integrated on the farmland with crops, is essential for two reasons, i) because canopy size decides the transmission of photosynthetically active radiation to the understorey crops and ii) that the impact of canopy management practices some times have drastic implications. The removal of aboveground portion of trees results in enormous decrease in photosynthesis and vigour of the trees, which is not desirable. There exists a positive correlation between the amount of intercepted incident solar radiation and total aboveground production of foliage and wood biomass. Thus, it is of utmost importance to have full understanding of the crown architecture and growth in order to design appropriate canopy management strategies. In agroforestry, tree canopy reduces and modifies the light available to the understorey crops. The main purpose of canopy management is to make multipurpose tree species suitable and compatible with the associated arable crops in any agro forestry system. Reports are available, where the positive and/or negative impacts of the presence of tree canopies on the performance of understorey vegetation have been indicated. The major management options for manipulating trees in agroforestry are based on the alternation of light (solar radiation) profile and moisture distribution. The retention of big tree with oversize canopy not only reduces productivity of understorey agricultural crops but also becomes less productive during the subsequent years.

CANOPY MANAGEMENT OPTIONS

The options for managing tree canopy are numerous e.g. pruning, coppicing, pollarding, lopping, partial crown removal etc. However, each one of these options has equal number of advantages and disadvantages. The impacts of such practices can some times have drastic implications as the removal of aboveground portion results in enormous decrease in photosynthesis and the plant may die (Thakur and Sehgal, 2003). In addition, the management of tree canopy is bound to have direct bearing on the root characteristics as well as growth, vigour and biomass production of the tree itself (Kang *et. al* 1981; Basappa, 1986; Dutt and Jamwal, 1987; Heering, 1995; Puri and Gargya, 1995; El-Fadl, 1997, Singh, *et. al.* 1998; Thakur, 2000; Thakur and Sehgal, 2001, 2003, 2004; Thakur and Singh 2003; Thakur *et al.* 2005).

CANOPY MANAGEMENT AND BIOMASS PRODUCTION

Canopy management practices like coppicing; pollarding and partial crown removal not only affects foliage and branch wood biomass production but also modifies physiological parameters and resource utilization pattern (Leverenge and Hinckley 1990; Wang *et. al* 1990; Sequeira and Gholz 1991; Law *et al.* 1992; Thakur and Sehgal 2003; Thakur and Singh 2003). Author's experience with canopy management methodology adopted for managing canopies of some of the important tree species of temperate region, indicated enormous influence of management options on vigour and biomass production potential. For example, *Grewia optiva* trees cut at 3.0 m and 4.5 m in comparison to that at 1.5 m cutting height produced substantially higher foliage and branchwood biomass. Trees cut at higher heights attained larger canopy size and exhibited better biomass production potential for longer duration. Four canopy management options in *Morus alba,* another important fodder and fuel tree species for short rotation forestry, namely i) no crown removal, ii) 25 per cent crown removal, 50 per cent crown removal and iv) 75 per cent crown removal not only affected radiation climate in the system but also influenced biomass productivity. No crown removal option in *Morus* produced higher annual foliage and branchwood biomass; whereas considerable decline in biomass production with decreasing canopy size was observed. Out of the four tree canopy management options tried; 75 per cent crown removal causing least adverse effects on growth and yield of associated agricultural crops seems suitable and compromised crown management practice. Canopies of most of the agroforestry tree species need to be managed, when four to five year old. Management practices strongly influence vigour and biomass production potential of individual tree species in the subsequent years. Some of the agroforestry tree species, namely *Grewia optiva , Morus alba, Celtis australis and Bauhinia variegata,* managed by subjecting to different cutting heights (stems cut to 0.5, 0.75, 1.0, 1.5 and 2.0 m) have shown significant effect of cutting heights on production of foliage and branch wood biomass up to five years after imposing canopy management treatments. Plants cut at lower heights produced lesser aboveground biomass in comparison to plants cut at higher heights. Decline in biomass production was observed with passing time in *Celtis australis and Bauhinia variegata,* whereas *Grewia optiva* and *Morus alba* were found to maintain vigour and biomass production ability for longer duration (Thakur 2002).

HEDGEROWS AS MANAGEMENT OPTION

Numerous fuel and fodder tree species maintained as hedgerows can be a suitable option for combining agricultural and medicinal herbs on the farmland successfully. Multipurpose tree species like *Leucaena leucocephala, Morus alba, Celtis australis, Bauhinia variegata, Grewia optiva, Robinia pseudoacacia, Salix* and important bushes like *Indigofera and Woodfordia*, etc. can suitably be maintained as hedgerows on the farmland / grassland and agricultural, medicinal and aromatic herbs, vegetable crops / grasses are successfully grown as alley components in the spaces between two hedgerows for diversification and greater benefits from the combinations. The emphasis, however, must be laid on i) the height of hedgerows; ii) plant spacing within the hedgerows; iii) alley width and iv) cutting frequency. Our findings indicate that suitable tree management options not only regulate biomass production ability and maintain vigour for extended period but also help agricultural crops to perform better under agroforestry interventions.

Planting geometry including tree orientation and tree spacing; row x row as well as plant x plant is another option to integrate desirable number of trees per ha with little or no adverse effects on the production efficiency of associated crops. Tree species like poplars, teak, sandalwood, shisham, Toona or fruit trees, proposed to be integrated as agroforestry trees on the farmland in combination with agricultural crops, need to be spaced appropriately in order to enhance compatibility and efficiency of the associated components. Suitable plant spacing, as chosen management practice, not only influences radiation climate, resource sharing pattern, below and aboveground interactions but also sustainability of the systems.

CANOPY ARCHITECTURE AND RADIATION CLIMATE

Canopy architecture, which includes size, structure, compactness, phyllotaxy distribution of functional leaf area is an important character that influences above as well as belowground biomass production. Light interception by a canopy and its efficiency of conversion into structural dry matter significantly affects production potential. Numerous investigations have demonstrated a strong linear relationship between intercepted photosynthetically active radiation (PAR) and aboveground dry matter production. So solar energy interception by continuous and discontinuous as well as evergreen and deciduous canopies form the basis for selection and management of tree species suitable for agroforestry system. It is generally believed that tree and crops compete strongly for light, however, it is also true that tree - crop combination substantially enhances light utilization ability of the system. So considering that other crucial factors are not limiting, the dry matter production by any crop is linearly related to total amount of solar radiation intercepted. The relationship can be conceptualized as the product of the following factors:

$$W = Sfedt$$

Where;

W = dry matter g m^{-2}

S = total solar radiation

f = incident light intercepted by the tree canopy

e = the amount of dry matter formed per unit radiation intercepted

t = the duration of crop growth.

It is clear from the equation that (f) in the equation has direct influence on the productivity potential and proper canopy management can result into positive interaction as well as efficient light utilization in any agroforestry system.

A model prepared on the basis of cumulative leaf area index (LAI) is often used to establish the relationship between canopy structure and PAR interception, e.g.

l_z/l_p = exp (-k* LAI)

Where;

L_z = total PAR transmitted through canopy

l_p = PAR incident at the top of the canopy

k* = represents the rate of decreased irradiance

LAI = leaf area index of the canopy

Yield reduction of arable crops in the range of 20-60 per cent is not uncommon due to shading caused by the presence of tree canopy. Different tree canopy management practices, undertaken at appropriate age and time, have been found to minimize yield reduction under agroforestry systems through reducing shade intensity to the agricultural crops (Bergez *et. al.* 1997; Semwal *et. al.* 2002; Thakur and Dutt 2003; Thakur and Singh 2003; Thakur and Dutt 2004; Thakur *et al* 2005). Since food crops are important for the subsistence farmers, any new agroforestry system will be acceptable to them only if crop yields are not unduly affected by the perennial species. How many trees ha^{-1} can be accommodated on the farm without adversely affecting the crop yield still remains inconclusive. Irrespective of the agro-climatic zones, different agroforestry systems are under serious pressures from two sides: increased demand for agricultural and forest products on the one hand and ecological and economic constraints that limit production on the other hand. In agroforestry, tree canopy reduces and modifies light available to the understorey crops with possible consequences for photosynthesis, water relations and photo-morphogenesis. So, canopy management practices can be adopted to exploit interactions between woody and non-woody (annual crops) components for the compatibility and success of any agroforestry system. Therefore, a better understanding of the interaction provides a strong scientific basis for improvement of traditional as well as evolving systems. Three distinct tree crop interaction zones may be distinguished in agroforestry systems:

a) A zone of light and root competition.
b) A zone of root competition, some area beyond the tree crown.
c) Open cropped area that was free from the interference of tree

here;

'The extent of the first two interactions depends on tree species, climate, soil, crown spread with age and size of tree. An understanding of these zones is essential for determining the optimum tree density and developing appropriate canopy management technology and maximizing resource use and productivity of the system. The major tree- soil-crop interactions that affect crop yields in agroforestry systems are related mostly to chemical aspects of soil fertility, soil physical properties, water relations and soil microclimate, especially shade'.

CONCLUSIONS

It may be summarized that the gains of agroforestry research with regard to canopy management till date are certainly impressive. However, in order to transform present day agroforestry into high tech scientific adventure in the future, there is need to continuously update technology in order to attain predictable understanding that will have wide applicability. Combination of traditional knowledge with scientific research is required for the development of improved technologies. Future agroforestry research is bound to have changed vision and strategies in order to make this alternate landuse more viable and profitable venture. Today we are equipped with adequate technologies to handle present day agroforestry, however, there is every probability that this package may not suffice to meet the challenges and problems in future. We don't have the comprehensive information on the advantages and disadvantages of different canopy management practices on various aspects related to growth, vigour, root characteristics, biomass production, longevity, etc. We have not yet completely understood the intricate resource sharing between different components of the system; extent of adverse impact on the production potential of crops, which undergo drastic change at the advent of canopy management. A great deal of work has been done on various aspects in agroforestry although the underlying physiological processes are yet to be understood completely. There is no specific data to establish any kind of relationship between canopy management practices and carbon sequestration. Quantification of above and belowground biological interactions, especially at tree-crop interface still remains inconclusive. It is important to workout the proportion of woody and arable components, which can be accommodated per unit area, in order to enhance yield advantage and benefits from the agroforestry systems by minimizing competition. Suitable management of tree canopies, permitting adequate transmission of photosynthetically active radiation beneath canopies has been found more critical for improving productivity of associated arable crops. Appropriate tree management practices not only regulate biomass production ability and maintain vigour for extended period but also help agricultural crops to perform better under agroforestry intervention. Diversification of farming systems by growing short rotation timber, paper, pulp, fuel and fodder tree species with high value cash crops on the farmland is a viable option for increasing income of farmers. So we need to continuously refine strategies and update the technology package based on the past experiences and failures to make agroforestry more viable, beneficial and widely acceptable option.

REFERENCES

Bassapa B (1986). Coppicing of silver oak (*Grevillea robusta). My Forest.* 22: 1-22

Bergez JE; Dalziel AJ; Duller C; Eason WR; Hoppe G and Lavender RH. 1997. Light modification in developing silvopastoral system in the UK: a quantitative analysis. *Agrofor. Systs.* 37:227-240.

Corlett JE; Black CR; Ong CK and Monteith JL (1992). Above and belowground interactions in a leucaena/millet alley cropping system. II. Light interception and dry matter production. *Agric. For. Meterol.* 60:73-91

Dutt AK and Jamwal U (1987). Effects of coppicing at different heights on wood production in *Leucaena. Leucaena* Res. Rep. 8:27-28.

El-Fadl MA (1997). Management of *Prosopis juliflora* for use in agroforestry systems in Sudan. *Trop. Forestry Rep. Helsinki.* 16:107

Gillespie AR, Jose S, Mengel DB, Hoover WL, Pope PE, Street JR, Biehle DJ, Stall T and Benjamin TJ (2000). Defining competition vectors in a temperate alley cropping system in the Midwestern USA. A Production Physiology. *Agrofor. Syst.* 48: 25-40

Heering JH (1995). The effect of cutting heights and frequency on the forage, wood and seed production of six *Sesbania sesbane accessions under irrigated conditions. Agrofor. Systs.* 30:341-350.

Kang BT; Wilson GF and Sipkens L (1981). Alley cropping with maize and *Leucaena* in Southern Nigeria. *Plant and Soil.* 63:165-169.

Khybri KL, Gupta RK, Ram S and Tomer HPS (1992). Crop yields of rice and wheat grown in rotation as intercrops with three tree species in the outer hills of western Himalayas. *Agrofor. Syst.* 17: 193-204

Law BE; Ritters KH and Othman LF (1992). Growth in relation to canopy light interception in a red pine (*Pinus resinosa*) thinning study. *For. Sci.*38:199-202.

Lawson TL and Kang BT (1990). Yield of maize-cowpea in alley cropping system in relation to available light. *Agric. For. Met.* 55: 347-357.

Leverenz JW and Hinckley TM (1990). Shoot structure, leaf area index and productivity of evergreen conifer stands. *Tree Physiol.* 6:135-149

Ong CK, Corlett JE, Singh RP and Black CK (1991). Above and belowground interactions in agroforestry systems. *For. Ecol. Manage.* 45: 45-57

Ong CK and Huxley PA (1996). Tree-crop interactions: A Physiological Approach. *CAB International, UK, 386 pp.*

Puri DN and Gargya GR (1995). Management of *Morus alba* and *Grewia optiva* for degraded lands. *Van Vigyan*. 33:109-113.

Rao MR, Nair PKR and Ong CK (1998). Biophysical interactions in tropical agroforestry systems. *Agrofor. Syst*. 38: 3-50

Semwal RL, Maikhuri RK, Rao KS, Singh K and Saxena KG (2002). Crop productivity under different lopped canopies of multipurpose trees in central Himalaya, India. *Agrofor. Syst*. 56: 57-63.

Sequeira W and Gholz HL (1991). Canopy structure, light penetration and tree growth in a slash pine silvo-pastoral system at different stand configurations in Florida. *The Forestry Chronical*. 67:263-267.

Singh C; Agarwal MC; Kumar N and Puri DN (1998). Biomass production of *Morus alba* under different management practices on degraded bouldry riverbed lands of Doon Valley. *Ind. Forester*. 124:252-260.

Thakur PS (2000). Comparative performance of multipurpose tree species growing at the degraded site. *Ind. Forester*. 126:895-900.

Thakur PS (2002). Effect of canopy management on vigour and biomass production potential in four agroforestry tree species from temperate region. *Ind. Forester*. 128:498-501.

Thakur, PS and Dutt V (2003). Performance of wheat as alley crop grown with *Morus alba* hedgerows under rainfed conditions. *Ind. J. Agrofor*. 5: 36-44

Thakur, PS and Dutt V (2004). Diversification of farming systems by combining commercial tree species with medicinal plants to boost economy of farmers under rainfed conditions. 4th *International Crop Science Congress, Brisbane, Australia, Pp 1-6.*

Thakur, PS and Dutt V; Sehgal S and Kumar R (2005). Diversification and improving productivity of mountain farming systems through agroforestry practices in northwestern India. *North American Agroforestry Conference, University of Minnesota, Rochester, USA, Abstract No* 25.

Thakur, PS and Sehgal S (2001). Effect of canopy management on root parameters in agroforestry tree species of temperate region. *Ind. J. Agrofor*. 2: 75-78

Thakur, PS and Sehgal S (2003). Growth, leaf gas exchange and production of biomass in coppiced and pollarded agroforestry tree species. *J. Trop. For. Sci*. 15: 432-440

Thakur, PS and Sehgal S (2004). Influence of canopy management on the performance of multipurpose tree species. *Ind. Forester*.130: 639-646.

Thakur, PS and Singh S (2003). Influence of canopy size modification of *Morus alba* on productivity of *Vigna mungo* and *Pisum sativum* in agroforestry system. *Ind. J. Pl. Physiol. (special vol.): 104-109.*

Wang YP; Jarvis PG; Isebrands JG; Dickson RE and Ceulemans R (1990). Influence of crown structure properties on PAR absorption, photosynthesis and transpiration in spruce: application of a model (MAESTRO). *Tree Physiol.* 7: 297-316.

33

Rural Women in Agroforestry in North Western Himalayas

VIDYA THAKUR
Department of Tree Improvement and Genetic Resources
Dr Y.S. Parmar University of Horticulture and Forestry
Nauni - 173 230 Solan (HP)

ABSTRACT

Women nearly constitute half of the population of the mountainous region of north-western Himalayas. Rural hill women with their multiple roles and functions in the society, form an integral part of sustainable development of hills. Irrespective of age and health, they have to perform domestic household and agriculture work. Collection of fodder, firewood and fuel wood from the forest areas for energy requirements and household cooking are also done by women. The present paper provides the background information about the women in hilly areas and their dependence on agroforestry for meeting daily requirements of household.

INTRODUCTION

More than ninety per cent of mountain population inhibit in rural areas. Women nearly constitute half of the population of the region having composite social profile. Mountain specificities which consists of habitats, societies and their interactions and adaptations can prove to be constraints as well as opportunities. Comparative economic growth or development and dynamism in social-economic structure are due to interlinking of various land use systems including agriculture, horticulture, forestry, animal husbandry and tourism. Rural women play a pivotal role in sustainable development of their areas and form an integral part of subsistence rural economy (Thakur, 2003). They are considered nucleus of hill farming because of their multiple functions/ role in the society. Hence this section of the society in the region is the main and basic asset. They depend on their surrounding environment for many of the basic needs, therefore they are inextricably linked with natural resources and environment, and thus have to cope with the harsh environment by developing mountain specific adaptations.

BACKGROUND

Rural hill women with their multiple roles and functions in the society, form an integral part of sustainable development of hills. Irrespective of age and health, they have to perform domestic household work. In addition to this she participates in various agricultural operations (sowing to harvest operations). Rearing of cattle is another important task being carried out by them, along with collection of fodder, firewood and fuel wood from the forest areas for energy requirements and household cooking. In contrast men contribute mainly in out door activities such as collection of off farm inputs, marketing of farm produces and other harsh agricultural operations. Secondly because of small and marginal land holding with meager production potential, there is dire need for additional non- farm family income. To cover up the same men have to migrate to adjoining urban areas in search of work to get supplementary income. Thus hill women in general have to work more than men irrespective of workload either as farm workers or just as a house wife, unless and until they are totally bedridden.

Hill women spend on an average of three to five hours per day for collection of fire and fuel wood, cow dung and crop wastes, in addition to other domestic chores. The nature and extent of their involvement in farming systems varies greatly from region to region. Other intraregional involvements depend mainly on varied farming systems, caste factor, socio-economic and socio culture conditions of the farm families (Swaminathan, 1985).

Women's drudgery starts long before the day break since childhood, when the little girl starts helping her mother in fetching water, fuel, fodder and over burdened with heavy workload right from the childhood, leads to health hazards. Thus many of the rural hill women are not healthy which may affect their family directly or indirectly.

AGROFORESTRY AND CHANGING SCENARIO

Traditional agroforestry practices i.e. growing fuel and fodder trees along the bunds of agricultural field is the common practice in the rural areas (Thakur, 1997). But about four decades ago concepts of traditional agroforestry were scientifically modified. Agroforestry is the comprehensive term for different forms of land use combining trees and shrubs with agricultural plants (agricultural or horticultural crops) and or live stock (pastoral components) on the same unit of land, either alternately or at the same time, using management practices suiting the socio-economic conditions of the people and the system is ecologically and economically viable. These land use systems are based on the natural richness of the species with the overall objective of increasing for productivity of the system by taking advantage of the perennial species and supply the rural population with food, fuel, fibre and fodder, without endangering or destroying the natural resources *vis-a-vis* reducing pressure on the forest wealth.

The agricultural prosperity of hill or mountain people solely depends upon the wider base of agroforestry approaches. They may be multipurpose forest tree production system (silviculture), agrisilviculture, silvipastoral, agrisilvipastoral and agrihortisilvipastoral system. This is because it uses the available land and resources in an optimal way and ensure the supply of food, fuel, firewood and small timber as well as environmental protection and amelioration of sites inspite of harsh environmental conditions. Thus agroforestry is the base of prosperity of rural people. There is a great potential of agroforestry approaches in the rural areas/hills, which can provide high income to the people along with reduction in drudgery's of women in the following ways:

- Agroforestry approaches help in easy availability of daily needs i.e. fuel, food and fodder etc.
- Upliftment of socio-economic status with high production and income generation.
- Introduction of high yielding multipurpose tree species along with high yielding suitable or ecologically compatible associated crops, vegetables, fruit crops and grasses etc.
- Introduction of best breeds of milch cattle for supplementing nutrition and income.
- Introduction of advanced scientific agro-techniques in addition to the scientific modification of existing one.
- Increase rate of girls education.
- Role of rural hill incomes in decision making in farm activities.

However, in these cases wide diversity according to the ecological compatibility, ecological and economic viability of trees, crops, vegetables, fruits crops and grasses must be maintained. Rural women because of their multiple roles/functions in the rural farming system are involved in the development of agroforestry approaches/models to the region because of the following reasons:

A. They are the managers of crop or germplasm of land races due to the following facts:

- Selection and storage of quality seeds of land races.
- Mixed and multiple cropping pattern are in the hands of rural women.
- They know about the best breed of cow and buffaloes.
- They are well versed with the environment and habitat requirement of land races i.e. crops, vegetables, fruits crop and grasses.
- They also well versed with the high quality fuel, firewood, fodder and fruits etc.
- They have good knowledge of indigenous medicinal and aromatic plants.
- They are also aware of the ecological compatibility of trees and crops, vegetables fruits crop and grasses etc.
- They know better about the natural resources, their uses and management as well.

B. They are the conservationists of agro-biodiversity, which is based on their (plants) usefulness (Thakur, 1997 and 2002). They are the real ecologist-cum-conservationist of par excellence as far as forests are concerned (Nair, 1991).

PROBLEMS OF RURAL WOMEN

In spite of women productive roles and continued contribution towards the sustainable development of the area, their role goes unnoticed and thus unrewarded because of i) lack of education, confidence and encouragement ii) shyness due to lack of exposure and iii) social taboos.

However, roles of rural women in identification and traditional management of natural resources and farming systems are still unnoticed and limited due to various socio-cultural barriers prevalent in rural society/community.

CONCLUSION

For the sustainable development of farming system in the hilly areas following suggestions are thus proposed.

- Involvement of rural women in decision making.

- Access of rural women to income for management of household activities.
- Recognition of invisible contribution of rural women in the society.
- Involvement of younger generation of rural areas in the conservation strategies of their rich cultural heritage with a blend of scientific development.

REFERENCES

Nair, G.R. (1991). Women as Guardians of forest wealth social welfare; Oct 4-5.

Swaminathan, M.S. (1985). Imparting a rural women user's perspective to agriculture, research and development IRRI.

Thakur, Vidya (1997) Gender, environment and sustainable livelihood. Final Report submitted to ICIMOD, Kathmandu (Nepal).

Thakur, Vidya. (2002). Gender Role in Conservation of Plant diversity. In: S. P. Vij, K. Konda, M. L. Sharma and A. Gupta (Eds) Plant diversity and conservation. Affiliated East-West Press Pvt. Ltd. New Delhi. pp. 331-334.

Thakur, Vidya. (2003). Apple cultivation- A boon or bane for hill women. *In*: K. K Jindal, R. Bawa and Tej Pratap (Eds) Apple farming and livelihoods in the Himalayas. Bishan Singh Mahendra Pal Singh Publication Dehradun. pp 339-342.

34

Bamboo based Agroforestry Systems in India and their Future Potential

D. R. BHARDWAJ AND KRISHAN CHAND

Department of Silviculture and Agroforestry, Dr. Y.S. Parmar UH&F
Nauni, Solan-173 230, Himachal Pradesh

ABSTRACT

The paper discusses distribution of bamboo, their uses and bamboo based agroforestry systems in India and their future potential. Bamboo, one of the fastest growing plant species is extremely versatile and has been traditionally put to a large number of uses. Application of modern technology and industrial processing has catapulted bamboo into a new global limelight. At present there is huge demand of bamboo based raw material *viz.*, timber, plywood, bamboo flooring, edible bamboo shoots, etc. in the international market and most of the international market of bamboo is captured by China. In India, most of the bamboo raw material comes from natural forest. To capture the emerging bamboo market we have to increase the bamboo cultivation outside the forest under different bamboo based agroforestry systems. Experiments have shown that bamboo based agrisilviculture systems is technically and economically feasible in our country under different agro-ecological conditions. The agriculture crops which can be profitably grown with bamboo are soybean, pigeon pea, ginger, turmeric, etc. Keeping in view the international scenario, ecological conditions and taking clues from China's success stories, some of the bamboo based agroforestry system components are also proposed.

INTRODUCTION

Bamboo is one of the fastest growing plant species, which occurs in a wide variety of climatic and edaphic conditions. It is extremely versatile, strong, renewable and environment friendly material. Most bamboo species produce mature fibre in 3 years, sooner than any tree species. Some bamboos grow up to one meter a day, with many reaching culm lengths of 25 metres or more when mature. Bamboo can be grown quickly and easily, and sustainably harvested in 3 to 5 years cycles. It grows on marginal and degraded land, elevated ground, along field bunds and river banks. It adapts to most climatic conditions and soil types, acting as a soil stabilizer, an ancient medicine, a food source, a critical element of the economy, integrally involved in culture and arts, an excellent alternative to wood and effective carbon sink and thus helping to counter the green house effect. Application of modern technology and industrial processing has catapulted bamboo into a new global limelight. China has emerged as the leader having an annual economic output of US $ 3 billion, which is growing at an annual growth rate of 30 per cent. India is second to China in bamboo diversity and hence potential for development of bamboo based industries in the country is very high. This industrial development will provide high foreign exchange, employment for million and thus stem rural workers flight to India's teeming cities.

India, China and Myanmar have 19.8 million hectares of bamboo reserves-80 percent of the world's bamboo forests. Out of this India's share is 45 percent, with nearly 130 different species of the plant, but only 4 per cent of its global market. The government would like to see its bamboo industry, concentrated in the northeast of the country, take 27 per cent of the world market by 2015. Bamboo occurs almost ubiquitously in the country, except in Kashmir and cover about 12.8 per cent of the forest area occupying over 9.57 million ha. Areas particularly rich in bamboo are the north-east region, Western Ghat and Andaman. About 130 species belonging to 24 genera have been reported (Sharma, 1987). Sixty six per cent of the growing stock of bamboos is available in the north-eastern states and the balance in rest of the country. Clump forming bamboos are 67.5 per cent of the total growing stock. Of all the clump forming bamboos, *Dendrocalamus strictus* is 45 per cent, *Bambusa bambos* 13 per cent, *D. hamiltonii* 7 per cent, *Bambusa tulda* 5 per cent, *B. pallida* 4 per cent and all other species put together 6 per cent of the total growing stock. *Melocanna bambusoides*, a non-clump forming bamboo, account for 20 per cent of the growing stock and found mostly in the north-eastern states. *D. strictus* is the most widely distributed bamboo occurring in most part of the country. *Bambusa bambos* is equally distributed widely in most deciduous forest and cultivated by the framers in northern India. *Arundinaria* and *Chimnobambusa* are two genera found in high altitude in the hills of the Western Ghat and outer Himalayas. *Ochlandra travancorica* in Kerala and *Oxytenanthera spp* in coastal Karnataka, Goa and Maharastra are the other commercially important species.

With use of bamboo as industrial raw material, rapid socio-economic transformation and industrialization in the country, bamboo gained importance as raw material not only for cottage industry but also for large industries of pulp and paper. The first bamboo-based paper mill was established in 1919 (Prasad and Gadgil, 1981) and subsequently several more came up to use bamboo as the basic raw material.

Bamboo resources in recent years, however, have shown perceptible decline. While the reasons being over-exploitation by the industries, other factors such as forest fires, grazing, shifting cultivation, invasion by weeds and gregarious flowering have also contributed substantially for the reduction of bamboo production and area in the country. As a consequence of which, the paper industries which earlier used to depend upon bamboo for 100 per cent raw material were compelled to manage with about 50 per cent only and the balance met with by softwoods still some pulp and paper industries have completely substituted bamboo with other raw material. Over and above, the country had to import substantial amount of foreign exchange to maintain the production level and quality standard. Small-scale and cottage industries were also affected due to the short supply.

India's bamboo based industries are likely to make a quantum jump if proper policies are put in place and implementation procedures are streamlined. Country's bamboo economy is expected to grow by over 15 per cent to touch Rs. 260,000 million by 2015. The National Mission on Bamboo Technology and Trade Development under the Planning commission has estimated that if proper encouragement is given to bamboo cultivation and its use, it can replace the projected import of timber to the tune of Rs. 300,000 million in the next 20 years i.e. 2025. The market size for bamboo plywood is to grow to Rs.5, 000 million from Rs. 2,000 million in 2001. The country exports about Rs. 1000 million bamboo flooring materials and another flooring material of Rs.1000 million is used domestically. It has been estimated that the total market size of bamboo flooring materials will rise to Rs. 19, 500 million by 2015. The demand for bamboo furniture is expected to grow to Rs. 32,650 million in 2015 from Rs. 3,800 million in 2001. By 2015 bamboo scaffolding requirement will rise to Rs. 8,610 million and for housing purposes the demand will rise to Rs. 11,630 million. The demand for road construction will rise to Rs. 2740 million and for bamboo grids the demand will be Rs. 1000 million. The demand for miscellaneous industry *viz.*, ice cream sticks, fire crackers, bamboo lathis and ladders will rise to Rs. 6000 million by 2004 (NABARD report, 2006).

The planning commission has estimated that the Indian bamboo shoots industry has the potential to grow at the rate of 25 per cent per annum and capture a market worth Rs.3000 million. A large potential export market exists for shoots in Japan, Hong Kong, Singapore and Thailand (NABARD report,

2006). The imported bamboo species for production of commercial bamboo shoots are *Melocanna baccifera, D. hamiltonii, D. gigantea, Chimnobambusa hookeriana, Bambusa balcooa* and *Bambusa nutans* (Bhat *et al.* 2003). Trials conducted for introduction of *D. asper* by different institution in India have revealed that this species can be successfully cultivated under subtropical and tropical conditions (personnel communication).

Under the above scenario the only viable option to meet out the above mentioned targets and to become self-reliant is to produce bamboo outside the existing forest area. Development of bamboo based agroforestry model, in this context, holds great promise in augmenting supply of bamboo products in our country without any adverse effect on agricultural land or environment.

Bamboo in agroforestry

Agroforestry is the integration of woody plants with other agricultural enterprises such as crop or livestock production to derive both economic and ecological benefits. Bamboo occurs in forest as well as raised in homesteads in many countries. In homestead, it is either found mixed with a large number of other species of tree or purely in patches (Krishnankutty, 1988). Many of the most useful bamboo species can occupy much the same ecological niche as tree, and are well suited for agroforestry. Bamboo has many advantages over trees such as a relatively short time for planting to harvest, the ability to sustainably provide building materials and edible products for many years or even decades, and versatility of use which outmatches most tree species. No other woody plant species compares with bamboos versatility in environmental conservation and commerce. It is an ideal replacement for both softwoods and hardwoods. Its growth rate is three times that of Eucalyptus and it mature in just three years. This is explicitly why there are large scale endeavours at national as well as international levels to promote bamboo under agroforestry system

Bamboo agroforestry allows for both short-term and long term (sustainable) economic development of both rural and urban locales, as income from seasonal harvests are coupled with practices that protect water, soil, and flora and fauna system. Among many prospective benefits, bamboo agroforestry systems can:

- Usefully connect farms, forests and communities to create vital, reciprocal resource relations-supporting a diverse "crop" portfolio of foods, fodder for domestic animals, fuels, medicines, organic recycling, building materials, ornament in the landscape, etc. and related economic stability.

- Complement landform and integrated water management efforts, especially where food, erosion and waste water treatment is expected.
- Create educational venues, recreational areas and wildlife habitat.
- Carbon sequestration benefits.

Bamboo based agroforestry systems in India

Agroforestry systems are the practice of mixed farming developed over centuries. Most of them are the long-term land management systems having a life cycle of more than one year. Moreover, these systems are the complex form of land management both ecologically and economically than other agricultural or forestry systems. Based on the components used in a given land management unit, a number of agroforestry systems are prevalent in India, of these the most commonly used in India are:

- *Agrisilviculture* : Agriculture + Forestry,
- *Sivilpastoral* : Forestry + livestock,
- *Agrisilvipastoral* : Agriculture + forest + livestock,
- *Agrihorticultural* : Agriculture + fruit trees,
- *Agrisilvihorticultural* : Agriculture+ forestry +fruit trees.
- *Hortipastoral* : Fruit trees + livestock.

Out of the above-mentioned agroforestry system bamboo can be grown in agrisilviculture, sivipastoral, agrisilvipastoral, agrisilvihorticultural systems. Under agrisilvicultural system, soyabean, niger, mustard, wheat, urd and arhar and some of the important crops which can be grown with bamboos. Shade crops like ginger, turmeric, cinnamon, etc. are some of the other commercial crops which can grow very well with bamboo plantations. In agroforestry systems where each plant receives individual care, bamboo shows promising results. The economic impact of the agroforestry with bamboo considerably influences general economic development. This system is especially important and significant for developing countries like India. Under this system because of use of various intercrops, products are obtained even in the early stages of plantations and the income is much higher than any other system.

Seshadri (1985) concluded that cultivation of soyabean (*Glycine max*) along with *Dendrocalamus strictus* was technically feasible and economically viable. Singh *et al.* (1992) studied the effect of *Bambusa nutans* Wall. ex Munro shade on the yield of some agricultural crops at mid hills of eastern Himalaya. He reported that bamboos are allowed to grow or are planted along farm boundaries or drainage lines and or uncultivated wastelands in the eastern Himalayan hills. Through experimentation he found that agriculture land near bamboos can be effectively utilized for growing ginger, turmeric, large cardamom, orchard grass and dinanath grass (*Pennisetum pedicellatum*) up to a distance of 11-15 m from the bamboo rows. Rice, finger millet, soyabeans, nandi setaria and fine stylo were suitable crops beyond this distance. Shanmughavel and Francis (2001) studied intercropping performance of four crops viz., pigeon pea, soyabean, ginger and turmeric with *Bambusa bambos* under Tamil Nadu conditions. He found intercropping of pigeon pea and soyabean to be more productive than ginger and turmeric. The land equivalent ratio (LER) of intercropping *B. bambos* with Pigeon pea and Soyabean was equivalent to that of 1.2 ha or 1.1 ha under monoculture. Jha *et al.* (2004) reported that intercropping of soyabean with *Melocanna baccifera* and *Dendrocalamus longispathus* is feasible on degraded Jhum land of Mizoram and gave better results than pure bamboo stands.

Wagh and Rajput (1994) studied the intercropping performance of bamboo with that of Mango, Cashewnut, Kokum and Rubber. He found that out of three traditional crops bamboo was found to be most profitable one. Jayasanker *et al.* (1997) carried out the profitability analysis of three bamboo species viz., *Bambusa bambos, Thyrostachys oliveri* and *D. strictus* grown in Kerala. He revealed that all the three species of bamboo yielded B-C ratio higher than one, indicating high profitability. Among the three species, *T. oliveri* showed better returns. Shanmughavel and Francis (1999) recorded higher annual net returns (Rs. 13,300) when Pigeon pea was intercropped in 1:1 rows at 3× 3 m spacing (250 plants/ha) in comparison to 1:2 rows spaced at 2× 2 m (500 plants/ha). Cost-benefit analysis of bamboo plantation based on *Dendrocalamus strictus* at Gual Pahari, Haryana revealed that this system yielded better economic returns (Rawat, *et al.* 2002). Tiwari (2001) conducted a study to determine the financial feasibility of bamboo based agroforestry system of Kheda district of Gujarat (India) using 7 management models. Results indicated that the profitability of bamboo was very high and that the crop was financially feasible even at very high discount rate and socio-economic factors are believed to be favourable to the domestication of bamboo as an agroforestry crops in the region. Singh (2002) suggested cultivation of bamboo along water springs as an agroforestry interventions for enhancing farmers income.

Singh *et al.* (1992) studied the impact of 25-30 years old *Bambusa nutans* clump raised in agrisilvicultural system on the chemical properties of soil. He found that available Phosphorus (P) increased whereas exchangeable K and Ca + Mg decreased with increasing distance from the bamboo row, soil pH and soil organic matter did not vary with distance. Patil *et al.* (2004) analysed the effect of bamboo based agroforestry system on soil profile and surface soil properties. He revealed that organic carbon content of these soils ranged from 0.43 to 0.72 per cent. Soil profile investigation showed that all of nutrients were increased in bamboo based agroforestry site. The organic carbon and organic matter content of these soils increased from 0.37 to 0.58 per cent and 0.63 per cent to 0.99 per cent, respectively. In addition to it bamboo based agroforestry system also increases the biodiversity under its habitat (Behari, *et al.* 2000).

Recent access to global market for various non-conventional products (edible fruits, herbal aromatics, cosmetics and medicines, spices, etc.) and bamboo, rattans and other palm group of trees of industrial importance will also provide opportunity to develop agroforestry based land use systems (Singh, 1999). There are many traditional bamboo based agroforestry practices existing in this region, which are economically viable but need in-depth understanding for agronomic improvement to enlarge the area under agroforestry system.

Bamboo based agroforestry system in China

In china, bamboo plantations can be seen not only in hills and slopes but also along agricultural fields, roadside, river and canal banks, homesteads, and in all possible areas one can see small plots of rice, soyabean, vegetable or bamboo being grown side by side. In hills the cultivation of bamboos is extensive and there are places known as 'Sea bamboo'.

In China, mostly bamboos are cultivated for bamboo shoots or bamboo timber. The main bamboo species is Moso bamboo (*Phyllostachys pubescence* or *P.praecox*). Initially for two to three years of plantation some agricultural crops like Soyabean or tubrose crops are planted as intercrop in bamboo plots forming a perfect model of agroforestry. The plantation are maintained with a culm density of about 2200-3000 culms per ha. To start with a spacing of 5m × 5m is given. These plots are intensively managed (Lou Yipin, 2000). Apart from soil working, fertilizers and manure are applied in high dosages for enhanced production. Some of the plots are irrigated through sprinkler method. For round the year shoot production i.e. to get winter shoot, spring shoots and summer shoots the soil is treated with Rice husk. Each plantation is scientifically thinned to maintain the crop density and mature culms are harvested at periodical intervals.

New bamboo shoot production is not uniform every year. A good production year is followed by an off year subsequently when shoot production is less. In good years, there is large production which is in excess of what is required for the crop density. So about 2/3rd of shoots are extracted for edible shoot production remaining are left for bamboo timber production. The off season ones are also generally meant for timber purpose. No culm over 6 year is retained. They number each culm every year and have record of all culms. 4th or 6th year culms are extracted for timber (Haridasan, 2000). This is a remarkable way to have a uniform bamboo production. This practice can be adopted for raising plantation of bamboos (especially monopodial) in India.

Apart from the above bamboo based agrisilviculture system Chinese scientist have recently developed an innovative chicken- bamboo system. In this system, chicken are allowed to roam freely during part of the day in the bamboo forest. Free-roaming chicken produces better quality meat having higher market value. This farming system leads to a complementary and multi-level utilization of resources (Jianbo and Wang, 2005). The system produces bamboo for construction, bamboo shoots and chicken meat, while improving or maintaining the soil fertility and preventing water and soil erosion, thus achieving significant social, economic and ecological benefits. The system is more intensive and requires continuous and higher labour inputs during the whole year. This is a positive development because it allows for increasing employment opportunities for growing number of unemployed people in the rural areas.

POTENTIAL BAMBOO BASED AGROFORESTRY SYSTEMS FOR INDIA

On the basis of burgeoning demand for bamboo culms and edible shoots in the international market, rising prices of pulses, recent access to global market for various non-conventional products (edible fungi, herbal aromatics, cosmetics, medicines and spices, etc.) and taking clues from the success story of bamboo cultivation in China, the following bamboo based agroforestry system components are suggested:

- Bamboo + crops
- Bamboo + crops + fish ponds
- Bamboo + edible fungi
- Bamboo + chicken + fish pond + earthworm rearing
- Bamboo + medicinal plants.

1. Bamboo + Crops

Because of ever increasing demand of quality culms and edible bamboo shoots in the international market, there is great scope for introduction of bamboo species under agrisilvicultural systems. The edible bamboo species for cultivation under this system are *Melocanna baccifera, D. hamiltonii, D. gigantea, Bambusa balcoa,* etc. In addition to it the exotic bamboo species whose introduction trials are showing promising results under subtropical and tropical conditions is *D. asper.* The culm and edible shoot of this species has got very good international value. For temperate climatic conditions of India, emphasis should be given for the introduction of *Phyllostachys pubescence.* The agro-niches of its natural habitat resembles with that of conditions prevailing under temperate Himalayan region. The various agricultural crops which can be used for raising with different bamboo species are soybean, pigeon pea, ginger and turmeric, etc. The continuous extraction of shoots for edible purposes will keep down the culm's density and thereby competition effect.

2. Bamboo +Crops + Fish pond

In the plains, particularly on lower and wetter fields where fish ponds are built, 1-3 rows of bamboo can be planted (spacing 8-9 m) along the pond banks, under which crops such as soyabean, pigeon pea, rye, etc. can be intercropped to establish the model. In this model, the function of food chain structure can be fully used i.e. the intercrops can be the fodder of fish, the fish waste and soil in the pond bottom can be dug out in winter as fertilizer for the bamboo.

3. Bamboo + Edible fungi

Edible fungi are a food with high vegetative protein content, and there are many species and varieties. Under bamboo plantation, edible fungi intercrops which can be explored are *Dictyphora sp., Pleurotus ostreatus,* and *Aluricularia auricular-judae.* The bamboo plantation site for producing edible fungi should be: warm, wet, and shady with thick and fertile soil.

4. Bamboo +Medicinal plants

In this system bamboo can be grown for timber and edible shoot production. Depending on topography, various species of medicinal plants can be selected and used in this model. Medicinal plants with high adaptability like *Aloe vera, Mentha* and *Ocimum sanctum* can be interplanted with bamboo species. This system can be adopted in the areas where Monkey's menance is there.

5. Bamboo + Chicken

Bamboo and chicken based agroforestry system holds scope for introduction under Indian conditions. This system will produce construction material, edible bamboo shoots and chicken meat and eggs, while improving and maintaining the soil fertility and preventing soil erosion. This system is labour intensive and thus will provide much needed employment opportunities.

BOTTLENECKS IN THE ADOPTION OF BAMBOO BASED AGROFORESTRY SYSTEM

Bamboo based Agroforestry systems are sustainable land use systems. Bamboo prevents erosion and improves soil fertility significantly. However, due to (Betser, 2000) the high underground competition with the intercropped crops, and the low price of the stem, farmers are not willing to expand the area under bamboo, despite the high demand of the products (Brian, 1998).

For successful integration of bamboo in the cropping system there is an urgent need for screening the species/varieties of bamboo which yields high value edible shoots. The soil-plant-water interaction in the bamboo-based agroforestry system is still a mystery. Understanding these interactions may allow us to scale up the bamboo agroforestry systems in our country.

REFERENCES

Behari, B; Rashmi Aggarwal; Singh AK; Banerjee, SK. (2000). Vegetation development in a degraded area under bamboo based agroforestry system. *Indian Forester* 126(7):710-720.

Betser, E. (2000). Rapid reconnaissance surveys in market research. Lecture notes in agroforestry tree selection. International centre for research in agroforestry (ICRAF), Nairobi, Kenya 12 P.

Bhatt, B.P. Singha, L.B.; Singh K; Sachan, MS. (2003). Commercial edible bamboo species and their market potential in three Indian tribal states of north eastern Himalayan region. *Journal of Bamboo and Rattan* 2(2):111-133.

Brian M. Belcher (1998). A production-to-consumption system approach: Lessons from the bamboo and rattan sectors in Asia. *In* Wollenberg and Ingles (Eds). Methods for the development and conservation of forest products for local communities. pp. 57-85.

Haridasan, K. (2000). Bamboo based socioeconomic revolution in rural south eastern china-relevance to north east India. *Arunachal Forests News*. 18(1&2):41-45.

Jayasankar, B; Anitha, V; Murleedharan, P.K. (1997). Economics of bamboo cultivation in homesteads agroforestry systems of Kerala. *In*: Iyenkar, P.K. (Ed.). Ninth Kerala Science Congress, Trivandrum, 27-29 January 1997. State committee on Science, Technology and Environment, Trivandrum: 22-24.

Jha LK, Lalnuntluaga; Marak C. (2004). Study on the growth performance of bamboo species of *Melocanna baccifera* and *Dendrocalamus longispathus* along with crop (*Glycine max*) in degraded Jhum land of Mizoram. *Indian Forester* 130 (9): 1071-1077.

Jianbo, JhoXingheng and Wang Daoze. (2005). Chicken Bamboo Farming in Southern China. Leisa Magazine 21.3, September 2005. pp 1-4

Krishnankutty, C.N. (1988). Bamboo resources in the homestead of Kerala. Proc. Of the international bamboo workshop, Nov 14-18, Cochin, India. Kerala Forest Research Institute and IDRC, Canada, pp 44-46.

Lou Yiping. (2000). Bamboo Silviculture and sustainable management in China. Ibid. 37-50.

NABARD report. (2006). Bamboo for integrated rural development. 1-11p. www.nabard.org/roles/ms/fw/bamboo.htm.

Patil, VD; Sarnikar, PN; Adsul, PB; Thengal PD. (2004). Profile studies, organic matter build-up and nutritional status of soil under bamboo (*Dendrocalamus strictus*) based agroforestry system. *Journal of Soils and Crops* 14(1): 31-35.

Prasad, S.N. and Gadgil, M. (1981). Conservation of bamboo resources of Karnatka. A technical report by working group on bamboo resources constituted by Karnataka state Council of Science and Technology, Karnataka.

Rawat, J.S; Singh, T.P; Rawat, R.B.S. (2002). Potential of bamboos in agroforestry in India. Proceedings of the National workshop on policy and legal issues in cultivation & utilization of bamboo, rattan and forest trees in private and community lands. Kerala forest research institute, Peechi: 38-44.

Seshadri, P. (1985). Intercropping of bamboo (*Dendrocalamus strictus* Nees) with soyabean (*Glycine max* L.) Herill an agroforestry study. Ph.D Thesis. Tamil Nadu Agricultural University, Coimbatore.

Shanmughavel, P; Francis, K. (2001). Intercropping trials of four crops in bamboo plantation. *Journal of Bamboo and Rattan* 1(1) vsp, Netherlands: 3-9.

Shanmughavel, P; Francis, K.(1999). Growth performance and economic returns of Pigeon pea in agroforestry. *Indian Journal of Agroforestry*. 22 (4) : 351-353.

Sharma, YML. (1987). Inventory and resources of bamboos. Proceeding of International workshop Singapore. pp. 99-120.

Singh, K.A. (1999). Resource management and productivity enhancement through agroforestry in the eastern hilly agro-ecosystems of India. *Indian Journal of Agroforestry*. 1(1): 63-72.

Singh, K.A.; Patiram; Singh, L.N; Roy, R.N. (1992). Effect of bamboo (*Bambus nutans* Wall. ex Munro) shade on the yield of some agricultural crops at mid hills of eastern Himalaya. *Indian Journal of Forestry* 15(4): 339-341.

Singh, K.A (2002). Agroforestry interventions in the farming systems of Eastern Himalayan region. Resource management perspective of Arunachal agriculture, 337-350.

Tiwari, D.D. (2001). Domestication of non-timber forest products (NTFPs): a case of bamboo farming in Kheda district, Gujarat, India. *Ind. For.*127 (7): 788-798.

Wagh, R and Rajput, J.C. (1994). Comparative performance of bamboo with the horticultural crops in konkan. Bamboo in Asia and Pacific. Proceedings of fourth international bamboo workshop, 27-30, November 1991, Chiangmai, Thailand. International Development Research Center, Canada FORSPA, Bangkok, Thailand: 85-86.

35

Potential and Strategies of Agroforestry Interventions for Rehabilitation of Mine Spoil Wastelands of Uttaranchal Himalaya

K.S. DADHWAL
Central Soil & Water Conservation Research & Training Institute
218, Kaulagarh Road, Dehradun-248 195 (Uttaranchal), India

ABSTRACT

Minerals are important raw materials for the economic development of any nation. However, unscientific mining for minerals poses a serious threat to the environment. Mine spoils from outer Uttaranchal Himalaya are highly variable from loose to hard and compact at places. Rehabilitation technology for degraded mine sites consist of several steps, each of them playing an important role. Rehabilitation of these spoils is imperative to meet our demand of fuel, fodder, timber, reduce soil erosion and improve environmental pollution. The paper presents the efforts of rehabilitation of mine spoils in Uttaranchal Himalayas both mechanical as well as vegetative measures. Agroforestry techniques used in rehabilitation of mine spoils has been given in this paper.

INTRODUCTION

Himalayan region which constitute about 54 million ha (16.5 %) area of the country is characterized by marginality, inaccessibility and fragility, compounded by high intensity erratic rainfall, steep slopes, large scale deforestation and faulty management practices. Natural resources in the region are under great stress due to mounting human and livestock populations. The increasing frequency of natural disasters, e.g. landslides, floods, droughts and deterioration of water bodies due to accelerated deforestation, conversion of marginal/forest land into agriculture and unscientific developmental activities e.g. mining, roads etc. further aggravate the problem. Soil erosion is the major constraint in crop production. High intensity rains in the monsoon season cause excessive runoff and soil loss. On erosion-prone steep slopes, about 50 per cent of the rainfall is lost as runoff, carrying with it 20 t ha^{-1} soil.

Though, minerals are important raw materials for the economic development of any nation, but, unscientific mining for minerals poses a serious threat to the environment (Dadhwal and Katiyar, 1995) and formation of mine spoil wastelands. Due to excessive use of explosives in mining operations, shear resistance of soil mass is reduced and shear stresses are increased. Due to unscientific mining in hilly terrains, slope height increases, resulting in slope instability. During heavy rains, mass movements are caused by mechanical activation of overland flow and sub-surface water (Haruyama, 1980). Therefore, due to mining, land degradation, pollution of air and water, droughts, floods, land slides and land slips, silting of reservoirs, chocking of bridges and disruption of communications are some of the serious environmental hazards. In outer north-west Himalaya of Uttaranchal there were about 62 mines (active and abandoned) which created land degradation and environmental problems; but in 1985, the Hon'ble Supreme Court of India banned quarrying in most of the mines keeping in view the eco-imbalances and other problems caused due to unscientific mining. At that time several kilometers of land in the outer north-west Himalaya was lying barren waste and abandoned after reckless quarrying leading to severe land degradation and environmental problems. Mine spoil/debris from these degraded lands disrupted roads, choked bridges, reduced the capacity of streams, silted reservoirs, polluted air and water, vanished the ecology and affected good agricultural lands. Therefore, the problem of mine spoil wasteland was very serious requiring dedicated and concerted efforts for its rehabilitation. Research efforts were attempted by Central Soil and Water Conservation Research and Training Institute, Dehradun from 1983 onwards and technologies were developed for stabilization and rehabilitation of mine spoil wastelands. Techniques developed, applied, their impact and future lines of work are briefed in this article.

FACTORS CONDUCIVE FOR CREATION OF MINE SPOIL WASTELANDS

- Unscientific blasting of more and more areas for mineral exploration.
- Lack of stringent rules and regulations governing environmental controls and precautions.
- Improper planning e.g. not installing diversions, vegetative waterways, etc. well ahead of the actual operation of mining.

DAMAGING EFFECTS OF MINE SPOIL WASTELANDS

- Compaction of spoil adversely affects plant establishment and growth by increasing resistance to root penetration and to the emergence of seedlings (Byrnes *et al.*, 1982).
- Limited plant growth due to lack of normal soil and available soil water.
- Reduced water holding capacity of the soil.
- Water pollution and environmental hazards.
- Slope aspect, soil water and shading by plants influences temperature of spoil material. For example, high temperature injury to plant seedlings is serious on black bituminous coal and black organic shales, but less on light coloured mineral shales and sand stones (Deely & Borden 1973).
- Spoil pH below 4.0 is usually toxic for plant growth.
- Over burden materials are low in organic matter and other nutrient contents.
- Due to low organic matter in most of the mine spoil wastelands, N is the primary limiting nutrient for plant growth. However, in most cases, a complete fertilizer (N, P and K) is needed for plant establishment.
- Reduced microbial growth and activity.
- Respiration and enzyme activities are hampered in mine spoil wastelands.
- Unreclaimed or poorly reclaimed lands often are barren because soil materials left on the surface after mining and reclamation does not support plant growth. This causes excessive erosion due to inadequate cover. Approximately 100 times more erosion has been reported from the abandoned mined wastelands compared to similarly located forest lands (USEPA, 1976).

POTENTIAL AND STRATEGIES FOR REHABILITATION

To define limitations for vegetal establishment, characterization of mines spoil wastelands is very important.

CHARACTERISTICS OF DEGRADED MINE SPOIL WASTELANDS

Mine spoils from outer Uttaranchal Himalaya are highly variable from loose to hard and compact at places. Older deposits are hard and compact and fresh one are loose and are composed of shales, cherts, loose rocks, schists and limestone mixed with soil. The mine spoil material (< 2mm) is sandy loam in texture, alkaline in reaction, very poor in nutritional status, highly calcareous, low in water retention characteristics and high in bulk density (Table 35.1) (Dadhwal *et al.*, 1991; Dadhwal *et al.*, 1992).

REHABILITATION MEASURES

Rehabilitation technology for degraded mine sites consist of several steps, each of them playing an important role. None of the measures in the chain should be missed. Some of the important steps emanated from research studies and experience are briefed as under (Dadhwal and Katiyar, 1990):

Mechanical measures

Mechanical measures are first line of defense and are very important to stop further damage. In highly degraded areas, where establishment of vegetation is very difficult due to moisture stress, movement of debris and rocky material on steep slopes, and absence of fertile soil, mechanical measures are must for controlling erosion, conserving soil particles and helping the vegetation to grow. Studies reveal that engineering structures e.g. cross barriers/check dams, log wood barriers, toe walls, gabions/loose check dams etc. arrest the movement of large amounts of loose rock and debris down the unstable and uneven steep slopes of the area and also through the drainage channels. Some of the important engineering measures, for stabilization of abandoned mined/wastelands suggested are as under:

- Planned placement of overburden in mountainous terrain is very important because it maximises the physical stability of the spoil. The elimination of outer fill slopes and control of outslopes is important for easier establishment of vegetation, spoil stability and for creation of configuration which is more compatible with the natural topography.
- During site preparation, position of water courses should be properly designed and assessed. Water control facilities e.g. channels, diversions and culverts, etc. must be properly designed and maintained. Construction of diversion drains is very much essential to avoid runoff water going to weak zones such as slide face and loose minespoil/debris dumps. Plots should be made depending upon topography.
- There must be a provision of road drains and culverts on mine haul roads. At the same time formation of paths and roads is also essential to transport plants, manures, water, fertiliser, etc.
- There should be construction of continuous contour trenches on minespoils and staggered contour trenches on denuded long slopes. These trenches are useful in water and finer particles retention and establishment of vegetation.

- Terracing of minespoil/debris deposited in heaps of steep slopes is advocated. Terracing will ease further to adopt biological means for spoil stabilisation.
- Measures like use of Geo-jute, Netlon, Geo-girds, Excelsior matting, mulching, etc. are also important to avoid rain drop impact on loose debris and also to facilitate the revegetation of the area. In steep slopes, the importance of these measures further increases. Mulching helps in moisture conservation and after decomposition in building nutritional status.
- Upto 40% slope, the construction of gabion drop structures in the main drainage channel is recommended.
- There must be a construction of silt detention basins in the main drainage channel wherever, technically feasible, normally up to 20 per cent slope.
- In small tributaries (first order streams) construction of logwood, brush wood and loose rock filled check dams is essential.
- Construction of stone filled log wood crib structures on steeply sloping, loose areas especially on slide faces is recommended. Studies indicate that these structures suit better than gabion drop structures where slope exceeds 40 per cent.
- Construction of chute spillways with cement concrete where the vertical fall is more than 3m is advocated.

The above mentioned strategies/measures are useful in grade stabilization, formation of bench terraces, retention of good soil, reduction in the steepness of the channel slopes, retention of runoff and thus better soil moisture status in the soil profile, increased subsurface flow instead of flash floods and facilitation in the establishment of vegetation.

Biological/Agroforestry interventions

Biological measures must be undertaken alone/simultaneously with the construction of engineering structures depending upon the site and situation. Measures adopted should have positive effect on soil properties and plant growth. The affected areas should be put to productive use, may be agriculture, forestry/agroforestry or recreation. Aesthetic value should be restored and adverse visual impact should be avoided. While taking biological means, economic and ecological considerations must be kept in mind. The following are some of the techniques emanated after research studies for the rehabilitation of abandoned limestone mined waste lands from Uttaranchal Himalaya:

1) Since minespoils of outer Himalaya are very poor in nutritional status (pH 8.0, organic carbon 0.13 %; available phosphorus 4 kg/ha; available potash 44 kg/ha; total nitrogen 0.02 %), minespoil/debris fertility management should be given due importance for the quick establishment of vegetation. Studies conducted for the minespoil from Sahastradhara (near Dehradun) reveal that 50 kg N/ha and 20 tonnes FYM/Leaf litter/ha to *Eulaliopsis binata* (bhabhar) gave better results. Similarly, di-ammonium phosphate @ 10 g/tree sapling gave encouraging results in the first year of establishment of MPTS.

2) Leaf litter/mulches are useful in moisture retention and fertility buildup. Leaf litter collected from the nearby forest (composed of grasses, wild Kadzu, *Sapium insignii* leaves etc.) after chopping incorporated in minespoil/debris gave better results for bhabhar growth and 20 tonnes/ha was found to be the ideal dose.

3) Vegetation establishment can be enhanced by placing and mixing of a normal soil with minespoil. One can have a normal soil from adjacent good areas. Placing normal soil upto a depth of 30 cm in the initial year of establishment was found good for *Eulaliopsis binata* (bhabhar) and *Chrysopogon fulvues* (gorda) grasses whereas in second year mixing treatment upto the same depth was found better over placing. Mixing 50:50 (soil: debris) performed well for the establishment of various tree, shrub and grass species. These studies suggest that if normal soil is available nearby then that can be used in pits or holes and can be mixed with debris before planting for better establishment of vegetation.

4) Select preferably leguminous plants for such degraded sites, which may add nitrogen. For example *Leucaena leucocephala* (Subabul) and *Pueraria hirsuta* (Kudzu) have root nodules which fixes nitrogen. The establishment of subabul and its growth is quick and can be used as a fuel and fodder (mixed with other grasses). Legumes can also be mixed with other plant species for re-vegetation of such affected sites.

5) While selecting plant species for rehabilitation of abandoned mined lands the following considerations should be taken into account:

   - Plant species selected should provide quick and immediate vegetative cover on these waste/barren and affected sites to avoid further damage. For immediate cover, grasses are best proposition. Grasses like *Chrysopogon fulvus* (gorda), *Eulaliopsis binata* (bhabhar), *Pennisetum purpureum* (hathighas), etc. have given good performance at Sahastradhara in Uttaranchal. *Bhabhar* is an industrial grass whereas other two can be used as animal fodder.

- Species selected should have economic and ecological values.
- Should add nitrogen (preferably).
- Since abandoned limestone sites are very rich in calcium carbonate content ($CaCO_3$ varies 4-80%), plant species selected should be lime loving.
- Plant species choosen should have better root soil binding characteristics.
- Should have aesthetic value.
- Plant speices chosen should be preferably native of that place.

6) Biological measures like contour wattling, mulching and planting can be adopted for such affected sites depending upon slope and other site characteristics.
7) In highly steep lands spraying over the dumps with seeds mixed with cowdung, hay, made into slurry, etc. is suggested.
8) Studies conducted under Mussoorie hills of Uttaranchal reveal that plant species like *Chrysopogon fulvus* (gorda), *Eulaliopsis binata* (bhabhar), *Pennisetum purpureum* (hybrid napier), *Pueraria hirsuta* (kudzu wine), *Saccharum spp* (kans grass), *Agave americana* (rambans), *Ipomoea carnea* (sadabahar), *Eucalyptus* hybrid (eucalyptus), *Bauhinia spp.* (kachnar), *Leucaena leucocephala* (subabul), *Vitex mnugundo* (semalu), *Salix tetrasperma* (salix/jalmala), *Acacia catechu* (khair), *Erythrina suberosa* (Jhingen), etc. have performed well and can be used for the rehabilitation of such abandoned ruined sites. The various agroforestry species coming under secondary succession alongwith evaluated and successfully reforested species in different degraded minesites of outer Himalaya are listed in Table 35.2 (Dadhwal *et.al.*, 1997).

Mechanical and biological measures applied has given encouraging results for the stabilisation of abandoned limestone mined lands from outer Himalaya over the time. Impacts of the technologies applied are enumerated as under:

IMPACT OF STRATEGIES APPLIED/MEASURES UNDERTAKEN

Impact on Vegetation

With continued biotic protection together with rehabilitation measures and protection against grazing and cutting of grasses and MPTs the retrogression and erosion has come to stand still. The vegetal cover has increased from 10 per cent to > 90 per cent due to reforestation over a period of 20 years.

Impact on minespoil characteristics

Impact of rehabilitation measures on some important characteristics over time was seen in treated area. On an average, after 18 years, pH of the spoil came down by 0.7 units (8.1 to 7.4), organic carbon increased from 0.13 to 0.45 per cent, whereas $CaCO_3$ (%) decreased from 54.6 to 29.5 and bulk density (Mgm^{-3}) from 1.63 to 1.45 which shows that planting of MPTs, shrubs and grasses over time have helped in improvement of minespoil characteristics (Table 35.3).

Impact on slope and runoff

The equivalent slope in the treated hilly terrain reduced from 38.3 per cent in 1983 to about 19 per cent in 1997 over a period of 14 years. Similarly, runoff and soil loss also decreased considerably over the time with mechanical and biological measures.

Impact on base flow

Because of the various treatment measures, the main drainage channels which used to dry up by the end of October, have now become perennial. The dry weather flow recorded in the main drainage channel was to the tune of 265 and 100 cum/day in the months of November and February, respectively. Farmers of the area have started utilizing this augmented flow for irrigation purposes.

Impact on water quality

Runoff samples collected from treated and untreated mined areas were analysed for calcium, magnesium, sulphates and bi-carbonates (Table 35.4). Runoff water from treated minesite was found within the permissible water quality limit (Juyal *et. al.* 1993).

Table 35.1. Important characteristics of minespoil wastes from outer Uttaranchal Himalaya

Characteristic	Average from 20 minesites
pH (1:2.5 spoil:water suspension)	8.0
Organic carbon (1%)	0.13
$CaCO_3$ (%)	61.6
Total N (1%)	0.02
Bulk density ($Mg\ m^{-3}$)	1.69
Water holding capacity (%)	16.6
Texture	Sl

Table 35.2. Prominent agroforestry species for rehabilisation of limestone minespoil wastes of outer Uttaranchal Himalaya with their economic use

Species	Economic Use
Trees	
*Acacia catechu (Khair)**	Wood for K, C, and F; B for tanning
Albizia chinensis (Siran)	F, Fd and species for saline and alkaline areas
A.lebbek (Siris)	F, Fd and species for saline and alkaline areas
Bauhinia purpurea (Gural)	Fd, wood for small timbers, B for tanning, flower buds as vegetable
B.retusa (Semla)	B, gum for sweets and medicines
B.variegata (Kachnar)	B for medicines; wood as F and small timbers
Boehmeria regulosa (Genthi)	Wood for making bowls and dairy utensils
Bombax ceiba (Semul)	Soft wood species, premature calyx as vegetable, cotton to stiff pillows and cushions, seeds as nourishing cattle feed
Cassia fistula (Amaltas)	Fruit as medicine, B for tanning/dyeing and wood for minor implements
Celtis australis (Khirik)	Fd, fuel and wood for handles and churnsticks
Dalbergia sissoo (Shisham)	F, Fd and timber; SC/afforestation
Dendrocalamus strictus (Bamboo)	Fd, SC
Erythrina suberosa (Madar)	Wood for sieve frames, SC
Eucalyptus hybrid *(Safeda)*	F, T, paper pulp
Ficus auriculata (Timla)	Fd, fruits (edible)
F.hispida (Kaksa)	Fd, yound fruits made into carry, B, FR for ropes
F.infectoria (Pilkhai)	Fd, yound fruits made into carry, B, FR for ropes
F.palmata (Pheru)	Fd, fruits (edible)
F.rumphii (Pilkhan)	Fd, fruits (edible)
F.semicordata (Khain)	Fd, fruits (edible), old leaves used in wood polishing
Grewia optiva (Bhimal)	Fd, FR, minor timber, F
Lannea coromandelica (Jhingan)	Fd, gum, SC
Leucaena leucocephala (Subabul)	Fd, F, N-fixation, SC
Mallotus philipinensis (Raini)	F/small timber, B for tannning, medicinal
Moringa oleifera (Sanjan)	Fd, fruit as vegetable, gum, medicine, SC

Contd...

Table 35.2. Contd...

Murraya exotica (Bilgar)	Wood used in making handles of implements and walking sticks, twigs used as tooth brushes
Ougenia oojeinensis (Sandan)	Fd, implements, furniture, gum
Phoeni acaulix (Khajoori)	Leaves for mat making, broom making, fruits (edible)
Salix tetrasperma (Jalmala)	Wood for gun powder charcoal
Toona cilata (Toon)	Fd, flowers fordye, bark medicinal, timber
Wendlandia exserta (Bathua)	Timber, agricultural implements
Shrubs	
Agave species (Sisal)	Ropes, leaf colloid used in detergent
Adhatoda zeylanica (Vasaka)	Leaves and twigs medicinal
Barleria cristata (Jhinti)	Roots/leaves medicinal, an antidote to snake bite
Carissa opaca (Karonda)	Fd, wood for combs, fencing and burning, fruit edible
Cocculus laurifolius (Tildhera)	Fuel
Euphorbia royaleana (Thor)	Live hedge fence, SC
Impmoea carnea (Besharam)	SC, fuel
Lantana camara (Lantana)	F, furniture
Mimosa himalayana (Al)	Fd, heartwood for tent pegs
Rumex nepalensis (Chulmora)	A weed on calcareous soil
Vitex negundo (Shimalu)	Branches for basket making, roots and leaves medicinal, SC
Wood fordia fruiticosa (Dhaula)	Wood for minor implements, flower for silk dye, and B for medicine
Grasses and Sedges	
Andropogon nardus (Andropogon)	SC
Apluda mutica (Apluda)	Fd
Arundo donax (Narkul)	SC, toys, pipe
Aristida cynantha (Arsitida)	SC, brooms
Bothriochloa intermedia (Bothriochloa)	Fd, SC
Chrysopogon fulvus (Gorda)	Fd, SC
Cynodon dactylon (Doob ghas)	Fd, SC

Contd...

Table 35.2. Contd...

Eragrostis nardoides (Eragrostis)	SC
Eulaliopsis binata (Bhabar)	Fb, paper pulp, SC
Heteropogon contortus (Speargrass)	Fodder, SC
Pennisetum purpureum (Napier grass)	Fodder, SC
Saccharum spontaneum (Kans)	SC, bufalo Fd, thatching
Thysanolaena maxima (Broom grass)	SC, broom
Eriophorum cosmosum (Pseudobhabar)	Fd, ropes, paper, SC
Herbs	
Argemone mexicana (Pili-katnari)	SC and medicinal
Euphorbia hirta (Dudhli)	Medicinal, weed of dry localities
Eupatorium adoratum (Eupatorium)	Medicinal, a weed in moist habitat
Galium rotundifolium (Galium)	A weed in moist habitat
Climbers	
Acacia pinata (Agai-bel)	Endemic to moist gullies/ravines
Atylosia scarabaeoides (Atylosia)	Fd, SC
Bauhinia vahlii (Malo)	Leaves as plates, cups, table cloth and raincaps; B for ropes, roasted pods and seeds edible
Peuraria hirsuta (Kudzu)	Fd, SC, Organic manure

C = Charcoal, F = Fuel, Fd = Fodder, FR=Fibre, K = Katha, B = Bark, SC = Soil Conservation

* Common/Vern. Names in parenthesis

Table 35.3 Changes in minespoil characteristics overtime (average of 14 locations).

Properties	Year						
	1985	1991	1992	1994	1995	1997	2002
pH (1:2.5)	8.1	7.9	7.7	7.6	7.6	7.5	7.40
Organic C (%)	0.13	0.18	0.26	0.39	0.4	0.42	0.45
Total N (%)	0.01	0.02	0	0	0	0.05	0.05
$CaCO_3$ (%)	54.6	34.6	34	32	31.8	31	29.50
Bulk density ($Mg\ m^{-3}$)	1.63	1.53	1.5	1.48	1.48	1.47	1.45

Table 35.4 Water quality parameters (ppm).

Site	Ca	Mg	SO_4	HCO_3
Treated oldmine	74	34	138	131
Untreated oldmine	188	39	240	98
Untreated fresh mine	389	120	756	77

FUTURE STRATEGIES

Several long-term studies have been conducted for the rehabilitation of mined lands in developed countries, but in India, barring few, very meagre information for rehabilitation of minespoils of various mines is available. Since the rehabilitation measures vary with the type of minespoil, site and physiographic conditions, the following future line of work is suggested for reclamation and rehabilitation of minespoil wastelands.

- Characterisation and classification of minespoil wastelands.
- To develop site specific soil working and planting techniques.
- To develop site specific fertility management programmes.
- For steep slope stabilisation, measures like Geojute, Netlon, Excelsior matting, etc. with planting of suitable vegetation are required to be studied.
- Select only such type of plant material (MPTs, shrubs, grasses, creepers, etc.) which provide immediate carpet, add nitrogen, have good root soil binding characteristics and are also ecologically and economically sound.

REFERENCES

Byrnes, W.R., McFee, W.W. and Steinhardt, G.C. (1982). *Soil Compaction Related to Agricultural and Construction Operations : Bull. Purdue Univ. Agric. Exp. Stan.* 397, West Lafayette, Indiana.

Dadhwal, K.S. and Katiyar, V.S. (1990). Measures for the reclamation of abandoned mined lands. *Himalaya (Man & Nature)*, 13 (9): 14-18.

Dadhwal, K.S. and Katiyar, V.S. (1995). Land degradation and environmental pollution by mining and land slides. *Bull. Indian Soc. Soil Sci.*, No.17, pp.114-126.

Dadhwal, K.S., Joshie, P. and Juyal, G.P. (1997). Prominent agroforestry species for reclamation of limestone minespoils of outer Himalaya. *Indian J. Soil Cons.*, 25(1): 33-40.

Dadhwal, K.S., Singh, B., Katiyar, V.S. and Dwivedi, V.K. (1991). Characteristics of limestone mined lands. *Indian J. Soil Cons.*, 19(3): 79-82.

Dadhwal, K.S., Singh, B. and Narain, P. (1992). Characteristics of limestone minespoil/ debris from outer Himalayas of Uttar Pradesh. *Indian Forester*, 118(9): 650-658.

Deely, D.J. and Borden, F.Y. (1973). *In* : *Ecology and Reclamation of Devastated Land* (Ed. R.J.Rutnik and G. Davis), Vol.1, p.69, Gordoh & Breach, New York.

Haruyama, M. (1980). *Proc. Int. Symp. Landslides*, New Delhi.

Juyal, G.P., Katiyar, V.S., Dadhwal, K.S. and Joshie, P. (1993). Rehabilitation of a steep mined area in Mussoorie hills. *Wasteland News*, 9(1): 57-91.

U.S.Environmental Protection Agency. (1976). *Erosion and Sediment Control*, Surface Mining in the Eastern U.S., Vol.I, EPA-625/3-76-006, U.S.Govt. Print. Office, Washington, D.C.

36

Tree Breeding — Status and Strategy for Agroforestry Species in India

V.K. GUPTA, R.V. KUMAR AND S.P. AHLAWAT
National Research Centre for Agroforestry, Jhansi - 284 003

ABSTRACT

In agroforestry and forestry sectors, little efforts have been made for tree improvement research in India. Planting of genetically poor quality of seedlings/propagules is one of the major reasons of low level impact of agroforestry systems in Indian Agriculture. Most of the agroforestry tree species have not been even gone through simple selection for superior types (plus trees). Exploration, collection and conservation of germplasm of woody agroforestry species, selection for fast growing trees, clonal propagation, seed orchards and other advances in tree breeding are still in fancy stage. At present, all the woody agroforestry species are propagated through collected seeds and/or propagules by unskilled labourers. It is well known that through tree breeding and establishment of seed orchards of improved genotypes, significant achievements towards the improvement of productivity of economic tree products is possible as happened in agricultural crops in India and industrial trees in USA and Europe. Even by small efforts of tree improvement, productivity of economic tree products can be raised by 20-30 per cent. In this paper breeding methodology, limitations, status and strategy of various agroforestry species has been discussed in special reference to *Leucaena leucocephala* (Subabul), *Prosopis cinereria* (Khejri), *Azadirachta indica* (Neem), *Populus* spp., *Dalbergia sissoo* (Shisham), *Acacia nilotica* (Desi babul), *Eucalyptus* spp. *Casuarina equisetifolia* (Casuarina), *Albizia lebbek* (Desi siris) and *Jatropha curcus* (Ratanjot).

INTRODUCTION

Pressure on natural resources in India is very high which is adversely affecting the forests as well as ecosystem of the country. Estimation of demand and supply of fuelwood and timber for house hold sectors shows a very wide gap. Average production of wood in India is 0.7 m^3 ha^{-1} yr^{-1} whereas, intensively managed poplar with superior clones yield 29-30 m^3 ha^{-1} yr^{-1} wood at the age of 5-7 years (Verma, 1993). These results give the reflection that there is a good possibility to meet the demand of fuelwood and timber for house hold sectors through large scale plantation of genetically superior high yielding varieties/clones in farmer's fields as well as in forest areas. Best example for increasing the productivity of per unit area per unit time is the plantation of genetically improved seedlings of poplar, eucalyptus and *Leucaena* in India. Superior clones of *Populus deltoides* with intensive management practices yielded mean annual increment (MAI) of 49.35 m^3 ha^{-1} yr^{-1} in Punjab which is perhaps the highest productivity under agroforestry system in the country (Singh *et al.*, 2001). At present, all the woody agroforestry species are propagated through collected seeds and/or propagules by unskilled labourers. Concept of plus tree selection, establishment of seed orchards and producing of quality seeds are still at experimental farms in India. Few sporadic efforts have been made for the improvement in isolated pockets at various research institutes of Indian Council of Forestry Research & Education, Dehra Dun; Indian Council of Agricultural Research, New Delhi and Agricultural Universities. These efforts are not coordinated one. Location trials have not been conducted. Site specific varieties have not been identified. Even by small efforts of tree improvement, productivity of economic tree products can be raised by 20-30 per cent (Chuntanaparb and Mac Dicken, 1991). In this paper breeding status and strategy of various agroforestry species has been discussed.

Following points should be taken into consideration for the success of any tree improvement programme of agroforestry species:

1. *Choice of species :* species should be well adapted at a specific site and suits in existing agricultural systems without significant reduction in yield of lower storey crops.

2. Economic value of tree products should be high so that B:C of the system should be high.

3. Selection criteria for improvement of agroforestry species should be decided before starting of the breeding programme so that desirable ideotypes can be selected in existing populations of a species in question in natural stands.

4. Knowledge of reproductive and floral biology of a species should be known

5. Economic and feasible methods for vegetative propagation should be standardized.

6. Technique and design to establish seed orchards of the species in question should be standardized
7. Large scale multiplication of promising genotypes should be done.

Ideotypes of Trees Desired in Agroforestry Systems

1. Trees should be fast growing with minimum annual increment of 20-30 m^3 $ha^{-1}yr^{-1}$
2. Tree should possess erect branches with minimum branch angle from the main stem so that maximum sunlight should be received by under storey crops.
3. Stem should be straight and cylindrical with clean bole height of 5-6 m at least.
4. High coppicing, lopping and pollarding ability.
5. Survive in harsh environment without watering.
6. Ability to provide quality green fodder during lean and dry period.
7. Ability to fix atmospheric nitrogen if possible
8. Having deep tap root system to avoid competition for soil resources with lower storey crops.

Brief review of the improvement work done in important woody agroforestry species

1. Subabul (*Leucaena leucocephala* (Lam.) de Wit)

Subabul is a self pollinated fast growing multipurpose leguminous tree species originated from Mexico and Central America (Brewbaker, 1987). Its potential as a fodder cum fuel tree in agroforestry systems in India is well known (Singh, 1986; Relwani, 1983; Deb Roy, 1988; Pathak, 1988) because of presence of large quantity of protein and vitamin A in their leaves, bear deep tap root system, erect growing habit and ability to improve soil fertility through fixation of aerial nitrogen and decomposition of protein rich leaves in the soil in a short duration. Average protein content varies from 22-30 per cent in their leaves and protein yield is around 3.63 t ha^{-1} yr^{-1} under cutting management systems (Brewbaker and Hutton, 1979). By simple selection in natural stands, Dr. Brewbaker and his team identified highly vigorous and straight trees which grew up to 20 m in southern Mexico and western Central America. By evaluating these selections, six outstanding varieties (K8, K132, K28, K29, K62) have been released. Out of these, K8 got maximum popularity.

Forage yield of improved variety K8 was 19.5 t ha^{-1} yr^{-1} against 10.5 t ha^{-1} yr^{-1}. of Hawaiian type (Brewbaker, 1987).

In CSIRO, Australia *Leucaena* breeding work started since 1950 for low mimosine content in their leaves and released the variety Conningham by hybridization of Peru X CPI 18228 followed by selection in 1976. In this variety, mimosine content is slightly less than cv Peru which was common variety in Australia. Scientists made efforts to further reduce mimosine content in same organisation. They could produce interspecific hybrids of *L. leucocephala* X *L. pulverulenta* which were quite vigorous and mimosine content of these hybrids were reduced further by 14 per cent (Bray, 1983) and increased forage yield by 42 per cent. These hybrids could not become popular due to lack of simple and economic technology to multiply vegetatively to retain the hybrid vigour or hybrids seed production.

In year 1984, suddenly *Leucaena* got setback due to wide spread of tiny homopterous insect called psyllids (*Heteropsylla cubana* craward) in whole of the leucaena growing areas of the world. Plants defoliated completely and growth hampered significantly. All the germplasm of *Leucaena* (more than 1000 of 17 *Leucaena* species) were tested against this insect. Two provenances viz. K636 and K584 of *L. leucocephala* were found tolerant and advocated to plant in psyllid infested areas. Other species of genus *Leucaena* resistant to psyllids were *L. diversifolia, L. collinsii, L. pallida* (Brewbaker *et al.*, 1989). Since all these species were not as vigorous as subabul so that effort were made to develop interspecific hybrids among these species rather than commercialise these. Breeding efforts in this direction resulted in interspecific hybrids KX1 (*L. diversifolia* X *L. pallida*), KX2 (*L. leucocephala* X *L. pallida*) and KX3 (*L. diversifolia* X *L. leucocephala*) were released by University of Hawaii which were quite resistant to psyllids and also vigorous.

1.1 Subabul Breeding in India

Looking to the importance of subabul in agroforestry systems as well as identifying limitations in this species in India, subabul breeding programme was initiated in 1975 in Indian Grassland and Fodder Research Institute (IGFRI), Jhansi. IGFRI collected 496 provenances of 14 *Leucaena* species.

Biometrical studies like genetic variability between and within species, correlation and path coefficient analysis (Gupta and Patil, 1988), gene action and hybrid vigour studies (Gupta and Patil, 1986) have been done at IGFRI. Studies revealed that a good amount of variability including mimosine content was available in genus *Leucaena* (Table 36.1). Interspecific hybrid vigour is quite significant but intervarietal vigour is absent (Table 36.2 & 36.3). In IGFRI through simple selection and hybridization, seven promising selections were developed namely Sp1, S10, S14, S15, S22, S23 and S24 (Pathak and Gupta, 1994). Selection S24 was quite promising and better than K8 by 29 per cent for bole yield and 65 per cent for leaf yield in summer (May) and 36 per cent for total biomass. It also had 48 per cent less pods than K8 and contributed 70 per cent for bole yield and 10 per cent for leaves against 64 per cent and 8 per cent by K8 to total biomass, respectively. This selection may be useful in northern India. Several interspecific hybrids were also prepared at IGFRI, Jhansi. Hybrids of *L. leucocephala* x *L. diversifolia* and *L. leucocephala* x *L. pulverulenta* were promising (Table 36.3). One of the interspecific hybrids of *L. leucocephala* x *L. pulverulenta* was superior over variety K8 by 88.9 % for fuelwood yield and 71.64 per cent for forage yield (Gupta, 1988). In other organisations only evaluation of received germplasm of leucaena from out side was done. In general, K8 was found superior over others in their trials.

1.2 Vegetaive Propagation

Grafting is relatively simple and historic technique for *Leucaena*. The best results obtained was in cleft, whip and tongue grafting rather than bud or T-shield grafts, using four-months seedlings as stock and taping with grafting tape. Several types of interspecific grafts were successful. Vegetative propagation through cuttings are successful in few locations. Hu (1983) obtained excellent rooting of cuttings under mist spray in Taipei greenhouse using 15 cm leafy twigs having terminal twigs intact in one year old trees. Leafless twigs from older trees failed to root. Micropropagation in *Leucaena* is successful but can not be considered for field application. Callus growth was achieved from germinating seeds and hormonal treatments were perfected leading to shoot differentiation and somatic embroids. Seed derived explants were carried out without callus formation through IBA and NAA media to vigrous growth shoots and with added ascorbic acid to healthy plantlets.

2. Khejri (*Prosopis cineraria*)

Prosopis cineraria is a far better indigenous species than the exotic for survival, adaptation, growth, production of fodder and fuel, increase the soil fertility and check desertification in Thar desert. Major limitation present in this species is that it is extremely slow growing attained 1 m height in 4 years. Due to its slow growth, large scale plantation of *P. cineraria* in Thar desert could not be done. Increasing demand of fodder and fuelwood in Thar desert could not be achieved unless fast growing provenances/ genotypes of this species would be identified. In this direction little efforts have been made before 1981 to study genetic variability available in nature particularly in

Rajasthan, its conservation and exploitation which are prerequisites to boost its growth and to increase the supply of fuel and fodder production. With this background intensive genetic improvement work on *P. cineraria* was initiated in 1983 to evolve fast growing genotypes by selection as well as to understand extent and pattern of genetic variability present in nature and its exploitation.

2.1 Survey

Eleven districts of Rajasthan were surveyed during 1984 and seeds from the best trees from each district were used to raise single tree progeny during August, 1984. During May and June, 1985, in second exploration trip, 7 districts of western Rajasthan, i.e. Jodhpur, Bikaner, Barmer, Jaisalmer, Churu, Sikar, and Nagour were surveyed. Seeds of seventy plus trees were selected. There was maximum variability for collar diameter and dbh. Coefficient of variation (CV) was maximum for collar diameter (65.7 %) followed by dbh (64.9 %), plant height (52.7 %) and crown diameter (48.14 %) (Table 36.4). Similar findings were reported by Jatasara and Paroda (1983) except number of branches per tree. Two types of branch angle namely acute and horizontal observed during present survey apparently suggested the presence of 2 different plant types in nature. Study also revealed predominance of the plant type with horizontal branches (69 %). Jatasara (1982) during his survey of the Thar desert also reported 2 distinct types *P. cineraria i.e.* one spineless with dropping twigs and a compact crown, the other with spines, non-dropping branches and a comparatively wide open crown. He also observed some intermediate ecotypes and suggested that these ecotypes can be crossed. The present study shows that wide variability exists in natural stands for all the characters studied. A wide range of variability was also observed in the percentage of crude protein, reducing sugars and total sugars of pods between trees within provenance and between provenances. Range was from 8.05-15.05 % for crude protein, 0.83-4.67 per cent for reducing sugars and 6.04-15.83 % for total sugar. In juvenile progenies, peroxidase isoenzymes showed large variation at some loci both within and between progenies.

2.2 Progeny Trials

During June, 1985, 70 single tree progenies (10 trees selected from each district) were sown in polythene bags and earthen pots in RBD design with four replications. The differences among progenies were significant only at 4- and 5- month age for plant height. First progeny trial of 10 plus trees was established during 1977 at the Central Research Farm, CAZRI, Jodhpur in randomised block design with three replications. These ten plus trees were identified at 2 locations i.e. Jodhpur (5 trees, JU-1 to JU-5) and Bikaner (5 trees, BK-1 to BK-5). The observations on fodder and other component characters were recorded at the age of 7.5, 8.5 and 9.5 years, respectively. There were significant differences among progenies at 2 locations during different years. Progeny number JU-5 has shown a consistently good performance over years. Both heritability and genetic advance as a percentage of the mean have moderate implying that selection of progeny is likely to be moderate. During May 1984, another survey of the arid region of Rajasthan was conducted and seeds of superior trees were collected. Of these, 11 half sib progenies were transplanted into the field in July-August, 1984 in randomised block design with 3 replications. Results revealed that there were significant differences among 11 progenies for tree height for all the 5 years. Progeny number 264 from Jodhpur ranked first throughout the study period. In CCSHAU, Hissar, provenances trial was established of about 8 provenances. Since trial was 2-4 years old thus no conclusion can be drawn (Toky, 1995). CCSHAU also prepared hybrids which are superior in relation to dbh, tree height and clear bole height.

2.3 Inter-relationship of Height of Mother Trees with Juvenile Progeny

Selection of plus trees from mature stands is generally based on various morphological characters like height, dbh, collar diameter, crown diameter, etc. and the presumption or rather than expectation is that the progeny from such trees will also have similar morphology and better growth. The correlation of 10 parents with their 7.5-year-old progenies have demonstrated positive correlations between various characters of parents and progenies. The correlation of dbh is the maximum (0.763). It revealed that the approach of direct selection from natural stands appear to be effective.

2.4 Seed Orchards

Some elite selections of *P. cineraria* attained growth height of 7.3 m. against 6.5 m. attained by *Acacia tortilis* (a fast growing species) in 6-years. Such identified elite trees were used to establish seed orchards. In the seedling seed orchards, there were 156 elite trees (mean height 3.54 m in 6-years). The best one among these are under further genetic exploitation. Clonal seed orchard was also established by preparing seedlings through air layering of elite trees. Using the seeds collected from the seedling seed orchard and the clonal seed orchard, 7000 saplings were raised in 1990-91 and supplied for multi-location trials.

2.5 Vegetative Propagation

There are several reports that *P. cineraria* has been propagated through root suckers (Rao, 1953) and by air layering (Rawat, *et al.*, 1982; Solanki *et al.*

1984, 1986) . There have been few reports that *P. cineraria* has been propagated through shoot cuttings (Arya and Tomar, 1989). The basic technique involved in air layering is removing the bark from 15mm girdle and applying the hormones to the upper end of the girdle. The girdle is then covered with a layer of moss or clay and wrapped in polythene. Solanki *et al*, 1986 showed that with either 100ppm IAA or with the proprietary rooting hormone Seradix B3, rooting was induced in 10-15mm branches but not 7mm branches, the best results being obtained with Seradix and clay. They also observed a distinct seasonal pattern in the success of air layering with the best treatments giving 100% success in July and August, a little rooting in January and February and no rooting between October and December and between March and May. *P. cineraria* is successfully propagated through tissue culture (Arya and Shekhawat. 1986; Solanki *et. al.*, 1991).

3. Neem (*Azadirachta indica* A. Juss)

Neem (*Azadirachta indica* A. Juss) belongs to the subfamily Meliodae of the family Meliaceae and is native to Indian subcontinent (Harm, 1940; Hooker 1973). It is distributed throughout south and south-east Asia, Australia, Africa and many other countries in central and south America including Caribbean, Puerto Rico and Virgin Islands. It has estimated population of 15 million trees and producing 5 lakh tons of neem seed annually in India (Venkateswarlu, 1996). In India, neem is found virtually in all states of the country, except the Himalayan region, due to wide range of adaptability of climatic, edaphic and topographical factors. It tolerates high temperature upto 49^0 C and thrives well in low rainfall areas (450 mm). In recent years, neem is getting much attention because of large number of liquid and powdered formulations of neem containing azadirachtin as a biological active gradient are being marketed in world as a nontoxic biological insecticides or chemicals. Neem is also used to synthesise medicinal, cosmeticals and animal health care products.

Realising its economic potential in world market and good adaptability in wide range of lands including degraded lands and climates, research work on tree improvement has been initiated in few research organizations i.e. National Research Centre for Agroforestry, Jhansi (Gupta *et al.* 1996b); Central Research Institute for Dryland Agriculture, Hyderabad (Venkateswarula, 1996); Tamil Nadu Agricultural University, Coimbatore (Shivasamy and Karivaratharaju, 1993); Arid Forest Research Institute, Jodhpur (Mishra, 1996); CCCS, Haryana Agricultural University, Hisar; Central Arid Zone Research Institute, Jodhpur; National Botanical Research Institute, Lucknow; Tropical Forest Research Institute, Jabalpur and Institute of Forest Genetics and Tree Breeding, Coimbatore.

3.1 Genetic Resources and Biodiversity

With the involvement of many organizations which are located in various neem growing areas, a vast number of germplasm of neem has been collected. Collected germplasm has large amount of variability for various morphological, growth and quality characters. National Research Centre for Agroforestry, Jhansi has collected 276 accessions in five exploration trips.These accessions belong to nine states viz. Uttar Pradesh, Madhya Pradesh, Maharastra, Andhra Pradesh, Orissa, Rajasthan, Haryana, West Bengal and Gujrat and include provenances as well as plus trees. In these collections, a wide range of variability was observed for 100-seed weight (11.00 - 33.04g), seed length (0.80 - 1.90cm), seed width (0.58 - 0.83cm), 100-kernel weight (4.43 - 18.73g), oil content (36.39 - 52.40%) on kernel basis, seed - kernel ratio, seed colours, seed shape, tree height, tree diameter, fruit ripening period and crown shape. The Forest Research Institute, Dehradun collected seeds from 20 sources from 8 states under Indo-Denish Project on tree and seed improvement. The test weight of 100 seeds varied from 10.5 - 34.7 g and leaf length from 2.3 cm to 7.2 cm. Arid Forest Research Institute, Jodhpur established provenances trial of 40 provenances belonging to 10 states and representing 10 agro-climatic zones of India. These germplams was collected under International Neem Net Work. The Forest College and Research Institute, Tamil Nadu collected the seeds of 149 plus trees from 18 neem growing districts and established the provenances trial. In the collected germplasm, 100-seed wt. ranged from 17.4 - 29.5 g and oil content varied from 40-44 per cent. Central Arid Zone Research Institute, Jodhpur collected the seeds of 754 CPTs in Rajasthan (Jindal and Satyavir, 1997). Good amount of variability was observed in the germplasm. The 100-seed weight ranged between 9.24 - 27.78g, 100-kernel weight varied from 3.5 - 13.51g, oil content varied from 25.17 to 55.09% and azadirachtin from 49 to 7572 ppm. Institute of Forest Genetics and Tree Breeding, Coimbatore collected the seeds of 442 CPTs from Tamil Nadu, Karnataka and Andhra Pradesh. The fruit length varied between 1.21 cm to 2.51 cm with average of 1.80 cm. The biggest seeds were observed from Thanjavur source measuring 2.04 cm in length. The oil content varied between 32.42% to 51.80%. The highest azadirachtin content was obseved in Kaur source recoding14525 ppm. Under Neem Network on Integrated Development of Neem financed by NOVOD Board, Gurgaon, more than 1000 accessions of neem collected by these organizations were evaluated for chemical composition. The finding revealed a large variation for azadirachtin and oil content and fatty acids (Kaushik, 2002). The azadirachtin content ranged from 0.02 - 1.6 %. About 27 samples showed more than 1 per cent azadirachtin. Similarly, oil content varied from 24 - 55 %. Oleic acid was found to be the major fatty acid and it ranged from 32 - 68 %. Palmitic acid ranged from 6 - 46 %, Linoleic from 3 - 36 % and Stearic 0.5 - 26 %. The azadirachtin and oil content were found to be affected by seed quality and moisture content. The state wise compilation of the average azadirachtin content reveals that southern peninsula states *viz*. Tamil Nadu, Karnataka and Andhra Pradesh have comparatively higher azadirachtin as compared to other states. The average content of fatty acids indicates that seed samples from Maharashtra, Delhi, Haryana, Chattisgarh, Rajasthan and Gujarat have high oleic acid (> 50 %).

3.2 Provenance and Progeny Trials

Though a large amount of germplasm of neem was collected by various organizations. However, provenances and progeny trials have not been established. Thus it is difficult to analyse that the variability existed in collected germplasm for various character attributes to genetical makeup of the material or the location effect. Only few organizations have established provenances and progeny trials.

At National Research Centre for Agroforestry, Jhansi, In a trial of 26 provenances collected from central India, a wide range of variability was observed for tree height (472.00 -662.13cm), collar diameter (12.03 - 17.82cm), dbh (10.20 - 13.90cm), seed length (1.16 - 1.58cm), 100-seed weight (19.11 - 24.96g), fruit bearing and oil content (35.04 - 51.42%) at the age of 6.5 year. On the basis of overall performance Bhopal provenance ranked first for growth and seed yield. This provenance also had oil content 45.71% which was 8.63 % higher than Jhansi provenance followed by Jabalpur, Damoh, Gulganj and Panna. Among these, Bhopal and Damoh was higher fruit producer and more straight than others, respectively. All these provenances belong to better soil with better rainfall areas (Gupta and Kumar, 2001). In progeny trial, twenty plus tree progenies of neem including check were evaluated for growth , straightness and seed yield. Results of 6.5-year-old trial revealed that PT- 24, PT- 22, PT- 15 and PT-14 were quite promising for growth parameters, straightness and seed yield. Among these, PT-24 was higher fruit producer and more straight than others. Results of two-years-old trial of 37 provenances conducted at Arid Forest Research Institute, Jodhpur showed that provenances of Indore, Ujjain, Palanpur, Rajkot and Ravinagar are quite good. Central Research Institute for Dryland Agriculture, Hyderabad identified four superior clones. These are one each from Anantpur and Hyderabad in Andhra Pradesh, Sabarkantha in Gujrat and Philibani in Orissa.

3.3 Reproductive Biology

To raise seed orchards and for genetic improvement of any species, a knowledge on floral and breeding systems are essential pre-requisites. There are some reports that neem is a cross pollinated species (Gill *et al.*, 1993; Venkateswarlu, 1996; Bisla *et al.*, 2002) based on preliminary information. To confirm the results of previous authors, National Research Centre for Agroforestry, Jhansi conducted a study on reproduction biology systematically and continuously for five years since 1994 and came to conclusion that neem may be a predominantly self pollinated tree species. Details of the study is as follows:

Flower bud initiation in neem starts in the 2nd week of March and continues up to last week of May with peaks between April 10-20, minor variation was observed due to variation in atmospheric temperature. Days taken from initiation of bud to opening of flowers ranged from 23-29 days with a mean 25.55 $\pm$ 0.70 days. The period of opening of flowers to fruit formation ranged from 15-20 days and duration of fruit formation to ripening of fruits was around 52 days. Anthesis starts between 6.30-7.00 p.m. and it completes in the late night. However, maximum buds (>90 %) open before 10.00 p.m. The dehiscence of anther starts before anthesis of flowers. It starts at 9.30 a.m. and continues up to 3.30 p.m. with maximum frequency between 12.30 - 1.00 p.m. in closed flowers which are going to be open during the night of the same day. Pollen stainability in 1 per cent acetocarbine ranged from 83.33-100.00% with average pollen size of 44.65μ. Generally pollens are of spheroidal type but few of them are of sub-spheroidal type. There was no pollen germination in sucrose solutions (5, 10, 20, 25 and 30%) with and without boric acid. However, germination was good on stigma but maximum germination takes place on the sides of stigma and not in the centre. Maximum pollen tube length observed after 36 hours of pollen dehiscence was 370μ. There are non-significant differences for pollen germination percentage (5.50 % - 10.75 %) on stigma between selfed and open pollinated flowers and also for growth of pollen tube. This indicates that selfing does not inhibit pollen germination and growth of pollen tube or absence of self incompatibility. Pollen germination starts from 4.00 p.m. in closed flower. Maximum frequency of fertilized ovules were observed after 36 hours of pollen dehiscence. There are minor differences in fruit setting between selfed and open pollinated flowers. Fruit setting in selfed flowers was 3.06 per cent while in open pollinated flowers per cent fruit setting was 4.15 per cent (Table 36.5). Growth attributes (plant height, collar diameter and diameter at breast height) and leaf characters recorded in selfed and open pollinated progenies at the age of 7.5 years also indicated that there was non-significant differences between progenies of selfed and open pollinated seeds for growth and leaf characters. Hence, it seems inbreeding depression is absent in this species. This study suggests that neem may be a predominantly self pollinated tree. To confirm the authenticity of results at micro-level, study on mode of pollination was also done through isozyme analysis with the help of gel electrophoresis. Under this study, estrase, acid phosphatase and peroxidase were analyzed in leaves of self as well as open pollinated progenies of same and other genotypes. Results showed uniform pattern of self and open pollinated progenies of the same genotypes and non-uniform pattern of progenies of other genotypes for estrase which further confirmed self pollinated behaviour of neem.

3.4 Vegetative Propagation

Neem can be propagated through cuttings in mist chamber, air layering, grafting and budding and through tissue culture. Out of these methods, air layering is cheapest , simple and highly successful.

Air layering : Rooting in air layered branches is 100 percent by application of 800ppm solution of IBA in rainy season (Gupta and Kumar, 2001). Airlayering can be done during March to September. These airlayered plants can easily be transferred in polybags without any mortality. Earlier Shanmugavelu (1967) also described vegetative propagation through air layering.

Branch Cuttings : IBA solutions of 100-200ppm gave very good response in mist chamber under 24 hrs treatment.

Tissue culture : Callus formation, somatic embroygenesis and direct multiple shoots induction from seedlings explant have been reported in neem (Rangaswamy and Promila, 1972; Sanyal *et al.*, 1981; Narayan and Jaiswal, 1985; Gautam *et al.*, 1993; Venkateshwarlu, 1996).

4. Poplar (*Populus* spp)

The history of poplar is quite old but it was largely confined to the Kashmir and the temperate arid zones. However, around 1950 with the import of large number of exotic clones and subsequent trial in north India by FRI, Dehra Dun, the usages pattern changed. Now these exotics (few clonal selections of *Populus deltoides*) are becoming an important part of agroforestry systems in some districts of Punjab, Haryana and Uttar Pradesh. This species is considered most successful and commercialised in agroforestry systems. The area planted with poplar during Eight Five Year Plan in north Indian states was approximately 23,000 ha per year. Presently 30 million trees of polar, producing 1,125 million m^3 industrial wood annually, are standing in Uttar Pradesh, Punjab, Hayana, Himachal Pradesh and Jammu and Kashmir in combination with Agricultural crops. The programme of introductions and research on poplars was initiated by Forest Research Institute, Dehradun when

24 clones were introduced from England (Mathur and Sharma, 1983). The successful attempts were made by FRI, Dehra Dun to raise genetically fast growing clones of poplars. Initially 'IC' clones were considered to be the best performers, however, in 1978, these clones lost vigour and resistance to pests and diseases by the 10th cycle of clonal cuttings. Since then several species, clones and cultivars of exotic poplar have been introduced, multiplied and tested by FRI, Dehra Dun; State Forest Department, Haldwani (Uttarachal); Dr. Y.S. Parmar University of Horticulture and Forestry, Solan (Himachal Pradesh) and WIMCO Seedling Ltd. - Rudrapur (Uttaranchal).

Among the successful exotic poplars introduced in india, the most widely planted species is *P. deltoids*. It is grown on a large scale in the terai region of Uttar Pradesh and the plains of Punjab, Haryana, Himachal Pradesh, Uttar Pradesh and Arunachal Pradesh. A large number of exotic clones of these species are grown in field plantation, e.g., Australian selections viz. G_3 and $G_{48,}$ American selections viz. D66, D121, D75 and D81 of *P. deltoides* and Italian selections viz. Lux 69/ 55 of *P. deltoides* and 72/ 58 of *P. deltoides x P. euramericana*, etc. Some of the new clones introduced recently which have shown promise in agroforestry systems are S7C4, S7C8, S7C15, S7C20, etc. Clones of 'L' series developed by Forest Deparment at Haldwani Research Centre (Uttaranchal) from open pollinated seeds of G48 and D121 mother plants have shown encouraging results and they are soon expected to become more popular among farmers than the existing clones under cultivation (G3 and G48). WIMCO Seedling Ltd. Rudrapur (Uttaranchal) has also come out with some clones namely Udai, Kranti, and Bahar which are also being planted by the farmers. However, clones developed by Uttaranchal Forest Department and WIMCO Seedling Ltd. have very narrow genetic base and as a result, none of the L-series clones could perform better than G_{48} in clonal field trial in Punjab (Sidhu, 1996). In order to explore the possibility of augmenting the genetic base of new clones, breeding orchard was established at Dehra Dun in 1998 with planting 40 clones (Singh *et al.*, 2001). Details are given in table 36.6. The aims of Poplar breeding at FRI are as follows:

1. Create wide spectrum of Poplar clones which are site specific.
2. To improve the quality of stem form and timber
3. To improve the growth rate
4. To develop greater tolerance to biotic and abiotic stresses
5. To develop timber suitable for diversified utilization
6. To increase the chance for intra and inter population mating
7. To increase the frequency of extreme genotypes

5 Shisham (*Dalbergia sissoo* Roxb.)

Sissoo is an important timber species belong to family Leguminaceae and subfamily Papilionaceae. It provides fuelwood, and fodder and enrich soil fertility. It provides important raw materials for a variety of wood based industries especially for furniture and building constructions. The wood is hard, heavy , strong, double elastic, seasons well and decay resistant.

5.1 Reproductive Biology

Extensive work on reproductive biology of sissoo was done at CCSHAU, Haryana (Bangarwa, 1996) which is given below :

1. Flower buds appear in the first week of March in northern India.
2. Flowers generally open (anthesis) between 10.30 and 13.30 hrs.
3. The anthers dehisce earlier in the day before anthesis.
4. Pollen viability varies from 56 per cent in March to 46 per cent in April.
5. Stigma become receptive six hours before anthesis and continue upto six hours after flower opening.
6. No difference in per cent fruit setting between isolated trees and trees in groups.
7. No fruit setting in emasculated flowers. However fruit setting takes place in emasculated flowers after artificial pollination.
8. Pod setting was around 40 per cent in natural condition.
9. Maximum pod setting occurs from last week of March to first week of April
10. *D. sissoo* is predominantly self fertilizing species.

Results were also confirmed by Tiwari and his team at G.B. Pant University of Agriculture and Technology, Pantnagar (Tewari, 2002).

5.2 Provenance and Plus Tree Progeny Trials

Exploration and seed collection work in *Dalbergia sissoo* was done by several organisations. Out of these, intensive work has been done by CCS Haryana Agricultural University, Hisar; National Research Centre for Agroforestry, Jhansi; G.B. Pant University of Agriculture and Technology, Pantnagar; Tropical Forest Research Institute, Jabalpur and Forest Research Laboratory, Kanpur (Table 36.7). A good amount of variability was reported by these organisations for tree height, clear bole height, straightness, crown shape , pod characters and dbh. National Research Centre for Agroforestry,

Jhansi made intensive survey of Bundelkhand region and collected seeds of 32 plus trees. Among these, PT-6 and PT-2 of Mahurani were fast growing and highly straight with mean annual increment of 15.53 $m^3ha^{-1}yr^{-1}$ and 12. 73 $m^3ha^{-1}yr^{-1}$, respectively (Gupta *et al.*, 2003). CCS Haryana Agricultural University did intensive survey of northern India (Rajasthan, Haryana and Punjab) and south western region of Nepal during January - February, 1998 in order to select plus trees of sisham based on good height, diameter, clear bole, stem straightness, aggressive apical dominance, compact crown, low branch angle and resistant to pest and diseases. Five progenies of plus tree viz. PT28 and PT12 from Haldwani, PT-283 from Gonda, PT-84 from Nepal and PT-9 from Tanakpur were found most promising for straightness, basal diameter, total height and unforked height. Heritability in broad and narrow sense were high for straightness and unforked height (Bangarwa, 2002). Similarly other organisations also conducted progeny trials results are awaited.

5.3 Vegetative Propagation

D. sissoo can be propagated easily in ordinary nursery by

1. Cuttings (100 ppm, 150 ppm IBA)
2. Air layering (800 ppm IBA)
3. Root suckers
4. It can also be propagated by tissue culture

6. Desi Babul *(Acacia nilotica* (L.) Wild ex Del.*)*

Acacia nilotica is a versatile indigenous multipurpose tree species of India. It is a good source of food, fodder, small timber and tannin. Its fast growth, good coppicing ability, fixing atmospheric nitrogen and wide range of adaptation has gained attraction of the people of south east Asia and Africa. It is found throughout semi-arid regions of the country from Punjab in north to Tamil Nadu in south. It is available in almost all the farmer's fields.

6.1 Reproductive Biology

Systematic work on reproductive biology was conducted at Punjab Agricultural University, Ludhiana (Dhillon and Khajuria, 1994) and National Research Centre for Agroforestry, Jhansi (Ahlawat *et al.*, 2003). Flowering in *A. nilotica* starts from June and continue upto December. It reaches at peak during September-October. Flowering proceeds from lower crown to top of tree as well as from the basal part of young shoots and continue with the growth of twigs in the axes of each leaf. Number of inflorescences varies from 22 - 40 with mean value of 32.2 (Ahlawat *et al.*, 2003). The number of anthers per flower varies from 90 to 140. Anthesis is observed during night between 0.30 to 1.30hrs. Anther dehiscence occurs at early morning. However, pollen grains were not found in mass due to sticky nature of anthers and pollen grains. The pollen grains are found in the form of compound grains of 12 to 16 pollens in the multiple of four. The individual pollen grain of such a group is polyhedral and short wedge shaped each presenting a flat angular usually square shape face to out side with the body of the cell tapering abruptly inwards, towards the centre of the group (Ahlawat *et al.*, 2003). Pollen size varies from 31.25x37.5μ to 43.75x50μ and mean size of 38.75x39.63μ. The size of compound grain (group of pollens) varies from 100x106.25μ to 118.25x11.825μ and mean value of 110.41x111.36μ. Based on stainability test, pollen viability was more than 90%. Pollens germinated after 4 to 24 hrs on medium having 10-15 % sucrose. Carpet remains folded even on third day after anthesis. Flower drop was such a high that ground under the trees looked like yellow carpet during June to October. Still mode of pollination of *Acacia nilotica* is not known clearly. It is considered to be highly cross pollinated species.

6.2 Provenances and Progeny Trials

Provenances/progeny trials were conducted by CCS Haryana Agriculture University, Hisar; Forest Research Institute, Dehradun; Tropical Forest Research Institute, Jabalpur, JNKVV, Jabalpur, which are given in table 36.8.

6.3 Vegetative Propagation

Branch cuttings : Application of IBA 1000ppm stimulates rooting in branch cuttings in month of June and rainy season (Batt, 1995; Khajuria, 1995). Gurumurthy *et al.* (1994) also got success to induce rooting (70 %) in branch cuttings by application of 100ppm IBA during August- September in hard wood. While in soft wood, it was 60% using 5000ppm IBA containing 0.1 per cent Emason during June - July.

Air Layering : As compare to branch cuttings, air layering is more successful and assured technique for cloning elite materials to develop seed orchards by application of 1000ppm IBA (Raj, 1991; Khajuria, 1995; Ahlawat *et al.*, 2003).

Tissue culture : Micropropagation in *A. nilotica* is also possible (Mathur and Chandra, 1994).

7. Eucalyptus (*Eucalyptus* spp.)

Eucalyptus species were introduced in India around 1790. Plantation of Eucalyptus in man made forests or under agroforestry systems is common in the country. It is an exotic species and its edaphic requirements have been studied since 1950. Large scale plantation of this genus have been established all over the country. It was planted intensively under agroforestry systems in Punjab, Haryana and western UP. However, their growth and establishment was poor due to lack of quality planting materials. Tree Improvement work was first initiated at FRI, Dehra Dun and two hybrids were developed in 1970 i.e. FRI 4 (*E. tereticornis x E. camaldulensis)* and FRI 5 (*E. camaldulensis x E. tereticornis*). These hybrids gave 109% more wood volume than Mysore Gum (Mysore hybrid) (Venkatesh and Sharma, 1977). However, these two hybrids could not become poplar for plantation due to lack of sufficient number of clones of these hybrids. Later on through tissue culture, several plus trees of *Eucalyptus tereticornis* have been multiplied at TERI, New Delhi; NCL, Pune; ITC Bhadrachalama Paper Board. These clones are giving 40-60% higher wood volume than the seedlings raised through seeds. Farmers are purchasing these seedlings even on higher rate. In Institute of Forest Genetics and Tree Breeding, Coimbatore, 17 Australian provenances received from Australia Tree Seed Centre, CSIRO, Australia were tested. It was concluded that the provenance Paluma, Ravenshoe, Gadgara and Herberton were superior in volume growth. M/s Bhadrachalam Paper Board Ltd. a wood based industry located in Khamman District in Andhra Pradesh did pioneering work in identification of phenotypically superior Eucalyptus trees in the plantation raised by APFDC and AP Forest Department, Andhra Pradesh, for taking up for vegetative propagation and field trials. They have carried out extensive tests of several clones in different agro-climatic zones and finally selected a few clones which were found to be promising with respect to their growth performance and adaptability to wide range of edaphic and climatic conditions. These clones are clone 1, 3, 4, 6, 7, 8, 10, 27, 32, 52, 71, 93, 99, 100, and 128. These clones are extensively grown in all the eucalyptus growing areas. (Lal, 2001).

7.1 Vegetative Propagation

Now a days, simplest technique for mass multiplication of superior trees is putting growing shoots of elite trees in the mist chamber for rooting. After formation of roots, these seedlings are transferred in root containers in shade house. Later on transferred in ordinary nursery for hardening. It can also be multiplied easily through tissue culture technique. Rooting is also possible with 100-150ppm solutions of IBA in ordinary nursery during rainy season.

8. *Casuarina equisetifolia* Linn

Casuarina equisetifolia is an ideal tree crop of coastal region. The tree is tall, fast growing, salt tolerant, wind resistant, adaptable to moderately poor soils because of its capacity to form root nodules and fix atmospheric nitrogen like a legume. Casuarina trees have soil ameliorative effects (Patel and Singh, 2000). It was introduced in India during the late 19th century (Pinyopusarerk and Heuse, 1993) and is now planted in more than 50,000 ha (Gurumurthi and Rawat, 1992) with short rotation of 3 to 4 years. The yield variation for wood per tree within a plantation ranges from 4 -35 kg in five-year-old plantation, suggesting that there is a tremendous scope for identifying superior performers (Gurumurthi and Rawat, 1992). Only few organizations have been working on tree improvement of *Casuarina equisetifolia* in India i.e. Forestry Research Station, Tamil Nadu Agricultural University, Mettupalayam; Regional Forest Research Centre, Rajahmundry, Andhra Pradesh and Institute of Forest Genetics and Tree Breeding, Coimbatore. Regional Forest Research Centre, Rajmundhari established provenances trial of 40 clones, progeny trial of 16 wind pollinated families and family-in-provenance trial of 100 wind pollinated families which is a part of an international provenance trial. Provenances of Malaysia and Thailand generally showed superior growth. However, land-race population often demonstrated superior stem straightness (Prasad, 1998). Progeny trial of three sources was established by Forest Research Station, Tamil Nadu Agricultural University, Mettupalayam and reported high heritability for tree height, root collar diameter, and number of branches (Jambulingam, 1990). IFG & TB, Coimbatore also established clonal seed orchards of elite trees. In the same organization, the international provenance trial has been established with seeds of 40 provenances obtained from Australia Tree Seed Centre, CSIRO, Australia. Many provenances of international collections were found to perform better than the local source. The Gede Forest Resat provenance from Kenya shows promising growth.

8.1 Vegetative Propagation

Many reports are now available with regards to vegetative reproduction of *C. equisetifolia*. Somasundaram and Jagadeesan (1977) rooted *C. equisetifolia* cuttings taken from young lateral shoots with heel smeared with seradix 'B'. Husain and Ponnuswamy (1980) rooted *C. junghuhniana* shoot and root suckers under mist chamber conditions. Kan and Hu (1983) initiated rooting of cuttings of *C. junghuhniana* and *C. equisetifolia* under green house condition. The highest success in vegetative propagation (91%) of *C. equsetifolia* was achieved by using cladode cuttings (Gurumurthi and Bhandari, 1988; Gurumurthi and Rawat, 1992).

9. Desi Siris (*Albizia lebbek*)

Exploration team of NRCAF, Jhansi, made survey of 36 districts of UP, MP and Rajasthan. During exploration, we observed high variability for leaves, pods, tree canopy, diameter at breast height, tree height, number of seeds per pod, 100-seed weight, retention of leaves during dry period. There were few trees which were entirely different than normal trees which are straight types. These trees attended the height more than 18 m against the normal tree of 10-12m.and very straight with black stems. These trees were named as giant tree . Seeds of these trees were collected. Progeny performance yet to be tested. In CCSHAU, Hisar, provenances trial of 12 provenances was conducted and found that provenance of Jammu was better than others for early growth parameters (Toky, 1995) (Table 36.9).

10. Ratanjot (*Jatropha curcus*)

Steep rising in the price of petroleum products forced the Government of India to think an alternative source of petroleum products. A more resilient source has been found in *Jatropha curcas*, an oil bearing, drought hardy shrub with ecological advantages and already being successfully grown and harvested as a biofuel in countries like Nicaragua. Even in India, Railways got success to run some trains by blending diesel and Jatropha oil.

The genus *Jatropha*, family Euphorbiaceae is a morphologically diverse genus comprising 160-175 species of shrubs, rhizomatous shrubs, herbs and small trees. About nine species of *Jatropha* have been recorded in India. Out of these, important ones are, *Jatroppha curcas* Linn., *J. gossypifolia* Linn., *J. glandulifera* Roxb., *J. multifida* Linn., *J. nana* Dalz. and *J. podagrica* (Anonymous, 1959). *Jatropha* is a wild growing hardy plant well adapted to arid and semi arid conditions. It has low fertility and moisture demand and can come up on stony, gravelly or shallow and even calcareous soil. *Jatropha curcas*, commonly known as Physic nut, Ratanjot, Chandrajot, Jamalgota, Jangli arandi, or Kala aranda is a native of Mexico and tropical South America. The plant is reported to have been introduced into Asia and Africa by Portuguese as an oil yielding plant. Now it is occurring almost through out India and in Andaman Islands in semi wild condition. It is reported to be cultivated in Central and Western India in states like Rajasthan, Madhya Pradesh, Maharashtra and Gujarat (Kumar *et al.* 2004). Similarly, there are reports of its cultivation in Andhra Pradesh and Tamil Nadu.

10.1 Flowering and fruiting

Jatropha carcus is monoecious and protandrous. Flowering occurs in the hot and the rainy season. The ratio of male to female flowers is 29:1. Both flower sexes open synchronously. The sexual system facilities geitonogamy and xenogamy. The flower visitors include bees, ants, thrips and flies. Bees and flies effect geitinogamy and xenogamy while ants and thrips effect only geitonogamy. By predominantly cross pollinating, the plant throws open the possibilities for geitonogamy. In line with this, the plant might be selectively eliminating the growing offspring, especially those originated from geitonogamy, in order to allocate resources for the xenogamous (Solomon *et al*., 2002). Fruits are set in winter when the shrub is leaf less or it may produce several crops during the year if soil moisture is not limiting and temperature is significantly high. Each inflorescence yields a bunch of approximately 10 or more ovoid fruits measuring approximately 2.0-3.5 cm long and 1.5-2.5 cm wide. It has nearly 422 fruits kg^{-1} and the 100 seed weight is of around 63.01 g (Kumar *et al*., 2004).

10.2 Seed and Oil yield

There is enough variability in fruit and seed characters of *Jatropha curcas* (Kumar *et al*. 2004). The range of oil and seed yields can be attributed to a variety of factors such as climate, intensity of management, site quality, size of plant and genetics (Raina and Gaikwad, 1987). *Jatropha curcas* seed contain 40-50 per cent of a semi-drying oil and seven per cent moisture (Chaudhary and Joshi, 1999; Patil *et al*., 2003). The seed yield, seed cell to kernel ratio and kernel oil content of *Jatropha curcas, J. gossypifolia* and *J. podarica* were compared. The seed yields are 5 t, 500 kg and 400 kg per ha annually for the three species, respectively, with seed cell to kernel ratio of 44:56, 35:65 and 25:75. The kernel oil contents are 47-52 per cent, 29 per cent and 54 per cent, respectively (Raina and Gaikwad, 1987).

10.3 Genetic improvement

The work of genetic improvement of this species is still in budding phase. Recently National Oilseeds and Vegetable Oils Development Board, Ministry of Agriculture, Govt. of India approved National Network on Integrated Development of Jatropha and Karanja for 25 organisations to collect the germplasm and evaluate the germplasm for morphological, seed and fruit characters, seed yield and oil quality. It is expected that more than 10,000 germplasm of *Jatrpha curcas* will be collected and evaluated for various economic characters. Some good varieties are expected to be released.

At National Research Centre for Agroforestry, Jhansi exploration trips were carried out during November 2003 and 2004 to collect the germplasm of *Jatropha curcas* from various parts of country. A total of 86 accessions were collected from different parts of Madhya Pradesh (Ratlam and Jhabua), Gujarat (Dahod, Panchmahal, Godhara and Palanpur), Bundelkhand region of Uttar Pradesh (Lalitpur and Madawara region) and from areas of Rajasthan (Udaipur, Rajsamand, Pali, Bhilwara and Chittorgarh). A wide range of variability for fruit and seed characters have been observed. The fruit length varied from 2.10 to 3.31 cm (2.76+ 0.017cm), fruit width from 1.20 to 2.50 cm, (2.14+ 0.013cm)), seeds from 1 to 3 seeds/fruit ((2.83+ 0.020), seed length from 14.0 to 21.0 mm ((17.93+ 0.028mm), seed width from 8.0 to 14.50 mm (10.91+ 0.012mm), seed thickness from 4.5 to 12.0 mm(8.61+ 0.017mm) and average 100 seed weight from 50.58g to 82.10g(66.67+ 0.838g). The initial growth of seedlings in the nursery was fast and seedlings attained the height of 44 to 58 cm and collar diameter of 0.85 -1.20 cm after 75 days of sowing. The initial growth of plants in the field is very fast as they attained the height from 90 to 142 cm, collar diameter from 3.8 to 4.9 cm and branches from 1 to 8 after four months of plantation. Germplasm from Ratlam is proving their superiority over others at initial stage of growth (Kumar *et al.*, 2004).

METHODOLOGY AND STRATEGY FOR IMPROVEMENT OF MPTS FOR AGROFORESTRY

The basic principles and practices of tree improvement are same as they apply for industrial trees. Superior phenotypes (plus trees) are selected from the best populations and their breeding potential should be evaluated in clonal or progeny tests on typical sites with typical managements. For immediate gain in 6-8 years (short cycle) following methodology may be successful:

1. Identify only 2-3 tree species for each zone according to the need of adaptability, accessibility and economic value of the species.
2. Identify areas where species in question are naturalised or native.
3. Identify areas where growth of stands of a species to be improved are good. These areas should be marked (seed producing zones) to fulfil immediate needs of quality seeds. Seedlings raised by these quality seeds would be expected to be better than the trees raised by seeds of unknown sources. Try to select the seed producing zones from areas where plantation would be done. Expected gain would be 10-12 per cent for production traits and 3-5 per cent for quality traits.
4. Whole of the area, where natural stands of a species is present, should be divided into 20 - 30 groups based on geographical, soil and climatic factors.
5. During exploration tour, based on eye judgement and measurements, best trees at a particular site should be identified (Plus trees) among selected trees (candidate trees) of the same age group within surrounding of 20 m. Number of plus trees should be more in the areas where performance of a species in question is good and minimum selection should be done from the poor areas.

6. Permanent spots should be marked in records so that we can use same trees to collect the seeds to raise nursery after confirmation of genetic potential through progeny tests. Genetic gain expected in the selected population would be 15-20 per cent.
7. Elite trees will be identified based on the results of Elimination Trial (ET), Growth Trials (GT) and Provenances Trials (PT). Rest of the trees should be removed from the fields. This area can also be used as a seed orchard for supplying of quality seeds. It can be said first generation seed orchard.
8. Clones or seed progenies of best trees should be used to establish 1.5 generation seed orchards to supply good quality of seeds. By using these seeds, improvement would be quite high.

Team of experts of All India Coordinated Research Project on Agroforestry have identified agroforestry woody species and organisations for intensive tree improvement programme. These are presented in table 36.10.

LIMITATIONS

1. Choice of tree species for a particular zone is still difficult task
2. Lack of adequate knowledge for long duration breeding trials of tree species
3. Lack of adequate skilled personnel
4. Lack of Co-ordinated efforts at national level
5. Long duration breeding cycle of tree species
6. Most of the species are open pollinated required many years to get genetically stabilized populations
7. Vegetative propagation methods are not available for most of the tree species
8. Reproductive biology of tree species is not studied thoroughly
9. There is no seed certification standards
10. Absence of seed producing and certification agencies

Table 36.1. Range of variation in *Leucaena* germplasm at IGFRI, Jhansi, India

Character	Range
Leaf length (cm)	5.3-17.9
Pinna number	5.1-42.3
Middle pinna length (cm)	3.8-7.7
Leaflet number per pinna	6.0-104.5
Leaflet length (cm)	0.4-3.9
Leaflet width (cm)	0.1-2.8
Seed number per pod	7-28
100-seed weight (g)	1.2-12.5
Plant height (cm) 3yr. old	60-641
Collar dia. (cm) 3yr. old	2.0-9.1
Mimosine in leaves (%DM)	1.2-5.4
Fodder yield/tree(DM kg) 3yr. old	0.07-4.1
Fuelwood yield/tree(DM kg) 3yr. old	0.15-24.5
Days to first flowering	91-982

Source: Gupta *1988*

Table 36.2. Hybrid vigour in *Leucaena leucocephala* for forage and fuelwood yields

	No. of F1s	Range of heterosis[a]		Average heterosis[a]		Mean performance[b]	
		Forage yield/pl.	Fuelwood yield/pl.	Forage yield/ pl	Fuel yield/pl .	Forage yield/pl .	Fuel yield/pl.
Trial-1	15	-81.2 to 67.5	-60.9 to 86.8	-16.3	-6.0	-53	-45
Trial-2	10	-33.9 to 54.8	-32.5 to 70.2	-2.9	-0.5	-26	-27
Trial-3	24	-68.1 to 85.0	-69.1 to 71.0	-18.3	-31.1	-47	-60

Source: Gupta and Patil, 1988 a= relative to better parent, b= relative to best parent

Table 36.3. Heterosis (%) over mid (MP) and best parent (BP) in interspecific hybrids of *Leucaena* species

Crosses	Collar diameter		DBH		Plant height	
	Mid parent	Better parent	Mid parent	Better parent	Mid parent	Better parent
L. lanceolata x *L. leucocephala*	5.9	-20.6	24.1	-9.5	4.8	-8.3
L. macrophylla x *L. leucocephala*	7.8	-23.5	0.0	25.2	3.8	-10.6
L. macrophylla x *L. lanceolata*	13.4	13.4	20.0	15.6	8.1	-3.6
L. shannoni x *L. leucocephala*	25.6	-10.0	13.3	-20.7	-21.1	-3.6
L. leucocephala x *L. pulverulenta*	61.9*	41.7*	128.1*	99.1*	-4.1	-6.5
L. leucocephala x *L. diversifolia*	94.3*	51.5*	86.1*	59.5*	23.8	8.3

Source: Gupta, 1990 *= significant at 5%, DBH= Dia. at breast ht.

Table 36.4. Mean and range of variation in seventy selected plants of *P. cineraria* in natural stands (10 each from 7 districts)

Trait	Range	Mean±SE	SD	CV
Height(m)	4.6-30.4	11.9± 0.75	6.28	52.7
Forking height(m)	1.2-04.6	3.12± 0.10	0.84	26.9
DBH(cm)	14.3-149.5	44.4± 3.4	28.8	64.9
Collar dia.(cm)	15.9-171.8	51.7± 4.0	33.8	65.7
No. of branches	2-9	4.5± 0.21	1.81	39.9
Canopy dia.(m)	1.20-9.72	4.6± 0.54	2.20	48.1
Pod length(cm)	6.4-18.0	12.2± 0.26	2.49	20.4
No. of seeds/pod	3-19	9.6± 0.39	3.33	34.8

Source: Solanki, 1996

Table 36.5. Percent fruit setting in controlled vs open pollinated

No. of trees	Selfed			Open pollinated		
	No. of flowers bagged	No. of fruits developed	Fruit setting (%)	No. of flowers bagged	No. of fruits developed	Fruit setting (%)
1	1093 ± 2.68	6 ± 0.10	0.55	545 ±5.16	31 ± 1.13	5.69
2	542 ± 1.57	16 ± 0.32	2.93	863 ± 4.60	43 ± 0.40	4.98
3	438 ± 2.15	9 ± 0.28	2.32	840 ± 2.32	36 ± 0.45	4.28
4	587 ± 5.19	36 ± 0.85	6.13	457 ± 4.30	23 ± 0.20	5.03
5	378 ± 4.60	30 ± 1.77	6.84	538 ± 4.14	28 ± 0.32	5.20
6	794 ± 4.20	34 ± 2.33	4.57	810 ± 5.20	38 ± 0.35	4.69
7	429 ± 5.10	16 ± 0.47	3.31	404 ± 4.99	23 ± 0.30	5.69
8	604 ± 5.33	11 ± 0.18	1.92	1077 ± 4.01	36 ± 0.52	3.34
9	636 ± 6.21	18 ± 0.50	2.83	905 ± 6.18	21 ± 0.22	2.32
10	428 ± 2.27	6 ± 0.32	1.40	835 ± 6.02	23 ± 0.40	2.75
Mean			**3.06**		**4.15**	

Source: Gupta *et al.* 1996a

Table 36.6. Clones of Populus developed in India which have been planted in the breeding orchard

Clone	Source	Percentage	Sex
L-200/84	State Forest Department, Haldwani	G48 x G3	Female
L-34/82	State Forest Department, Haldwani	G48 x G3	Female
Udai	WIMCO Seedlings Ltd., Rudrapur	G48 x G3	?
# 36	Univ. of Horticulture & Forestry, Solan	Open-pollinated selection U.S.A.	?
U.D.-88	Univ. of Horticulture & Forestry, Solan	Open-pollinated selection U.S.A.	?
P.D.-1	Univ. of Horticulture & Forestry, Solan	Open-pollinated selection U.S.A.	?
100-7	Univ. of Horticulture & Forestry, Solan	Open-pollinated selection U.S.A.	?
103	Univ. of Horticulture & Forestry, Solan	Open-pollinated selection U.S.A.	?
U.D.-55	Univ. of Horticulture & Forestry, Solan	Open-pollinated selection U.S.A.	?
98	Univ. of Horticulture & Forestry, Solan	Open-pollinated selection U.S.A.	?

Note : "?" indicates that sex is not known at this stage. These clones are included on the basis of outstanding performance in the field.

Source: Singh *et al.*, 2001

Table 36.7. Seeds of provenances and plus trees of *D. sissoo* collected by different organisations

Name of the organisation	No. of collections	Best plus trees/ provenances
CCSHAU, Hisar	106 Plus trees	PT- 9 (Tanakpur) PT-280, PT-12 (Haldwani) PT-283 (Gonda) PT- 84 (Dangarhi, Nepal)
NRCAF, Jhansi	32 plus trees collected from Bundelkhand region	PT-6, PT-2 (Mauranipur) MAI of best tree 3.08m for ht. and 5.14cm for cd
G.B.P UA&T, Pantnagar	170 from northern states	-
TFRI, Jabalpur	10 provenances	Raipur provenance
FRI, Dehradun	70 Plus trees	-
Forest Research Lab., Kanpur	20 Provenances	Mauranipur provenance

Table 36.8. Seed of provenances and plus trees *of A. nilotica* collected by different organisations

Name of the organisation	No. of collections	Best plus trees/ provenances
CCSHAU, Hisar	21 plus trees	North west and central India are better than southern India provenances
JNKVV, Jabalpur	10 provenances	not known
PAU, Ludhiana	39 provenances	"
FRI, Dehradun	20 provenances	"
AFRI, Jodhpur	28 provenances	Shivpuri, Hastinapur provenances
IFGT, Coimbatore	18 provenences	Hissar provenance
State Forest Research Laboratory, Kanpur	9 provenances	Banaskantha, Akola, Adilabad
NRCAF, Jhansi	29 provenance and 31 Plus trees	Results are awaited

Table 36.9. Seeds of provenances and plus trees of *A. lebbek* collected by different organisations

Name of the organisations	No. of collections	Best plus tree/ provenance
CCSHAU, Hisar	12 plus trees	Jammu provenance
IFG & TB, Coimbatore	13 provenance	Shimoga prov. from Karnataka Salem provenance from Tamil Nadu
NRCAF, Jhansi	147 provenances	Results are awaited

Table 36.10 Agroforestry woody species and organisations for intensive tree improvement programme

SNo.	Species	Centre
1.	*Populus deltoides/ P. ciliata*	Hissar, Ludhiana, Pantnagar, Solan
2.	*Alnus nepalensis*	N.E. Hills, Solan
3.	*Ailanthus excelsa*	Bhubaneshwar, Mettupalayam, Fatehpur, Sekhawati, Avikanagar, Nagpur, Dehradun, Ranchi
4.	*Azadirachtica indica*	Hissar, Ludhiana, Jhansi, Mettupalayam, S.K. Nagar, Nagpur, Hyderabad
5.	*Dalbergia sissoo*	Pantnagar, Ludhiana, Hissar, Jhansi, Pusa, Jammu, Jabalpur
6.	*Prosopis cineraria*	Hissar, Jodhpur, Fatehpur, S.K. Nagar
7.	*Anthocephalus cadamba*	Jorhat, Pantnagar, Jhargram
8.	*Acacia nilotica*	S.K. Nagar, Ludhiana, Nagpur, Hissar, Rahuri, Parbhani, Makhdoom, Jabalpur
9.	*Tamarindus indica*	Mettupalayam, Hyderabad, Rahuri, Parbhani
10.	*Embilica officinalis*	S.K. Nagar, Jhansi, Faizabad, Raipur, Kanpur, Ranchi

FUTURE STRATEGY

1. National network should be established for exploration, collection, multiplication, distribution and conservation of germplasm of important MPTs for agroforestry species
2. In first phase, genetic improvement work should be started at national level in *Populus deltoidis/P. ciliata, Alnus nepalensis, Ailanthus excelsa, Azadirachta indica, Dalbergia sissoo, Prosopis cineraria, Anthocephalus cadamba, Acacia nilotca, Tamarindus indica, Embilica officinalis, Casuarina equisetifoia,* Eucalyptus spp. and bamboo spp.
3. All the germplasm should be conserved *in situ* as well *as ex situ*
4. Mode of pollination and reproductive biology should be studied of important tree species
5. Simple and feasible techniques should be standardised for vegetative propagation
6. Seed production zones should be identified to fulfil immediate need of the farmers
7. Clonal seed orchards should be established
8. Seed certification system should be started in tree species also
9. Responsibility of multiplication of quality seeds/ propagules of MPTs should be given to NSC and SSC under the guidance of experts.
10. Data base of all the information related to tree improvement of agroforestry species should be prepared.

ACKNOWLEDGMENT

Authors are grateful to Director, National Research Centre for Agroforestry, Jhansi for providing good library facilities to consult literature to prepare this document as well as giving the permission to send the lecture text.

REFERENCES

Ahlawat, S.P., Gupta, V.K. and Kumar, R.V. (2003). Phenology and reproductive biology of *Acacia nilotica*. Abstract, National Symposium on Agroforestry in 21st century held from February 11-14, 2003 at Ludhiana, India. pp.69.

Anonymous (1959). *Jatropha curcas*. Wealth of India, Raw Materials H.K. Vol.V, CSIR, New Delhi, pp.293-298.

Arya, H.C. and Shekhawat, N.S. (1986). Clonal multiplication of tree species in the Thar desert through tissue culture. *Forrest Ecology & Management* 16:*201-208*.

Arya, S. and Tomer, R. (1989). Study of rooting response in juvenile stem cuttings of *Prosopis cineraria* (L). *Macbride, Jeevanti* 7: *65-68*.

Bangarwa, S. Kulvir. (1996). *Sissoo Breeding*. Systematic Printers, Hissar, Haryana, India.

Bangarwa, S.Kulvir. (2002). Plus tree selection and progeny testing for establishment of first generation seed orchard in *Dalbergia sissoo* Roxb. *Indian J. Agroforestry 4: 122-131.*

Batt, R.S. (1995). Field selection and raising of clonal bank of *Acacia nilotica* (L.) ssp *indica. M.Sc. Thesis*, Punjab Agricultural University, Ludhiana, India.

Bisla, S.S., Dhillon, R.S., Chillar, S., Hooda, M.S. (2002). Phenology and breeding system of Neem (*Azadirachta indica* A. Juss). In proceedings of World Neem Conference, November 27-30, 2002, Mumbai, India. pp.6

Bray, R.A. (1983). High yield from *Leucaena* hybrids. *Leucaena Research Report 4:1-2.*

Brewbaker, J.L. (1987). Leucaena a multipurpose tree genus for tropical agroforestry. In: *Agroforestry a Decade of Development.* (Eds. H.A. Steppler and P.K.R. Nair). ICFRI, Nairobi Kenya.pp289-323.

Brewbaker, J.L. and Hutton, E.M. (1979). Leucaena-A versatile tropical legume. In : *New Agriculture Crop* (Ed. G.A. Ritchie). AAAS selection symposium westview Press, Colorado, USA. pp 207-259.

Brawbaker, J.L., Sorenson, C.T. and Wheeler, R.W. (1989). New tree crops from interspecific Leucaena hybrid. In : *Leucaena Psyllids Problems and Management* (Eds. B. Napombeth & K.G. MacDicken). Funny Publishing Limited Partnership, Bangkok, Thailand. pp. 99-104.

Chaudhary, D.C. and Joshi, D.N. (1999). *Jatorpha curcas* a mutipurpose species for economic prosperity and wasteland development. *Advances in Plant Sciences Research in India* 9:35-39.

Chuntanaparb, L.K.G. and Mac Dicken. (1991). Tree selection and improvement for agroforestry. In: *Biophysical Research for Asian Agroforestry* (Eds. M.E. Avery, M.G. R. Cannel and C.K. Ong). Winrock International, USA and South Asia Books, USA.

Deb Roy, R. (1988). Forage production through silvopasture system. In : *Pasture and Forage Crop Research A State of Knowledge Report* (Ed. Punjab Singh). Range Management Society, Jhansi. pp. 305-312.

Dhillon, G.P.S. and Khajuria, H.N. (1994). Floral biology of *Acacia nilotica* ssp. *indica* under sub-tropical conditions. *J. Res. Punjab Agric. Univ.* 31: *421-426.*

Gautam, V.K., Nada, K. and Gupta, S.C.. (1993). Development of shoots and roots in anther derived callus of *Azadirachta indica* A.Juss- A medicinal tree. *Plant Cell Tissue and Organculture 34: 13-18*

Gill, B.S., Singhal, V.K. and Maninder, K. (1993). Studies on genetic systems of N.W. Indian Neem. Abstract. World Neem Conference, February 24-28, 1993, Bangalore, India.

Gupta, V.K. (1988). Breeding research on subabul (*Leucaena leucocephala*) in India. In: *Pasture and Forage Research: A state of knowledge report* (Ed. P. Singh) . Range Management Society of India, Jhansi, India.

Gupta, V.K.. (1990). Genetics and breeding of *Leucaena leucocephala* (Lam) de Wit. In: *Multipurpose Tree Species for Agroforestry Systems* (Ed. P.S.Pathak)

Gupta,V.K. and B.D. Patil. (1986). Heterosis and combining ability estimates in *Leucaena leucocephala* (Lam de Wit). *Indian J Tree Sci. 5: 67-73.*

Gupta, V.K. and Patil, B.D. (1988). Genetic variability and path analysis in subabul (*Leucaena leucocephala*). *Indian J. Agric. Sci.* 58:*483-484.*

Gupta, V.K. and Kumar, R.V. (2001). Genetics and breeding of *Azadirachta indica, Dalbergia sissoo, Acacia nilotica* and *Anogeissus* spp. Annual Report, NRCAF, Jhnasi, India.

Gupta, V.K., Kumar, R.V., Ahlawat, S.P. and Datta, A. (2003). Performance of plus tree progenies of sisham (*Dalbergia sissoo* Roxb.). Abstract, National Symposium on Agroforestry in 21st century held from February 11-14, 2003 at Ludhiana, India. pp.68.

Gupta, V.K., Solanki, K.R., Gupta, R. and Datta, A. (1996a). Reproductive biology of Neem (*Azadirachta indica* A. Juss). *Range Mgmt. & Agroforestry17(2) : 187-192.*

Gupta, V.K., Solanki, K.R., Gupta, R. and Dwivedi, P. (1996b). Provenances trials and reproductive biology of neem (*Azadirachta indica*) in Central India. *In: Tree Improvement for Sustainable Tropical Forestry* (Eds. M.J. Dieters, A.C. Matheson, D.G. Nickles and C.E. Hardwood). Proc. QFRI-IUFRO Conference at Colaundra, Queensland, Australia, Oct. 27-Nov.2. pp. 81-82.

Gurumurthi, K. and Bhandari, H.C.S. (1988). Induction of rooting in cladode cuttings of *Casuarina equisetifolia. Current Science* 57 : 958-959.

Gurumurthi, K., Jayachandran, C.K. and Thirunavankkarasu, M. (1994). Rooting trial in branch cuttings of *Acacia nilotica* (L) Del. *Indian J. Forestry* 17: 112-118.

Gurumurthi, K. and Rawat, P.S. (1992). Rooting of *Casuarina equisetifolia* cuttings as influenced by season and auxin. *NFTRR* 10: *137-140.*

Harm, H. (1940). Miliaceae In: *Die NaturlichenPflanzenfamilien* (Eds. A.. Engier and Prant). : 1-172.

Hooker, J.P. (1973). *Flora of British India. Vol. I.* Periodical Expert, New Delhi.pp.540-544.

Hu, Ta Wei. (1983). Leucaena research in Tiwan. In: *Leucaena Research in the Asia-Pacific Region*. IDRC, Ottawa, Canada.

Husain, A.M.M. and Pannuswamy, P.K. (1980). Propagation of *C. junghuniana* by planting shoot and root sucker. *Indian Forester* 106 : 298-299.

Jambulingam, R. (1990). Recent developments in research on *Casuarina* in Tamil Nadu. Proceedings of the Second International Casuarina Workshop, Cairo, Egypt. January 15-20. : 45-54.

Jatasara, D.S. (1982). Germplasm collection of *Prosopis cineraria* (L.) MacBride from India desert. *Forage Research* 8 :1-9

Jatasara, D.S. and Paroda, R.S. (1983). Evaluation of *Prosopis cineraria* germplasm from Thar desert in situ.Transaction of the Indian Society of Desert Technology and University Centre of Desert Studies 8: 138-142

Jindal, S.K. and Satya Vir. (1997). Phenology and breeding system and seed production in Neem (*Azadirachta indica* A. Juss). *Van Anusandhan* 9: *17-19.*

Kan, W.H. and Hu, J.W. (1983). Vegetative propagation of *Casuarina* by leafy twig cuttings under mist spray. Taiwan Forestry Research Institute Bulletin, No.390, pp.8.

Kaushik, N. (2002). Prospecting chemical diversity of Neem (*Azadirachta indica*) trees growing in India for value addition. In proceedings of World Neem Conference, November 27-30, 2002, Mumbai, India. pp. 41-43.

Khajuria, H.N. (1995). Improvement of *Acacia nilotica* (Willd) for ex Del. *Annals of Biology:* 12(2) : 302-305.

Kumar, R.V., Ahlawat, S.P., Gupta, V.K., Datta, A. and Handa, A.K. (2004). Variability in fruit and seed characters of Jatropha curcus. In Proceeding of International Conference on " Biofuels: perspective and prospects" Pub. Winrock International, New Delhi, Indiapp.247-248.

Lal, Piare. (2001). Registration of clones and certification of clonal planting stock. *Indian Forester* 127

Mathur, I and Chandra, N. (1984). Induced regeneration in stem explants of *Acacia nilotica. Current Science*: 52(18) : 882-883.

Mathur, R.S. and Sharma, K.K. (1983). Poplar in India. *Indian Forester* 109:591-631.

Mishra, R.N. (1996). Neem improvement research at Arid Research Institute. Jodhpur. *Indian Forester* 121: 997-1002.

Narayan, P. and V.S. Jaiswal. (1985). Plantlet regeneration from leaflet callus of *Azadirachta indica* A. Juss. *J. Tree Sci.* 4: *65-68*

Patel, N.L. and Singh, S.P. (2000). Effect of different tree species on site amelioration. *Indian J. Forestry* 23 : *192-196.*

Pathak, P.S. (1988). Ecology and potential of subabul in agroforestry systems. In: *Pasture and Forage Crop Research- A state of knowledge report* (Ed. P.Singh). Range Management Society of India, Jhansi.

Pathak, P.S. and Gupta, V.K. (1994). Evaluation of new selections of *Leucaena leucocephala* on dry degraded lands . In: *Agroforestry System Degraded Lands* (Eds. P.Singh, P.S. Pathak and M.M. Roy). Oxford & IBH Publishing Co. Pvt. Ltd. New Delhi, India.

Patil V, Singh, K and Bhandari, T.N. (2003). Experience and observations about Jatorpha plantations (1986 - 2003). Agroforestry and Wasteland Development Foundation, Nasik, Maharastra, India. : 2-10.

Pinyopusarerk, K. and Heuse, A.N. (1993). *Casuarina* : an annotated bibliography. International Centre for Research in Agroforestry (ICRAF), Nairobi, 298p.

Prasad, N.S. (1998). Genetic control of growth and form in early age tests of *Casuarina equesetifolia* in Andhra Pradesh, India. *Forest Ecology and Management* 110 : 49-58.

Raina, A.K. and Gaikwad, B.R. (1987). Chemobatany of Jatropha species in India and further characterization of " Curcus oil". Journal of the Oil Technologists Association of India. 19(4):81-85

Raj, D. (1991). Air layering in *Acacia nilotica. Indian Forester* 117 : *156.*

Rangaswamy and Promila. (1972). Morphogenesis of the adult embroy of *Azadirachta indica* A. Juss. *Pflanzen Physiol. 67: 377-379*

Rao, H.S. (1953). Vegetative propagation and forest tree improvement. *Indian Forester* 79: *176-183*

Rawat, T.S., Sharma, H.C. and Pundir, J.S.P. (1982). Air layering a success in propagation of Khejri (*Prosopis cineraria*). *Udanika* 5:43- 44.

Relwani, L.L. (1983). Kababul (*Leucaena leucocephala*) - A renewable source of fuel, timber and pulp. In : *Leucaena leucocephala* in India (Eds. R.N. Kaul, M.G. Gogate and N.K. Mathur) Vangaurd Pree, Dehradun, India.

Sanyal, M., Banergy, M. and P.C. Dutta, (1981). *In vitro* hormone induced induced chemical and histological differentiation of stem callus of *Azadirachta indica* A. Juss. *Indian J. Expt. Bio.* 19: *1067 - 1068*

Shanmugavelu, K.G. (1967). A note on the air layering of Eugenia (*Eugenia jambalana* Lam) and neem (*Azadirachta indica* A. Juss). *South Indian Hot.* 15: *70-71.*

Shivasamy, M and Karivaratharaju, T.V. (1993). Seed quality variation in relation to ecological variation in neem Abst. World Neem Conference, 24th - 28th February, 1993, Bangalore, India. pp. 62

Sidhu, D.S. (1996). Economic returns of cultivation of genetically superior Poplar clones in Punjab (India). Proceedings of 20th session of the International Poplar Commission. Vol. 1 : 451-459.

Singh, N.B., Kumar, Dinesh, Rawat, G.S. and Gupta, R.K. (2001). Establishment of breeding orchard of *Populus deltoides* BARTR. *Indian Forester* 127: 3-8.

Singh, P. (1986). Silvipastoral systems for wastelands management analysis. IGFRI, Jhansi, Silver Jublee.

Solanki, (1996). Recent advances in genetic improvement of *Prosopis ceneraria* in India. In: *Tree Improvement for Sustainable Tropical Forestry* (Eds. M.J. Dieters, A.C. Matheson, D.G. Nickles and C.E. Hardwood). Proc. QFRI-IUFRO Conference at Colaundra, Queensland, Australia, Oct. 27-Nov.2. : 81-82.

Solanki, K,R., N.L. Kacker and S.K. Jindal. (1984). Propagation in *Prosopis cineraria* (L.) Mac Bride by air layering. *Current Science* 53: *1166 – 1167*

Solanki, K.R., Kacker, N.L. and. Jindal, S.K. (1986). Air layering in *Prosopis cineraria* (L.) Mac Bride. *Indian Forester* 112: *202-207.*

Solanki, K.R., Kackar, N.L.,Singh, Manjit and Jindal, S.K. (1991). Collection and propagation of desirable genotypes of *Prosopis cineraria* (L.) Druce, *Indian Forester (Special Issue).*

Soloman, A.J., Raju and Ezradanam, V. (2002). Pollination ecology and fruit behavior in a monoecious species Jatropha curcas L. (Euphorbiaceae). Department of Environmental Science, Andhra University, Visakhapatnam, india. *Current Science* 83(11): 1395-1398

Somasundaram, T.R. and Jagadeesan, S.S. (1977). Propagation of *C. equisetifolia* (Forst) by planting shoots. *Indian Forester,* 103 *: 735-738.*

Tewari, S.K., Shubhanjana and Pandey, S.B.S. (2002). Reproductive biology of *Dalbergia sissoo* (Sisham). *Indian Forester* 128 *: 336-340.*

Toky, O.P. (1995). Genetic diversity in arid and semiarid Indian trees and its role in sustainable agriculture. Paper presented on National Seminar on "Community forestry: Biodiversity of forest species". May 2-3. Indian Society of Tree Scientists.

Venkateshwarlu, B. (1996). Tissue culture technology for propagation of elite neem trees. *SAIC, News Letter, March,1996* : 1.

Venkatesh, C.S. and Sharma, V.K. (1977). Rapid growth rate and higher yield potential of Heterotic Eucalyptus species Hybrids FRI-4 and FRI-5. *Indian Forester* 103 (12) : 795-802.

Verma, (1993). Studies on growth and performance of *Populus deltoides* in Punjab. *M.Sc. Thesis.* Punjab Agricultural University, Ludhiana. pp.154.

37

Agroforestry for Sustainability — Soil and Water Conservation Strategy

SATYANANDA PATRA[1] AND CALVIN FOX[2]

[1] Center for Drought Management, Orissa University of Agriculture and Technology, Bhubaneswar, Orissa

[2] Asian Rural Life Development Foundation (India) Raipada, Kalinga, Phulbani, Orissa

ABSTRACT

Contour hedgerow technology has different names in different regions. Sloping Agriculture Land Technology (SALT) has alley farming with hedgerow barrier technology. It is basically an agroforestry system in which nitrogen - fixing trees or shrubs are planted very closely in rows and cash crops grown in between the alleys. The SALT has been considered as an alernative to shifting cultivation. Forest dwellers are mostly tribals and shifting cultivation have been their practice for food production. The slash and burn has caused enormous damage to the forest composition. The forests on the hills and on the mountains have suffered a lot due to "Podhu" cultivation. The similar situations have been observed also in most of the Asian country. The hilly area has been considered under the hedgerow system for soil and water conservation. The introduction of the agroforestry systems in the farmers fields and in the wastelands have brought potential forest farming. The young leaves of some hedge row plants provide good fodder for livestock. Hedgerows can also be managed to provide fuelwood, if necessary. The hedgerows and their trimming provide mechanical and biological barriers and help minimise soil erosion by reducing surface runoff velocities, leading to higher deposition of soil sediments and living source of nutrient cycling. The food and cash production has been observed to be the most essential components in four types of SALT models.

INTRODUCTION

Soil is the result of the gradual weathering of rocks, and minerals. Soil formation is a slow process, it takes place at the rate of 2.5 centimeters per century. Topsoil is rich and fertile because of its organic matter content. Plants and animals die, decay, disintegrate, and are incorporated in the soil, making the soil fertile and capable of supporting the growth of food crops. Soil erosion has been well recognized as direct cause of declining soil fertility and land productivity. Many technologies have been developed and promoted to prevent soil and water conservation, of which terracing has been advocated as the main soil conservation approach for sloping agricultural lands. Bench terraces can greatly reduce erosion, increase infiltration, and make the land more easily managed. Many traditional agricultural systems have developed extremely effective terraces that ensure the long-term stability of the land. However, terrace promoting as the panacea to all erosion problems on sloping agricultural land may be questionable. There are many reasons why this is not so. Integrated watershed management is another mountain stabilizing technique. However, due to various reasons, the problem of soil erosion has been getting worse in the region. Soil erosion and resultant soil fertility decline have threatened the food security of the subsistence farming in the region because food crops production on sloping lands leads to a serious and often drastic decline in the productive potential of the land. Although many farmers are quite aware of this situation, they persist in farming sloping land in order to meet the family' s subsistence food requirements. Soil conservation practices are seldom used. In contrary, many farmers practice up and down farming, which leads to severe soil erosion, and few of them practice recommended contour farming which can help reduce soil loss. Therefore it is important to analyse why farmers frequently reject soil conservation practice even though they are aware that adoption of such practices is necessary to protect future productivity of land resources.

Sloping Agricultural Land Technology (SALT) is a soil conservation-oriented farming system and it is being increasingly studied and tested in many countries of the world. The SALT system can adequately conserve soil and water, enrich soil, and facilitate farming without terracing. In the management of the sloping agricultural lands, contour hedgerows of nitrogen fixing plants have proved effective in improving soil fertility and preventing the farmland from further degradation. The contour hedgerows of nitrogen fixing plants are thickly planted along the contour lines. They not only act as the effective barrier of soil erosion, but also can considerably improve the soil fertility through the bioactivities of the symbiosis and by addition of large amount of fresh green manure or mulch materials to soil. The foliage of vigorous perennial legumes may accumulate 100-600 kg of nitrogen per ha. These pruning are nitrogen rich and also contain sufficient other nutrients, such as potassium and phosphorus. It is reported that dry leaves of *Gliricidia sepium* contain 3-4 per cent of nitrogen, 3.5 per cent of potassium and that litters of *Erythrina poeppigian* contain 2.3-3.6 per cent of nitrogen and its fresh leaves contain 4.1-4.9 per cent of nitrogen (Glover, 1986). (Table 37.1) The foliage of *Leucaena* contains about 4 percent nitrogen. The contour hedgerows not only provide nutrients for the growth of the companion crops in the alleyways but can also improve supply of phosphorus and potassium through their deep root systems from deep layers of the soil. Though the biomass production of the hedgerows is different in different agro-ecological zones, the testing and demonstration of the contour hedgerow systems in the HKH region indicates that it can provide annually 2012 tonnes fresh materials per ha. The very low nutrient soils with a content of soil nitrogen of 0.061 per cent, and organic matter of 0.66 per cent has been greatly improved.

THE SALT SYSTEM

SALT is a package technology on soil conservation and food production, targeting different soil conservation measures in just one setting. Basically, SALT is a method of growing field and permanent crops in 3 m to 5 m wide bands between contoured rows of nitrogen fixing trees. The nitrogen fixing trees are thickly planted in double rows to make hedgerows. When a hedge is 1.5 to 2 m tall, it is cut down to about 75 cm and the cuttings (tops) are placed in alley-ways to serve as organic fertilizers.

SALT : An Agroforestry Scheme

SALT is a diversified farming system which can be considered agroforestry since rows of permanent shrubs like Mango, Guava, citrus and other fruit trees are dispersed throughout the farm plot. The strips not occupied by permanent crops, however, are planted alternately to cereals (corn, upland rice, sorghum, etc.) or other crops (sweet potato, melon, pineapple, castor bean, etc.) and legumes (soybean, mung bean, peanut, etc.) .This cyclic cropping provides the farmer some harvest throughout the year. SALT also includes planting of trees for timber and firewood on surrounding boundaries. Examples of tree species for "boundary forestry" in SALT are mahoganies, casuarinas, sesbanias, cashewnuts, etc.

History of SALT

SALT was developed on a marginal site in Kinuskusan, Bansalan, Davao del Sur by the Mindanao Baptist Rural Life Center (MBRLC). In 1971, MBRLC started to employ contour terraces in its sloping areas. Dialogues with local upland farmers acquainted the Center with farm problems and needs which gave the Center the impetus to work out a relevant and appropriate upland farming system.

From testing different intercropping schemes and observing ipil-ipil-based farming systems in Hawaii and at the Center, the SALT was finally verified and completed in 1978. While it was still in the developing stage, the following guideleines were considered essential. The system must :

- adequately control soil erosion.
- help restore soil structure and fertility.
- be efficient in food crop production.
- be applicable to at least 50 percent of hillside farms.
- be easily duplicated by upland farmers with the use of local resources and preferably without taking loans.
- be culturally acceptable.
- have the small farmer as the focus and food production as the top priority.
- require minimal labour, and
- be economically feasible.

Various Forms of SALT

There are several forms of SALT, and a farmer may wish to use the SALT system in several variations.

SALT-1 (Simple Livelihood Land Technology) is a conventional SALT project where hedgerows of different nitrogen fixing trees and shrubs are established on the contour lines. The manure from the animals is utilized as fertilizer both for agricultural crops and the forage crops. There are three variations of SALT that have been developed at the Mindanao Baptist Rural Life Center (MBRLC). These are as follows:

SALT-2 (Simple Agro-Livestock Land Technology) is a small livestock-based agroforestry where preferably dairy goats are reared with a land use of 40 per cent for agriculture, 20 per cent for forestry and 40 per cent for livestock.

SALT - 3 (Sustainable Agroforest Land Technology) is a cropping system in which a farmer can incorporate food production, fruit production, and forest trees can be marketed. The farmer first develops a conventional SALT project to produce food for his family and possibly food for livestock. On another area of land he can plant fruit trees such as rambutan, durian, and lanzones between the contour lines. The plants in the hedgerows will be cut and piled around the fruit trees for fertilizer and soil conservation purposes. A small forest of about one hectare will be developed in which trees of different species may be introduced.

SALT-4 (Small Agrofruit Livelihood Technology): The system's general objectives are to produce food, increase income and practice soil conservation in a limited sloping land (one-half hectare). Marketing of agricultural crops has been cited by most farmers as one of the biggest problem as most uplands have no road to market the produce. Because of this, the MBRLC introduced the growing of high value crops like fruits. Fruits, after all, can be easily marketed, not mentioning they can stand the rigors of spoiling unlike the easily perishable vegetables and other perennial crops. In cases where fruits easily rot, farmers may resort to planting those fruits that can easily be marketed right in their farm or neighboring areas.

The importance of the sloping agricultural lands in the developing countries will be increasing with population growth and diminution of the agricultural lands due to land degradation and urbanization development. The main problems in the sloping agricultural lands includes serious soil loss and nutrient leaching. Soil erosion is a main cause of soil fertility decline. The present common and effective way of soil conservation and fertility management of the sloping agricultural lands is to construct terrace. Terracing has played a very important role in controlling soil loss and improving land productivity, but limitations are also present. They include large quantity of investment in construction and maintenance, instability, not applicable to sandy and coarse texture soils. Planting of contour hedgerows can effectively reduce soil loss from the sloping agricultural lands. The systematic placement of the hedgerows can reduce soil loss very considerably. The aggressive taproot system helps break up compacted subsoil layers, both improving the penetration of moisture into the soil and decreasing surface runoff. Our experience shows that in a very degraded soil the surface runoff from the plots with hedgerows is only 37-47 per cent of that from the traditional sloping farmland (Tang 1998). The data from the tropics in Africa and the Philippines indicates that soil loss with contour hedgerows is only 2.7-1.6 percent of that from the traditional farming practice (Kiepe, 1995; Palmer 1996). The results from the HKH region indicate that from the second year of contour hedgerow planting, 80-95 per cent of the soil loss can be reduced. The normal farming activities will eventually lead to the formation of terrace between two hedgerows, hence we call it bioterrace. Compared with the conventional bench terrace, the bioterrace has numerous advantages including low investment, stability, capability of improving soil fertility, and providing fodder or fuelwood. In the remote rural regions, where fertilizer is unobtainable or too expensive, growing of nitrogen fixing plants for manure by establishing contour hedgerows may prove attractive and practical. In addition, many herbaceous legumes can be planted as cover crops to improve soil fertility and to conserve soil because cover crops can shield the soil surface from erosion by water and wind.

Clearance of forests and other types of vegetation in the mountains has been regarded an important cause of soil erosion. The combined result of vegetation clearance and soil erosion results in large areas of wasteland. Afforestation and / or reforestation have been the main activity of rehabilitation of degraded mountain ecosystems. However, due to soil erosion, such areas are usually low in nutrients and poor in soil properties. Planting of trees is often unsuccessful. Though there may be various causes of such failure, selection of inappropriate species could be an important one. Many plants cannot be easily established at the very degraded and infertile habitats. In such case, many legumes are valuable for initial plantings in bare grounds, wastelands, or sand dunes because they have the ability to thrive on poor soils and in areas with long dry seasons. Once established, they can create conditions that induce other species to grow well. Thus they act as pioneers, preparing difficult sites for farming or forestry. Many nitrogen fixing plants could be prime candidate for restoring forest cover in watersheds, slopes, and other lands that have been denuded. Successful examples are common. In ICIMOD project on rehabilitation of site in the middle hills of Nepal, *Tephrosia candida* grows very well in eroded soil where no other grasses can grow. In Ningnan County of the southwestern China, planting of nitrogen fixing plants improved the soil condition. The soil moisture of the plots planted with eucalyptus and legumes was 6.9-8.2 per cent higher than that of the bare land. The obvious improvements are found with soil organic matter and total nitrogen. Contents of organic matter of the soil with planting of nitrogen fixing legumes were 29.7-79.9 per cent. The contents of the total nitrogen in the plots with cultivation of mixed eucalyptus and nitrogen fixing plants and pure nitrogen fixing plants are 5.9-35.3 per cent higher than that from bare land plot and plot of *Eucalyptus camaldulensis* (Chen, 1991). Use of Lespedeza in rehabilitating degraded red soil in Jiangxi of China is also successful (Wang and Cai, 1993).

SOIL CONSERVATION AND SALT

Enormous data shows that properly planted and managed hedgerows of nitrogen fixing plants can adequately control soil erosion. The data from the birthplace of SALT shows that the annual soil loss for SALT is only 0.9-1.7 per cent of that from non-SALT systems and the runoff from the SALT system is only 7.7 per cent of that from non-SALT system (Palmer, 1996). The data from the HKH region show that contour hedgerows function effectively as soil erosion barriers from the second year of planting and their efficiency in controlling soil loss from SALT system is 4.7-22.5 per cent of non- SALT systems, and as much as 37-47 per cent of the runoff water is reduced by hedgerow systems owing to a prolonged infiltration.

SOIL FERTILITY MAINTENANCE

Research indicates that soil erosion is a major contributing factor to soil fertility decline because large quantity of soil nutrients is lost with soil. Therefore reduction of soil loss is an important mechanism for maintaining soil fertility. In addition, in SALT system, hedgerows also provide large amount of mulch materials or green manure. In the tropical condition, the tested nitrogen fixing plants can provide from 28.5 to 51.9 tonnes per ha of fresh materials and 5.4-16.1 tonnes of the dry matter (Palmer, 1996). This nitrogen and protein rich foliage can improve soil fertility effectively. On farm research data indicate that, on a very degraded soil, compared to the control, the content of the total nitrogen of the soil with hedgerows increased by 14.3-58.2 per cent and 69.3-133.9 per cent in the second and third years respectively after hedgerow planting. Similarly the organic matter increased 4.7-32.3 per cent and 19.8-55.5 per cent in the second and third years respectively (Tang Ya, 1998). The above data show cearly that this technology is effective in maintaining soil fertility.

CROP YIELD AND SALT

Crop yield is the final products of the farming activities. Whether or not a technology can enhance land productivity could be an important influencing factor for technology extension. In Philippines, the results showed that corn yield was 0.5-1.0 t/ha in the traditional farming system, while the corn yield was 2.0-2.5 t/ha, 3.0-4.0 t/ha in the crop-based SALT system and animal husbandry based SALT system, respectively. The annual income of the later two systems is 2.4 and 5.3 times of traditional farming system (Palmer, 1996). The results from the HKH region indicate that although the hedgerows occupy about 15-25 per cent of the land surface, crop yields from areas with hedgerow plantations are higher than that in the control areas. In two sites, it was observed that the crop yields of maize increased by 17-51 per cent in the second year and 30-60 per cent in the third year (Tang Ya, 1998).

WATER CONSERVATION AND SALT

The total water collected in the Non-SALT basin was remarkably higher (38,111liters) than that collected in the SALT basin (2, 974) liters). This indicates less surface runoff in the SALT system, meaning more water is being placed back into the soil, ensuring less erosion and greater water availability to crops.

The total sediment load collected in the Non-SALT was drastically higher (1, 968 kg) than the SALT treatment (17kg), thus showing NF agroforestry systems such as SALT 1 to be excellent erosion control system. The combination of mulch and barrier used in SALT is the key to good control in upland farming systems. Although true that grasses such as vetiver (*Vetiveria zizanodies*) and engineering structures may be better as pure barriers to soil erosion, the barrier is but one aspect of erosion control. Management of the cover can play just as critical a role in controlling soil erosion as the barrier itself.

GENERAL EVALUATION OF THE TECHNOLOGY

The above discussion clearly indicates that SALT has great potential for environmental conservation while providing option for sustainable mountain agricultural development. This technology is easy to apply once farmers are trained. An important principle is to locate the contour lines, which is similar to that for bench terracing, with which many farmers in the region are familiar. The second important principle is to plant the hedgerows along the contour lines with proper nitrogen fixing plants (NFPs), which is new to many of the farmers in the region. The real challenge may be selection of the right NFPs and availability of seeds, which is usually not easy for farmers.

Some believe that this technology is labour intensive. Hedgerows are pruned every month during the growing season in tropics to prevent shading of the crops. When this technology was applied in the subtropics, especially in the HKH region, the situation is different. Because the temperature is lower in these areas and over 70-90 per cent of the rainfall is received during the monsoon season, the growth of the hedgerows is not so fast as in the tropics. The pruning is done 2-4 times especially during the monsoon. In some cold areas, the hedgerows need pruning once a year only, which is much less than that in the tropics. In addition, farmyard manure has been the most important organic fertilizer and its transport to farmland is a burden to the local farmers. The on-site provision of green manure is much less labour consumptive compared to transportation of farmyard manure to the field.

The Aravallis are among the oldest hill systems in the country. In the state of Rajasthan, the Aravallis fall in the semi-arid zone. Though the region gets enough rainfall (500-700mm) in the monsoons, the sloping land result in fast runoff, a problem compounded by deforestation. Soil erosion, degradation and small sized landholdings remain the major limitations for agriculture. The Sloping Agricultural Land Technology (SALT) program was envisaged for 10 predominantly poor villages in the Udaipur district of Rajasthan, falling entirely in the Aravallis. A total of 500 hectares of land was treated with SALT techniques in these villages. The 415 villagers who are part of the project are mainly small and marginal farmers. The specialty of the SALT program is that it has produced results even in a water scarce area. In 2002, a drought year, farmers in Kaccher report better maize yields by using a combination of liquid manure and mulching techniques. In the same village, pasturelands protected through social fencing are now producing enough fodder for the entire village.

Agricultural productivity in Northeast India, especially in the upland areas is severely constrained due to the terrain. Further, agricultural production has shown declining trends due to the reduced jhum cycle consequent to increasing pressures of population and the resultant demands on land. The overall result is rapid deterioration of the land and water resources and environmental degradation. Several agencies have attempted to find alternatives to ameliorate the situation, but none have been able to develop an acceptable, convincing model. SALT offers a viable alternative which can prove to be a feasible, low cost but simple innovation to shifting cultivation.

The fundamental aim of SALT is to ensure soil and water conservation, enhance soil fertility and agricultural productivity (including animal husbandry and fisheries) and improve income generation. At the larger scale, it aims to transform shifting cultivation, restore and rehabilitate degraded land with low external inputs. The basic technology applied is the introduction of nitrogen fixing species as hedgerows with the twin aim of providing a bio-slope stabilizer to promote soil and water conservation and improve soil fertility through biological nitrogen fixation and mulching of the plots with green manure generated from hedgerow pruning. Two important objects - soil and water conservation and soil fertility enhancement- are achieved by a simple, low cost intervention, with little cost to the farmer. Other methods for slope stabilization, soil and water conservation and soil fertility enhancements are much more expensive, both in terms of labour and costs. Hedgerows require minimal labour and costs. Even the most marginal farmer can afford the introduction of hedgerows in his plot. For the upland farmer of Northeast India, such an easy, simple and low cost intervention would be easily acceptable.

Establishing hedgerows

On each prepared contour line, make two furrows at a distance of 50 cm apart. Plant one seed per centimeter. Planting must be done at the start or during rainy season. To avoid washing out of newly-planted seeds cover them with mulching materials. Newly-planted hedgerows must be weeded and cultivated at least once a month or more, if necessary. Nitrogen fixing species are important because they manufacture their own nitrogen. Species which can be used for hedgerows include *Gliricldia sepium, Leucaena leucocephala, L. diversifolia, Calliandra calothyrsus, Indigofera tysemani* and those that are locally grown in the area.

One of the main problems, which have led to erroneous results within alley cropping research in many areas, is the inclusion of non-nitrogen fixing plants to be used as vegetative barriers within a hedgerow. Two widely-accepted facts support this claim : 1) not all legumes are nitrogen fixing, and 2) some non legumes are nitrogen fixing. The word legume basically means "pod bearing plant", and the most common examples are beans and pulses. Many plants in the legume family are nitrogen fixing; however, many are not.

The guidelines for a good species to be used in NF agroforestry systems as contour hedgerows are as follows:

1. A nodulating, nitrogen fixing plant.
2. Coppies readily under heavy pruning (12 times per year).
3. Grows well in the MBRLC double hedgerow system.
4. Produces at least 25 metric tons of fresh biomass/hectare/year based upon five meter double hedgerow intervals for approximately 2000 linear meters per hectare.
5. Readily sets seed for farmer reproducibility.
6. Able to be reproduced asexually.
7. Resistant to insects and diseases.
8. Able to grow well when planted very thinly.
9. Adaptable to a wide variety of soil and climates.
10. Deep/tap rooted.
11. Grows into a tree if left unattended.
12. Usable for forage.
13. Multi-purposed (i.e. fencing, fuel, feed, etc.).

The *Calliandras* are excellent hedgerow species but have a problem with seeding at lower altitudes. The MBRLC is at 350 m altitude, and the *Calliandras,* based upon local experience, need at least 500 to 600 m. Planting of nitrogen fixing plants to supply fuelwood will contribute cosiderably to the conservation of forests. Legummes and nonlegumes can be planted in the marginal lands. Planting of *Amorpha fruticosa* in the terrace risers in Gansu of China can produce 2340 to 2625 kg. fuelwood a year. Nitrogen fixing plants can be managed for other uses or application. Many nitrogen fixing plants are good timber products. Legumes can also be managed to produce gums and tannin. *Acacia mearnsii* is a good candidate for tannin production.

The leguminous woody fodder is advantageous over grasses and non-legumes as they have higher contents of crude protein and calcium while other nutrients are more or less similar (Table 37.1). The development of the woody fodder can be combined with development of the wastelands. This may be attractive because this approach will provide benefits to the local people in the course of rehabilitation of the wasteland. Failure of the earlier approaches of wasteland development might be due to the more timber-oriented reforestation approach and local needs were not well considered. The HKH region is rich in legume diversity. There will be a very bright prospect for the development of woody fodder in the region.

Though SALT is effective in controlling soil erosion, improving soil fertility, and enhancing crop yield, its adoption by farmers is limited. Many reasons contribute to this limited adoption, but lack of direct and visible cash income is probably a major one. Technologically, this technology has shown that after adoption of this technology, about 80-94 per cent of soil loss can be reduced and increased crop yield was also observed. In a long-term basis, the benefits of this technology are believed to be much higher compared to the traditional farming practices, because the traditional up-down farming will lead to serious soil and nutrient loss and land has to be abandoned due to very low soil fertility after several years' of continuous farming. While in the hedgerow farming, in addition to green mulch or green manure from prunings of the hedgerow, soil loss is greatly reduced and so is the nutrient loss, and soil fertility is improved. Land productivity under the contour hedgerow system will be maintained. But direct and visible cash incomes are lacking.

In order to enhance the direct and visible benefits of the hedgerows, efforts have been made in the project sites in China. A local cash plant, mulberry tree was planted between two rows of the double hedgerows. Mulberry trees are an important cash plant in the project region in China and are planted in almost all possible marginal lands and risers of terraced land. The conventional management of the mulberry plantation is that it is fertilised every time for each of the four times of silkworm rearing, meaning four times of fertiliser application. In the project area, supply of mulberry leaves has been a key issue for sustainability of sericulture in Ningnan Country. It has been found that the mulbery trees planted within the hedgerows grow very well and better than those received regular fertilisers, and the leaf yield of mulberry is about four times of the conventional plantation (Yuan *et al* 1999), implying more leaf yield as mulberry with less inputs. The inclusion of mulberry trees within the double hedgerows has played important role in convincing local farmers to see the benefits and to adopt this technology, because the hedgerow-mulberry system is very similar to the riser-mulberry plantation system in the region.

ADVANTAGES OF SALT FARMING AGROFORESTRY

The advantages of SALT are that it is a simple, applicable, low-cost, and timely method of farming uplands. It is a technology developed for Asian farmers with few tools, little capital, and little learning in agriculture. Contour lines are run by using an A-frame transit that any farmer can learn to make and use. A farmer can grow varieties of crops he is familiar with and old farming patterns can be utlized in the SALT system.

If farmers leave the SALT farm, like some tribal groups do, the nitrogen fixing trees and shrubs (NFTS) will continue to grow and overshadow the crop area. By the time the land is reverted to cultivation, the soil has been enriched already by the large amount of NFTS leaves and there is no erosion to content with.

- Choice of models for farmers according to local suitability-agroforestry/agrohorticultural/agropastoral and fisheries.
- Effective and efficient land-use optimization.
- Potential for increased cropping phase to more than two years
- Low external inputs
- Potential high economic returns
- Gradual transformation to sedentary agriculture
- Long term conservation and optimal use of land and water resources.
- As per provision system of upland framing, SALT has certain good qualities over both the traditional techniques of slash-and-burn and conventional terrace farming.
- SALT is efficient in food crop production and restore soil fertility.
- SALT is culturally acceptable because the farming techniques are in harmony with Asian beliefs and traditional practices.
- SALT has the small family as the focus and food production as the top priority. Fruit trees, forest and other crops are secondary priority.
- SALT is economically feasible and ecologically sound.
- The SALT farm can easily revert back to forest land if left unfarmed.

Table 37.1. Nutrient composition of some common fodder plants (% of absolute dry matters)

Plant Species	Crude Protein	Crude Fat	Crude Fibre	Crude Ash	N-free Extract	Ca	P
Grasses /nonlegumes							
Setaria vividis	12.1	2.0	26.5	13.8	45.6	0.43	0.33
Pennisetum flaccidum	9.6	2.3	20.0	9.4	58.7	0.25	0.29
Arthraxon prionodes	7.2	2.0	32.5	9.6	48.7	0.86	0.18
Digitaria cruciata	14.2	2.2	26.4	14.5	42.7	0.50	0.38
Eragrostis ferruginia	9.7	2.3	35.9	7.4	44.7	0.49	0.09
Andropogon yunnanensis	11.6	2.3	30.7	11.5	43.9	0.31	0.23
Tripogon filiformis	10.1	2.2	34.8	11.5	41.4	0.70	0.14
Mariscus umbellatus	8.9	1.4	30.1	10.4	49.2	0.47	0.30
Grewia optiva	17.20	-	19-22	10-14	42-49	2.7-3.6	0.14-2.24
Celtis australis	14-15	-	20	12-18	45	3.5-4.9	0.18
Legumes							
Sophora vicifolia	22.5	2.9	8.9	6.8	48.9	1.06	0.30
Bauhinia fabri	21.3	4.0	18.9	7.0	48.8	1.31	0.24
Lespedeza dahurica	16.1	1.8	31.7	7.4	43.0	1.42	0.21
Indigofera bungeana	17.3	2.2	32.8	7.2	40.5	1.40	0.17
Campylotropisis poliantha	17.1	3.5	25.0	7.8	46.6	1.84	0.18
Kummerowia striata	14.8	-	26.0	11.9	45.9	1.37	0.15
Campylotropis sp.	10.4	1.6	43.2	4.1	39.6	0.72	0.15
Acacia sp.	19.6	2.7	23.9	6.7	47.9	1.37	0.21
Bauhinia variegata	11.16	-	25.33	-	-	1.8-4.1	0.2-0.4
Dalbergia sissoo	17	-	22	9	49	-	-
Leucaena leucocephala	20	-	10.17	5-10	50	3	0.2

Source: CIB and SPGI 1984; (Prakash and Hoching 1986)

CONSTRAINTS

The heavy pressure of human population for its various requirements may not allow good land to be put under fodder production to the extent it has been estimated. In such a situation to meet the increased fodder requirement there is no way except that of improving productivity of the fodder crops through scientific management of the permanent pasture and grazing lands. Woody fodder is an important source basis of livestock development. Development and use of woody fodder is common in the South Asia but very rare are in the Himalayan region. Although calculations of the economic viability of SALT technology have been done and high rate of return asserted, as shown by the data from it birthplace (Palmer, 1996), actual costs are sometimes underestimated, because some, such as time and skill required and availability of planting materials are overlooked. When a technology is being promoted, a overestimation of economics usually happens. But what farmers will believe is not what they are told but what they see. Despite effective in soil conservation and improvement of soil fertility, increased crop yield, and various positive effects on soil physical and chemical properties, the use of SALT by farmers is limited. Farmers do not operate as independent decision- makers, but rather are subjected to and influenced by a variety of factors. Thus, decision to adopt and use conservation practices, is made not only in the context of personnel and family circumstances, but also in response to government policies, institutional arrangements, community attitudes, customs and free market forces. Evidence suggests that farmers perceive that soil erosion causes on-site losses, yet most of them do not invest in measures to control it. Innovations proposed to the farmers must bring him advantage, must be acceptable to him, and must be low risk for economic loss.

REFERENCES

Annual Report (2002-2003) Action for food production, New Delhi

Chaudhry, M.A. (1995). Soil erosion and conservation. *Soil Science* National Book Foundation Islamabad.: 509-537

Chen, Keming (1991). Layout and demonstration study of fuelwood in Ningnon country, Sichuan China. In Huang Zhilia and wei Taichang (editors), Research on Development of integrated Rural Energy of Nigeran country in subtropical Hot and Dry Valley, *Chengdu Science and Technology University*, Press 1991.

Das, D.C. (1985) Problems of soil erosion and land degradation in India. *In* lead papers, National Seminar, *Soil Conservation and Water Management*, September 17-18, 1985, New Delhi pp1-24

Glover, M. (1986) NFT highlights, NATA *86-06*. Winrock International.

ICMOD (1998) News Letter No. 32 winter

Kerr, J.M. and Sanghi, N.K. Indigenous soil and water conservation in India's semi-arid tropics : 255- 289.

Kiepe, P. (1995) No Runoff, No soil Loss oil and water conservation in hedgerow barrier systems. *Tropical Resource Management Papers* 10. Wageringer Agricultural University, The Netherlands.

Mac Dicken, K.G. (1994) Selection and Management of nitrogen-fixing trees, Winrock international, USA, and Food and Agriculture Organization, Bangkok, Thailand.

NFT High light : A publication of the Nitrogen fixing Tree Association 1010 Holomua Road, Paia, Hawaii, 967796744, USA.

Palmar, J. (1995) Slopping Agricultural Land Technology (SALT) Nitrogen Fixing Agroforestry for sustainable soil and water conservation. A publication of the Mindanao Baptist Rural Life Center.

Palmer, J. (1996) Sloping Agricultural Land Technology (SALT) - Nitrogen fixing agroforestry for sustainable soil and water conservation. A publication of the Mindavao Baptist Rural life Center, Davao, Philippines, 59pp.

Parkash, R and Hocking, D. (1986) Some Favourite Trees for Fuel and fodder. Society for Promotion of Wastelands Development, Abhay Rastogi at Prabhat offset Printers Private Limited, Meerut, India.

Partap, T and Watson (1994) Slopping Agricultural Land Technology (SALT): A reservative option for sustainable mountain forming. ICIMOD occasional paper No. 23 Kathamandu, Nepal: International centre for Integrated mountain Development.

Swarup, S and Tewari, S.C. (1998) Farmers dependent on forest for fodder, fuel and timber in hills of Himchal Pradesh. *In* Khosla, P.K. and R.N. Sehgal (editors). *Trends in Tree Sciences*, India Society of Tree Scientist, New Delhi.

Tang Ya, (1998) Bioterracing and soil conservation : Experience with contour hedgerow planting in parts of the Hindu Kush- Himalayan region. *Issues in Mountain Development* 98/7. International center for Integrated mountain Development Kathamandu, Nepal.

Tang Ya (1999) Factors influencing farmers adoption of soil conservation programme in the Hindu Khush Himalayas.

Training Manual : G.B. Pant Institute of Himalayan Environment & Development, N.E. Unit, Itanagar.

Wang Mingzhu and Cai Shenkun (1993) Lespedeza- Pioneer plants for revegetation in bare red soil. *In* Wang Mingzhu *et al* (editors) Research on Ecosystems of the Red soils. Vol. Z-pp. 55-58. Jiangzi Science and Technology Press.

Watson, H.R. (1997) The Development and spread of SALT and its modification in the Philippines and their Asia Countries. Paper presented it Regional workshop in Appropriate Technologies for soil conserving farming system project. Organised by ICIMOD in collaboration with chengdu Institute of Biology, Chinese Academy of Science. 8-11 July 1997, Ningnan Sichana Presina. P.R. China

Youns, A. (1984) Agroforestry for soil conservation, UK ; CAB International

Yuan, Yuanliang, Sun Hai and Tang Ya. (1999). A preliminary study of economic benefit of inclusion of mulberry within contours hedgerows of nitrogen fixing plants *Eco-agricultural Research.*

Photo 11.1. Populus deltoides + sugarcane based agroforestry systems

Photo 11.2. *Tectona grandis* based agrisilviculture system at farmer's field (mustard is an agricultural component)

Photo 11.3. *D. sissoo+M. indica* based agrisilvihorticulture system developed by a farmer in North eastern tarai zone. *Pisum sativum* is agricultural crop

Photo 11.4. A five year old *C. equisetifolia* based Agrisilviculture system.

Photo 19.1. Arrowroot Intercropped in *Acacia mangium* Alleys

Photo 19.2. Stylosanthes hamata In Hortisilvipastoral System. (Guava+ Gmelina +Stylo).

Photo 26.1 Mango + Cowpea based agrihorticulture agroforestry system

Photo 26.2 Mango + *Sorghum* based agrihorticulture agroforestry system

Species Index

K

L

M

N

O

P

Q

R

S

U

V

W

Z

Subject Index

F

G

H

M

R

S